Laser Annealing
of Semiconductors

CONTRIBUTORS

B. R. Appleton

Pietro Baeri

J. C. Bean

Salvatore Ugo Campisano

A. G. Cullis

Gaetano Foti

J. F. Gibbons

C. Hill

S. S. Lau

James W. Mayer

J. M. Poate

Emanuele Rimini

T. W. Sigmon

Frans Spaepen

David Turnbull

Martin F. von Allmen

C. W. White

J. S. Williams

S. R. Wilson

D. M. Zehner

Laser Annealing
of Semiconductors

Edited by
J. M. POATE
Bell Laboratories
Murray Hill, New Jersey

JAMES W. MAYER
Department of Materials Science
Cornell University
Ithaca, New York

 1982

ACADEMIC PRESS

A Subsidiary of Harcourt Brace Jovanovich, Publishers

New York   London
Paris   San Diego   San Francisco   São Paulo   Sydney   Tokyo   Toronto

RECID = 89-1

ACADEMIC PRESS, INC.
111 Fifth Avenue, New York, New York 10003

United Kingdom Edition published by
ACADEMIC PRESS, INC. (LONDON) LTD.
24/28 Oval Road, London NW1 7DX

Library of Congress Cataloging in Publication Data
Main entry under title:

Laser annealing of semiconductors.

    Includes bibliographical references and index.
    1. Semiconductor industry--Laser use in--Congresses.
2. Semiconductors. I. Poate, J. M. II. Mayer, James W.,
Date   .
TA1673.L34   1982   621.3815'2   82-8816
ISBN 0-12-558820-8   AACR2

Contents

Chapter 4. Heat Flow Calculations

Pietro Baeri and Salvatore Ugo Campisano

Chapter 5. Supersaturated Alloys, Solute Trapping, and Zone Refining

C. W. White, B. R. Appleton, and S. R. Wilson

Chapter 6. Microstructure and Topography

A. G. Cullis

Chapter 7. Epitaxy by Pulsed Annealing of Ion-Implanted Silicon

Gaetano Foti and Emanuele Rimini

Chapter 12. **Silicides and Metastable Phases**

Martin F. von Allmen and S. S. Lau

Chapter 13. **Factors Influencing Applications**

C. Hill

List of Contributors

Numbers in parentheses indicate the pages on which the authors' contributions begin.

B. R. Appleton (111), Solid State Division, Oak Ridge National Laboratory, Oak Ridge, Tennessee 37830

Pietro Baeri (75), Istituto di Struttura della Materia, Università di Catania, 95129 Catania, Italy

J. C. Bean (247), Bell Laboratories, Murray Hill, New Jersey 07974

Salvatore Ugo Campisano (75), Istituto di Struttura della Materia, Università di Catania, 95129 Catania, Italy

A. G. Cullis (147), Royal Signals and Radar Establishment, Malvern, Worcestershire WR14 3PS, England

Gaetano Foti (203), Istituto di Struttura della Materia, Università di Catania, 95129 Catania, Italy

J. F. Gibbons (325), Stanford Electronics Laboratories, Stanford University, Stanford, California 94305

C. Hill (479), Plessey Research (Caswell) Limited, Allen Clark Research Centre, Caswell, Towcester, Northamptonshire NN12 8EQ, England

S. S. Lau (439), Department of Electrical Engineering and Computer Sciences, University of California, San Diego, La Jolla, California 92093

James W. Mayer (1), Department of Materials Science, Cornell University, Ithaca, New York 14853

J. M. Poate (1, 247), Bell Laboratories, Murray Hill, New Jersey 07974

Emanuele Rimini (203), Istituto di Struttura della Materia, Università di Catania, 95129 Catania, Italy

T. W. Sigmon (325), Stanford Electronics Laboratories, Stanford University, Stanford, California 94305

Frans Spaepen (15), Division of Applied Sciences, Harvard University, Cambridge, Massachusetts 02138

David Turnbull (15), Divisison of Applied Sciences, Harvard University, Cambridge, Massachusetts 02138

Martin F. von Allmen (43, 439), Institute of Applied Physics, University of Bern, CH-3012 Bern, Switzerland

C. W. White (111, 281), Solid State Division, Oak Ridge National Laboratory, Oak Ridge, Tennessee 37830

J. S. Williams (383), Department of Communication and Electronic Engineering, Royal Melbourne Institute of Technology, Melbourne, Victoria 3000, Australia

S. R. Wilson (111), Semiconductor Group, Motorola, Inc., Phoenix, Arizona 85008

D. M. Zehner (281), Solid State Division, Oak Ridge National Laboratory, Oak Ridge, Tennessee 37830

Preface

In the past five years there has been a remarkable display of interest in the laser annealing of semiconductors. Interest in this field developed in the period 1977–1978 with workshops in Albany, New York, and in Catania, Italy, followed by symposia at the Materials Research Society meetings in Boston from 1978 to 1981. By July 1981 there were over 800 publications in this field. The contributors to this volume have been involved in the field since its emergence in 1977.

The subject deals with the materials science of surfaces that have been subjected to ultrafast heating by intense laser or electron beams. The time scale is such that layers can melt and recrystallize in a few hundred nanoseconds. This rapid resolidification of semiconductors has led us into novel regimes of phase formation and has served as a means of removing implantation damage. It was this latter aspect that stimulated the field.

We believe that the phenomena of not only the energy deposition and heat flow but also of the basic crystal growth processes are sufficiently well understood to warrant this volume. The chapters follow a logical sequence from basic concepts to device structures.

We should like to thank our fellow contributors for their efforts. They revised their chapters as necessary to fit within a common framework. We especially acknowledge our friends in Catania for their efforts.

Dawn and Betty, as always, gave us unflagging support.

Chapter 1

Introduction

J. M. POATE

Bell Laboratories
Murray Hill, New Jersey

and

JAMES W. MAYER

Department of Materials Science
Cornell University
Ithaca, New York

I. Directed Energy Processing of Semiconductors

A new field of materials science—directed energy processing—has emerged in the past five years for the processing and modification of the surface layers of semiconductors. Directed energy sources such as lasers or electron beams are used to heat the surface. The unique temporal and spatial control exercised over the heat flow by these beams allows formation of quite novel structures and alloys. For example, surface layers can be melted and solidified in exceedingly short times to produce metastable alloys. The dimensions of the layers that can be modified by the incident beams are just those required by Si integrated circuit technology.

1

The materials and dimensions of the Si technology are beautifully illustrated in the transmission electron microscopy photograph in Fig. 1, which is a cross section through an actual insulated-gate field-effect transistor (IGFET) test structure. The source–drain regions were fabricated by As implantation followed by an annealing treatment in O_2 at 900°C for 35 min. The resulting n^+ junction depth (a in Fig. 1) is 1000 Å, with the same

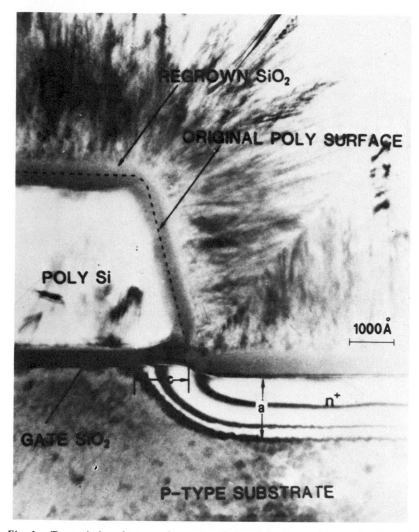

Fig. 1. Transmission electron microscopy photograph of a vertical section through an IGFET structure at the edge of a gate. The parameters a and c, respectively, are the implanted junction depths and lateral penetration of the implanted junction following the anneal. [From T. T. Sheng and R. B. Marcus, *J. Electrochem. Soc.* **881,** 128 (1981).]

lateral spread (c in Fig. 1) under the gate as measured from the position of the original implantation. The gate contact is 3500 Å of polycrystalline Si on a 250-Å SiO$_2$ gate oxide. The thicknesses of the active regions are

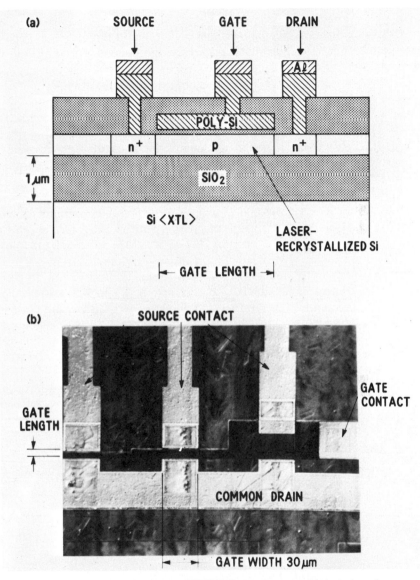

Fig. 2. (a) Cross section of an *n*-MOSFET test structure fabricated on laser-recrystallized poly-Si, 0.5 μm thick, on amorphous SiO$_2$. (b) Optical micrograph of the test structure showing the drain and source contacts for three gate lengths of 3, 6, and 30 μm. (From G. K. Celler, Bell Laboratories.)

confined to the outer micrometer of the Si structures—a distance compatible with directed energy processing.

The first demonstrations of the utility of directed energy processing came from the annealing of implantation damage using lasers. Thus the term "laser annealing" was coined. It is something of a misnomer, as most of the annealing mechanisms are in fact due to liquid or solid phase recrystallization. The field has progressed far beyond the annealing of implantation damage, as demonstrated in Fig. 2. It is possible to recrystallize amorphous or polycrystalline Si on amorphous substrates by laterally sweeping a Si melt puddle with a moving laser or particle beam. Very large grains or even single-crystal Si can be produced by this technique, which may lead to a new generation of devices. Figure 2a shows a schematic cross section of an *n*-channel metal–oxide–semiconductor field-effect transistor (MOSFET) fabricated in a 0.5-μm-thick film of Si. The Si layer was deposited in the form of small-grain (0.1 μm) polycrystalline Si on 1 μm of amorphous SiO_2 which had been thermally grown on a Si wafer. The Si layer was then heated by a scanning continuous-wave (cw) Ar^+ laser and recrystallized into grains with average lateral dimensions of 5 μm. The actual test devices have gate oxides of 270-Å thickness with channel lengths from 100 to 0.3 μm and channel widths from 120 to 20 μm. An optical micrograph (Fig. 2b) shows a test structure with different channel lengths of 3, 6, and 30 μm and constant channel widths. The electron surface mobilities are found to increase with decreasing channel length and approach that of devices in single-crystal Si. The success of these devices results from the ability to produce very large grain Si layers on an amorphous, insulating substrate. They could not be fabricated without the laser recrystallization techniques.

There is little doubt that the rapid expansion of this field is due to the driving force of the semiconductor industry. New processing techniques are needed to produce structures on the submicrometer scale. Much of the research interest, however, centers around the fact that the directed energy techniques allow the exploration of new realms of materials science. Spaepen and Turnbull in Chapter 2 present the concepts of solidification and crystallization pertinent to the present subject.

II. Energy Deposition and Heat Flow

The facility of rapidly heating and cooling surface layers without heating the bulk depends on the pulse duration time τ and the coupling depths of the heat source. These parameters for both laser and electron-beam sources are discussed by von Allmen in Chapter 3. For Q-switched solid state lasers the output pulses range from 10^{-9} to 10^{-7} sec, whereas for mode-locked lasers pulses as short as 10^{-12} sec can be generated. The cooling or quench rates of the surface layers using these pulsed sources will be in the range 10^9-10^{14}°C/sec. Longer irradiation times can be achieved by the use of continuous sources with scanned spots. The fastest irradiation times will be $10^{-5}-10^{-6}$ sec, with cooling rates less than 10^9°C/sec.

The coupling of lasers to materials is very sensitive to the laser wavelengths and the state of the material. For example, the optical absorption length α^{-1} of 0.5-μm photons in crystalline Si is on the order of 1 μm. On the other hand, the coupling, or energy loss, of electron beams to solids does not depend, for all practical purposes, on the state of the material but is just a simple function of the incident energy. In spite of this, lasers are generally used in pulsed annealing because of their wide availability.

We can illustrate the concepts of lasers heating by the use of order-of-magnitude calculations. Baeri and Campisano in Chapter 4 present detailed heat flow calculations. The rate at which heat is dissipated in a solid is determined by the heat diffusivity $D = \kappa/C\rho$, where κ is the thermal conductivity, C is the specific heat, and ρ is the density. For the laser heating of Si we assume that $\kappa \sim 0.2$ J/cm sec K, $\rho = 2.2$ g/cm, and $C = 1$ J/g K, giving $D \sim 0.5$ cm²/sec. For τ of 10^{-7} sec therefore, the characteristic heat diffusion length $(2D\tau)^{1/2}$ is ~3 μm. Figure 3 illustrates schematically

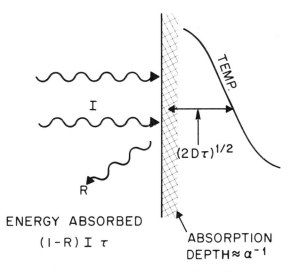

Fig. 3. Temperature distribution in laser-irradiated Si in which the absorption distance α^{-1} is less than the heat diffusion length $(2D\tau)^{1/2}$. For $\tau = 10^{-7}$ sec, $(2D\tau)^{1/2} = 3$ μm.

the case where $\alpha^{-1} < (2D\tau)^{1/2}$. The absorbed energy is simply the laser power density multiplied by the pulse length $I\tau$ minus the reflected energy $IR\tau$. The average temperature rise in this $(2D\tau)^{1/2}$-thick layer is therefore given by

$$\Delta T = (1 - R)I\tau/C\rho(2D\tau)^{1/2}$$

If zero reflectivity is assumed, then an absorbed energy density of 1 J/cm² will be required to raise this layer to the melting point.

The heating and cooling rates for the case where $\alpha^{-1} < (2D\tau)^{1/2}$ are given simply by $\Delta T/\tau$. It is the cooling rate that ultimately determines the composition and structure of the irradiated material. Probably one of the parameters of greatest interest is the velocity of recrystallization of the molten zone. This velocity v can be estimated as illustrated schematically in Fig. 4. The heat liberated at the advancing interface is given simply by $\Delta H_m\,\rho v$, where ΔH_m is the enthalpy of melting. This heat must be balanced by the heat flow ($\kappa\,\partial T/\partial z$) into the substrate. The temperature gradient can be calculated to first order by $T_m/(2D\tau)^{1/2}$. The velocity of the interface is about 3 m/sec. These estimates agree quite well with velocities obtained by detailed calculations. The computations of Baeri and Campisano give $v \sim 2$ m/sec for this regime.

Measurements of the liquid–solid interface velocity have been made by utilizing the metallic conductivity of molten Si. After the photoconductive response decays, the sample conductance is determined primarily by the thickness of the molten layer. Thermally generated carriers in the solid make only a small ($\simeq 10\%$) contribution because of the strong thermal gradients near the interface. Figure 5 shows transient conductance of a

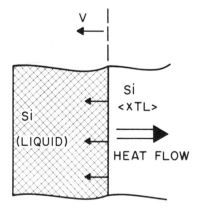

Fig. 4. Schematic illustrating the velocity of the liquid–solid interface, which is determined by the balance between the latent heat liberated and its heat flow into the substrate (heat liberated per second equals heat flow: $\Delta H_m\,\rho v = \kappa\partial T/\partial z$).

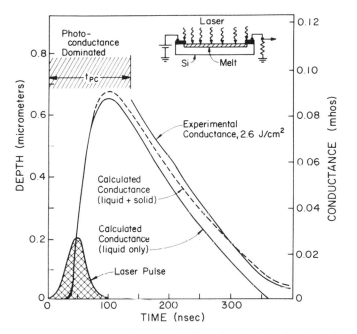

Fig. 5. Transient conductance of Au-doped Si irradiated with a Q-switched laser at 2.6 J/cm². The shaded area indicates the photoconductance-dominated time as determined from below-melt-threshold irradiations at 0.2–0.8 J/cm². Calculated conductance curves based on a thermal heat flow model are shown for the molten metallic Si (solid line) and the liquid and solid (dashed line); thermally generated carriers add a contribution equivalent to about 0.05 μm. [Adapted from G. Galvin, M. O. Thompson, J. W. Mayer, R. B. Hammond, N. Paulter, and P. S. Peercy, *Phys. Rev. Lett.* **48**, 33 (1982).]

Au-doped Si sample during irradiation with a Q-switched ruby laser at 2.6 J/cm². For room temperature irradiations at 1.9–2.6 J/cm², the interface velocity was found to be 2.8 m/sec, in close agreement with calculations (dashed line in Fig. 5) based on a thermal model of heat flow similar to that described in Chapter 4 by Baeri and Campisano.

III. Interfaces and Surfaces

Some of the more interesting manifestations of surface processing result from the motion of the crystallizing interface. Figure 6 shows Si crystallization rates as a function of interface temperature. The various heating schemes are shown with the time required to crystallize 1000 Å indicated in parentheses. The solid phase rates are obtained from measurements of

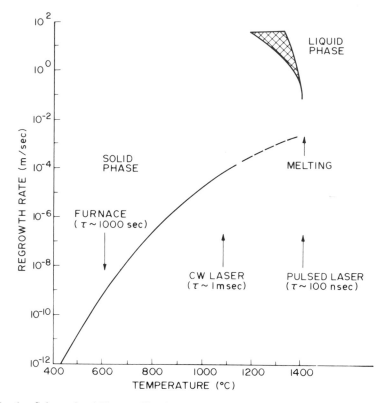

Fig. 6.   Schematic of Si crystallization rates (in meters per second) versus temperature. The solid phase data come from measurements of the regrowth of amorphous Si on Si. The liquid phase data come from calculations of the velocity of the liquid–solid interface in laser-induced surface melting.

the regrowth of amorphous Si layers on Si; most of the data come from low temperature furnace measurements as discussed in Chapter 10 by Gibbons and Sigmon. In the furnace measurements the temperature gradients are essentially zero, and the rate of regrowth depends upon the temperature dependence of bond breaking and the atomic configuration of the interface. Because of its higher free energy, amorphous Si is thermodynamically unstable in the presence of crystalline Si and tends to regrow epitaxially, layer by layer, on the underlying crystal. The interface velocity is an exponential function of temperature, and the plot represents an extrapolation to higher temperatures. It is broken in the region beneath the melting temperature because it is believed (Spaepen and Turnbull, Chapter 2) that the amorphous Si melts at considerably lower temperatures than crystalline Si.

Once the Si is melted, however, there is a sharp jump in the crystallization velocity because of the much greater mobility of the atoms in the liquid and the very high temperature gradients present in laser melting. In the case of conventional liquid phase crystal growth (Czochralski for example) the temperature gradients are, by comparison, low. There the rate of extraction of the seed crystal from the melt is determined by the rate of extraction of latent heat through the seed crystal. Growth rates, typically 10^{-5} m/sec, are much slower than the recrystallization velocities of the liquid surface layer produced by pulsed irradiation. The hatched region in Fig. 6 shows the velocities which are experimentally attainable using pulsed irradiation. The recrystallization velocity can be varied, for example, by means of the temperature gradient and laser pulse length (Baeri and Campisano, in Chapter 4). The recrystallization velocity is ultimately determined by the undercooling of the melt (Spacpen and Turnbull, in Chapter 2); the higher the undercooling, the greater the velocity. Definitive estimates of Si undercooling do not yet exist, and the hatched region just indicates plausible values. The upper limit to the recrystallization velocity represents the case where the quench rate is so fast that an amorphous layer is produced. It is remarkable that the velocity of the recrystallizing interface, whether from the amorphous or liquid phase, can be varied over at least 14 orders of magnitude.

An important way of tracking the motion of the interface is to observe the location of impurities after the interface has passed. The behavior for the solid phase is shown schematically in Fig. 7 with data from Chapter 10 (Gibbons and Sigmon). Arsenic has been implanted in Si to form an amorphous layer which has been subsequently recrystallized with a cw Ar laser. The laser conditions are such that the layer recrystallizes in less than 10^{-2} sec. Measurements show that passage of the amorphous crystal interface through the As distribution does not cause any perceptible movement of the As atoms except that they are located on Si lattice sites after crystallization. The lack of motion is simply due to the fact that there is not sufficient time for any significant As diffusion to occur. For example, the diffusivity of As in Si at 1300°C is 10^{-12} cm²/sec, and the diffusion lengths $\sqrt{Dt}$ in 10^{-2} sec will therefore only be 10 Å. Moreover, the exceedingly short diffusion distances mean that second-phase precipitation will be suppressed and metastable solid solutions can be formed. This behavior is very different than that occurring in conventional furnace annealing (~900°C for 30 min), where diffusion distances of 1000 Å are common and equilibrium solid solubilities are not exceeded. The consequences and advantages of rapid solid phase crystallization are discussed in Chapter 10 by Gibbons and Sigmon.

One of the more striking indications of the possibilities of pulsed laser

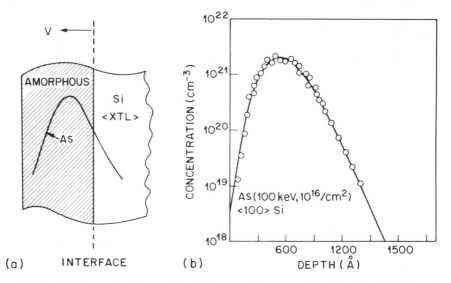

Fig. 7. (a) Schematic of the motion of the amorphous crystal interface for As-implanted ⟨100⟩ Si annealed with a cw Ar laser. (b) The As distribution. The As distribution is not measurably perturbed by the solid phase epitaxy. —, Implanted; ○, cw laser annealed.

irradiation as a unique tool for surface processing came from early experiments showing that dopants could be incorporated on lattice sites at concentrations far in excess of solid solubilities. This subject is discussed fully by White, Appleton, and Wilson in Chapter 5. Figure 8 shows their example of pulsed ruby laser irradiation of As implanted in Si at laser energies of 1–2 J/cm² and $\tau \sim 15$ nsec. Under these laser conditions the Si layer melts and recrystallizes at interface velocities of 3–4 m/sec. The interface is moving at such a high velocity that there is a good probability that impurities will be trapped. This process can be imagined as a competition between the velocity of the interface and the diffusive velocity of the impurities in the liquid at the interface. Consider the recrystallization over a distance of 5 Å. Typical impurity diffusion coefficients in liquid Si are $\sim 10^{-5}$ cm²/sec. The effective diffusive velocity of the impurity away from the interface can therefore be ~5 m/sec. But this is the same sort of interface velocity encountered in surface melting, and there is a good chance that the impurities will not escape the moving front. Once they are engulfed on lattice sites, they will not diffuse in the solid because quench rates are so fast.

All the As is trapped, and the resulting depths profile can be uniquely fitted with a near-unity interfacial segregation coefficient k' as compared to the equilibrium value of 0.3; where $k' = C_l/C_s$, the ratio of impurity concentrations at the liquid–solid interface. Solid solubilities are exceeded by a factor of 3–4. The situation is different for relatively insoluble impurities such as Bi, where the equilibrium concentrations are exceeded by a factor of 500. However, not all the Bi is trapped, and some is zone refined to the surface. These trapping phenomena are velocity and orientation dependent and make a fascinating extension of existing near-equilibrium crystal growth and solubility data. Because of the phenomenon of constitutional supercooling, at certain velocities and impurity concentrations the interfaces become unstable, resulting in the incorporation of dopants in cellular arrays. Examples of such microstructures that can result from segregation at the interface are shown in Chapter 6 by Cullis. Chapter 6 gives many examples of the fascinating microstructures produced by surface irradiation.

The final act of solidification is the freezing of the surface and incorporation of any zone-refined impurities at the surface. The surface structures produced by melt quenching can be quite different than those produced by conventional surface treatments. Zehner and White in Chapter 9 review this rapidly developing field.

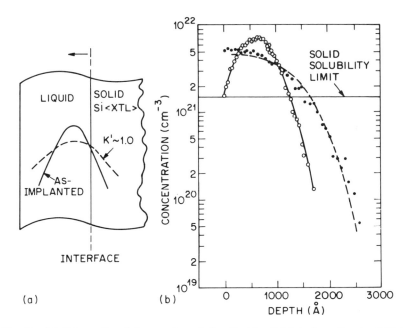

Fig. 8. (a) Schematic of the motion of the liquid–solid interface for As (100 keV, $6.4 \times 10^{16}/cm^2$)-implanted $\langle 100 \rangle$ Si annealed with a pulsed ruby laser. (b) The As distribution. The As redistributes in the liquid phase, but is not segregated by the interface. —, Implanted; ●, laser annealed; --, calculated ($k' = \sim 1.0$).

IV. Epitaxy and Alloying

Much of the earliest interest in laser annealing arose from the fact that lasers could be used to regrow epitaxially amorphous, implanted layers of Si. This subject is reviewed by Foti and Rimini, in Chapter 7, who compare the processes and remaining defect structures with those produced by furnace annealing. Essentially perfect single-crystal material can be produced with good electrical properties. The situation is different for the laser annealing of implanted GaAs, as reviewed by Williams in Chapter 11. Although the structure as measured by channeling or diffraction techniques is good, the electrical characteristics are not. Defect structures in the binary material, on the atomic scale, are apparently produced by the rapid motion of the interface.

An attraction of laser irradiation is that, in the melt phase, epitaxy is not very sensitive to interface cleanliness, as discussed by Poate and Bean in

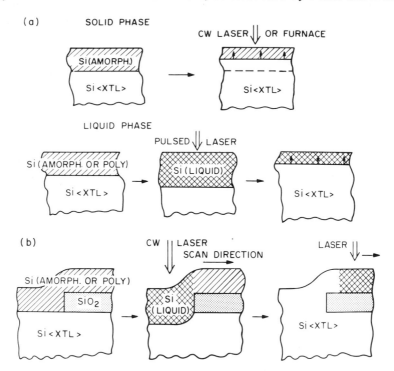

Fig. 9. Schematic of the two geometries for epitaxy. (a) Vertical epitaxy, the regrowing interface, in both the solid and liquid phases, moves in a fashion planar to the surface. (b) Lateral epitaxy, the melt puddle formed by a cw laser is scanned. Lateral epitaxy thus occurs at the trailing edge of the melt puddle.

Chapter 8. Silicon epitaxy can therefore be achieved with deposited films, whereas it is difficult to obtain epitaxial regrowth of deposited layers in the solid phase not only because of the importance of the interface conditions but also because gaseous impurities can be incorporated in the deposited amorphous layers. The vertical epitaxy of amorphous layers is shown schematically in Fig. 9, where we use the term "vertical" to distinguish this mode of crystallization from lateral growth.

One of the most exciting developments of this subject is that the lateral epitaxy of deposited Si layers has been realized. As shown schematically in Fig. 9, the molten zone can be moved controllably across the sample. In this example the Si substrate is used as a seed, and epitaxy occurs up and over the amorphous SiO_2 layer. The SiO_2 does not melt, and lateral sheets of single-crystal Si result. Nucleation can be controlled, for example, by shaping the melt puddle without recourse to substrate seeding. The subject of lateral epitaxy is discussed by Poate and Bean in Chapter 8.

One of the most crucial aspects of the fabrication of semiconductor devices is the alloying of metal layers for contact formation. Alloy contacts have been fabricated using lasers, and the results for Si are discussed by Gibbons and Sigmon (Chapter 10) and by von Allmen and Lau (Chapter 12) and for the compound semiconductors by Williams (Chapter 11). The fact that the quench rates are so fast can be utilized to fabricate unusual materials. In Chapter 12 von Allmen and Lau discuss the phase formation in the context of the equilibrium phase diagrams and the attainable quench rates.

V. Surface Crystallization and Alloying—Perspectives

The driving force behind much of this work is to produce new structures and materials or more efficient processing technologies for semiconductor device fabrication. The techniques are quite complementary to planar device technology whether for annealing or alloying. Indeed the possibility of treating the surface while the bulk remains cold appears to be an ideal direction for the technology. It is not possible to state now where the greatest use will occur. The device possibilities and limitations of directed energy processing are discussed by Hill in Chapter 13.

Whatever their ultimate utility, there is no doubt that these sophisticated surface heating techniques are opening new areas of crystal growth and alloying. Liquid phase quench rates and crystallization velocities have been considerably extended. Novel metastable phases and surface structures have been fabricated.

List of Symbols

The following symbols are used generally throughout the book; expanded lists of symbols are given in Chapters 2 and 4.

C_l, C_s	concentration of impurities in liquid, solid
C, C_p	heat capacity or specific heat (J/g K)
D	diffusion coefficient of heat impurities (cm²/sec)
$\Delta H_c, \Delta H_m$	enthalpy of crystallization, melting
I	power density (W/cm²)
k', k_i	segregation coefficient at the interface
R	reflectivity
T, T_m	temperature, melting temperature
v	velocity
α	absorption coefficient (cm⁻¹)
κ	thermal conductivity (W/cm K)
ρ	mass density (g/cm³)
τ	pulse duration

Chapter 2

Crystallization Processes

FRANS SPAEPEN and DAVID TURNBULL

Division of Applied Sciences
Harvard University
Cambridge, Massachusetts

I. Introduction

In this chapter we discuss the melting and solidification that may occur during laser annealing. Although crystallization is the thermodynamically preferred mode of solidification, a melt may under certain conditions solidify to an amorphous solid.

In laser annealing, as well as in a number of other microprocessing methods, a small volume of material is energized and then quenched. Usually, the energized state is one in which the atomic mobility is quite high. In the quench it becomes thermodynamically metastable, or even unstable, and may then transform to a different phase. In the latter part of the quench the material usually goes through a temperature regime in which it becomes frozen into a single configuration deriving from the intermediate state (Turnbull, 1981a). We suppose that configurational freezing has occurred when the average time for localized atomic rearrangements becomes long compared with the period of observation or use of the material.

Usually the magnitude and concentration of the energization is so great that the departures from equilibrium during the quench are very wide, so that formation of any one of a considerable variety of states becomes thermodynamically possible. Thus the outcome of the quench will be determined by the kinetically preferred course of structural evolution within this range of thermodynamic possibilities.

Typically the volume of material energized is very small. Also, the thermal gradients produced in the material in both the energization and quenching steps are extreme in magnitude compared with those achieved in normal materials processing. For example, when materials are energized by laser pulses in the picosecond range, the thermal gradients may be five to eight orders of magnitude greater than those typical in ordinary processing (Bloembergen, 1979).

Among the most important of the processes that occur in laser annealing is the formation of thin molten overlays on crystal surfaces and the crystallization of molten or amorphous solid overlays. These processes are effected heterogeneously by the movement of crystal–melt or crystal–amorphous-solid interfaces. This interfacial motion occurs in a sequence of at least two steps. For example, a crystal grows into a pure amorphous phase by an interfacial rearrangement followed by transport of the liberated heat away from the interface. Melting occurs by the reverse sequence. Impurities rejected in the interfacial process also must be transported away.

The question of the existence of a crystal in a state that is metastable relative to its melt is a much debated one. However, it is well established that crystals such as quartz (Ainslie *et al.*, 1961; Uhlmann, 1980), cristobalite (Ainslie *et al.*, 1961), albite (Uhlmann, 1980), and P_2O_5 (Cormia *et al.*, 1963), which melt to very viscous fluids, can be superheated far into their metastable ranges without internal melting. Thus the issue has mainly to do with the nature of melting to fluids with low viscosity. As observed, such melting is always heterogeneous; i.e., it is effected by the movement of crystal–melt interfaces. The thermodynamic crystallization point is the temperature at which two highly dynamic opposing processes, the forward and reverse motions of a planar crystal–melt interface, are exactly in balance so that the interface is stationary. If the interface moves forward the melt in contact with it must be undercooled; if it moves backward, the crystal at the interface must be, to some degree, superheated. The superheating must be greater the greater the speed of this reverse motion. Presumably, the regime of crystal metastability will be terminated at some temperature T_i where the shear modulus goes continuously to zero. Calculation of this instability point, at which homogeneous melting may be possible, is the object of many microscopic theories of melting. However, since the shear moduli of most crystals are quite large and fall slowly with T at T_m, the metastability ranges should extend generally to very large superheatings. We should expect that the interfacial superheatings will be quite large during ultrarapid laser melting.

II. Concepts of Crystal Growth

A. *General Formulation*

The velocity v of the crystal–melt interface may be expressed as (Turnbull and Cohen, 1960; Hillig and Turnbull, 1956; Turnbull, 1962)

$$v = fk_i\lambda[1 - \exp(\Delta G_c/RT_i)] \qquad (1)$$

where f is the fraction of interfacial sites at which rearrangement can occur, k_i is the frequency and λ is the displacement per rearrangement (the approximate interatomic distance), ΔG_c is the free energy of crystallization per gram atom, and T_i is the interfacial temperature. If the effective rearrangements were confined to ledges on a terraced interface, f would be the fraction of ledge sites. The thermodynamic factor $1 - \exp(\Delta G_c/RT_i)$ in the expression reduces the forward rate by the reverse rate. When the departure from equilibrium is not large, $RT_i \gg |\Delta G_c|$, and v reduces to

$$v \cong -fk_i\lambda \frac{\Delta G_c}{RT_i} \cong -\frac{fk_i\lambda(\Delta S_c)(T_m - T_i)}{RT_i} \qquad (2)$$

where ΔS_c is the entropy of crystallization per gram atom and T_m is the thermodynamic equilibrium temperature.

The flux $\dot{Q}$ of crystallization heat from the interface is given by

$$\dot{Q} = -\kappa(\mathrm{grad}\ T)_i = v\ \Delta H_c/\bar{V} \qquad (3)$$

where κ is the thermal conductivity of the transporting medium, $(\mathbf{grad}\ T)_i$ is the thermal gradient at the interface, ΔH_c is the heat of crystallization per gram atom, and $\bar{V}$ is the volume per gram atom. A lower limiting estimate of the thermal gradient can be obtained by assuming a linear temperature profile between the interface, at temperature T_i, and the heat bath around the specimen, at temperature T:

$$(\mathbf{grad}\ T)_i = (T - T_i)/d$$

where d is the characteristic distance between the interface and the heat bath; in conventional annealing experiments d is on the order of the specimen size; in laser annealing d is on the order of the thickness of the heated layer.

Equation (3) can then be rewritten

$$v = - (\kappa\bar{V}/\Delta H_c)(T - T_i)/d = k_h\lambda(T - T_i)/T_m \qquad (4)$$

where k_h is a characteristic frequency associated with the heat flow process:

$$k_h = \kappa\lambda^2/\Delta s_c\, d \qquad (5)$$

where Δs_c is the entropy of crystallization per atom; the atomic volume has been approximated by λ^3. The relation between the rates of interface advance and heat removal is obtained from Eqs. (1) and (4):

$$v = - [\kappa\bar{V}(\mathbf{grad}\ T)_i]/\Delta H_c = fk_i\lambda[1 - \exp(\Delta G_c/RT_i)] \qquad (6)$$

or in linearized form for both rates [Eqs. (2) and (4)]:

$$v = k_h\lambda(T - T_i)/T_m = - fk_i\lambda(\Delta S_c/R)(T_m - T_i)/T_i \qquad (7)$$

The interface temperatures can now be determined from this relation:

$$[(T_m - T_i)/(T_i - T)](T_m/T_i) = (k_h/fk_i)(R/\Delta S_c) \qquad (8)$$

Since for most materials $(\Delta S_c/R) \sim 1$, the interface temperature T_i (with $T < T_i < T_m$) is determined primarily by the ratio of the two characteristic frequencies k_h and fk_i.

If $fk_i \gg k_h$, then $T_i \to T_m$ and the interface motion is said to be heat flow limited, by which we mean, as shown in Fig. 1a, that $T_m - T_i$, *while nonzero*, is small compared to the difference $T_i - T$, driving the heat flow. In the alternative limit, $k_h \gtrsim fk_i$, $T_i \to T$, and the interface motion is considered to be interface limited; i.e., $T_m - T_i$ is very large (see Fig. 1b) compared with $T_i - T$. Figures 1c and 1d show schematic diagrams of the temperature profile for a liquid layer cooled from the crystal side for both growth regimes.

This general formalism and the conditions for heat-flow- or interface-

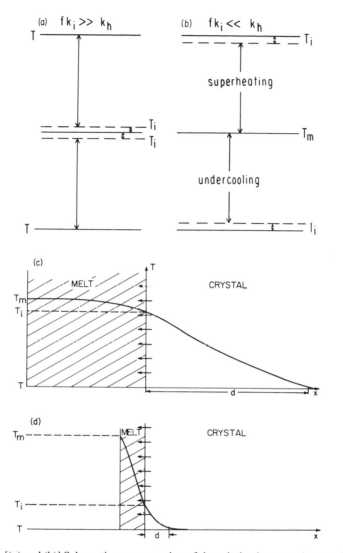

Fig. 1. [(a) and (b)] Schematic representation of the relation between the transformation temperature T_m, the interface temperature T_i, and the ambient temperature T for (a) the heat flow-controlled and (b) interface-controlled regimes. [(c) and (d)] Schematic diagrams of the temperature profile during (c) heat-flow-limited and (d) interface-limited crystal growth into a liquid layer cooled from the crystal side.

limited crystal growth will now be discussed more quantitatively for three different systems: pure metals, metallic alloys, and covalently bound materials.

B. Pure Metals

When free of heterophase impurities, pure metal melts exhibit a very high resistance to crystal nucleation, but, once formed, the crystal–melt interface moves with a high velocity even at quite small undercoolings. The reduced undercooling ΔT_r° at the onset of measurable homophase crystal nucleation (Turnbull, 1956, 1981b) in some liquid metals is on the order of 0.25–0.30 or larger, corresponding to a scaled surface tension $\alpha \sim 0.6$ (see Section III).

Metal crystals growing in their own melts develop dendritic morphologies at small undercoolings, indicating that the growth is diffusion-controlled, with the growth site fraction approaching unity ($f \approx 1$). We have proposed that the rate of the interfacial process in the growth of crystals into pure metal melts may be limited only by the frequency of collision of atoms z_1, per interface site, from the melt on the crystal face (Turnbull, 1974; Turnbull and Bagley, 1975; Spaepen and Turnbull, 1976). Then k_i would be equivalent to z_1, for which an upper limiting value can be obtained from the speed of sound v_s:

$$k_i = z_1 \lesssim v_s/\lambda$$

Typical values for metals give $k_i \approx 10^{13}$ sec^{-1}. If we choose as a criterion for heat flow control a difference between T_m and T_i of less than 1 K, it follows from Eq. (8) that this growth regime occurs when $k_h < 10^{10}$ sec^{-1}, since $(T_m/T_i) \sim 1$ and $(T_i - T) \sim 10^3$ K. For metals, $\kappa \approx 10^2$ W m^{-1} K^{-1} and $\Delta s_c \sim k_B$ (Boltzmann's constant). From Eq. (5) it follows then that heat flow-controlled flow occurs when $d > 2 \times 10^5 \lambda \approx 60$ μm. This condition is satisfied for bulk specimens. The crystallization of thin laser-melted layers, however, is largely interface-controlled, even for pure metals.

Walker (quoted in Chalmers, 1964) and Colligan and Bayles (1962) measured the speed v of dendrite tips of Ni or Co in their pure melts as functions of ambient undercooling ΔT_r [$=(T_m - T)/T_m$]. The speeds were quite well characterized and increased roughly as $(\Delta T_r)^2$ to $\Delta T_r \sim 0.1$, where $v \sim 50$ m/sec. Owing to the curvature of the dendrite tips and the high thermal gradients the interfacial undercooling ΔT_{ri} [$=(T_m - T_i)/T_m$] in these experiments must have been far below the ambient. From Eq. (2) and the parameters discussed above, the upper limiting growth speed can be estimated as

$$v \lesssim v_s \beta \, \Delta T_{ri} \tag{9}$$

where $\beta = \Delta S_c/R$.

Using the estimate $v_s \sim 2000$ m sec^{-1} for Ni and setting $\beta \sim 1$ we calculate that the growth rate measured by Walker at $\Delta T_r = 0.1$ is consistent with

$$\Delta T_{ri} \geq \Delta T_r/4 = 0.025$$

and that the thermal gradient $|\mathbf{grad}\ T|_i$ would have been on the order of 10^7–10^8 K m^{-1}.

We expect that the upper limiting rate of the interface motion in melting also will be given by Eq. (9). Thus measurement of this interface speed should permit estimation of the lower limit of interfacial superheating, i.e.,

$$\left| \Delta T_{ri} \right| \geq \left| \frac{v}{v_s\beta} \right|$$

C. Metallic Alloys

If interface movement requires the redistribution of impurity, by either rejection by one phase or ordering, then k_i must scale with k_D, the frequency of diffusive transport in the interfacial region, rather than with z_1 (Spaepen and Turnbull, 1976). Since a highly correlated set of motions is required for diffusion in condensed systems, we expect that k_D will lie well below z_1 and fall more sharply with decreasing temperature. It can be estimated from the diffusivity:

$$k_i = k_D = 6D/\lambda^2$$

For typical values of the diffusivity in metallic liquids, $D \approx 10^{-8}$ m^2 sec^{-1}, this gives $k_i \approx 6 \times 10^{11}$ sec^{-1}. Since the other parameters are similar to those for pure metals, the heat flow-controlled regime can be said to occur when $k_h < 1.6 \times 10^9$ sec^{-1} or $d > 3.6 \times 10^6\lambda \approx 1$ mm.

The quench rate of a molten layer (thickness d) on a cold substrate can be estimated in terms of the quantities introduced above:

$$\dot{T} = \dot{Q}\bar{V}/C_p d \tag{10}$$

where $\dot{Q}$ is the heat flux out of the melt into the substrate and C_p is the molar specific heat (~ 25 J K^{-1} mole^{-1} for metals). The heat flux can be written

$$\dot{Q} = -\kappa(\mathbf{grad}\ T) \approx \kappa(T_m - T)/d \tag{11}$$

which gives in Eq. (10):

$$\dot{T} = \kappa\bar{V}(T_m - T)/C_p d^2 \tag{12}$$

In order to quench alloy melts into the glassy state, quench rates on the order of $\dot{T} \gtrsim 10^6$ K sec^{-1} are necessary. This means that, according to Eq. (12), $d \lesssim 200$ μm. It seems therefore that the epitaxial crystal growth process, which must be prevented to form a glass, is always interface limited.

D. Covalently Bound Materials

In covalently bound materials (Si, Ge, SiO$_2$, GeO$_2$, B$_2$O$_3$), the molecular transport frequencies—e.g., k_D or k_η, the frequency of viscous flow—are relatively small even at T_m. For example, the self-diffusivity in Si at its melting point is $D = 4 \times 10^{-16}$ m^2 sec^{-1} [average value estimated from data listed in Hirvonen and Anttila (1979)]. This lets us estimate $k_i \approx 2 \times 10^4$ sec^{-1}. The crystallization process is therefore interface limited if $k_h/f \gtrsim 70$ sec^{-1} [see Eq. (8)]. Equation (5) shows that this condition is satisfied for $fd \lesssim 250$ m (using $\kappa \approx 10$ W m^{-1} K^{-1}, $\Delta s_c = 3.6$ k_B), which is always the case. The interfacial temperature is therefore always close to the ambient temperature.

As the undercooling increases from zero, v rises to a maximum and then falls on a course approaching an exponential decrease with falling T at temperatures well below the maximum. This behavior is consistent with a thermally activated interface frequency k_i. In the framework of transition state rate theory k_i may be expressed (Spaepen and Turnbull, 1979; Fratello et al., 1980)

$$k_i = n_r \nu_r \exp(\Delta S_1'/R) \exp(-\Delta H'/RT)$$

where ν_r is the normal frequency of the internal motion leading to interfacial rearrangement, $\Delta S_1'$ and $\Delta H'$ are, respectively, the entropy and enthalpy of activation, and n_r is the number of rearrangements per activation. Often a pair of dangling half-bonds are formed by the activation, and they may, with little additional activation, migrate over several interfacial sites before recombining, thus effecting n_r rearrangements. This process is quite analogous with that occurring in the free-radical mechanism of homogeneous chemical reactions.

The thermodynamic factor becomes $1 - \exp(n'\Delta G_c/RT)$, where $\Delta G_c/\bar{N}$ is the free energy of crystallization per atom ($\bar{N}$ is Avogadro's number) and n' the number of atoms crystallized per rearrangement. In an earlier paper (Fratello et al., 1980) we mistakenly used n_r instead of n' in the exponential. In most chain reactions of interest here $n' \ll n_r$. If $n' \gg 1$, the exponential may approach unity at rather small undercoolings.

In fused silica the frequency of homophase crystal nucleation is below measurable levels under all conditions so far tested (Ainslie *et al.*, 1962). Indeed, there has been no demonstration of the actual occurrence of homophase crystal nucleation in any covalent glass-forming melt. When crystallization occurs in these materials, it is nucleated by heterophase impurities. This high resistance to crystal nucleation is partly accounted for by the high activation energies of interfacial rearrangement, but it also indicates, if simple nucleation theory is valid, that the scaled interfacial tension α of these materials is $\gtrsim \frac{1}{2}$ (see Section III).

E. Conclusions for Thin Molten Overlays

Laser or electron beam heating of a surface results in the formation of a molten overlay whose thickness depends primarily on the total energy fluence. For typical laser irradiation experiments the fluence is ~1 J cm^{-2}, resulting in a layer thickness, hence a characteristic distance d, for the temperature gradient on the order of 1 μm. In the previous subsections we have shown that the regrowth in this case is interface limited for all materials. It is therefore important to include the undercooling at the interface, $T_m - T_i$, in all model analyses of these experiments. The dependence of the crystal nucleation frequency and the crystal growth velocity on the undercooling is therefore discussed in the following sections.

III. Concepts of Nucleation

In laser annealing molten or amorphous solid overlays often withstand, for some time, large undercoolings without crystal nucleation. We now discuss the origin of this resistance in terms of the simple or "classical" nucleation theory (Turnbull, 1956, 1981b). This theory also provides the basis for interpreting f, the growth site fraction, which appears in Eq. (1).

A. Bulk Nucleation

The central assumption of the simple theory is that the reversible work of nucleus formation may be found by extrapolating the Gibbs theory of capillarity into the microscopic regime. This work is a maximum,

$$W^* = 16\pi\sigma^3/3(\Delta G_v)^2 \qquad (13)$$

when the radius of the nucleus reaches a critical value

$$r^* = -2\sigma/\Delta G_v \tag{14}$$

where σ is the interfacial tension, here assumed to be isotropic, between the crystal and amorphous phase and ΔG_v is the Gibbs free energy of crystallization per unit volume of crystal. Kinetic analysis in combination with fluctuation theory leads to an expression for I, the steady state nucleation frequency per unit volume, having the form

$$I \cong Ak_i \exp[- 16\pi\sigma^3/3(\Delta G_v)^2 k_B T] \tag{15}$$

where A is a constant which can be specified by the kinetic analysis and k_i is the temperature-dependent exchange frequency already defined. It is convenient to reform Eq. (15) in terms of scaled variables as follows (Turnbull, 1964, 1969):

$$I \cong Ak_i \exp[- 16\pi\alpha^3\beta/3(\Delta T_r)^2 T_r] \tag{16}$$

where $\alpha = -(\bar{N}\bar{V}^2)^{1/3}(\sigma/\Delta H_c)$, $\bar{N}$ is Avogadro's number, ΔH_c is the enthalpy of crystallization per gram atom, $\beta = -\Delta S_c/R$, ΔS_c is the entropy of crystallization per gram atom (assumed to be T-independent), $T_r = T/T_m$, and $\Delta T_r = (T_m - T)/T_m$.

The variation in I with ΔT_r for different choices of the parameters α and k_i has been displayed in other papers (Turnbull, 1964, 1969; Spaepen and Turnbull, 1976). As ΔT_r increases from zero, I rises steeply but does not reach experimentally measurable values until the undercooling exceeds some "onset" level ΔT_r°, which increases with α.

In laser annealing large undercooling or superheating may be attained in periods so short that the density of structural fluctuations, leading to nucleus formation, may not have reached the steady state value assumed in the derivation of Eq. (15). Calculation of the period, the "transient" τ_{tr}, needed to establish the steady distribution is a difficult problem. However, a lower limiting value for τ_{tr} may be estimated, as pointed out by Hillig (1962), from the relation

$$\tau_{tr} \geq i^*/k_i \tag{17}$$

where i^* is the number of atoms in the nucleus.

B. Surface Nucleation and Roughness

When the crystal growth rate is interface controlled, the crystal takes on a form in which it is bounded by planes normal to the directions of slowest growth. Usually these are the planes on which the atomic packing is most dense. Consider a terrace bounded by a ledge on such a plane. The ledge positions will be the strongest binding sites and preferred as growth sites to isolated positions on the terrace. The configurational entropy of the system can be increased, but at a large energy cost, by the breakup or "roughening" of the interface into a multiterraced structure with a high density of ledge sites.

For a densely packed face of a crystal in contact with a dilute fluid, Burton *et al.* (1951) showed that the energy of interface disordering is so large that the form of the interface should, at equilibrium, remain virtually smooth and perfect to temperatures approaching T_m. At some higher temperature a roughening transition would occur. This high resistance to interface disordering reflects the relatively large atom displacements, e.g., movement of an atom from a position within to one on the terrace, needed to produce new configurations.

In the smooth interface regime, growth of the perfect crystal is effected by the nucleation and growth of terraces on the densely packed faces. Application of simple nucleation theory leads to the condition that to expand rather than collapse the radius of a circular terrace must exceed the critical value

$$r^* = - \sigma_e/\Delta G_v \tag{18}$$

where σ_e, assumed to be isotropic, is the work per unit area of forming the vertical edge of the terrace. A kinetic analysis analogous to that used for the three-dimensional nucleation problem leads to an expression for the steady frequency per unit area I_s of terrace nucleation having the form

$$I_s = A_s k_i \exp(\pi \lambda \sigma_e^2/\Delta G_v k_B T) \tag{19}$$

where λ is the ledge height. This equation predicts that substantial departures from equilibrium, increasing with σ_e, will be required for measurable rates of terrace nucleation, hence of crystal growth on perfect densely packed planes. The substantial growth rate usually observed under near-equilibrium conditions reflects operation of the Frank (1949) screw dislocation mechanism. In this mechanism growth occurs by addition of material at the edges of steps of the spiral ramp structures formed around the emergence points of screw dislocation in the interface. The spacing ΔR between the growth steps is proportional to r^*:

$$\Delta R = K_s r^* \tag{20}$$

and the fraction of growth (also ledge) sites is

$$f \cong \lambda/R \cong \lambda/K_s r^* \tag{21}$$

The form of f, with departure from equilibrium, on the close packed faces

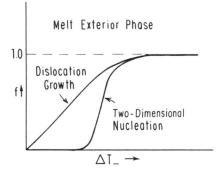

Fig. 2. Fraction of potential growth sites in the interface f versus undercooling ΔT_- for two regimes of crystal growth.

of perfect and dislocated crystals is shown schematically in Fig. 2. After the roughening transition f should approach unity.

When the average crystallographic orientation of the interface plane deviates somewhat from that of a low index face, it is thermodynamically favorable (Herring, 1951) for the interface to break into terraces which expose the low index plane, and monatomic ledges, which may serve as growth sites. When the orientation deviation is described by a single angle θ of inclination of the average to the low index orientation, the growth site fraction f will increase as $\sin \theta$.

The microscopic theory for the equilibrium structure of a densely packed crystal face in contact with its melt or a concentrated fluid is still not well developed. However, inferences on this structure, as well as on the growth mechanism, can be drawn from the morphology of crystals growing, near equilibrium, within their melts. In particular, polyhedral growth forms indicate sharp interface structure and interface-controlled growth, while dendritic morphologies imply a rough interface and diffusion-limited growth. Jackson (1958) noted that the morphologies exhibited in growth from the melt correlate quite well with the magnitude $|\Delta S_c|$ of the crystallization entropy. In particular, the morphology tends to be dendritic when $|\Delta S_c| \leq 2R$ and polyhedral when $|\Delta S_c| \geq 2R$.

IV. Phase Transformations of Elemental Silicon

A. *Free Energies of the Phases*

The differences in free energy between the three phases of the elemental semiconductors [the covalently bound crystalline (c) and amorphous (a) phases, and the metallic liquid (*l*)] are illustrated in Fig. 3. Whereas the

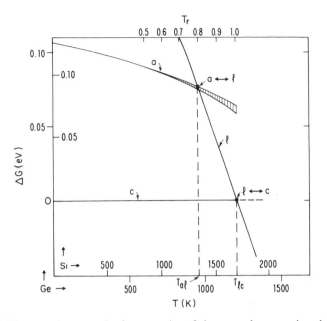

Fig. 3. Differences between the free energies of the amorphous semiconductor (a) or liquid (*l*) phase, and the crystalline (c) phase, for Si (inner scales) and Ge (outer scales). The curves have been calculated for Ge and rescaled for Si according to the melting temperatures and moduli.

thermodynamic properties of the crystalline and liquid phases are well known, those of the amorphous phase can only be estimated.

The curve in Fig. 3 is based on a calculation for a-Ge (Spaepen and Turnbull, 1979); its heat of crystallization and specific heat have been measured (Chen and Turnbull, 1969) in a limited temperature range; the calculation also uses a model-based estimate (Spaepen, 1974) of the residual entropy of the tetrahedrally coordinated random network. A similar calculation has been made by Bagley and Chen (1979).

No such measurements have been reported so far for a-Si. Therefore, as a simplest approximation, the curves for Si are obtained by scaling the temperature axis according to the melting point of the crystals $[T_{lc}(Ge) = 1210 \text{ K}; T_{lc}(Si) = 1683 \text{ K}]$. Since the most important contribution to the heat of crystallization is the bond-bending strain energy in the amorphous phase, the energy axis is scaled according to the crystalline shear moduli (Huntington, 1958). $[c_{44}(Ge) = 67 \text{ GPa}; c_{44}(Si) = 79 \text{ GPa}]$. Direct measurements of ΔH_{ac} for Si are necessary to make this estimate more precise.

In the free-energy diagram in Fig. 3 the a → *l* transition has been drawn as one of first order, thermodynamically. This seems plausible in view of the intrinsically different short-range order in the two phases [the respective coordination numbers are $Z(a) = 4$ and $Z(l) \cong 12$, for example]. There is no direct experimental evidence for the first-order nature of this transition. However, the necessity for including a finite latent heat of melting ΔH_{al} in the heat flow in a crystallization analysis of pulsed heating experiments on a-Si overlays can be taken as indirect evidence of a first-order transition (Baeri *et al.*, 1980).

Figure 3 shows that the amorphous semiconductor phase melts at a considerably lower temperature than the crystalline one: $T_{al}(Ge) = 970$ K and $T_{al}(Si) = 1350$ K.

B. Solid Phase Epitaxial Regrowth

When an a-Si overlay on a single-crystal Si substrate is crystallized by furnace annealing or by long laser or electron beam pulses ($t > 10^{-3}$ sec), the regrowth process occurs entirely in the solid state: both the crystal and the overlay are covalently bound semiconductors. The process, therefore, corresponds to the one discussed in Section II,D: it is interface controlled by the multiple rearrangements of dangling bond pairs following a bond-breaking activation event. This has been shown explicitly from structural models of the a-Si–c-Si interface (Spaepen and Turnbull, 1979): (i) interfaces without broken bonds are energetically favored, hence necessitating the activation step for crystallization (Spaepen, 1978); (ii) breaking of one bond creates sufficient topological freedom for multiple crystallization rearrangements; the two half-bonds have been shown to propagate along a ⟨110⟩ ledge on a {111} interfacial plane (Spaepen and Turnbull, 1979).

Csepregi *et al.* (1978) have measured the orientation dependence of the regrowth velocity v. Figure 4 shows their data as a function of the angle θ between the growth direction and the [111] direction in the crystal substrate. As they point out, and as discussed in Section IIIB, the proportionality between v and $\sin \theta$ indicates that growth occurs by a (111) layering mechanism. Most likely, this layering is governed by the motion of ⟨110⟩ ledges, since these are the densest directions in the {111} planes.

In order to explain the data in Fig. 4 completely, however, it is necessary to use two different ledge velocities $v_l = v/\sin \theta$; for orientations deviating from the [111] toward the [110] direction, $v_{l,1} = 52$ Å/min; for orientations deviating toward the [100] direction, $v_{l,2} = 122$ Å/min. This difference can be explained by the different nature of the ledges involved. Figure 5a shows that an interface plane (hkl), deviating from [111] toward [110], is made up of (111) planes ending at [1$\bar{1}$0] ledges of the type AA'

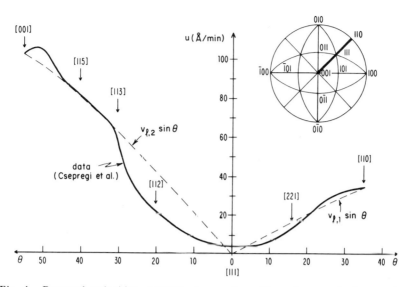

Fig. 4. Regrowth velocities at 550°C of amorphous Si overlays on single-crystal substrates of different orientations as a function of the angle between the regrowth direction and the substrate [111] orientation. (Data from Csepregi *et al.*, 1978.)

shown in Fig. 6. On the other hand, as Fig. 5b shows, an interface plane (*hkl*) deviating toward [100] is made up of (111) planes ending at [1$\bar{1}$0] ledges of the type BB′ in Fig. 6. Figure 6 is a view of one of the (111) planes; in the [1$\bar{1}\bar{1}$] direction (i.e., into the substrate). It clearly demonstrates the difference between the two types of [1$\bar{1}$0] ledges: the AA′ ledge has only one bond connecting to the amorphous phase, whereas the BB′ ledge has two. It is plausible to assume that this would make the latter type more mobile. If, as a result of this higher mobility, the BB′ ledge facets into slower AA′ segments (see Fig. 5b), the overall ledge would still have a higher velocity than a straight AA′ ledge, since kink nucleation would not limit the velocity of the faceted ledge.

The finite velocity of the growth in the [111] direction (~5 Å/min in Fig. 4) cannot be explained by homogeneous nucleation of new layers. The growth velocity in this regime can be written as (Hillig, 1966)

$$v = \lambda I_s^{1/3} v_l^{2/3} \tag{22}$$

where I_s is the surface nucleation frequency of Eq. (19). Since the rearrangements associated with the formation of a layer nucleus are the same as those for ledge motion, the frequency k_i can be written as

$$k_i = 2\pi v_l / \lambda \tag{23}$$

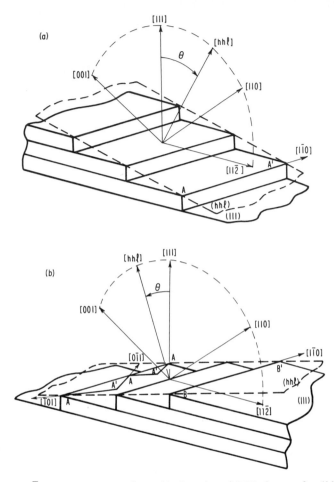

Fig. 5. (a) [1$\bar{1}$0] Ledge structure, formed by layering of (111) planes, of an (hhl) a-Si–c-Si interface for a substrate orientation deviating from [111] in the direction of [110]. (b) Same as (a), but for substrate orientations deviating in the direction of [100]. Notice that the BB′-type ledge can break up into AA′AA′ segments.

Combining Eqs. (9), (22), and (23) gives then for the growth velocity:

$$v \propto v_l \exp(\pi\lambda\sigma_e^2/3 \; \Delta G_v k_B T) \tag{24}$$

Csepregi *et al.* (1978) observed that the activation energy for growth in the [111] direction was the same as for all other directions. Since in all these other cases the temperature dependence of v is governed by v_l only, homogeneous nucleation can be the mechanism for growth in the [111]

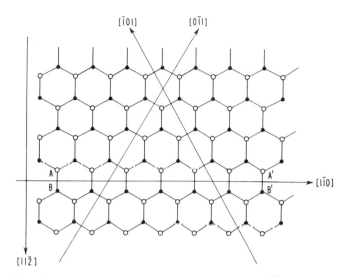

Fig. 6. Top view of a (111) plane, showing the two types of [1$\bar{1}$0] ledges, AA' and BB'. The BB'-type ledge has two bonds connecting to the overlaying amorphous phase. O, Bond along [$\bar{1}\bar{1}\bar{1}$]; ●, bond along [111].

direction only if the exponential factor in Eq. (24) is almost temperature independent in the range of the observations. However, an estimate (Spaepen, 1982) of the temperature dependence of this factor, based on a model-derived value of 0.4 J m^{-1} for σ_e, gives an activation energy of 0.4 eV, which is not negligible compared to the overall activation energy of 2.3 eV. Furthermore, the magnitude of v (5 Å/min) requires a nucleation frequency $I_s \approx 10^{14}$ m^{-2} sec^{-1}, using the value of $v_l = 122$ Å/min derived above. A calculation (Spaepen, 1982) of the homogeneous nucleation frequency, based on the model value of σ_e, gives $I_s \approx 10^7$ m^{-2} sec^{-1} only.

The alternative explanation for the finite value of v along [111] is growth on a fixed number of nuclei occurring with a concentration of N per unit area in the interface. This gives for the growth velocity

$$v = \lambda N^{1/2} v_l \tag{25}$$

It is clear that in this regime v has the same activation energy as v_l, hence as in all the other orientations of growth. The observed magnitude would require a nucleus concentration $N = 10^{16}$ m^{-2}. The nature of these nuclei is as yet unknown. However, only a few degrees of deviation from the precise [111] orientation would create a sufficient number of interface ledges to account for this concentration.

C. Liquid Phase Epitaxial Regrowth

If an a-Si overlay can be brought to its melting temperature T_{al}, before
solid state crystallization takes place, the overlay will melt by the motion
of an a–l interface from the surface inward, and if the layer melts completely, it will regrow epitaxially by the motion of a c–l interface outward.

Extrapolation of the fastest solid state regrowth velocity (i.e., along
[100]) to $T_{al} = 1350$ K gives $v = 4 \times 10^{-5}$ m sec^{-1}. Crystallizing an overlay,
with a typical thickness of 1 μm, completely at this temperature would
therefore take 2×10^{-2} sec. It is clear that laser and electron beam pulses
can heat the overlay rapidly enough to permit liquid phase epitaxy.

Since the liquid phase is metallic in nature, the atomic rearrangements
governing the motion of the c–l and a–l interfaces are probably more like
those in pure metals (collision) than those in purely covalent systems
(bond breaking). Since it is observed, however, that crystals grown from
the melt exhibit {111} facets, the fraction of sites f available for crystallization at the c–l interface is much less than in a metallic system, where
$f \approx 1$; it is probably again determined by the ledge structure of the interface.

Figure 7 summarizes the various regimes for crystal growth or melting
when the melt is the exterior phase. If the growth sites fraction f is con-

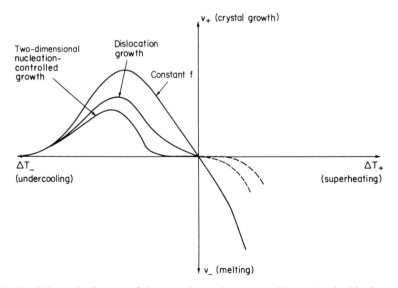

Fig. 7.   Schematic diagram of the crystal growth (v_+) or melting (v_-) velocities in various
regimes for a Si crystal surrounded by its melt. The dashed lines represent melting of a thin
layer on a flat Si crystal.

stant, the growth velocity increases linearly with undercooling ΔT_-. This would be the case for metals ($f \approx 1$) and for regrowth of a Si overlay away from the [111] orientation ($f < 1$ but constant). An upper limit of the slope can be estimated from Eq. (9) (i.e., assuming $f = 1$):

$$\frac{d(v)}{d(\Delta T_i)} = v_s \frac{\beta}{T_m} \approx 10 \quad \text{m sec}^{-1} \text{ K}^{-1}$$

If f is not too small, a lower limit can be obtained from Eq. (2) by assuming that the attachment frequency is determined by the liquid diffusivity ($k_i \approx 10^{11}$ sec^{-1}; see Section II.C):

$$\frac{d(v)}{d(\Delta T_i)} = k_i \lambda \frac{\beta}{T_m} = 0.06 \quad \text{m sec}^{-1} \text{ K}^{-1}$$

If the undercooling becomes large enough ($\Delta T_i > T_m/2$), the crystal growth velocity would decrease with increasing ΔT_i because of a decrease in k_i, which would eventually result in glass formation; this is of little practical importance for elemental Si.

The curves in Fig. 7 for dislocation and two-dimensional nucleation-controlled growth apply to the regrowth of Si in the [111] direction but are difficult to obtain quantitatively. Since during melting of a *three*-dimensional Si crystal surrounded by its melt new ledges can always be formed at the crystal edges, $f \approx 1$ for this process (solid curve in Fig. 7). The formation of a *two*-dimensional molten layer on a [111] crystalline surface, however, is limited by the presence of ledges and would therefore probably be described more accurately by curves similar to those for dislocation or nucleation-controlled growth (dashed curves in Fig. 7).

The orientation dependence of the regrowth velocity of Si is also reflected in the anisotropy of the effective impurity distribution coefficient, k (see also Section V). Baeri *et al.* (1981) showed that, for the same regrowth velocity, Bi impurity was incorporated more during regrowth along [111] than along [100]. This could be explained by the difference in ledge velocity for the two cases: the number of ledges controlling growth along [111] is probably much smaller than the intrinsic number of ledges forming the (100) interface; for the same normal regrowth speed, the ledge velocity for [111] regrowth must therefore be much higher than for [100] regrowth, resulting in increased impurity trapping by the faster ledges.

D. Formation of a-Si by Melt Quenching

Liu *et al.* (1979) have demonstrated that a-Si is formed by 30-psec laser pulse irradiation of a single-crystal surface. At the uv wavelength used in the experiment, such a pulse melts a surface layer a few hundred angstroms thick. Since the thermal gradients in this case are extremely steep, the regrowth of the c-Si is very fast. However, since in this regime all regrowth is interface controlled (see Section II.B), the interface temperature T_i can deviate considerably from the melting temperature $T_{lc} = 1683$ K. If $T_i < T_{al}$, it becomes thermodynamically possible to form the amorphous phase from the melt at the interface. Therefore, topological perturbations in the crystal growth process are not corrected in this regime but continue to grow as the amorphous phase.

Using Eq. (8) with $T_m = T_{lc}$, $T_i = T_{al}$, and $T = 300$ K, the condition for formation of the amorphous phase becomes

$$(k_h/fk_i)(R/\Delta S_c) > 0.4 \tag{26}$$

Using the definition of the heat flow frequency, Eq. (5),

$$k_h = \kappa\lambda^2/\Delta s_c d > 0.1fk_i$$

Since the melt is metallic, $fk_i \approx 10^{12}$ sec^{-1} and therefore, $d \lesssim 200$ Å. This is precisely the thickness one would estimate from the wavelength and the duration of the laser pulse (Bloembergen, 1979). The quench rate in this regime can also be estimated from Eq. (12) and gives $\dot{T} \approx 10^{13}$ K sec^{-1}.

V. Nonequilibrium Impurity Incorporation during Crystal Growth

A. Thermodynamic Conditions

The thermodynamic conditions for nonequilibrium distribution of impurity during crystallization have been set forth by Baker and Cahn (1971). The essential requirement for forming an infinitesimal amount of crystal, in which the mole fraction of the minor constituent B is x_s, from a two-component melt of B in solvent A is that the accompanying Gibbs free-energy change per mole:

$$\Delta G = (1 - x_s)\,\Delta\mu_A + x_s\,\Delta\mu_B \leq 0 \tag{27}$$

where

$$\Delta\mu_A = \mu_s^A - \mu_l^A \qquad \text{and} \qquad \Delta\mu_B = \mu_s^B - \mu_l^B$$

where μ_s^A, μ_s^B are the chemical potentials of A and B, respectively, in the forming crystal and μ_l^A, μ_l^B are the chemical potentials of A and B, respectively, in the melt where the mole fraction of B is x_l. When the dilute solution approximations apply, the composition dependences of $\Delta\mu_A$ and $\Delta\mu_B$ may be expressed as

$$\Delta\mu_A = RT \ln \frac{(1 - x_s)(1 - \bar{x}_l)}{(1 - x_l)(1 - \bar{x}_s)} \tag{28}$$

$$\Delta\mu_B = RT \ln(x_s\bar{x}_l/x_l\bar{x}_s) \tag{29}$$

where $\bar{x}_l$ and $\bar{x}_s$ denote the equilibrium mole fractions of B in the liquid and crystal, respectively, at the temperature T.

A schematic representation of the isothermal variation in G with x for the liquid and crystalline solutions is shown in Fig. 8. The molar decrease in free energy accompanying crystallization is proportional to the length of the line cd. We note that, at the melt composition x_l, $\Delta G \le 0$ for any crystal composition lying between the points a and b on the diagram. At all interfacial melt compositions between $\bar{x}_l$ and x_0 some impurity rejection across the interface must attend crystallization, and in this regime the crystal growth rate cannot exceed the rate at which the impurity is transported away from the interface. However, diffusionless growth is possible when $x_l \le x_0$, where x_0 is the melt composition at the point 0 of intersection of the liquid and crystal curves.

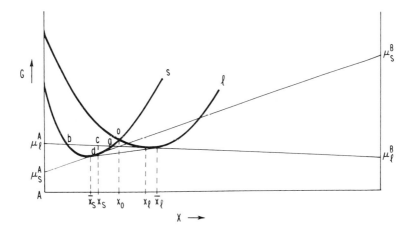

Fig. 8. Schematic diagram of the free-energy differences and chemical potentials for discussion of the impurity-trapping problem. See text.

B. Kinetic Analysis

The kinetic analysis of the interface motion in a two-component system is a difficult problem that has been treated from various perspectives by a number of authors (Cahn et al., 1980; Jackson et al., 1980; Wood, 1980; Hillert and Sundman, 1977; Turnbull, 1980). An instructive review and critique of the various analyses was presented by Cahn et al. (1980). Here we attempt no complete analysis, but we discuss general features of the problem and some limiting cases.

At steady state the velocity of the planar interface must satisfy simultaneously an equation, like Eq. (1), for the interface kinetics, one for heat transport, and one for impurity transport, as well as the thermodynamic conditions. The impurity drift speed v_B along the concentration gradient $\partial c/\partial y$ normal to the interface may be expressed as

$$v_B = -\frac{M_B}{\bar{N}}\frac{\partial \mu_B}{\partial y} \qquad (30)$$

where M_B is the diffusive mobility of the impurity and $\bar{N}$ is Avogadro's number. For dilute solutions $M_B \approx D_l/k_B T$ and

$$\frac{\partial \mu_B}{\partial y} \approx \frac{RT}{c}\frac{\partial c}{\partial y}$$

hence

$$v_B = -\frac{D_l}{c}\frac{\partial c}{\partial y} \qquad (31)$$

We expect $|\partial c/\partial y| \leq c/\lambda$, so $v_B \leq D_l/\lambda$. With $D_l \sim 10^{-5}$ cm² sec⁻¹ and $\lambda \sim 10^{-8}$ cm, v_B should not exceed a value $v_B^0 \sim 10$ m/sec, and for gradients of the usual magnitude it would be one or more orders of magnitude below this maximum. The upper limiting speed of the interface process given by Eq. (9) would exceed the maximum v_B at all $\Delta T_{ri} \geq 0.05$.

From an alternative viewpoint the frequency k_D of diffusive jumps should be v_B^0/λ and that, z_1, of interface jumps $\sim v_s/\lambda$. Using $v_s \sim 2000$ m/sec and $v_B^0 \sim 10$ m/sec we see that at equivalent driving free energy the interface process should be on the order of 200 times faster than the diffusive process. From this result it seems that the diffusionless growth mode, with complete solute trapping, is likely to predominate whenever x_0 substantially exceeds x_l.

Now suppose that $x_l > x_0$, so that the composition must change across the moving interface. When the composition x_l' in the melt at the interface equals that, x_l, in the bulk, ΔG of crystallization is, as we have noted, proportional to the length cd at the solid composition x_s. As the interface moves, the impurity is rejected into the melt, so that x_l' comes to exceed x_l with attendant development of a concentration gradient at the interface proportional to $x_l' - x_l$. The interface velocity will evolve toward a steady state value in which the rates of interface motion, heat transport, and impurity migration into the melt are matched. To complete the solution of the problem we need, in addition to these relations, the net solvent–impurity exchange frequency across the interface. It is in the microscopic theory for this frequency that the differences between the various treatments mainly arise. Here we summarize an analysis due to Aziz (1982).

Assuming that there is no net adsorption of impurity in the interfacial region and that the diffusivity of impurity across the interface in either direction is constant at D_i, Aziz obtains the following expression for the net rate of change $\dot{x}_s$ of the impurity content of the solid at the interface:

$$\dot{x}_s = -(D_i/\lambda^2)[x_s(t) - x_l'k_e(T_i)] \qquad (32)$$

where $k_e(T_i)$ is the equilibrium distribution coefficient of impurity at the interfacial temperature T_i. This relation leads to the upper limiting value D_i/λ^2 for the frequency of solvent–impurity exchange across the interface.

A monolayer of composition x_l' is instantaneously solidified at t = 0. Solute is then allowed to diffuse back into the liquid until the next monolayer is added at t = λ/v, trapping the remaining solute into the bulk of the solid where diffusion is negligible. Solution of the differential equation with the boundary condition $x_s = x_l'$ at $t = 0$ leads to

$$x_s(t) = x_l'\{k_e(T_i) + [1 - k_e(T_i)] \exp(-D_it/\lambda^2)\} \qquad (33)$$

and setting $t = \lambda/v$,

$$x_s = x_l'\{k_e(T_i) + [1 - k_e(T_i)] \exp(- D_i/\lambda v)\} \qquad (34)$$

This equation combined with an equation, such as Eq. (1), for the interface speed and appropriate solutions to the heat and impurity transport equations can, with x_l and T_s given, be solved in principle for v, T_i, x_l', and x_s.

Actually, the results of the analyses are usually presented in terms of variations in the ratio x_s/x_l' with imposed interface speed v. In Aziz's analysis with D_i/λ taken to be 10 m/sec (i.e., $D_i \sim D_l$) the transition from $x_s/x_l' = k_e$ to unity occurs over one order of magnitude in v with $v \sim 10$ m/sec at midrange and 30 m/sec for complete trapping. The midrange velocity v_m would be reduced substantially by choosing $D_i < D_l$. It increases somewhat with increasing $k_e > 0.1$ but shifts rather little with decreasing $k_e < 0.1$. The latter result reflects the effect of setting an upper limit on the solvent–impurity exchange frequency.

An analysis due to Baker, reported by Cahn *et al.* (1980), results in an expression for x_s/x_l' which simplifies to

$$\frac{x_s}{x_l'} = \frac{[\exp(-\lambda v/D_l)]/k_i}{1 + [\exp(\lambda v/D_i)]\ln(k_e/k_i)} + \frac{k_e[1 - [\exp(-\lambda v/D_l)]/k_i]}{1 + [\exp(\lambda v/D_l)]\ln k_i} \quad (35)$$

where k_i is the equilibrium distribution coefficient of impurity in the interfacial region (i.e., the possibility of impurity adsorption at the interface is allowed for) and D_i is the diffusivity in the interfacial layer. When $k_i = k_e$ and $D_i = D_l$, the expression becomes

$$\frac{x_s}{x_l'} = \frac{(v\lambda/D_l) + \ln k_e}{(v\lambda/D_l) + [(\exp(-\lambda v/D_l)/k_e]\ln k_e} \quad (36)$$

It gives results approaching those of Aziz (1982) as $k_e \to 1$ but predicts midrange velocities one-half to two orders of magnitude greater than those in Eq. (34) with decreasing $k_e < 0.1$. For example, at $k_e = 0.1$ it gives $v_m \sim 40$ m/sec. This disparity in the predictions of the two equations seems to arise because Baker does not impose an upper limit on the solvent–impurity exchange frequency.

C. Microsegregation and Stability of the Interface

Cahn *et al.* (1980) have also reviewed the application of theories of dendrite formation to the problem of rapid solidification. They point out that the classical theory of constitutional undercooling (Tiller *et al.*, 1953), which takes into account only the depression of the melting point due to impurity rejection, predicts increasing dendritic instability with increasing interface velocity. The more complete theory of Mullins and Sekerka (1964), which also takes capillarity effects into account, shows that, although the classical theory is essentially correct at low interface speeds, it greatly overestimates the tendency toward dendrite formation in the rapid solidification regime: at high interface speeds, the wavelength of the instability decreases, which leads to an increased stability against dendrite formation with increasing interface velocity. Specific calculations by Cahn *et al.* (1980) for the solidification of doped Si show that the tendency toward dendrite formation, in a given temperature gradient, is greatest for interface velocities of about 1 cm sec^{-1}; at lower velocities stabilization occurs because of a decrease in constitutional undercooling; at higher velocities it occurs because of increased capillary stabilization.

In these calculations, as in the original theory by Mullins and Sekerka (1964), it is assumed that solute rejection can be described by the equilibrium distribution coefficient k_e. However, since the actual distribution coefficient during rapid solidification can be considerably larger than k_e (see Sections V.A and B), this impurity trapping effect will lead to enhanced stability against dendrite formation over the predictions of the equilibrium theory. Narayan (1981) has used this approach, in conjunction with empirical relationships for the dependence of k on the interface velocity, to describe his observations of dendrite formation in laser-annealed implanted Si.

Acknowledgments

We thank M. J. Aziz and J. M. Poate for helpful discussions on their research and for their suggestions for improvement of this chapter. Our research in this area has been supported in part by the Harvard University M. R. L., under contract DMR-79-23597, and by the Office of Naval Research, under contract N-00014-77-C-0002.

List of Symbols

a_η	proportionality constant scaling k_η and η
c	concentration (mole m^{-3})
C_p	heat capacity at constant pressure (J K^{-1} $mole^{-1}$)
ΔC_p	difference in heat capacity at constant pressure (J $mole^{-1}$)
$C_p(l)$	heat capacity at constant pressure of liquids (J $mole^{-1}$)
d	characteristic distance from the temperature gradient (m)
D	diffusion coefficient (m^2 sec^{-1})
D_i	impurity diffusion coefficient at the interface (m^2 sec^{-1})
D_l	diffusion coefficient in the liquid (m^2 sec^{-1})
f	fraction of sites in the interface at which rearrangement can occur
ΔG	Gibbs free-energy difference (J $mole^{-1}$)
ΔG_c	Gibbs free-energy change per mole in crystallization (J $mole^{-1}$)
ΔG_{ac}	Gibbs free-energy difference between amorphous and crystalline phases (J $mole^{-1}$)
ΔG_{lc}	Gibbs free-energy difference between liquid and crystalline phases (J $mole^{-1}$)
ΔG_v	Gibbs free-energy change per volume in crystallization (J m^{-3})
ΔH_c	enthalpy of crystallization (J $mole^{-1}$)
ΔH_{ac}	enthalpy difference between amorphous and crystalline phases (J $mole^{-1}$)
ΔH_{al}	enthalpy difference between amorphous and liquid phases (J $mole^{-1}$)
ΔH_{mv}	enthalpy of melting per volume (J m^{-3})
$\Delta H'$	activation enthalpy (J $mole^{-1}$)
i^*	number of atoms in the critical nucleus
I	nucleation frequency (m^{-3} sec^{-1})
I_s	surface nucleation frequency (m^{-2} sec^{-1})
k	rate constant of reaction (sec^{-1})
k_B	Boltzmann's constant (J K^{-1} $atom^{-1}$)
k_c	impingement frequency (sec^{-1})
k_D	frequency of configurational rearrangement for diffusion (sec^{-1})

k_e — equilibrium impurity distribution coefficient

k_η — frequency of configurational rearrangement for viscous flow (sec^{-1})

k_h — frequency characteristic of heat flow (sec^{-1})

k_i — frequency of configurational rearrangement for crystallization (sec^{-1}) *or* equilibrium impurity distribution coefficient at the interface

M_B — mobility of the B atom (N sec m^{-1})

n' — number of atoms crystallized per rearrangement

n_r — number of rearrangements resulting from a single activation

$\bar{N}$ — Avogadro's number

$\dot{Q}$ — heat flux (J m^{-2} sec^{-1})

r^* — radius of the critical nucleus (m)

Δs_c — entropy of crystallization per atom (J K^{-1} atom^{-1})

ΔS_c — entropy of crystallization (J K^{-1} mole^{-1})

$\Delta S'$ — activation entropy (J K^{-1} mole^{-1})

$\Delta S'_1$ — activation entropy for single event (J K^{-1} mole^{-1})

t — time (sec)

T — heat bath temperature (K)

T_g — glass transition temperature (K)

T_i — temperature of the interface (K)

T_{ir} — reduced interface temperature, $\equiv T_i/T_m$

T_{al} — transition temperature between amorphous and liquid phases (K) (i.e., amorphous phase melts)

T_{lc} — transition temperature between liquid and crystalline phases (K) (i.e., crystal melts)

T_0 — upper limiting temperature for diffusionless growth

T_m — melting temperature (K)

ΔT — undercooling, $\equiv T_m - T$ (K)

ΔT_i — interface undercooling, $\equiv T_m - T_i$ (K)

T_r — reduced temperature, $\equiv T/T_m$

ΔT_r — reduced undercooling, $\equiv \Delta T/T_m$

ΔT_{ri} — reduced interface undercooling, $\equiv \Delta T_i/T_m$

v — crystal growth velocity (m sec^{-1})

v_B — drift speed of impurity atom B (m sec^{-1})

v_B^0 — upper limit for the drift speed of impurity atom B (m sec^{-1})

v_l — ledge velocity (m sec^{-1})

v_m — midrange velocity for impurity trapping (m sec^{-1})

v_s — speed of sound (m sec^{-1})

$\bar{V}$ — molar volume (m^3 mole^{-1})

W^* — work to form a critical nucleus (J)

x_s — mole fraction in the solid phase

x_l — mole fraction in the liquid phase

$\bar{x}_s$ — equilibrium mole fraction in the solid phase

$\bar{x}_l$ — equilibrium mole fraction in the liquid phase

x'_l — mole fraction in the liquid at the interface

x_0 — limiting mole fraction for impurity trapping

y — position coordinate (m)

z_l — collision frequency of liquid atoms (sec^{-1})

Z — coordination number

α — scaled surface tension, $\equiv -(\bar{N}\bar{V}^2)^{1/3}(\sigma/\Delta H_c)$

β — ratio of the entropy of crystallization and the gas constant, $\equiv \Delta S_c/R$

η	viscosity (N sec m^{-2})
θ	angle between the normal and the [111] direction
κ	thermal conductivity (W m^{-1} K^{-1})
λ	interatomic distance (m)
λ_D	jump distance for diffusion (m)
μ_B	chemical potential of component B (J mole^{-1})
μ_s^A	chemical potential of component A in the solid phase (J mole^{-1})
μ_l^A	chemical potential of component A in the liquid phase (J mole^{-1})
$\Delta\mu^A$	chemical potential difference of component A between liquid and solid phase, $\equiv \mu_s^A - \mu_l^A$ (J mole^{-1})
ν_r	normal frequency of reaction mode (sec^{-1})
σ	surface tension, surface free energy (J m^{-2})
σ_e	surface tension of ledge (J m^{-2})
τ_{tr}	time constant for nucleation transient (sec)

References

Ainslie, N. G., MacKenzie, J. D., and Turnbull, D. (1961). *J. Phys. Chem. Glasses* **65**, 1718.

Ainslie, N. G., Morelock, C. R., and Turnbull, D. (1962). *In* "Symposium on Nucleation and Crystallization in Glasses and Melts" (M. K. Reser, G. Smith, and H. Insley, eds.), p. 97. Am. Ceram. Soc., Columbus, Ohio.

Aziz, M. J. (1982). *J. Appl. Phys.* **53**, 1158.

Baeri, P., Foti, G., Poate, J. M., and Cullis, A. G. (1980). *Phys. Rev. Lett.* **45**, 2036.

Bacri, P., Foti, G., Poate, J. M., Campisano, S. U., and Cullis, A. G. (1981). *Appl. Phys. Lett.* **35**, 800.

Bagley, B. G., and Chen, H. S. (1979). *In* "Laser–Solid Interactions and Laser Processing" (S. D. Ferris, H. J. Leamy, and J. M. Poate, eds.), p. 97. Am. Inst. Phys., New York.

Baker, J. C., and Cahn, J. W. (1971). *In* "Solidification" (T. J. Hughel and G. F. Bolling, eds.) p. 23. Am. Soc. Met., Metals Park, Ohio.

Bloembergen, N. (1979). *In* "Laser Solid Interactions and Laser Processing" (S. D. Ferris, H. J. Leamy, and J. M. Poate, eds.), p. 1. Am. Inst. Phys., New York.

Burton, W. K., Cabrera, N., and Frank, F. C. (1951). *Philos. Trans. R. Soc. London* **243**, 299.

Cahn, J. W., Coriell, S. R., and Boettinger, W. J. (1980). *In* "Laser and Electron Beam Processing of Materials" (C. W. White and P. S. Peercy, eds.), p. 89. Academic Press, New York.

Chalmers, B. (1964). "Principles of Solidification," p. 114. Wiley, New York.

Chen, H. S., and Turnbull, D. (1969). *J. Appl. Phys.* **40**, 4212.

Colligan, G. A., and Bayles, B. S. (1962). *Acta Metall.* **10**, 895.

Cormia, R. J., MacKenzie, J. D., and Turnbull, D. (1963). *J. Appl. Phys.* **34**, 2239.

Csepregi, L., Kennedy, E. F., Mayer, J. W., and Sigmon, T. W. (1978). *J. Appl. Phys.* **49**, 3906.

Frank, F. C. (1949). *Discuss. Faraday Soc.* **5**, 48, 67.

Fratello, V. J., Hays, J. F., Spaepen, F., and Turnbull, D. (1980). *J. Appl. Phys.* **51**, 6160.

Herring, W. C. (1951). *Phys. Rev.* **82**, 87.

Hillert, M., and Sundman, B. (1977). *Acta Metall.* **25**, 11.

Hillig, W. B. (1962). *In* "Symposium on Nucleation and Crystallization in Glasses and Melts" (M. K. Reser, G. Smith, and H. Insley, eds.), p. 77. Am. Ceram. Soc., Columbus, Ohio.

Hillig, W. B. (1966). *Acta Metall.* **14,** 1868.
Hillig, W. B., and Turnbull, D. (1956). *J. Chem. Phys.* **24,** 914.
Hirvonen, J., and Anttila, A. (1979). *Appl. Phys. Lett.* **35,** 703.
Huntington, H. B. (1958). *Solid State Phys.* **7,** 213.
Jackson, K. A. (1958). *In* "Growth and Perfection of Crystals" (R. H. Doremus, B. W. Roberts, and D. Turnbull, eds.), p. 319. Wiley, New York.
Jackson, K. A., Gilmer, G. H., and Leamy, H. J. (1980). *In* "Laser and Electron Beam Processing of Materials" (C. W. White and P. S. Peercy, eds.), p. 104. Academic Press, New York.
Liu, P. L., Yen, R., Bloembergen, N., and Hodgson, R. T. (1979). *Appl. Phys. Lett.* **34,** 864.
Mullins, W. W., and Sekerka, R. F. (1964). *J. Appl. Phys.* **35,** 444.
Narayan, J. (1981). *J. Appl. Phys.* **52,** 1289.
Spaepen, F. (1974). *Philos. Mag.* **30,** 417.
Spaepen, F. (1978). *Acta Metall.* **26,** 1167.
Spaepen, F. (1982). To be published.
Spaepen, F., and Turnbull, D. (1976). *In* "Rapidly Quenched Metals" (N. J. Grant and B. C. Giessen, eds.), p. 205. MIT Press, Cambridge, Massachusetts.
Spaepen, F., and Turnbull, D. (1979). *In* "Laser–Solid Interactions and Laser Processing" (S. D. Ferris, H. J. Leamy, and J. M. Poate, eds.), p. 73. Am. Inst. Phys., New York.
Tiller, W. A., Jackson, K. A., Rutter, J. W., and Chalmers, B. (1953). *Acta Metall.* **1,** 428.
Turnbull, D. (1956). *Solid State Phys.* **3,** 225.
Turnbull, D. (1962). *J. Phys. Chem.* **66,** 609.
Turnbull, D. (1964). *In* "Physics of Non-Crystalline Solids" (J. W. Prins, ed.), p. 41. North-Holland Publ., New York.
Turnbull, D. (1969). *Contemp. Phys.* **10,** 473.
Turnbull, D. (1974). *J. Phys. (Orsay, Fr.)* **35**(C-4), 1.
Turnbull, D. (1980). *J. Phys. (Orsay, Fr.)* **41**(C-4), 109.
Turnbull, D. (1981a). *Metall. Trans.* **12A,** 695.
Turnbull, D. (1981b). *Prog. Mater. Sci.,* Chalmers Anniversary Volume, p. 269.
Turnbull, D., and Bagley, B. G. (1975). *In* "Treatise on Solid State Chemistry" (N. B. Hannay, ed.), Vol. 5, p. 513. Plenum, New York.
Turnbull, D., and Cohen, M. H. (1960). *In* "Modern Aspects of the Vitreous State" (J. D. MacKenzie, ed.), Vol. 1, p. 38. Butterworths, London.
Uhlmann, D. R. (1980). *J. Non-Cryst. Solids* **41,** 347.
Walker, J. L. (1964). *In* "Principles of Solidification" (B. Chalmers, ed.), pp. 114, 122. Wiley, New York.
Wood, R. F. (1980). *Appl. Phys. Lett.* **37,** 302.

Chapter 3

Fundamentals of Energy Deposition

MARTIN F. VON ALLMEN

Institute of Applied Physics
University of Bern
Bern, Switzerland

I. Introduction

Laser beams as well as kiloelectron volt energy electron beams are currently being used in a whole range of new applications—from growing crystals to forming metallic glasses, and from depositing films to purifying surfaces. All these actions are ultimately produced by heat into which a smaller or larger part of the beam energy is transformed by various coupling mechanisms. These mechanisms determine not only the amount of heat created but also its spatial and temporal distribution. They should therefore be considered important factors in designing an experiment.

The technology for producing high-intensity beams of directed energy is a development of the past 20 years. Since the first demonstration of a

pulsed ruby laser by Maiman in 1960, lasers have become mature and
reliable sources of energy and are used on a routine basis in a number of
industrial processing techniques. Today, lasers covering a broad range of
wavelengths, pulse characteristics, and output powers are available on the
market. Table I lists some characteristics of the lasers most commonly
used in the processing of electronic materials. Laser radiation, in addition
to its high intensity, has some features not found in ordinary thermal
radiation: It is highly directional, and it is coherent, i.e., capable of self-
interference. The directionality enables the experimenter to concentrate
the beam energy with the aid of lenses to a spot of any size, down to little
more than one wavelength. Self-interference is the basis of holography but
is usually detrimental in laser processing: mutual interference of scattered
parts of a beam causes the intensity to vary from point to point on an
irradiated surface.

Sources of high-intensity electron beams suitable for material processing became available only a few years ago. Current electron beam anneal-

TABLE I

TYPICAL CHARACTERISTICS OF SOME COMMERCIAL HIGH-POWER LASERS[a]

Laser	Active species	Wave-length (μm)	Output characteristics	Average power or pulse energy	Mode of excitation
Gas lasers					
Argon	Ar^+	0.33–0.52	cw/pt	20 W	LE, LP
Krypton	Kr^+	0.41–0.8	cw/pt	<10 W	LE, LP
Nitrogen	N_2	0.337	Pulsed (5 nsec)	5 mJ	TE, AP
Excimer	ArF, KrF, etc.	uv	Pulsed (10 nsec)	0.2 J	TE, LP
Carbon monoxide	CO	5–6	cw or pulsed (μsec)	50 W, 0.1 J	TE or LE
Carbon dioxide	CO_2	10.6	cw or pulsed (nsec or μsec)	10–1000 W, 1–100 J	TE or LE, LP or AP
Solid state lasers					
Ruby	Cr^{3+}	0.694	Pulsed	10 J	Flash lamp
YAG or glass	Nd^{3+}	1.06	cw or pulsed (nsec or msec)	10–100 W, 1–10 J	Flash lamp
Liquid lasers					
Dye	Various	uv and visible	cw or pulsed	5 W, 1 J	Laser or flash lamp

[a] cw, Continuous wave; pt, pulse train; LE, longitudinally excited; TE, transversally excited; LP, low pressure; AP, atmospheric pressure.

ing machines work with electron energies in the 5–100 keV range and are capable of delivering up to several tens of joules per pulse. Electrons, in contrast to photons, interact as individual particles and do not show interference. However, the mutual repulsion of the electrons in a beam can also produce self-induced variations in local energy density on an irradiated surface by space charge effects.

This chapter reviews the main mechanisms of interaction of photon and electron beams with condensed matter from the point of view of energy deposition. Finally, the influence of coupling phenomena on the final sample temperature is discussed.

Beam intensities of interest range up to about 10^8 W/cm^2, sufficient to heat and even melt the surface region of the irradiated material. Even though the basic mechanisms of absorption are well known, coupling phenomena in this high-flux regime are not as well understood as they are at low intensities.

II. Laser Beams

A. *Optical Properties of Materials*

1. GENERAL

The response of matter to high-intensity optical radiation has been intensively studied over the past 20 years. A wealth of unusual effects has been discovered, and new insights into the properties of matter thereby obtained. Still, the basis for understanding high-intensity radiation effects are the phenomena observed with ordinary light—summarized as optical properties.

Electromagnetic radiation with wavelengths ranging from uv to ir interacts exclusively with electrons, since nuclei are too heavy to respond significantly to the high frequencies ($\nu > 10^{13}$ Hz) of the electromagnetic field. Moreover, the photon energies of interest are too small to affect core electrons. Therefore, the optical properties of matter are largely determined by the energy states of its valence electrons (bound or free). Generally, bound electrons respond only weakly to the electromagnetic wave and mainly affect its phase velocity. Free electrons are able to be accelerated, i.e., to extract energy from the field. However, since the external field periodically changes its direction, the oscillating electrons reradiate their kinetic energy unless they undergo frequent collisions with the atoms. In this case, energy is transmitted to the lattice and the external field is weakened. Reradiation of energy is the cause of reflection.

Since optical wavelengths λ are large compared to atomic distances, the response of a homogeneous material can be described in terms of averaged macroscopic quantities, such as the complex refractive index $\hat{n} = n + ik$. The real part n gives the ratio of the phase velocities in vacuum and in the material, whereas the imaginary park k describes the damping of the light wave. The two parameters n and k (which are, of course, functions of the wavelength) completely describe the response of the material to the light wave. However, if one is mainly interested in energy deposition, it is more convenient to use an equivalent set, namely, the reflectivity R and the absorption coefficient α or its inverse, the absorption length d. These variables are related to n and k by the well-known relations (for normal incidence)

$$R = [(n - 1)^2 + k^2]/[(n + 1)^2 + k^2] \qquad (1)$$

and

$$\alpha = 1/d = 4\pi k/\lambda \qquad (2)$$

If light of *intensity* I_0 (in watts per square centimeter) is normally incident on a sample, then the *power density* (in watts per cubic centimeter) absorbed at depth z is given by

$$\Phi(z,t) = I_0(t)[(1 - R)/d]e^{-z/d} \qquad (3)$$

Yet another quantity used to characterize a beam is the *fluence* F (in joules per square centimeter), which is the time integral of the intensity I_0.

In the following, we briefly summarize the behavior of R and d or α for various classes of target materials at low light intensities. For a more extensive discussion of optical properties the reader is referred to standard textbooks.

2. METALS

Metals, because of their large density of free electrons, have very small absorption lengths (on the order of 10 nm) over the whole optical spectrum. The reflectivity is high above a certain critical wavelength. Below the critical wavelength λ_{cr}, which is in the visible or uv part of the spectrum, the reflectivity decreases sharply. As an example, Fig. 1 shows R and α for Al and Au at room temperature. The critical wavelength of Al is below 100 nm, while that of Au is at about 600 nm (the yellow color of Au is due to its low reflectivity in the green and blue regions). The critical wavelength is the wavelength at which the light frequency equals the *plasma frequency* ν_p of the free-electron plasma and is given by

$$c/\lambda_{cr} = \nu_p = (1/2\pi)\sqrt{Ne^2/m_e\epsilon_0} \qquad (4)$$

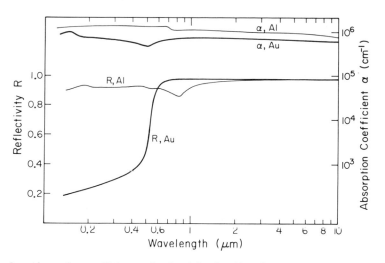

Fig. 1. Absorption coefficient and reflectivity for Al and Au evaporated films at room temperature. (Data from AIP Handbook, 1972.)

where N and m_e are the effective density and mass of the free electrons, respectively, c is the velocity of light, and $\epsilon_0 = 8.854 \times 10^{-12}$ F/m.

At longer wavelengths, the reflectivity of metals tends toward unity. If the period of the light wave is long compared to the average time t_c between two collisions of an electron with the lattice (practically for $\lambda \gtrsim 10$ μm), the classical Hagen–Rubens formula can be used (Mott and Jones, 1958):

$$R = 1 - 0.036/\sqrt{\sigma_0 \lambda} \tag{5}$$

where σ_0 is the dc conductivity per microhm-centimeter and λ is in micrometers. The appearance of the conductivity in the expression for an optical constant reflects the fact that absorption of light in a metal is caused by the same collisions that produce its electrical resistance. Values of R and d for various metals at selected laser wavelengths are given in Table II.

3. SEMICONDUCTORS AND INSULATORS

In materials with a completely filled valence band, photons, depending on their energy, can induce two kinds of electronic transitions, namely, *intraband* (bound-bound) and *band to band* (bound-free). Further, resonant coupling to high-frequency phonons (by means of bound electrons) is observed.

TABLE II

OPTICAL ABSORPTION LENGTHS *d* AND REFLECTIVITIES *R* OF EVAPORATED METAL FILMS,
SEMICONDUCTORS, AND CRYSTALLINE INSULATORS AT ROOM TEMPERATURE FOR
DIFFERENT WAVELENGTHS[a]

	0.25 μm		0.5 μm		1.06 μm		10.6 μm	
Material[b]	*d*	*R*	*d*	*R*	*d*	*R*	*d*	*R*
Al	8 nm	0.92	7 nm	0.92	10 nm	0.94	12 nm	0.98
Ag	20 nm	0.30	14 nm	0.98	12 nm	0.99	12 nm	0.99
Au	18 nm	0.33	22 nm	0.48	13 nm	0.98	14 nm	0.98
Ge (c)	7 nm	0.42	15 nm	0.49	200 nm	0.38	>1 cm	0.36
Ge (a)	10 nm	0.48	50 nm	0.47	1 μm	0.42	>1 cm	0.34
Si (c)	6 nm	0.61	500 nm	0.36	200 μm	0.33	1 mm	0.30
Si(a)	10 nm	0.75	100 nm	0.48	1 μm	0.35	>1 cm	0.32
GaAs (c)	6 nm	0.6	100 nm	0.39	70 μm	0.31	>1 cm	0.28
KCl	>1 cm	0.05	>1 cm	0.04	>1 cm	0.04	>1 cm	0.03
SiO$_2$	>1 cm	0.06	>1 cm	0.04	>1 cm	0.04	40 μm	0.2

[a] Data on metals and insulators are mainly from AIP Handbook (1972); data on semiconductors were collected from various sources and are meant as typical values only. In particular, the data on amorphous semiconductors vary considerably depending on purity and method of preparation.
[b] c, Crystalline; a, amorphous

If the photon energy $h\nu \ll E_g$ (usually true for semiconductors in the ir and for insulators in the visible region), only intraband and phonon absorption are possible. Coupling in this regime is weak. The absorption lengths *d* range typically from centimeters to meters but may decrease sharply as a function of intensity or temperature, as discussed in Section II.B. Basically, this is an unfavorable regime for beam heating.

When the photon energy approaches the gap energy, the degree of absorption increases because band-to-band transitions become available. Two types of materials have to be distinguished, namely, materials where the wave vectors of the valence band maximum and the conduction band minimum coincide (*direct band materials*, e.g., GaAs) and materials where this is not the case (*indirect band materials*, e.g., Ge and Si). For absorption of a photon with $h\nu \approx E_g$ in the latter type of material, the simultaneous absorption or emission of a phonon providing the extra crystal momentum is required. This is illustrated in Fig. 2. It can be seen that direct transitions are also possible in indirect gap materials, provided the photon energy is sufficient (e.g., $h\nu > 3.4$ eV for Si).

The two kinds of band-to-band transitions lead to different dependences of the absorption coefficient on photon energy in the region of the band

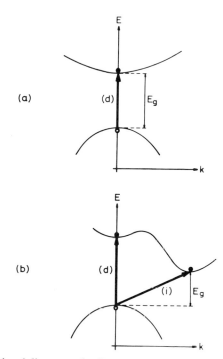

Fig. 2.   Schematic band diagrams of a direct gap material (a) and an indirect gap material (b). The heavy arrows symbolize direct (d) and indirect (i) transitions.

gap. A simplified treatment yields (see, e.g., Wooten, 1972)

$$\alpha_d = A(h\nu - E_g)^{1/2} \qquad \text{for} \quad h\nu \geq E_g \tag{6}$$

for direct transitions, and

$$\alpha_i = B \frac{(h\nu - E_g + E_p)^2}{\exp(E_p/kT) - 1} \qquad \text{for} \quad E_g - E_p < h\nu < E_g + E_p \tag{7}$$

or

$$\alpha_i = B \left[\frac{(h\nu - E_g + E_p)^2}{\exp(E_p/kT) - 1} + \frac{(h\nu - E_g - E_p)^2}{1 - \exp(-E_p/kT)} \right] \qquad \text{for} \quad h\nu > E_g + E_p$$

for indirect transitions. Here A and B are constants and E_p is the phonon energy. In the first expression of Eq. (7), only phonon absorption is allowed, whereas in the second phonon emission is also energetically possible. Equations (6) and (7) allow only for single-phonon processes and neglect the detailed structure of the state density at the band edges. (For a more detailed discussion, see Wooten, 1972.)

At photon energies significantly exceeding E_g the optical response of insulators and semiconductors tends toward that of metals. Figure 3 shows the situation for crystalline Si: For $\lambda > 1$ μm the absorption coefficient α is low, while R is still relatively high because of the large refractive index ($n = 3.5$ at 1 μm). The weak absorption band above $\lambda = 6$ μm is due to phonon absorption (Spitzer, 1967). Below $\lambda = 1$ μm both R and α increase because of band-to-band absorption. In the near-uv region, the reflectivity decreases again, as the light frequency exceeds the free-carrier plasma frequency, in analogy with the behavior of metals.

Insulators such as oxides and ionic salts behave in similar ways, except that the gap energies generally correspond to wavelengths in the vacuum uv. The refractive indices and consequently the reflectivities in the visible region are lower than those of Si or Ge in the ir because of a smaller lattice polarizability. As an example, Fig. 4 shows R and α for crystalline quartz (SiO_2). Some typical values for R and d can also be found in Table II.

B. Beam-Induced Changes in Optical Properties

1. GENERAL

So far, we have briefly reviewed optical properties at low light intensities. At sufficiently high intensities, the response of matter to light is modified by a number of mechanisms, a thorough discussion of which

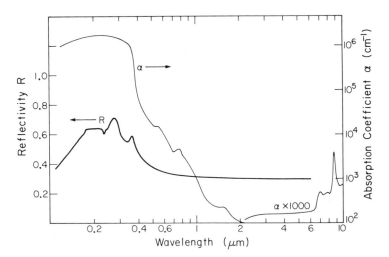

Fig. 3. Absorption coefficient and reflectivity of crystalline Si at room temperature. (Data in part from AIP Handbook, 1972.)

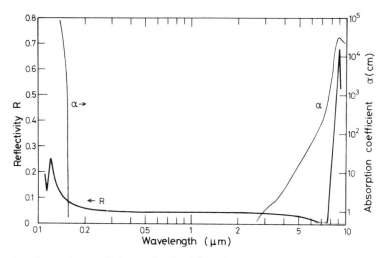

Fig. 4. Absorption coefficient and reflectivity of quartz at room temperature. (Data in part from AIP Handbook, 1972.)

could easily fill a book by itself. We therefore limit our discussion to an outline of such effects that directly affect energy deposition. These effects can conveniently be subdivided into two categories, namely, effects due to carrier excitation and effects due to lattice heating. While the former category is of practical significance only for nonmetals, the latter applies to any material.

2. CARRIER EXCITATION

When a beam of photons of energy $h\nu > E_g$ is absorbed in a semiconductor or an insulator, excited carriers are generated at a rate $1/t_e = I\sigma/h\nu$, where σ is the relevant absorption cross section of the carriers. At high absorbed intensities I, this rate may approach or exceed the rate of carrier relaxation. As a result, the energy distribution of the carriers is significantly changed as compared to that defining the optical properties of the material at low light intensities.

Relaxation of hot carriers is a complex phenomenon and still a subject of active research. However, for the sake of simplicity we may assume that only two mutually independent mechanisms are important, namely, carrier–lattice collisions (at a rate $1/t_c$) and recombination (at a rate $1/t_r$). Usually, $1/t_c > 1/t_r$. Thus, in the order of increasing power density two regimes can be distinguished: In the regime $1/t_e > 1/t_r$ a significant den-

sity of free carriers builds up, which then dominate the optical properties
[optically induced free-carrier absorption (FCA), plasma effects]. If even
$1/t_e > 1/t_c$, a hot carrier plasma is created, which is no longer in thermal
equilibrium with the host lattice.

   a. Optically Induced Free-Carrier Absorption.   Absorption of photons
with $h\nu > E_g$ in a nonmetal always leads to free carriers. However, only
above a certain intensity does their density become sufficient to produce a
measurable contribution to total absorption. Under some simplifying assumptions, an effective absorption coefficient α_{eff} can be defined as the
sum of an absorption coefficient α_0 characterizing normal band-to-band
absorption and a coefficient for FCA. The latter is the product of the
density N_c of carrier pairs and a FCA cross section σ_{FC}; thus

$$\alpha_{eff} = \alpha_0 + N_c\sigma_{FC} \tag{8}$$

The value of σ_{FC} is independent of N_c for not too high carrier densities, but
it depends on the nature of the dominant scattering mechanism (phonon
scattering or ionized impurity scattering (Rosenberg and Lax, 1958;
Schumann and Phillips, 1967)) and it increases with the wavelength. For
$\lambda = 1.06\ \mu m$, Svantesson (1979) found $\sigma_{FC} = 5.1 \times 10^{-18}\ cm^2$ in high-
purity Si at room temperature. Between 200 and 400 K, σ_{FC} was proportional to the absolute temperature (Svantesson and Nilsson, 1979).

   The attenuation of a light beam inside the material is described by α_{eff},
while the carrier density created is equal to the density of photons absorbed by band-to-band absorption only. An expression for the resulting
carrier density for short times, when recombination or diffusion of carriers
can be neglected, was given by Gauster and Bushnell (1970). Based on
their results, α_{eff} for $z \ll 1/\alpha_0$ can be expressed as

$$\alpha_{eff} \cong \alpha_0 + (\sigma_{FC}/h\nu) \int_0^t \Phi(0,t')\,dt' \tag{9}$$

where Φ is given by Eq. (3) with $d = 1/\alpha_0$. For Si, Eq. (9) is expected to
be a reasonable approximation for nanosecond laser pulses (for longer
pulses optically induced FCA does not seem to play a role).

   An absorption length that increases as a function of time and pulse
intensity makes beam heating strongly nonlinear. This is illustrated by
Fig. 5, which shows the result of a numerical calculation of Nd laser-
induced (1.06-μm) FCA in Si (Lietoila and Gibbons, 1981). It is obvious
from the figure that the contribution from FCA drastically reduces the
pulse energy necessary for surface melting. Also indicated in the figure is
the effect of the temperature dependence of the band gap.

   An additional complication in the description of heating in the present
regime arises from the fact that only photons absorbed by FCA are avail-

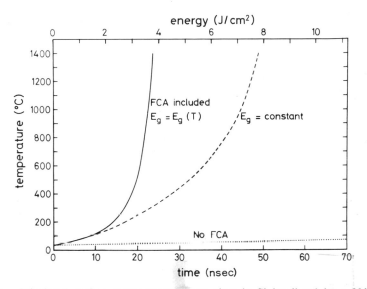

Fig. 5. Calculated surface temperature versus time in Si irradiated by a Nd laser Q-switched pulse. The dotted curve results if only lattice absorption is considered. The dashed line indicates the influence of the decrease in band gap with temperature, and the solid line, in addition, allows for FCA. (From Lietoila and Gibbons, 1981.)

able for lattice heating instantaneously (on a nanosecond time scale), while band-to-band absorption releases heat only via recombination, thus with a delay on the order of t_r that again depends on N_c (Dziewior and Schmid, 1977).

b. Carrier Avalanche Multiplication. The effect considered now also results in the production of free carriers, but it does not rely on band-to-band absorption and may therefore occur even at photon energies much smaller than the band gap. This mechanism is often responsible for damage in optical materials. Above a certain threshold intensity, available free carriers (created, maybe, by thermal excitation of shallow impurity levels) can acquire enough energy to create additional carriers by impact ionization. The result is an avalanche multiplication of free carriers similar to the one occurring in the dc breakdown of dielectrics (Holway and Fradin, 1975). For long wavelengths (generally for $1/\nu > t_c$) the threshold intensity corresponds to a rms electric field inside the material close to that necessary for dc breakdown (Bloembergen, 1974). At shorter wavelengths the threshold intensity increases approximately as $1/\lambda^2$. Further, it is found that surfaces tend to have lower thresholds than bulk material (Bloembergen, 1974; Bass and Fradin, 1973). As a consequence of the carrier avalanche, the absorption coefficient increases suddenly by orders of magnitude once a certain threshold intensity is reached. This usually results in explosive mechanical damage.

    c.   Effects of a Hot, Dense Carrier Plasma.   At sufficiently high intensities $1/t_e > 1/t_c$ and the rate of energy gain by the carriers exceeds the rate of energy loss to the lattice. The result is a splitting up of carrier and lattice temperatures (Yoffa, 1980a). Figure 6 shows a schematic energy diagram for the electrons in a semiconductor, illustrating this regime. A typical collision rate $1/t_c$ is 10^{12} sec^{-1}, whereas $1/t_e$ for beam intensities of 10^8 W/cm^2 may be comparable or even higher. A large density of hot carriers therefore builds up before the lattice is appreciably heated. Collisions between carriers establish thermal equilibrium among them on a time scale of 10^{-14} sec. Auger processes and other recombination mechanisms establish equilibrium between electrons and holes but do not affect the total carrier density. After the pulse ends, the hot carriers thermalize with a time constant t_c and recombine within typically $t_r = 10^{-9}$ sec (Dziewior and Schmid, 1977).

    As far as coupling is concerned, the effect of the plasma is ultimately to limit the rate of local energy deposition in two ways:

    (i)   the carrier plasma frequency increases as $\sqrt{N_c}$; at a sufficiently high carrier density, light is reflected from the plasma (for reflection of visible light a density of about 10^{20} cm^{-3} is necessary);

    (ii)   diffusion of hot carriers away from the absorbing surface effectively increases the heated volume; Yoffa (1980b) has calculated that for Si this diffusion can be characterized by an effective diffusivity of about 100 cm^2/sec for typical laser annealing parameters. Under such circum-

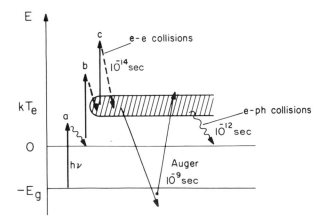

Fig. 6.   Schematic energy diagram of the electrons in a semiconductor under intense irradiation, showing band-to-band absorption (a) and FCA (b and c).

stances, expression (3) for the absorbed power density would no longer be a good approximation for the distribution of the produced heat flux.

The effects of a dense, hot plasma on the lattice are not yet completely understood. Van Vechten (1980) has argued that removal of a significant number of electrons from bonding states by intense laser radiation should weaken the covalent bonds in amorphous Si enough to produce a state of vanishing shear stress at lattice temperatures far below the melting point ("cold fluid"). He proposes that recrystallization of ion-implanted Si (see Chapter 7 by Foti and Rimini) by Q-switched pulses occurs from this cold fluid rather than from a normal melt. The necessary diffusion of vacancies out of the recrystallizing zone is claimed to be helped by screening of the electrostatic interaction of vacancies with other defects by the dense carrier plasma. Perhaps the most controversial point in this theory is the explanation of the long period (100 nsec or more) of enhanced reflectivity upon irradiation (Auston et al., 1979). This phenomenon has been observed by many workers; it is what one expects from normal melting of the Si (see Table IV and Fig. 10). Van Vechten and his associates ascribe the enhancement in reflectivity to the carrier plasma. However, the observed duration of enhanced reflectivity, as well as the assertion that the lattice stays "cold" during regrowth, requires an increase in the carrier relaxation time t_c by many orders of magnitude. The explanation proposed for this increase involves screening of the electron–phonon interaction by the plasma. The adequacy of this mechanism has been put in doubt (Dumke, 1980). A recent experimental investigation of the wavelength dependence of enhanced reflectivity revealed the characteristics expected for ordinary metallic molten Si rather than those of an expanding carrier plasma (Nathan et al., 1980).

3. THERMAL EFFECTS

The phenomena discussed now can play a role at moderate light intensities only if the total pulse energy is sufficient to raise the temperature of the solid. The optical properties can then change for two basic reasons: (a) at elevated temperatures more phonons are present that can interact either with the photons or with the carriers; and (b) the band structure of the solid may change with the temperature.

a. Phonon-Assisted Transitions. Phonons are necessary for light absorption in indirect semiconductors at photon energies close to the band gap, as discussed in Section II.A.3. As the temperature of the material is raised, more phonons are present and the probability of phonon absorption strongly increases, as evident from Eq. (7). An example for this regime is the absorption of Nd laser light ($h\nu = 1.17$ eV) in crystalline Si. Figure 7 shows the absorption coefficient as a function of sample temperature for crystalline and amorphous Si, measured with a low-intensity laser beam (Siregar *et al.*, 1979). [An additional effect contributing to the increase in $\alpha(T)$ in crystalline Si is the rather strong decrease in the band gap with temperature.] Note that for amorphous Si α changes only slightly with temperature because momentum conservation plays no role in the disordered lattice.

b. Phonon Scattering. Phonons are also required to transfer energy from the carrier system to the lattice. Therefore, the carrier–lattice collision frequency $1/t_c$ increases with the lattice temperature. A practical consequence of this is that the reflectivity of most metals decreases with temperature. As an example, Fig. 8 shows the quantity $1 - R$ versus temperature for Al and Au. The curves were calculated from electrical dc resistivity data. In addition to the steady increase in $1 - R$ with temperature, an abrupt increase is observed for most metals upon melting (Gubanov, 1965).

c. Thermally Induced Free-Carrier Absorption. This effect is similar to optically induced FCA discussed earlier, except that the carriers are thermally instead of optically excited. It is therefore also possible for $h\nu \ll E_g$. An effective absorption coefficient formally identical to the one

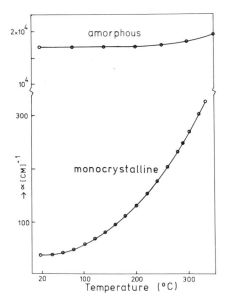

Fig. 7. Measured absorption coefficient in Si at $\lambda = 1.06$ μm as a function of sample temperature. (From Siregar *et al.*, 1979.)

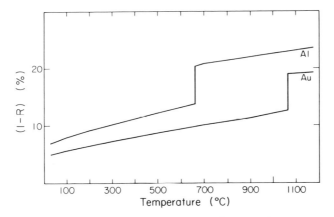

Fig. 8.   For bulk Al and Au, $1 - R$ at $\lambda = 1$ μm as a function of temperature calculated using data on dc resistivity from AIP Handbook (1972) and Gubanov (1965).

in Eq. (8) can be defined, where α_0 now stands for the residual lattice absorption (due to intraband or phonon coupling) and N_c is a function of temperature (Kittel, 1976):

$$N_c = N_0 + 4\left(\frac{2\pi kT}{h^2}\right)^{3/2} \cdot (m_e m_h)^{3/4} \cdot \exp\left[-\frac{E_g(T)}{2kT}\right] \qquad (10)$$

Here m_e and m_h are the electron and hole effective mass, respectively, and N_0 is the concentration of ionized impurities. Thermally induced FCA governs the absorption of ir radiation in Ge and Si and is often referred to as thermal runaway. As an example, Fig. 9 shows measured transmission

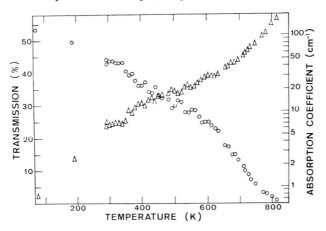

Fig. 9.   Measured transmission (O) and absorption coefficient ($\triangle$) as a function of sample temperature for a 290-μm-thick Si wafer at 10.6 μm. (From Siregar *et al.*, 1980.)

and the absorption coefficient as a function of the sample temperature of crystalline Si for CO_2–laser radiation ($\lambda = 10.6 \ \mu$m) (Siregar *et al.*, 1980).

    d.   Thermally Induced Changes in Band Structure.   Under this heading a number of phenomena can be summarized that produce either an increase or a reduction in the degree of absorption. A few examples may illustrate the point.

    (i)   The gap energy of most semiconductors decreases with temperature. This affects primarily the absorption coefficient at photon energies close to the gap energy (see Fig. 5).

    (ii)   A number of semiconducting materials, including Si, become metallic upon melting (Shvarev *et al.*, 1975). This induces a strong increase in α and R. As an example, Fig. 10 shows a time-resolved reflectivity measurement of Si during irradiation with a Q-switched laser pulse. The reflectivity, monitored with a He–Ne laser beam ($\lambda = 633$ nm) is seen to stay enhanced for a time of about 100 nsec (Auston *et al.*, 1979). This is on the order of magnitude of the expected lifetime of a molten surface (Surko *et al.*, 1978).

    (iii)   A decrease in α and R is observed when amorphous semiconductors crystallize. For example, the ratio of the room temperature absorption coefficients of amorphous and crystalline Si is about 10^4 (see Fig. 7 and Table II).

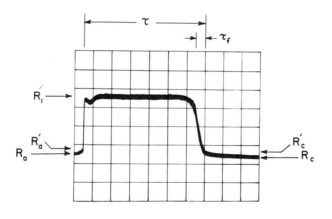

100 nsec / div.

Fig. 10.   Time-resolved reflectivity (measured at $\lambda = 630$ nm) of Q-switched laser-irradiated Si (Auston *et al.*, 1979). Reflectivities of room temperature amorphous Si, Si just prior to melting, metallic Si, hot crystalline Si, and room temperature Si are characterized by R_a, R'_a, R_l, R'_c, and R_c, respectively. τ designates lifetime of a molten surface.

e. Evaporation Effects. Yet another large class of phenomena is observed if the sample reaches temperatures where intense evaporation takes place. Vapors, due to the localization of the electrons, are basically transparent insulators. However, in dense vapors substantial scattering or defocusing of the beam can result in an actual decrease in intensity at the sample surface. Moreover, optical breakdown by avalanche multiplication of carriers is observed in dense vapors at light intensities above 10^7 W/cm^2 (Bergelson *et al.*, 1974). The process is very similar to the avalanche breakdown in solids mentioned earlier. The threshold intensity is again roughly proportional to $1/\lambda^2$ but decreases strongly in the presence of easily ionizable impurities such as dust particles (Smith, 1977). The same mechanism is responsible for gas breakdown in the atmosphere surrounding the sample that occurs at sufficient intensities even without evaporation of sample material.

The presence of an ionized gas layer can lead to complete shielding of the sample surface (Stürmer and von Allmen, 1978). Alternatively, a drastic decrease in the apparent reflectivity of metals has been observed in the presence of a vapor plasma (von Allmen *et al.*, 1978). A more detailed discussion of this regime cannot be given without reference to the dynamics of evaporation and gas expansion, which is beyond the scope of this volume.

4. SUMMARY OF BEAM-INDUCED ABSORPTION PHENOMENA

We have briefly reviewed a number of beam-induced coupling effects that are of practical significance for beam heating. The main features of these effects are summarized in Table III.

As we mentioned earlier, the list of self-induced effects of high-power radiation known today is by far not exhausted by our discussion. Other important effects include

(i) phenomena produced by an intensity-dependent refractive index, such as self-focusing, or generation of harmonics of the laser wave inside the solid (nonlinear optics),

(ii) scattering of laser radiation from laser-driven coherent phonons (stimulated scattering), and

(iii) structural modifications created by laser-produced shock waves inside an irradiated solid.

These are only a few examples. These effects, although being intensively studied today, seem to be of less practical importance in present beam annealing techniques.

TABLE III

SUMMARY OF SELF-INDUCED OPTICAL COUPLING EFFECTS[a]

Effect	Mainly applies to		Results in
	Material	Spectral range	
Optically induced FCA	Semiconductors, insulators	$h\nu \gtrsim E_g$	$d(F){\downarrow}$
Carrier avalanche	Semiconductors, insulators	$h\nu < E_g$	$d(F){\downarrow}$
Hot plasma	Semiconductors	$h\nu \gtrsim E_g$	$D{\uparrow}, R{\uparrow}$
Phonon-assisted transitions	Indirect semiconductors	$h\nu \approx E_g$	$d(T){\downarrow}$
Phonon scattering	Metals	$\lambda > \lambda_{cr}$	$R(T){\downarrow}$
Thermally induced FCA	Semiconductors	$h\nu < E_g$	$d(T){\downarrow}$
Variations in $E_g(T)$	Semiconductors (Si, Ge)	$h\nu \approx E_g$	$d(T){\downarrow}$ if $E_g(T){\downarrow}$
Semiconductor–metal transition	Some semiconductors (Si, Ge)	Any	$R{\uparrow}, d{\downarrow}$ upon melting
Crystallization	Amorphous Si, Ge	$h\nu \approx E_g$	$d{\uparrow}$
Vapor breakdown	Any metal	Any	Shielding $R{\downarrow}$

[a] F, Fluence; T, temperature, R, reflectivity, d, absorption length, D, thermal (carrier) diffusivity; ↑, increase, ↓, decrease.

C. Interference

Laser-irradiated samples often reveal a microscopic surface structure consisting of rings, ripples, corrugations, and the like, particularly if surface melting is induced during irradiation (Hill, 1980). Some of these structures result from interference effects (see Affolter *et al.*, 1980, and references therein), while others are due to material-related thermal or mechanical phenomena (Vitali *et al.*, 1979). Only the first category of effects will be briefly discussed here, because it is related to a fundamental property of laser radiation, namely, coherence. Laser light has a very narrow bandwidth, since it results from an undamped oscillation. Radiation with a bandwidth of $\Delta\nu$ is said to be coherent over a length $l_{coh} = c/\Delta\nu$, the coherence length. Two parts that split off a coherent beam and travel along different paths can produce a persisting interference pattern if their path lengths differ by less than l_{coh}. For typical continuous-wave (cw) gas lasers l_{coh} can be several meters, whereas for solid state Q-switched lasers it is usually on the order of 1 cm. This means that the conditions for interference are practically always given.

Interference patterns result if part of the laser beam is scattered from an obstacle and interferes with the remainder of the beam. If the obstacle is a simply shaped macroscopical object such as an aperture or an edge, interference leads to the familiar diffraction patterns described in every textbook on optics. Apart from diffraction effects (which are easily eliminated) interference phenomena may result for the following reasons:

(i) *Scattering from small particles*, such as dust particles in the atmosphere or small protrusions on the sample surface. If the scatterer is small enough for scattering to be isotropic, we have interference of a spherical wave with a plane wave. The resulting pattern consists of concentric circles (for normal incidence) or of a system of ellipses (for an incidence angle $\varphi \neq 0$) and is described by

$$r_n = n\lambda[G + \sqrt{1 + H(1 - G^2)}]/(1 - G^2) \tag{11}$$

where $G = \cos \vartheta \sin \varphi$ and $H = 2h/n\lambda \cos \varphi$. Here r and ϑ are the polar coordinates of the interference maxima, h is the distance of the scattering center from the sample surface, and n is an integer. Some familiar interference patterns described by this formula are shown in Fig. 11.

(ii) *Scattering from a diffuser*, such as a diffusing screen or a diffusely reflecting surface. The patterns of this type are called speckling patterns and consist of a random distribution of bright spots with a mean diameter of (Goldfisher, 1965)

$$\Lambda = \lambda a/B \quad \text{for} \quad a \gg B \tag{12}$$

where B is the beam diameter and a is the optical distance between the diffuser and the sample.

(iii) *Multimode interference*. This effect is not related to scattering of any kind but depends on the properties of the laser source. Whenever a laser oscillates simultaneously in several transversal resonator modes, then the intensity on a sample is found to fluctuate in space and time because of the interfering modes. The maximum intensity reached in a "hot spot" can be estimated from (Ryter and von Allmen, 1981)

$$I_{max}/\langle I \rangle = (r_m/r_{00})^2 \tag{13}$$

where $\langle I \rangle$ is the mean intensity on the beam axis, r_m is the useful output mirror radius, and r_{00} is the radius of the TEM_{00} mode at the position of the output mirror. An upper limit to the "lifetime" of the hot spot is given by the inverse of the frequency separation of adjacent resonator modes.

There are two obvious ways to reduce the detrimental effects of interference on beam heating:

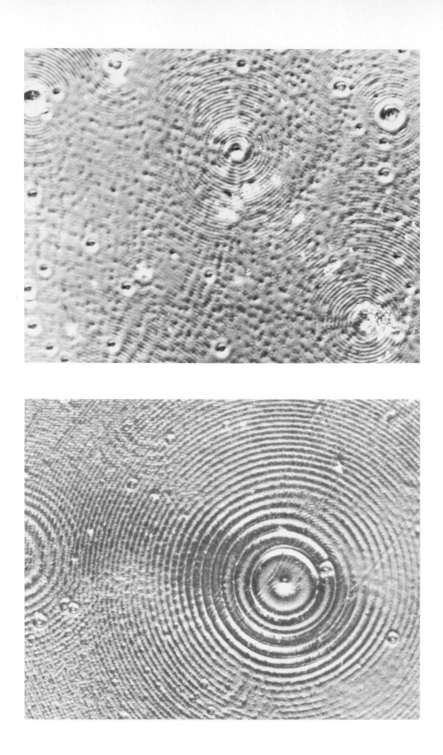

(i)   The physical size of the fringes or hot spots can be made small compared to the thermal diffusion length $2\sqrt{DT}$, where D is the thermal diffusivity of the sample material; or

(ii)   the lifetime of the hot spots can be reduced such that their energy is too small to produce a material effect. This can be achieved by "scrambling" the phases in the laser beam with a suitable device (Cullis and Webber, 1980).

We shall further discuss the sensitivity of the sample temperature to variations in intensity in Section IV.

III.   Electron Beams

A.   Electron–Solid Interactions

Electron beam annealing has been found to produce results very similar to those obtained with laser beams of the same absorbed intensity and pulse duration. This is not surprising, since both techniques ultimately rely on the heat produced. The coupling mechanisms, however, are rather different.

From the point of view of a high-energy electron (of a typical energy between 10 and 100 keV), matter consists of nearly free electrons and nuclei. A beam of electrons incident on a solid therefore undergoes two kinds of encounters. *Nuclear collisions*, due to the large mass ratio, are essentially elastic and merely change the direction of the beam electron. (Emission of bremsstrahlung is unimportant as far as energy deposition is concerned.) Energetic electrons may also displace atoms from their lattice sites if the transmitted energy exceeds the displacement energy (about 2.4 eV for Si, corresponding to an electron energy of about 125 keV).

Electronic collisions are inelastic and gradually slow down the primary electron, its energy being transformed into excitation energy and kinetic energy of secondary electrons. The excited secondary electrons subsequently thermalize and recombine in a similar fashion as in the case of light absorption. Emission of secondary electrons occurs mainly for beam electron energies below 1 keV; the kinetic energies of the secondary electrons are, however, too low (<50 eV) to influence energy deposition sig-

---

Fig. 11.   Interference fringes on the surface of laser-irradiated palladium silicide (normal incidence). The equidistant fringes in the upper micrograph are produced by scattering centers at the surface [$h = 0$ in Eq. (11)]; the fringes in the lower figure are due to scattering from particles in the atmosphere in front of the sample surface.

nificantly (Bishop, 1974). The nuclear collisions for beam electrons of energy E above 5 keV have been found to be adequately described by a screened Rutherford scattering cross section, which is proportional to E^{-2},

$$\frac{d\sigma}{d\Omega} = \left\{ \frac{e^2[Z - f(\vartheta)]}{16\pi\epsilon_0 E \sin^2(\vartheta/2)} \right\}^2 \tag{14}$$

where Z is the target atomic number, ϑ is the scattering angle of the electron, and the factor $f(\vartheta)$ accounts for screening of the nucleus by the electron shells. The inelastic electron–electron interactions involve a very large number of collisions with electrons of different energy states before the incident electron is sufficiently slowed down to be absorbed. This process may be approximated by a continuous energy loss of the beam electron along its path, described by the Bethe formula:

$$-\frac{dE}{ds} = \frac{e^4}{8\pi\epsilon_0^2} \frac{N_A\rho Z}{AE} \ln\left(\frac{aE}{J}\right) \tag{15}$$

Where N_A is Avogadro's number, ρ is the mass density, and A is the atomic weight of the material. (The factor $N_A Z/A$ is equal to the density of all electrons in the solid.) J is an average ionization potential which depends on the electronic structure of the material, and $a = 1.166$. Note that the structural term J enters only weakly in the expression for the energy loss. An empirical relation between J (in electron volts) and Z is (Everhart and Hoff, 1971)

$$J = 9.76Z + 58.8Z^{-0.19} \tag{16}$$

The path s in Eq. (15) is determined by the nuclear collisions and is therefore different for every individual beam electron. A depth distribution of the total deposited energy can be determined numerically by the Monte Carlo method (Berger, 1963; Murata *et al.*, 1971). In these calculations, a large number of individual electron paths are simulated, and expression (15) is applied along each path. The energy deposited in each volume element of the material is then summed up and displayed as a function of depth to yield the so-called depth–dose function $\langle dE/dz \rangle$. As an example, Fig. 12 shows depth–dose functions in aluminum for various electron energies and normal incidence of the electron beam (Matsukawa *et al.*, 1974).

B. Coupling Parameters for Electron Beams

From the depth–dose function $\langle dE/dz \rangle$ an expression for the absorbed power density $\Phi_e(z)$ can be obtained, which is equivalent to the quantity Φ

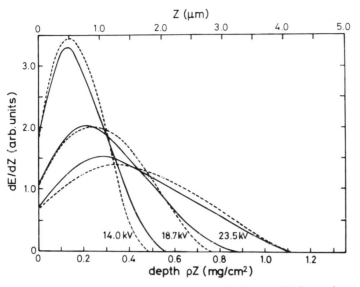

Fig. 12. Deposited energy as a function of depth in Al ($\alpha = 0°$) for various electron energies. Solid line, experimental; dashed line, Monte Carlo. (From Matsukawa *et al.*, 1974.)

given in Eq. (3) for light absorption. Like Φ, Φ_e may be characterized by two parameters, a penetration depth and a reflectivity, which we denote by d_e and R_e, respectively. For a beam of electron energy E and current density $i(t)$ we can write

$$\Phi_e(z,t) = [i(t)/e](1 - R_e)f_E(z/d_e) \tag{17}$$

where the depth distribution f_E is derived from a depth–dose function (dE/dz) valid at energy E. As for the *penetration depth* d_e, various definitions have been used in the literature. A suitable parameter is the extrapolated range according to Grün (1957), usually denoted by R_G. It is obtained from a depth–dose function by extrapolating the straight-line portion at the bottom end of the curve down to the axis, as shown in Fig. 13. The variation in R_G with electron energy follows a power law:

$$R_G = kE^b \qquad (\mu g/cm^2) \tag{18}$$

where k and b are constants. Hereby the depth is measured in units of ρz instead of z in order to make the range expression roughly independent of the material. Everhart and Hoff (1971) derived depth–dose curves from beam-induced conductivity measurements in thin SiO_2 layers and obtained $k = 3.98$ and $b = 1.75$ for Si.* They were able to fit the depth

* For comparison, in air $k = 4.57$ and $b = 1.75$.

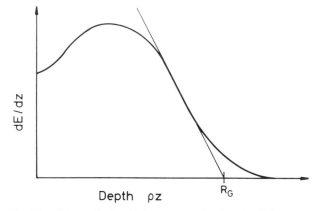

Fig. 13. Schematic depth–dose curve showing the Grün range.

distribution of deposited energy to a polynomial which they normalized to the absorbed energy:

$$\frac{R_G}{E(1 - R_e)} \frac{dE}{dz} = 0.60 + 6.21\left(\frac{z}{R_G}\right) - 12.4\left(\frac{z}{R_G}\right)^2 + 5.69\left(\frac{z}{R_G}\right)^3 \quad (19)$$

Note that the electron range often given in the literature is different from the range of the deposited energy used here. The two kinds of ranges may or may not be proportional, although both are usually described by an expression of the form of Eq. (18) (Feldman, 1960).

Part of the incident electrons experience head-on collisions with nuclei and are backscattered out of the sample before they have lost much of their kinetic energy. Backscattered electrons have energies ranging from 50 eV up to the primary energy E; in contrast to the secondary electrons they can significantly affect the fraction of the beam energy deposited. In analogy with the reflection of a light beam, a coefficient for *energy reflection* can be defined as

$$R_e = y \langle E_B \rangle / E \quad (20)$$

where y is the backscattering yield and $\langle E_B \rangle$ is the mean energy of the backscattered electrons. Figure 14 shows the quantity R_e for normal incidence as a function of target atomic number Z. It was calculated based on measured backscattering yields (Bishop, 1974) and using an empirical relation for $\langle E_B \rangle$ due to Sternglass (1954) and valid for beam energies between 0.2 and 32 keV:

$$\langle E_B \rangle / E = 0.45 + 0.002Z \quad (21)$$

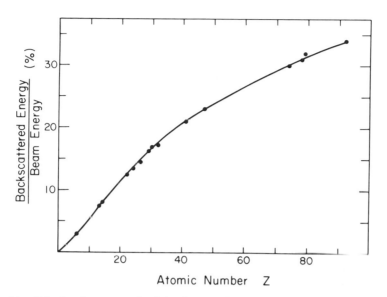

Fig. 14. Calculated energy reflectivity R_e as a function of target atomic number Z. Normal incidence, 30-keV electrons. Based on data for the backscattering yield as given by Bishop (1974).

The reflection coefficient R_e also depends on the incidence angle of the electron beam, because the backscattering yield increases significantly for nonnormal incidence (Bishop, 1974), as shown in Fig. 15.

Electron beams used for annealing applications are usually neither monoenergetic nor exactly collimated. In order to obtain effective coupling parameters for such a case, range and reflectivity relations have to be weighted with the energy spectrum and the angular distribution of the electron beam; they may also be obtained directly from a Monte Carlo simulation. As an example, Fig. 16 shows a calculated depth–dose function in Si for a commercial electron pulse generator (Greenwald *et al.*, 1979).

C.   Beam-Induced Coupling Effects

The parameters d_e (respectively, R_G) and R_e characterizing the degree of coupling of electron beams are far less sensitive to the temperature or the electronic structure of the target material than the parameters d and R pertinent to light. This is not surprising, because while photons interact only with valence electrons, energetic beam electrons can interact with almost every electron in the solid. The detailed energy states of the val-

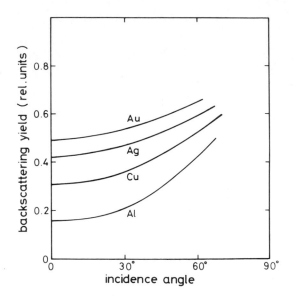

Fig. 15. Backscattering yield as a function of incidence angle of the electron beam. (From Bishop, 1974.)

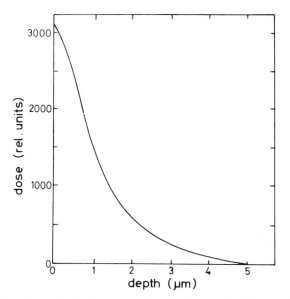

Fig. 16. Calculated depth distribution of deposited energy in Si for a pulse from a commercial electron pulse generator. (From Greenwald *et al.*, 1979.)

ence electrons and all parameters relating to it therefore have far less influence.

Beam-induced coupling effects can, however, result from the electrostatic charge accumulated during irradiation of poorly conducting materials. If the potential built up in the material reaches a sizable fraction of the beam accelerating potential, then part of the electron kinetic energy is lost by work done against the space charge field. Thus, a significant reduction in electron range as a function of time during irradiation has been observed in dielectrics, e.g., plastics (Gross and Nablo, 1967). The field strengths produced in electron-irradiated insulators can reach about 10^6 V/cm; they are limited by leakage currents within the material caused by beam-induced conductivity or by dielectric breakdown (Gross et al., 1973).

IV. Heating by Laser and Electron Beams

As discussed in the past two sections, absorption of both laser and electron beams mainly creates excited electrons which thermalize with a time constant of 10^{-11} sec or less. As thermal equilibrium between the electrons and the lattice is established, part of the energy is taken over by lattice vibrations or phonons. Heat is carried away from the region of absorption by the diffusion of phonons as well as carriers. This is usually described by an overall thermal diffusivity D. If we assume D to be constant, the heat conduction equation reads

$$\frac{\partial T}{\partial t} = D\nabla^2 T + \frac{A}{\rho C_p} \tag{22}$$

Here ρ and C_p are the density and specific heat of the material and A is the rate of heat production per unit volume. Solutions of Eq. (22) are discussed in detail in Chapter 4 by Baeri and Campisano. The point we want to stress here is the impact of coupling phenomena on sample temperature.

As long as D itself is not strongly affected by the beam (this may arise if a dense carrier plasma is created in a nonmetal—see section II.B.2), we may assume that the distribution of thermal energy production equals that of absorbed beam power, i.e., that $A = \Phi$, with Φ as given by Eq. (3) or (17). As discussed in the previous sections, the penetration depth d and/or the reflectivity R may change as a result of the irradiation, causing variations in the spatial distribution as well as the integrated value of A.

In most annealing procedures the sample temperature has to be kept within narrow limits to ensure the desired result. However, particularly in

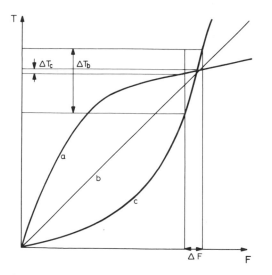

Fig. 17. Schematic temperature-versus-fluence curves, showing the temperature fluctuation ΔT caused by a fluctuation in fluence ΔF for various heating characteristics.

pulsed beam annealing, the process time is too short to allow any real-time control. Moreover, laser as well as electron beams have a finite reproducibility and contain lateral variations in intensity. The result is a variation in sample temperature, which should be kept as small as possible. The amount of these variations depends on the dynamics of the heating process. Figure 17 shows schematic temperature-versus-fluence curves illustrating this point: In curve b the temperature rises linearly with the pulse fluence, in curve a the rate of temperature rise decreases with fluence, and in curve c the reverse is true. The final temperature reached is the same in all cases, but obviously a given fluctuation in beam fluence (due either to lateral variations in intensity or to variations from shot to shot) causes a large variation in temperature in case c, whereas in case a it does not. From a practical point of view, clearly case a is the desirable one. Which one of the cases in Fig. 17 represents a given experimental situation depends in particular, on the behavior of d and R. For illustration let us consider two simple cases.

A beam of fluence F, incident on the surface of a sample of thickness L, creates heat in a surface layer of thickness d (the absorption length for light or the energy range d_e for electrons). During the heat pulse of duration t (thought to be fixed), heat diffuses over a dimension $\delta = 2\sqrt{Dt}$. (If the sample thickness L is smaller than δ, the following discussion applies with δ replaced by L.) Two limiting cases can be distinguished:

(i) If $d \ll \delta$, then heat is created practically only at the surface. The surface temperature at the end of the pulse is given by

$$T(0,t) = 1.13F(1 - R)/\rho C_p \delta \qquad (23)$$

The slope of $T(F)$ now depends mainly on the behavior of $R(T)$ and $C_p(T)$. Since C_p usually increases with temperature, $T(F)$ tends to saturate, corresponding to case a above, if $R(T)$ is constant. If, however, R strongly decreases with temperature (as in the case of light reflection of metals), then the slope of $T(F)$ is shifted toward case b or c.

(ii) If, on the other hand, $d \gg \delta$, then heat is created over a large depth. It can be shown that the surface temperature in this case is approximately

$$T(0,t) = F(1 - R)/\rho C_p d \qquad (24)$$

In this case, the behavior of d as a function of temperature or fluence is crucial. We have mentioned several mechanisms that lead to a strong decrease in the absorption length for light during irradiation. A decrease in d produces a temperature rise as in case c above, at least as long as the condition $d \gg \delta$ is fulfilled.

Two examples may illustrate how the discussed concepts work out in practical laser annealing experiments. Consider a sample consisting of several alternating thin layers of Au and Si on a glass substrate to be reacted by a 30-nsec Nd laser pulse (von Allmen *et al.*, 1980). If the top layer is Au, then $d \ll \delta$ and the decrease in $R(T)$ leads to an unfavorable situation. If the top layer is chosen to be, say, 30 nm of Si (acting also as an antireflection coating), then R even increases with temperature (Si turns metallic upon melting) and the heating curve resembles curve a in Fig. 17. The unfavorable decrease in the absorption length of Si plays no role here, since the Si layer in only 30 nm thick, corresponding to a case with $d \ll \delta$. The experiment confirmed that the latter configuration could be annealed easily, while in the former the control of pulse intensity was extremely critical.

Similar arguments apply for the case of pure crystalline Si heated by a short Nd laser pulse. Here the strong decrease in d during irradiation results in an extreme type-c heating curve (von Allmen, 1979). However, by simply preheating the sample to, say, 250°C (thus decreasing the initial absorption length), a sufficiently stable heating curve is achieved (von Allmen *et al.*, 1979). The same effect can be obtained if part of the pulse energy is converted to a higher photon energy by frequency doubling (Auston *et al.*, 1979) in order to create an appropriate density of free carriers.

TABLE IV

CHARACTERISTICS OF LASER AND ELECTRON BEAMS FOR SAMPLE HEATING

Characteristic	Laser beam	Electron beam
Particle energy	0.1–5 eV	5–100 keV
Interacting particles	Valence electrons, phonons	All electrons, nuclei
Material properties pertinent to coupling	Electronic structure	Density, atomic number
Depth distribution of absorbed energy	Exponential	Roughly Gaussian
Self-induced effects	Varying coupling parameters interference	Space charge effects
	$\lambda = 1.06 \ \mu m^a$	$E = 10 \ keV$
Typical absorption length		
Al	10 nm	1.5 μm
Au	13 nm	2 μm
Si (c)	200 $\mu m \rightarrow$ 13 nm	1 μm
Si (a)	1 $\mu m \rightarrow$ 13 nm	1 μm
Typical energy reflectivity		
Al	0.94 → 0.8	0.08
Au	0.98 → 0.8	0.31
Si (c)	0.33 → 0.7	0.08
Si (a)	0.35 → 0.7	0.08

[a] The arrows indicate variations in the coupling parameters upon heating from room temperature to the melt.

V.   Conclusions

To summarize this chapter, Table IV shows a brief comparison of the essential features of laser and electron beams relevant for material heating. It can be seen, that, as a general trend, electrons deposit their energy more efficiently than photons. The price for this advantage of electron beams is a usually more complicated hardware. On the other hand, lasers presently offer more flexibility and a wider range of parameters to chose from for a given purpose.

It is clear that, while from a physical point of view both beam technologies have advantages, a choice between them will also depend on other criteria—such as adaptability to a given industrial process or cost-effectiveness.

References

Affolter, K., Lüthy, W., and Wittmer, M. (1980). *Appl. Phys. Lett.* **36**, 559.

American Institute of Physics (AIP) Handbook (1972). 3rd ed. McGraw-Hill, New York.

Auston, D. H., Golovshenko, J. A., Simons, A. L., Slusher, R. E., Smith, R. P., Murko, C. M., and Venkatesan, T. N. C. (1979). *Appl. Phys. Lett.* **34**, 777.

Bass, M., and Fradin, D. W. (1973). *IEEE J. Quantum Electron.* **QE-9**, 890.

Bergelson, V. I., Golub, A. P., Nemchinov, S. P., and Popov, S. P. (1974). *Sov. J. Quantum Electron. (Engl. Transl.)* **3**, 288.

Berger, M. J. (1963). *In* "Quantum Mechanics" (B. Alder, S. Fernbach, and M. Rotenberg, eds.), Methods in Computational Physics, Vol. 1. Academic Press, New York.

Bishop, H. E. (1974). *In* "Quantitative Scanning Electron Microscopy" (D. B. Holts, ed.), p. 41. Academic Press, New York.

Bloembergen, N. (1974). *IEEE J. Quantum Electron.* **QE-10**, 375.

Cullis, A. G., and Webber, H. C. (1980). *In* "Laser and Electron Beam Processing of Electronic Materials" (C. L. Anderson, G. K. Celler, and G. A. Rozgonyi, eds.), p. 220. Electrochem. Soc., Pennington, New Jersey.

Dumke, W. P. (1980). *Phys. Lett. A* **78A**, 477.

Dziewior, J., and Schmid, W. (1977). *Appl. Phys. Lett.* **31**, 346.

Everhart, T. E., and Hoff, P. H. (1971). *J. Appl. Phys.* **42**, 5837.

Feldman, C. (1960). *Phys. Rev.* **117**, 455.

Gauster, W. B., and Bushnell, J. C. (1970). *J. Appl. Phys.* **41**, 3850.

Goldfisher, L. I. (1965). *J. Opt. Soc. Am.* **55**, 247.

Greenwald, A. C., Kirkpatrick, A. R., Little, R. G., and Minnuci, J. A. (1979). *J. Appl. Phys.* **50**, 783.

Gross, B., and Nablo, S. V. (1967). *J. Appl. Phys.* **38**, 2272.

Gross, B., Dow, J., and Nablo, S. V. (1973). *J. Appl. Phys.* **44**, 2459.

Grün, A. E. (1957). *Z. Naturforsch. A* **12A**, 89.

Gubanov, A. J. (1965). "Quantum Electron Theory of Amorphous Conductors." Consultants Bureau, New York.

Hill, C. (1980). *In* "Laser and Electron Beam Processing of Electronic Materials" (C. L. Anderson, G. K. Celler, and G. A. Rozgonyi, eds.), p. 26. Electrochem. Soc., Pennington, New Jersey.

Holway, L. H., and Fradin, D. W. (1975). *J. Appl. Phys.* **46**, 279.

Kittel, C. (1976). "Introduction to Solid State Physics," 5th ed., p. 228. Wiley, New York.

Lietoila, A., and Gibbons, J. (1981). *In* "Laser and Electron Beam Solid Interactions and Materials Processing" (J. F. Gibbons, L. D. Hess, and T. W. Sigmon, eds.), p. 23. North-Holland Publ., Amsterdam.

Matsukawa, T., Shimizu, R., Harada, K., and Kato, T. (1974). *J. Appl. Phys.* **45**, 733.

Mott, N. F., and Jones, H. (1958). "The Theory of the Properties of Metals and Alloys." Dover, New York.

Murata, K., Matsukawa, T., and Shimuzu, R. (1971). *Jpn. J. Appl. Phys.* **10**, 678.

Nathan, M. I., Hodgson, R. T., and Yoffa, E. J. (1980). *Appl. Phys. Lett.* **36**, 512.

Rosenberg, R., and Lax, M. (1958). *Phys. Rev.* **112**, 843.

Ryter, D., and von Allmen, M. (1981). *IEEE J. Quantum Electron.* **QE-17**, 2015.

Schumann, P. A., Jr., and Phillips, R. P. (1967). *Solid-State Electron.* **10**, 943.

Shvarev, K. M., Baum, B. A., and Gel'd, P. V. (1975). *Sov. Phys.—Solid State (Engl. Transl.)* **16**, 2111.

Siregar, M. R. T., von Allmen, M., and Lüthy, W. (1979). *Helv. Phys. Acta* **52**, 45.

Siregar, M. R. T., Lüthy, W., and Affolter, K. (1980). *Appl. Phys. Lett.* **36**, 787.
Smith, D. C. (1977). *J. Appl. Phys.* **48**, 2217.
Spitzer, W. G. (1967). *In* "Optical Properties of III-V Compounds" (R. K. Willardson and A. C. Beer, eds.), Semiconductors and Semimetals, Vol. 3, p. 17. Academic Press, New York.
Sternglass, E. J. (1954). *Phys. Rev.* **95**, 345.
Stürmer, E., and von Allmen, M. (1978). *J. Appl. Phys.* **49**, 5648.
Surko, C. M., Simons, A. L., Auston, D. H., Golvchenko, J. A., Slusher, R. E., and Venkatesan, T. N. C. (1979). *Appl. Phys. Lett.* **34**, 635.
Svantesson, K. G. (1979). *J. Phys. D* **12**, 425.
Svantesson, K. G., and Nilsson, N. G. (1979). *J. Phys. C* **12**, 3837.
Van Vechten, J. A. (1980). *In* "Laser and Electron Beam Processing of Materials" (C. W. White and P. S. Peercy, eds.), p. 53. Academic Press, New York.
Vitali, G., Bertolotti, M., and Stagni, L. (1979). *In* "Laser–Solid Interactions and Laser Processing" (S. D. Ferris, H. J. Leamy, and J. M. Poate, eds.), p. 111. Am. Inst. Phys., New York.
von Allmen, M. (1979). *In* "Laser–Solid Interactions and Laser Processing" (S. D. Ferris, H. J. Leamy, and J. M. Poate, eds.), p. 43. Am. Inst. Phys., New York.
von Allmen, M., Blaser, P., Affolter, K., and Stürmer, E. (1978). *IEEE J. Quantum Electron.* **QE-14**, 85.
von Allmen, M., Lüthy, W., Thomas, J.-P., Fallavier, M., Mackowsky, J. M., Kirsch, R., Nicolet, M.-A., and Roulet, M. E. (1979). *Appl. Phys. Lett.* **34**, 82.
von Allmen, M., Lau, S. S., Mäenpää, M., and Tsaur, B. Y. (1980). *Appl. Phys. Lett.* **36**, 205.
Wooten, F. (1972). "Optical Properties of Solids." Academic Press, New York.
Yoffa, E. (1980a). *Phys. Rev. B* **21**, 2415.
Yoffa, E. (1980b). *Appl. Phys. Lett.* **36**, 37.

Chapter 4

Heat Flow Calculations

PIETRO BAERI AND SALVATORE UGO CAMPISANO

Istituto di Struttura della Materia
Università di Catania
Catania, Italy

I. Introduction

Since the first results related to laser annealing of ion-implanted semiconductors were obtained, the number of published papers has increased enormously, as demonstrated by the large number of international conferences devoted to this subject (Rimini, 1978; Ferris *et al.*, 1979; Anderson *et al.*, 1980; White and Peercy, 1980; Gibbons *et al.*, 1981). All these results can be interpreted within the framework of a simple thermal

75

model (Baeri *et al.*, 1979; Wood *et al.*, 1980). The transformation of electronic excitation into heat was well established in the case of pulsed laser irradiation of metals (Ready, 1971), and the validity of such assumptions in relation to the irradiation of crystalline and amorphous semiconductors is given in Chapter 3. It can be assumed that the absorbed light intensity is instantaneously converted to heat which can diffuse according to the conventional heat diffusion equation.

A limit to the quantitative estimate of the effects of high-power laser irradiation is related to a lack of precise optical and thermal parameters of the materials typically used in such experiments. Common irradation conditions are pulse duration 1–100 nsec, energy density 0.5–5 J/cm² at ruby (0.693-μm) or Nd (1.03-μm) wavelengths. At such high irradiation power densities the absorption coefficient usually given in the literature may not be correct because of the possible occurrence of nonlinear effects. In the case of Nd irradiation of Si, the match of the photon energy with the energy gap of the semiconductor further complicates the quantitative evaluation. In fact, the absorption coefficient is strongly dependent upon temperature, presence of dopant and free carriers, crystal perfection, etc.

The absorbed energy produces a dynamical temperature variation, so that the temperature dependence of both optical and thermal parameters must be used in a reliable calculation. However, some of these parameters are known within a limited temperature range, and some extrapolation has to be used. Moreover, for the annealing of ion-implanted materials, the experiments deal with irradiation of an amorphous or heavily damaged surface layer. Thermal parameters of such materials are not well known, especially at high temperatures where crystallization effects prevent conventional measurements.

In this chapter all the numerical evaluations have been performed using the data given by Goldsmith (1961) or by Toulokian (1967), which concern the thermophysical properties of crystalline materials. For the absorption coefficient of both amorphous and crystalline semiconductors and its temperature dependence we adopted data by Sze (1969), von Allmen *et al.* (1979), and Bean *et al.* (1979). For liquid semiconductors the absorption coefficient was assumed to be (250 Å)⁻¹ in all cases.

Heating in an irradiated sample is a consequence of the balance between the deposited energy, governed by optical parameters of the sample and characteristics of the laser pulse, and the heat diffusion, determined by thermal parameters and the pulse duration. In the two limiting cases in which the optical absorption depth is small compared with the thermal diffusion length during the pulse, and vice versa, simple estimates of the temperature rise have been outlined by Bloembergen (1979). Some details will be given in Sections II and III. However, in many practical cases, exact calculations are needed, and complex analytical or numerical evaluations must be used. In particular, for phase transitions occurring during the irradiation, only numerical solutions are available. The mathematical problem is stressed in Section II, while in Sections III–V a set of numerical solutions is given in order to cover a broad range of cases of practical interest. The calculations presented in these sections agree well with experimental data. However, if more exact knowledge of the optical and thermal parameters becomes available, their numerical values could change, but the trends will still be valid. Simple scaling rules are given for this purpose. Applications to Ge and GaAs are presented in Section VI. The problem of diffusion and segregation of impurities during the liquid transient produced by the laser pulse is considered in Section VII. The effect of a very long pulse duration is outlined in Section VIII. Appendixes A and B report details of the numerical method used to solve the heat and mass transport equations.

II. The Mathematical Problem

A. Analytical Solution

As discussed in Section I, we assume that the laser light is converted to heat instantaneously. The time and space temperature evolution can be evaluated by means of the conventional heat diffusion equation, adding a source term which depends on time and space (Baeri *et al.*, 1979). We assume for simplicity that the laser beam travels along the z axis, it is uniform in the x,y plane, the target composition is homogeneous in this plane, and structural changes, if any, occur in the z direction only. The cross section of the laser beam is assumed to be much greater than the heated sample thickness, so that edge effects are negligible.

Under these conditions we deal with a unidimensional problem and the heat equation becomes

$$\frac{\partial T}{\partial t} = \frac{\alpha}{\rho C_p} I(z,t) + \frac{1}{\rho C_p} \frac{\partial}{\partial z}\left(\kappa \frac{\partial T}{\partial z}\right) \qquad (1)$$

where $I(z,t)$ is the power density of the laser light at depth z and time t, T is the temperature, and ρ, C_p, κ, and α are the density, specific heat, thermal conductivity, and absorption coefficient of the sample, respectively. For a homogeneous absorbing medium the power density will be

$$I(z,t) = I_0(t)(1 - R)\, e^{-\alpha z} \qquad (2)$$

where $I_0(t)$ is the output power density from the laser and R is the target reflectivity.

Equation (1) can be solved analytically (Carslaw and Jaeger 1959) under the following assumptions:

(1)   no temperature dependence of thermal (κ, C_p) and optical (α, R) parameters,
(2)   no phase change induced by the laser pulse,
(3)   no inhomogeneity of the sample along the z axis (no dependence of κ, C_p, and α on z),
(4)   constant laser power, and
(5)   infinite sample thickness.

The exact solution is

$$T(z,t) = \{(2I_0/\kappa)(Dt)^{1/2}\ \mathrm{ierfc}|z/2(Dt)^{1/2}|-(I_0/\alpha\kappa)\ e^{-\alpha z}$$

$$+(I_0/2\alpha\kappa)\exp(\alpha^2 Dt - \alpha z)\ \mathrm{erfc}|\alpha(Dt)^{1/2} - z/2(Dt)^{1/2}|$$

$$+(I_0/2\alpha\kappa)\exp(\alpha^2 Dt + \alpha z)\ \mathrm{erfc}|\alpha(Dt)^{1/2} + z/2(Dt)^{1/2}|\} (1 - R) \tag{3}$$

If the light absorption length (α^{-1}) is very small compared with the heat diffusion length $\sqrt{Dt} = \sqrt{\kappa t/\rho C_p}$, the solution can be simplified:

$$T(z,t) = [2I_0(Dt)^{1/2}/\kappa]\ \mathrm{ierfc}\ [z/2(Dt)^{1/2}](1 - R) \tag{4}$$

and for $z = 0$, i.e., at the sample surface,

$$T(0,t) = (2I_0/\kappa)(Dt/\pi)^{1/2}(1 - R) \tag{5}$$

Equations (3)–(5) can be useful in many cases. In particular, from Eq. (5) it follows that the energy density required to bring the surface to the melting point is proportional to the square root of the pulse duration and is independent of the exact value of the absorption coefficient.

B. Numerical Calculations

In the semiconductor case, the thermal and optical parameters are temperature dependent (Goldsmith, 1961; Toulokian, 1967). Moreover, they show strong variations with the phase (amorphous, crystal, or liquid). So they depend in an explicit (from the sample structure) or implicit (through the temperature change) way on z and t. On the other hand the source term is not simply given by Eq. (2) if the absorption coefficient is z dependent, as in the case of layered samples. Because of this intriguing dependence of thermal and optical parameters on z and t and the release of the latent heat during possible phase transitions, Eq. (1) cannot be solved analytically for a general case. Appendix A gives an outline of the numerical method used for the solution of Eq. (1).

The numerical calculations for Si were performed using values given in Figs. 1 and 2 and Table I. For liquid Si we use $R = 0.7$ and $\alpha = 250$ Å^{-1}. The numerical method details the time evolution of the temperature at several depths. As an illustration the temperatures evaluated at the surface, at 0.35 and 1.0 μm, respectively, are shown in Fig. 3 for a 1.7-J/cm^2, 10-nsec ruby laser pulse. The surface layer reaches the melting point at about the maximum power of the pulse and then melts. A maximum temperature of 2200 K is reached at about 18 nsec. During cooling it remains liquid at 1700 K up to 180 nsec. The temperature remains constant at the melting point because of the absorption and release of the latent heat ΔH_m. For layers located at 0.35 and 1 μm the material remains solid at any time and the maximum temperature decreases with depth. Quenching rates and temperature gradients are estimated to be on the order of 10^9 K/sec and 10^7 K/cm, respectively. If the energy is high enough, melting starts from the sample surface and propagates inside the sample. The result of the calculation is a planar melt front starting from the surface and propagating inside the sample in a time comparable with the laser pulse duration. A maximum molten thickness is reached, depending on the energy density of the pulse, and a typical rate of its increase is 0.6 μm/J.

The kinetics of the melt front is shown in Fig. 4 for the case in Fig. 3. The melting proceeds with a planar front at a velocity on the order 10

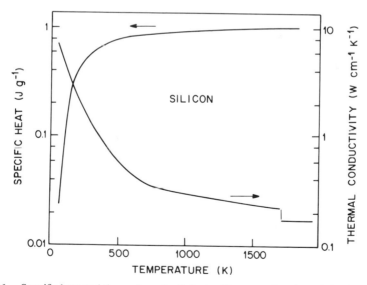

Fig. 1. Specific heat and thermal conductivity in silicon as a function of the temperature (Goldsmith, 1961). The same curves have been used for amorphous material.

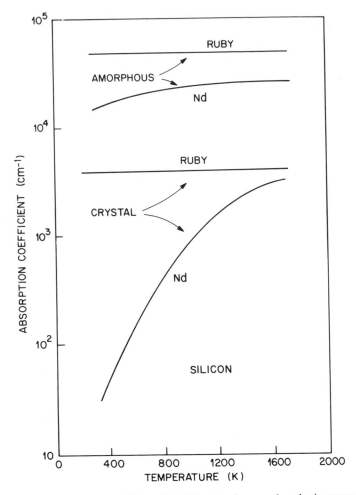

Fig. 2. Light absorption coefficients for Nd and ruby wavelengths in amorphous and crystalline silicon as a function of the temperature.

m/sec, while the solidification requires a time that can be on the order of 10τ for an energy density of a few joules per square centimeter. The front velocity during solidification is on the order of a few meters per second, and details are given in Section V.

III.   The Role of the Absorption Coefficient

The heat flow calculations, as outlined in the previous section, can be easily accomplished for a square pulse and temperature-independent

TABLE I

THERMAL AND OPTICAL PROPERTIES OF SILICON

	Crystal	Amorphous
ρ (g/cm³)	2.33	2.33
ΔH_m (J/g)	1790	1250
T_m (K)	1700	1190
R (ruby)	0.34	0.40
R (Nd)	0.32	0.37

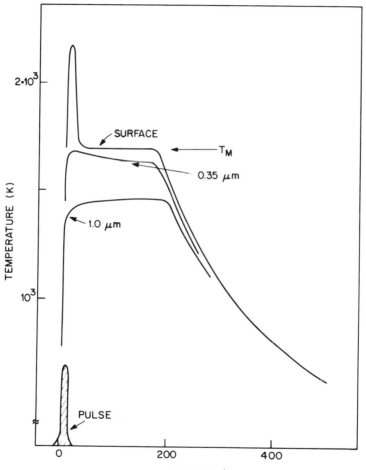

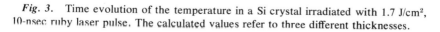

Fig. 3. Time evolution of the temperature in a Si crystal irradiated with 1.7 J/cm², 10-nsec ruby laser pulse. The calculated values refer to three different thicknesses.

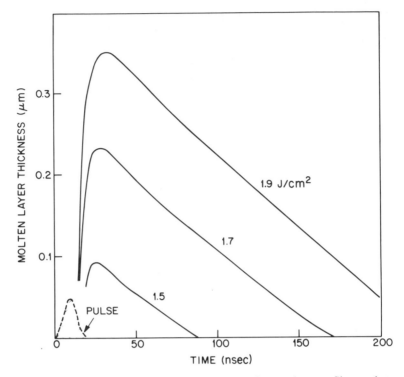

Fig. 4. Kinetics of the melt front for a 10-nsec ruby laser pulse on a Si crystal at several energy densities.

thermal parameters of the material under investigation. From Eq. (3) it is possible to evaluate the threshold energy density E_{th} required to warm the sample surface from room temperature (RT) up to the melting point. Results of such calculations for Si and for several values of the pulse duration are given in Fig. 5. The following parameters have been used: $\kappa = 0.28$ W cm^{-1} K^{-1}; $D = 0.17$ cm^2 sec^{-1}; $T_m = 1700$ K; R $= 0.35$. The κ and D values have been chosen intermediate between the corresponding values at room temperature and T_m. This choice is due to the requirement of constant parameters in the analytical solution.

 For the pulse duration range considered here, the heat diffusion length is greater than the absorption length for $\alpha \gtrsim 10^5$ cm^{-1}. Within this range E_{th} scales as $\tau^{1/2}$ [see Eq. (5)], and the numerical result is sensitive to the shape of the pulse. In the range where $\alpha^{-1} \gg \sqrt{D\tau}$ the result of the calculations is insensitive to the time duration of the pulse and is not appreciably affected by its shape. A square or a Gaussian pulse will give rise, for example, to the same results. The arrows in the figure indicate the α

values corresponding to Nd and ruby laser light (1.06, 0.53, and 0.69 μm) absorbed by amorphous or crystalline material.

In the case of amorphous Si and for ruby, Nd, and Nd ($\lambda/2$, double frequency) wavelengths the threshold energy density for surface melting scales as $\tau^{1/2}$. In the crystalline material there is a small dependence of E_{th} on τ for $\lambda = 0.69$ and 0.53 μm. For $\lambda = 1.06$ μm, E_{th} is completely independent of τ. In the last case, however, the calculated E_{th} value is strongly affected by the large variation in α with the temperature; E_{th} ranges from ~100 to ~2 J/cm^2 for α ranging from values of RT to T_m, respectively. Nonlinear effects and free carrier absorption may be of some importance in this case, and the correct numerical evaluation requires precise knowledge of α at various intensities and temperatures.

In the next section we do not detail the physical processes leading to a

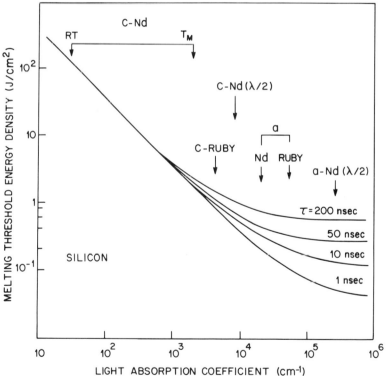

Fig. 5. Energy density threshold for surface melting of a Si sample as a function of the light absorption coefficient and for several durations of a square pulse. The arrows indicate the values of the absorption coefficient for Nd, ruby, and Nd ($\lambda/2$) in crystalline and amorphous material.

given absorption coefficient; instead values generally measured at small light intensities are adopted. Because of the large value of the absorption coefficient at the ruby wavelength in amorphous Si, the assumption is in this case correct.

IV. Amorphous Silicon

The coupling of the laser energy with amorphous material is different from that with crystalline material because of its greater absorption coefficient, leading to a different threshold for surface melting, as discussed in connection with Fig. 5. Moreover, experimental work by Chen and Turnbull (1969) has shown that the enthalpy of the melting of amorphous Ge is lower than the corresponding value for crystalline Ge. Theoretical estimates by Bagley and Chen (1979) and by Spaepen and Turnbull (1979) point out that both the enthalpy of melting and the T_m of amorphous Si and Ge can be depressed by 20–40% compared to single-crystal values. Electron irradiation experiments on amorphous Si (Baeri *et al.*, 1980a, 1981a) give results in agreement with such an assumption.

Results of computation of the maximum melt front penetration as a function of the energy density are reported in Fig. 6 for a single crystal and for amorphous silicon irradiated with a ruby laser pulse of 10- or 50-nsec duration. For amorphous Si we adopted T_m and ΔH_m 30% lower than the corresponding values for crystalline Si.

The almost linear relationship between maximum molten thickness and energy density in the crystalline material is not appreciably affected by the pulse duration, as pointed out in Fig. 5. Instead, in the amorphous material the threshold for surface melting scales approximately as $\tau^{1/2}$, and its smaller value, with respect to the crystal, is mainly due to its higher absorption coefficient.

For ion-implanted materials or films deposited on a crystal substrate, the thickness of the amorphous layer can be either smaller or larger than the absorption length α^{-1} in the amorphous material. In the first case only a fraction of the entire energy is used to heat and melt the surface, the remaining fraction being distributed over a greater depth because of the smaller value of α in the crystal. The energy density threshold for surface melting will then be intermediate between those for the all-amorphous and the all-crystal cases and will depend on the amorphous layer thickness. In Fig. 7 are reported the results of numerical computation of E_{th} as a function of the thickness of the amorphous layer in the case of Si irradiated with ruby and Nd pulses of 10-nsec duration. An energy density of 1.5

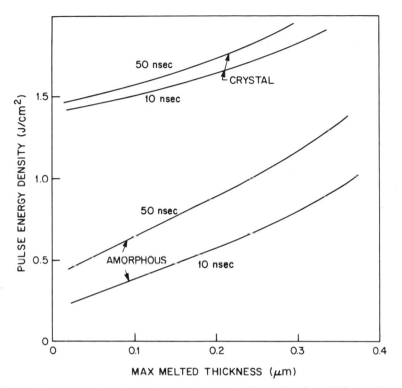

Fig. 6. Energy density of a ruby laser pulse required to melt a given thickness of crystal or amorphous Si. Two different pulse durations are used.

J/cm^2 is required to melt an all-crystal sample at the ruby wavelength, while for Nd no estimate is given. For a sample with a 500-Å-thick amorphous surface layer, 0.68 and 1.55 J/cm^2 are required to reach the melting point at the surface for ruby and Nd laser pulses, respectively. At very large amorphous layer thicknesses, saturation values of 0.38 and 0.28 J/cm^2 for E_{th} are obtained at λ_{Nd} and λ_{ruby}, respectively. Exceeding the threshold for surface melting, the liquid–solid interface propagates inside the sample. The kinetics of the melting and recrystallization processes is shown in Fig. 8 for a 0.2-μm-thick amorphous layer on a Si crystal irradiated with a 50-nsec ruby laser pulse. The threshold for surface melting occurs at about 0.5 J/cm^2, and the thickness of the molten layer increases with the energy density up to about 0.8 J/cm^2 when all the amorphous Si is melted but the underlying crystal is not. For energy densities in the range 0.8–1.15 J/cm^2 the amorphous side of the interface is molten but the liquid front does not penetrate into the crystal. To overcome this interface,

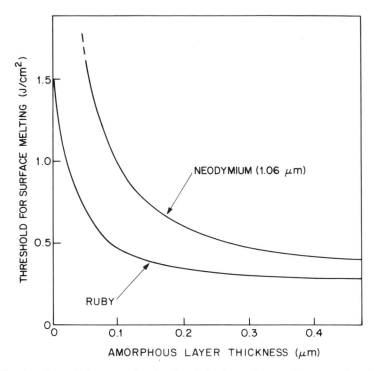

Fig. 7. Computed energy density threshold for surface melting as a function of the amorphous layer thickness of Si on a single-crystal substrate for irradiation with a 10-nsec ruby or Nd laser pulse.

pulses of greater energy density are required. In detail, at 1.25 J/cm² all of the amorphous layer is melted, and its temperature is higher than the T_m of the crystal, so that part of the contiguous crystal substrate is melted too. During the subsequent solidification, liquid phase epitaxy results in a good single-crystal structure. At 1.0 J/cm² all of the amorphous layer is molten, but the temperature at the interface does not reach the crystal melting point. A large amount of residual disorder is expected in this case in the crystallized layer. The assumed difference in the melting temperature and enthalpy between crystalline and amorphous material implies the formation of a stagnant liquid layer of thickness equal to that of the amorphous one. Calculations reported in Fig. 9 for the energy-density–maximum-molten-thickness relationship for a 0.2-μm-thick amorphous Si layer on a crystal substrate irradiated with a 50-nsec ruby laser pulse illustrate the effect of the above assumptions. For comparison, the calculated curve assuming the same values of T_m and ΔH_m for amorphous and crystalline material is also reported. The discontinuity at 0.2 μm in the energy density corresponds to the amount of energy required to raise the temperature of the interface to the crystal melting point. The energy density required to melt just the interface of the underlying crystal, so as to obtain good epitaxial regrowth, is almost independent of the assumed T_m value for the amorphous Si and only slightly dependent on the assumed value of ΔH_m.

The threshold energy densities for good annealing for several combinations of amorphous layer thickness, laser pulse duration, and wavelength are reported in Table II. The values show a minimum mainly as a result of the difference in the absorption coefficient between amorphous and crystalline material. Thin layers require a large amount of energy to be annealed because most of the incident power is diluted in the crystal substrate. For large thicknesses the annealing threshold energy density increases because of the large amount of material to be melted. This trend

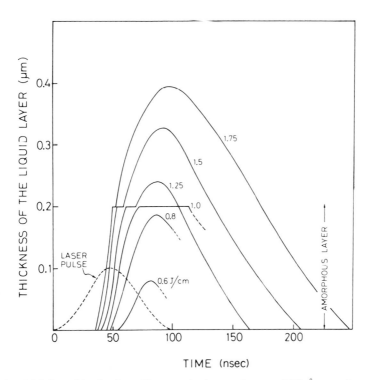

Fig. 8. Melt front kinetics for a 50-nsec ruby laser pulse on a 2000-Å amorphous silicon layer on a silicon crystal. The melting temperature and melting enthalpy of the amorphous Si have been assumed to be 1190 K and 1260 J/g, respectively.

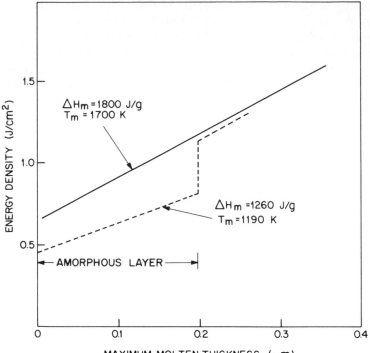

Fig. 9. Energy-density–maximum-molten-thickness relationship for a 50-nsec ruby laser on a 0.2-μm silicon amorphous layer on a silicon crystal with the same thermal parameters as in Fig. 8 (dashed line) and assuming for the amorphous layer the same thermal parameters of the crystal (solid line).

TABLE II

THRESHOLD[a] FOR GOOD ANNEALING OF
AMORPHOUS Si LAYERS

Amorphous layer thickness (Å)	10 nsec		50 nsec	
	Ruby	Nd	Ruby	Nd
500	0.7	1.7	1.05	2.2
1000	0.65	1.3	1.05	1.6
2000	0.7	1.2	1.15	1.5
4000	1.3	1.5	1.6	1.9

[a] In joules per square centimeter.

is more evident for Nd wavelength irradiation, where the change in the absorption coefficient between crystalline and amorphous material is greater than for ruby wavelength irradiation.

V. Liquid–Solid Interface Velocity

In this section we will treat the solidification process, neglecting any atomistic description. The velocity of the liquid–solid interface is determined by the rate of heat extraction from the interface into the bulk. This rate in our case depends on the temperature gradient just behind the interface $(\partial T/\partial z)_i$, and a simple relationship correlates v, κ, and $(\partial T/\partial z)_i$:

$$v = \frac{\kappa}{\Delta H_m \rho}\left(\frac{\partial T}{\partial z}\right)_i \tag{6}$$

where κ is the heat conductivity in the interface region, i.e., at T_m. The temperature gradient in the liquid phase is assumed to be zero. The only way to change the value of v is through the term $(\partial T/\partial z)_i$, as the other parameters are constant and determined by the material under investigation. In the case of silicon, as pointed out in the previous sections, a simple estimate of this term gives a value of about 10^7 K/cm. The resulting velocity is then on the order of several meters per second. The liquid–solid interface velocity plays an important role in the understanding of the impurity redistribution following laser annealing of ion-implanted crystals (Baeri *et al.*, 1980c, 1981b,c). It has been shown that heavily supersaturated solid solutions can be formed in ion-implanted semiconductors and metals (White *et al.*, 1980).

The velocity of the liquid–solid interface is a critical parameter in determining the supersaturation ratio and the percentage of surface accumulation, as is discussed in a next section. It is therefore interesting to evaluate the role of the external parameters in determining the recrystallization velocity.

The temperature distribution is determined by the competition between the rate of energy deposition and the rate of heat diffusion. The liquid–solid interface velocity is then a function of the pulse duration and of the thermal diffusivity of the bulk. The last parameter can be varied by changing the substrate composition, e.g., by depositing films on metallic or insulating substrates. In the case of silicon wafers the thermal diffusivity can be increased by more than one order of magnitude by changing the substrate temperature from 600 to 77 K. The kinetics of the liquid–solid interface for a 15-nsec ruby laser pulse on 1000-Å-thick amorphous silicon and for 300- and 600-K crystal substrate temperatures is shown as an example in Fig. 10. The main difference is related to the different solidification velocities and for the same maximum molten thickness is a factor of 2 smaller at 600 K compared with room temperature substrate irradiation.

The liquid–solid interface velocity can be controlled also by the rate of heat deposition, i.e., by the pulse duration. In Fig. 11 the interface velocity is reported as a function of the maximum molten thickness for different time durations of a ruby laser pulse. Different energy density scales are associated with each pulse duration. The case considered is for a Si sample with a 0.1-μm amorphous layer. For irradiation near the threshold energy density, melting and solidification occur in a time comparable with the pulse duration, and the overall temperature distribution is determined also by the rate of heat deposition. The temperature gradient is determined by the heat diffusion in a time comparable with the laser pulse, so we can assume

$$\left(\frac{\partial T}{\partial z}\right)_i \sim \frac{T_m}{\sqrt{2D\tau}} \sim \frac{2 \times 10^3}{\sqrt{\tau}} \quad (\text{K/cm}) \tag{7}$$

where τ is the pulse duration in seconds. For high-energy densities the melting process proceeds for a long enough time that the solidification

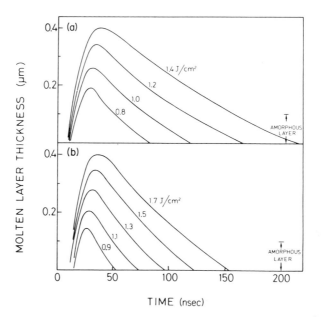

Fig. 10. Kinetics of the melt front for a 15-nsec ruby laser pulse on a 1000-Å amorphous layer on a silicon crystal at (a) 600 K and (b) 300 K substrate temperature.

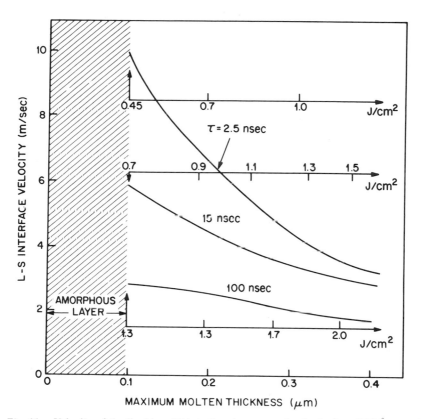

Fig. 11. Velocity of the liquid–solid interface (average value in the last 1000 Å near the surface) during the regrowth as a function of the molten layer thickness produced by ruby laser pulses of different time durations on a 1000-Å-thick amorphous layer on a silicon single crystal. Only pulses above the annealing threshold are evident. Different energy scales are associated with each curve.

starts after the end of the laser pulse and the temperature distribution is unaffected by the way the sample was heated.

Numerical calculations and experimental *in situ* reflectivity measurements by Auston *et al.* (1978) have shown that the surface remains melted for a time interval proportional to E^2. By substituting this time interval for τ in Eq. (7), we obtain the temperature gradient and thus the solidification velocity that is proportional to E^{-1}. The maximum recrystallization velocity occurs for irradiation near the annealing threshold (see Fig. 11). These maximum velocities are reported in Fig. 12 as a function of the laser pulse duration in the range 0.5–100 nsec and for different substrate temperatures. For pulse durations exceeding 10 nsec the calculated data follow the

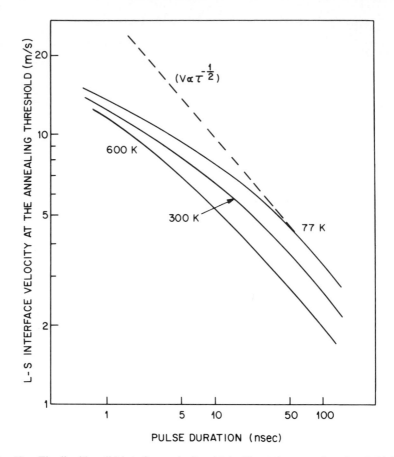

Fig. 12. The liquid–solid interface velocity obtainable at the annealing threshold for a ruby laser pulse on a 1000-Å amorphous layer on a silicon crystal at 77-, 300-, and 600-K substrate temperatures as a function of the time duration of the laser pulse.

$\tau^{-1/2}$ trend, while for shorter pulses the interface velocity seems to saturate. In this case we can in fact neglect heat diffusion and assume

$$\left(\frac{\partial T}{\partial z}\right)_i \sim T_m \frac{\alpha}{e} \qquad (8)$$

where α is the absorption coefficient of the laser light and $e = 2.718$. Using $\alpha \simeq 10^5$ cm^{-1}, intermediate between that of solid and that of liquid silicon, we obtain $v \sim 20$ m/sec. This value should then be an upper limit even for very short laser pulses. Any physical process that increases the absorption coefficient will increase this limiting value. The calculated velocity scales as $\tau^{-1/2}$ for $\tau \geq 10$ nsec at each considered temperature, and the values increase continuously with decreasing temperature as a result of the variation in thermal diffusivity.

VI. Other Semiconductors

In previous sections we presented results of numerical and analytical calculations referring to laser-irradiated silicon. The outlined procedure can be applied to other materials as well. In this section we consider some results concerning the irradiation of Ge and GaAs samples.

The optical and thermal parameters for these semiconductors that we have used in computations are given in Table III. The temperature dependence of the thermal conductivity for both Ge and GaAs is reported in Fig. 13.

For photon energy greater than the band gap (0.76 eV for Ge and 1.4 eV for GaAs) the light absorption length is smaller than the heat diffusion length for $\tau \geq 10$ nsec. Under these conditions the energy density threshold for surface melting is approximately proportional to $\tau^{1/2}$. The validity of such an estimate is better in the actual case of irradiation of amorphous layers because of their higher absorption coefficients with respect to the underlying crystal. For the same reason the threshold for surface melting is not appreciably affected by the thickness of the amorphous layer. The calculated annealing thresholds for a variety of cases in Ge and GaAs are given in Table IV.

TABLE III

PROPERTIES OF Ge AND GaAs

	Ge		GaAs	
Parameter	Crystal	Amorphous	Crystal	Amorphous
T_m (K)	1232	850	1510	1510[a]
ΔH_m (J/g)	411	300	548	548[a]
R (ruby, Nd)	0.4	0.4	0.35	0.35[a]
α (cm^{-1})				
Ruby	1×10^5	2×10^5	3×10^4	1×10^5
Nd	2×10^4	2×10^5	—	—
Nd ($\lambda/2$)	2×10^5	2×10^5	1×10^5	—
C_p (J/g K)[b]	0.33	0.33	0.36	0.36

[a] Assumed the same as for crystal.

[b] Average between room temperature and T_m.

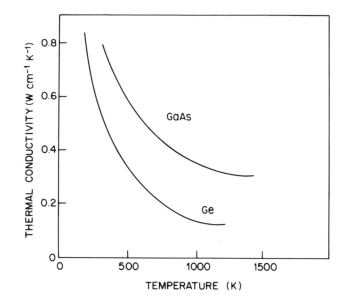

Fig. 13. Thermal conductivity as a function of temperature for Ge and GaAs.

Irradiation of GaAs with Nd pulses requires a high-energy-density threshold for surface melting because of the small absorption coefficient of the 1.06-μm wavelength in the crystalline material. In the case of amorphous layers the lack of precise knowledge of the absorption coefficient prevents any reasonable estimate. In any case the E_{th} value for surface melting is expected to be strongly dependent upon the amorphous layer thickness. It must be noted, however, that if the absorption coefficient is

TABLE IV

ANNEALING THRESHOLD[a]

Amorphous layer thickness (Å)	10 nsec		50 nsec	
	Ruby	Nd	Ruby	Nd
Ge				
1000	0.3	0.35	0.45	0.5
3000	0.55	0.6	0.8	0.9
GaAs				
1000	0.45	—	0.7	—
3000	0.7	—	1.1	—

[a] In joules per square centimeter.

great enough ($\alpha \geq 10^5$ cm^{-1}), the threshold for surface melting does not depend appreciably on its value (see, for example, Fig. 5). This result, justified by the approximations leading to Eqs. (4) and (5), is of great importance in many practical cases where the absorption coefficient is high but known only with a large degree of uncertainty.

As pointed out for Si, for a molten layer thickness on the order of 10^{-5} cm, the maximum molten thickness increases proportionally to $E - E_{th}$. The velocity of the liquid–solid interface during the solidification and for a near threshold irradiation can be scaled from material to material by scaling the thermal parameters involved. [See Eq. (6) and Section V.]

For Ge and GaAs using thermal diffusivity values of 0.11 and 0.20 cm^2/sec, respectively, and other parameters given in Fig. 13 and Table III, the solidification velocities are greater then the corresponding values for Si by a factor of 1.35 for Ge and of 1.8 for GaAs. In both cases this faster regrowth is mainly attributable to the smaller enthalpy of melting of these semiconductors compared to silicon.

VII. Impurity Redistribution

A. Theoretical Approach

The formation of a liquid layer implies a redistribution of the implanted impurities over distances comparable with the implantation range. The diffusion coefficient in liquid Si is on the order of 10^{-4} cm^2/sec. In addition the presence of a moving liquid–solid interface affects the depth distribution of the impurities characterized by different solubilities in the two adjacent phases. Because of the high crystallization velocities and quenching rates, thermodynamical equilibrium concepts do not generally apply. In this section we give a method for determining the impurity depth profile and correlating it with the nonequilibrium parameters (Baeri et al., 1980b; Campisano et al., 1980).

The impurity redistribution is governed by the diffusion equation

$$\frac{\partial}{\partial t} C_l = \frac{\partial}{\partial z} D_l \frac{\partial}{\partial z} C_l \tag{9}$$

where $C_l(z,t)$ is the impurity concentration in the liquid phase at depth z and time t. Boundary conditions given by the presence of the surface and interface are:

$$\left. \frac{dC_l}{dz} \right|_{z=0} = 0 \tag{10}$$

$$\frac{dC_l}{dz}\bigg|_{z=\text{interface depth}=z_i(t)} = 0 \tag{11a}$$

$$\frac{dC_l}{dz} + (1 - k')vC_l[z_i(t)] = 0 \tag{11b}$$

Conditions (11a) and (11b) apply to equal solubility in the two adjacent phases and in the presence of segregation effects, respectively. The segregation coefficient is defined as

$$k_0 = C_{s,i}/C_{l,i} = C_{s,o}/S_{l,o} \qquad \text{at} \quad v = 0$$

$$k' = C_{s,i}/S_{l,i} \qquad \text{at} \quad v \neq 0$$

where the subscripts i and 0 refer to the interface and to the bulk, respectively, and the subscripts s and l indicate liquid and solid. In conventional crystal growth, where the solidification velocity is on the order of 10^{-4} m/sec or less, k' coincides with k_0. As shown before, the crystallization, after laser irradiation, can reach a value of several meters per second, and then k' can be significantly different from k_0.

The diffusion equation with boundary condition (11b) was solved analytically by Tiller *et al.* (1953) for a semi-infinite medium and for a flat initial distribution. The solution has a simple form as a function of the parameter $\xi = vz/4D_l$.

At large values of the reduced thickness ξ the concentration in the solid reaches the original value in the liquid as a result of the continuous pileup of dopant at the liquid–solid interface. This steady state condition is obtained after an ''initial transient'' which extends up to $\xi = 1/k'$. Equation (9) with boundary conditions (10) and (11b) cannot be solved analytically for a nonconstant initial distribution, as in the case of ion implantation. For this purpose it is convenient to use the standard method of solving parabolic differential equations. Appendix B reports in detail the numerical method adopted to solve the mass transport equation.

B. Numerical Calculations

The results of a numerical method applied to the initial constant distribution of impurities indicate that, for a finite liquid thickness, the amount of surface accumulation is determined by the value of k' and by the parameter $\xi = vz_l/4D_l$, where z_l is the initial thickness of the liquid layer. To evaluate the relative influence of these two parameters the percentage of impurities accumulated in the last 10% of the initial liquid thickness is reported in Fig. 14. Saturation at a 90% level arises from the fact that only the fraction of impurities segregated is calculated. Here and

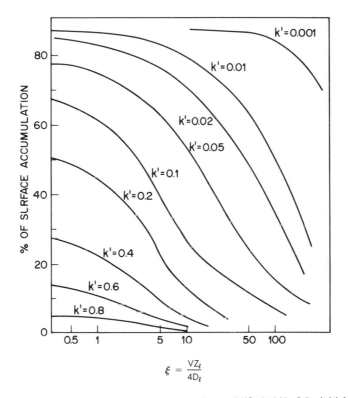

$$\xi = \frac{VZ_l}{4D_l}$$

Fig. 14. The surface fraction of impurities in the last solidified 10% of the initial thickness z_l of a liquid layer as function of its reduced thickness for several values of the segregation coefficient k'. The starting distribution is constant, and 10% of the total amount of impurities initially in the liquid layer has been subtracted from the result.

in the following ξ represents the initial reduced thickness of the liquid layer.

So far only a constant distribution in the liquid has been used as the initial boundary condition, but usually this is not the case. In ion-implanted samples, for example, we encounter, to a first approximation, Gaussian distributions centered at depths z_m ranging between 10^2 and 10^4 Å from the surface and with FWHM, which are typically $\Delta z_m \sim z_m/2.5$. The value of ξ in these cases ranges between 1 and 100. Figure 15 shows the solution of Eqs. (9)–(11) for an initial Gaussian distribution centered at 500 Å, $\Delta z_m = 170$ Å FWHM, and for a liquid layer of initial thickness $z_l = 1000$ Å. The velocity of the liquid–solid interface is 200 cm/sec, the diffusion coefficient in the liquid is 10^{-4} cm^2/sec, and the segregation coefficient is $k' = 0.1$. Calculations have been made at several times showing

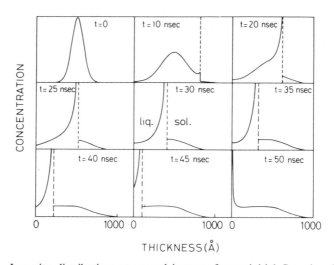

Fig. 15. Impurity distributions at several instants for an initial Gaussian distribution embedded in a liquid layer 1000-Å thick ($t = 0$) during solidification with a planar liquid–solid interface (dashed line) moving with 2 m/sec velocity from the right- to the left-hand side of each square.

the progressive deformation of the impurity profile inside the liquid and the progressive formation of the final distribution in the solid. The dashed vertical line represents the positions of the liquid–solid interface. Impurities are rejected at the freezing interface and build up to high concentrations just ahead of the interface. The freezing of the last layer results in surface segregation of the impurities that were in the final liquid. The concepts developed in connection with Fig. 14 regarding the dependence of the amount of impurity surface accumulation on the parameter ξ can be applied with good approximation to most of the practical cases of ion-implanted dopants. To a first approximation the value of z_l must be replaced by $z_m + \Delta z_m$. The maximum extent of the liquid layer z_l does not practically affect the amount of surface accumulation. Figure 16 shows the numerical solution of Eqs. (9)–(11) for an initial Gaussian-shaped distribution (a) centered at $z_m = 300$ Å, with $\Delta z_m = 170$ Å, assuming a diffusion coefficient of 10^{-4} cm²/sec, a velocity $v = 200$ cm/sec, a segregation coefficient $k' = 0.2$, and an initial thickness of the liquid layer of 500, 1000, and 2000 Å for cases (b), (c), and (d), respectively. Very little difference is observed in the surface peak, but the distribution inside the sample broadens with the initial molten layer thickness. Within the range of practical values of z_l, z_m, and Δz_m, the critical parameter determining the

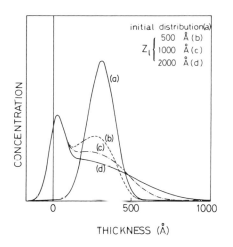

CONCENTRATION

initial distribution(a)

$$Z_l \begin{cases} 500 \text{ Å (b)} \\ 1000 \text{ Å (c)} \\ 2000 \text{ Å (d)} \end{cases}$$

(a)
(b)
(c)
(d)

0   500   1000

THICKNESS (Å)

Fig. 16. Modification of an initially Gaussian distribution after solidification of a liquid layer with three different thicknesses.

percentage of surface accumulation is the implantation depth z_m and not the maximum molten thickness z_l.

The final dopant profile depends on several parameters: the implant distributions characterized by z_m and Δz_m, the segregation coefficients and diffusivities of the implanted species (k', D_l), the velocity of the solid–liquid interface, and the melt thickness (v, z_l) associated with the irradiation conditions. In many cases one is interested in the fraction of dopants retained inside the sample or, equivalently, the percentage of surface accumulation. The availability of the set of curves shown in Fig. 14 for flat initial distributions should be useful. We now show by some examples that the percentage of surface accumulation for Gaussian profiles can be determined also by the parameter $vz/4D_l$ and by the k' value. Moreover a good estimate of the impurity fraction accumulated at the surface is given by the data in Fig. 14 with z replaced by $2z_m$. As a first example we report in Fig. 17 a set of numerical solutions for $z_m = 0.15$ μm, $D_l = 10^{-4}$ cm²/ sec, $v = 2$ m/sec, and for several k' values—0.2, 0.06, and 0.02. The amount of surface accumulation increases with decreasing k' and is 7, 36, and 72%, respectively. These values agree with those (8, 42, and 68%) obtained from Fig. 14 for the reduced thickness $\xi = 15$. For a fixed k' value the percentage of surface accumulation changes by varying the ξ value. In Fig. 18 we report a set of numerical solutions for $z_m = 0.05$ μm, $D_l = 10^{-4}$ cm²/sec, $k' = 0.2$, and several values of v—6, 2, and 0.67 m/sec. The amount of surface accumulation increases with decreasing liquid–solid interface velocity: 7, 35, and 50%, respectively. The corresponding values of ξ are 15, 5, and 1.7, and the surface accumulation from

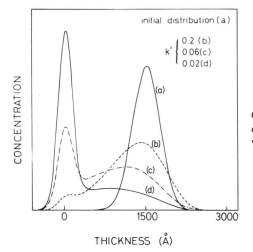

Fig. 17. Modification of an initial Gaussian distribution after solidification of the liquid layer for three different values of the segregation coefficient.

Fig. 14 results in 8, 28, and 43%. As a last case we consider the changes in surface accumulation with the implantation range, and the results are reported in Fig. 19. The assumed conditions are $v = 2$ m/sec, $k' = 0.06$, $D_l = 10^{-4}$ cm²/sec, and $z_m = 0.05$ and 0.15 μm. The calculated surface accumulations are 56 and 33%, respectively, and the ξ values are 5 and 15, to which correspond surface accumulations of 63 and 41% as obtained from Fig. 14.

In all cases the calculated surface peak is very sharp, but a depth resolution of 100 or 300 Å has been used to display the results in a form

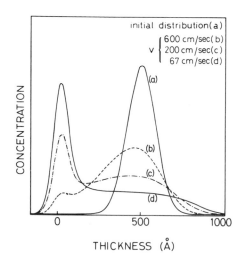

Fig. 18. Modification of an initially Gaussian distribution after solidification of the liquid layer with three different velocities and for the same value of the segregation coefficient.

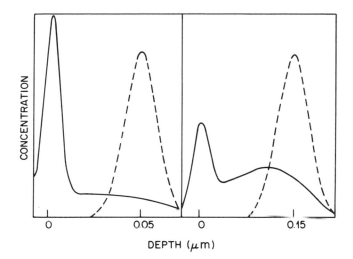

Fig. 19. Modification of two initial Gaussian distributions (dashed line) with different implantation ranges after solidification of the liquid layer with the same velocity and the same segregation coefficient.

more suitable to be compared with backscattering spectra. We conclude that for the same values of ξ and k', although different D_l, v, or z_m values are used, the profiles exhibit the same surface accumulation which, indeed, is very well approximated by those calculated from Fig. 14.

VIII. Continuum Lasers

By increasing the laser pulse duration the deposited energy is smeared out by the diffusion process. For $\alpha^{-1} \ll \sqrt{D\tau}$ the surface temperature increases with the square root of time during the laser pulse, as shown by Eq. (5). In Fig. 20 we show the energy density required to reach the melting temperature on the Si sample surface for a pulse duration in the 10^{-6}–10^{-2} sec range. For irradiation at this threshold value a temperature in the range $0.9T_m$–T_m is maintained for a time interval on the order of 0.2τ.

This temperature–time combination can be enough to activate processes in the solid phase. Williams *et al.* (1978) have shown that a time of about 10^{-5} sec is enough to regrow 500 Å of amorphous Si at a substrate temperature close to the melting point. This extrapolation to high temperature of the low-temperature data is reported in Fig. 21 where the time required to regrow 500 Å of amorphous Si is plotted as a function of the

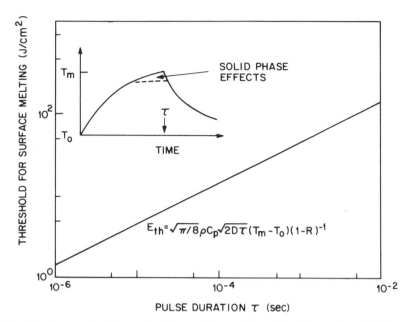

Fig. 20.   Energy density required to reach the melting point at the surface of a Si sample irradiated with a square pulse as a function of the pulse duration. The time dependence of the temperature during the pulse is schematically shown. A constant reflectivity of 0.4 and parameters averaged between room temperature and T_m have been used.

substrate temperature. From these data and calculations reported in Fig. 20 we obtain that, for $\tau > 10^{-5}$ sec, solid phase effects are important for irradiations near the melt threshold value.

To obtain the required power output a focused cw laser can be used. For the available commercial lasers, a typical size should be 50–100 μm in beam diameter. Under these conditions, however, the source can no longer be considered planar. Transverse heat flow must be taken into account, the problem is not unidimensional, and the above calculations are no longer valid.

Under these conditions for a cw irradiation a steady state temperature

$$\Delta T = P/\sqrt{2}\pi a\kappa$$

is reached as a result of the balance between heat absorption and diffusion. In this geometry the relevant parameter governing the temperature rise is the ratio P/a, i.e., absorbed power/beam radius. The steady state temperature is reached after a transient time on the order of $C_p a/\kappa$ (Auston *et al.*, 1978). In Fig. 21 is also reported the linear power density P/a required to reach a given temperature. The mathematical problem of heat

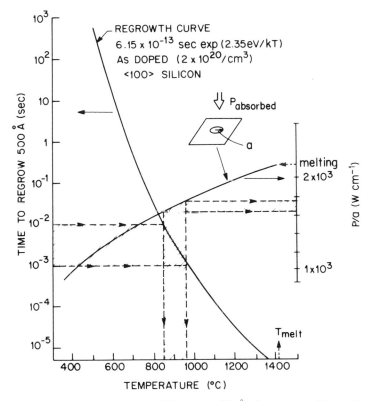

Fig. 21. The time required to epitaxially regrow 500 Å of amorphous Si as a function of the substrate temperature. Also plotted is the calculated absorbed laser power per spot radius (see inset) as a function of the steady state temperature attained at the center of the spot. The hatched areas define the temperature regions and laser power required to regrow thermally the 500-Å amorphous layer. The amorphous Si is doped with As at 2×10^{20} atoms/cm³.

flow (Carslaw and Jaeger, 1959) has been solved by Lax (1977, 1978) for the case of cw laser irradiation. The occurrence of melting can also be taken into account analytically (Kokorowski *et al.*, 1981). Detailed calculations for the case of an elliptical laser beam are reported in Chapter 10.

Appendix A. Heat Transport

The target or a suitable portion of the whole target is divided into slices of thickness Δz. Each layer is labeled with a positive integer i, and by

$i = 1$ we indicate the slice that lies at the irradiated surface of the target.

Assuming as zero time the start of the laser pulse, we must define at each time t and layer i the following quantities:

(i) the temperature T_{b_i} of each layer,

(ii) the structure; i.e., we must define if the ith layer is crystalline, amorphous, or liquid. This can be accomplished by a suitable variable FLAG$_i$ which may assume the value 0 (crystalline), 1 (amorphous), or 2 (liquid).

(iii) Moreover the layer may contain solid and liquid in equilibrium if the absorbed energy has been sufficient to reach the melting temperature but not sufficient to melt all of the layer because of the latent heat of fusion of the material. So we define a quantity FF$_i$ which is the fraction of the material that has been melted according to the balance between the absorbed and the latent heat.

Thermal and optical parameters are in general defined for each layer by suitable functions of T_{b_i}, FLAG$_i$, and FF$_i$. For the first layer we must take into account also the reflectivity $R(T_{b_1}, \text{FLAG}_1, \text{FF}_1)$ in order to compute the power incident at time t:

$$I_1 = I_0(t)(1 - R) \tag{A1}$$

The power density incident reaching the layer is evaluated by

$$I_i = I_{i-1}\, e^{-\alpha_{i-1}\, \Delta z} \tag{A2}$$

where α_{i-1} means $\alpha(T_{b_{i-1}}, \text{FLAG}_{i-1}, \text{FF}_{i-1})$

Knowing T_{b_i} at time t, we must compute the temperature T_{a_i} at the time $t + \Delta t$, where Δt is a suitable time interval. This is accomplished for each layer i as follows: The energy absorbed from the laser pulse during time Δt by the ith layer is

$$\Delta Q_{\text{abs}} = I_i(1 - e^{-\alpha_i \rho \Delta z})\, \Delta t \tag{A3}$$

The energy transferred by thermal diffusion from the nearest-neighbor layers, per unit area, is

$$\Delta Q_{\text{diff}} = \left| \kappa_- \frac{T_{b_{i-1}} - T_{b_i}}{\Delta z} + \kappa_+ \frac{T_{b_{i+1}} - T_{b_i}}{\Delta z} \right| \Delta t \tag{A4}$$

where

$$\kappa_- = (\kappa_{i-1} + \kappa_i)/2 \quad \text{and} \quad \kappa_+ = (\kappa_{i+1} + \kappa_i)/2. \tag{A5}$$

The new temperature can be calculated by

$$T_{a_i} = T_{b_i} + (\Delta Q_{\text{abs}} + \Delta Q_{\text{diff}})/C_p \rho\, \Delta z \tag{A6}$$

Equation (A6) cannot be used if $T_{b_i} = T_m$, or $T_{b_i} < T_m \leq T_{a_i}$, or $T_{a_i} \leq T_m < T_{b_i}$. If so, melting or solidification processes occur and the latent heat must be included in the computation. The quantity

$$\Delta Q' = \Delta Q_{abs} + \Delta Q_{diff} - (T_m - T_{b_i})C_p\rho\,\Delta z \qquad (A7)$$

represents the energy available for melting the layer (if $T_{b_i} \leq T_m$) or arising from the solidification (if $T_{b_i} > T_m$). Thereafter the quantity

$$\Delta FF = \Delta Q'/\Delta H_m \rho\,\Delta z \qquad (A8)$$

is the melted or solidified fraction of the layer during the time Δt. The new value $FF_i(t + \Delta t) = FF_i(t) + \Delta FF$ can now be calculated, and if $0 \leq FF_i \leq 1$, we set $T_{a_i} = T_m$ and the computation follows for the subsequent layers. If $FF_i > 1$, we set $FF_i = 1$ and the quantity $\Delta Q'' = (FF_i - 1)\rho\,\Delta z\,\Delta H_m$ is calculated; if $FF_i < 0$, we set $FF_i = 0$ and the quantity $\Delta Q'' = FF_i\rho\,\Delta z\,\Delta H_m$ is calculated. $\Delta Q''$ is the heat available for varying the temperature after complete melting or solidification of the layer has occurred.

Finally, the new temperature

$$T_{a_i} = T_m + \Delta Q''/C_p\rho\,\Delta z \qquad (A9)$$

can be calculated.

The computation procedure is then applied to the subsequent layers. When the computation of the T_{a_i} values is performed for all the layers, we set $T_{b_i} = T_{a_i}$ for each i; the time t is then incremented by the value Δt, and the computation starts again for all layers. The increment Δt is related to the depth interval Δz by the stability relation

$$\kappa\,\Delta t/C_p\rho\,(\Delta z)^2 < \tfrac{1}{2} \qquad (A10)$$

The boundary conditions are the following:

(i)   the temperature distribution in the target at time $t = 0$,

(ii)   no loss of heat at any time from the illuminated surface,

(iii)   for a bulk sample $\lim_{z\to\infty} T$ = constant at all times.

For assumption (ii) one must note that both blackbody radiation and heat exchanged with the surrounding gaseous medium are negligible. If the target has a limited thickness, condition (ii) must be imposed on the rear surface also. Condition (iii) is approximated by T_N = constant at every t, where T_N is the temperature of the last Nth layer at a depth Z. The thickness Z must be great enough so as to neglect the heat flux through the rear surface for all the times of interest in the solution of Eq. (1).

Appendix B.   Mass Transport

The sample is divided into N slices of thickness δz labeled progressively from 1 to N starting from the surface.

The initial concentration $c(t,z_i)$ at $t = 0$, where z_i is the depth inside the sample at which is located the ith layer, $z_i = (i - \frac{1}{2}) \delta z$, is calculated for each layer. For all the layers the flux F of particles is then computed on both sides by considering the concentration gradients:

$$F_{(i-1)\to i} = D_l[c_l(t,z_{i-1}) - c_l(t,z_i)]/\delta z \qquad (B1)$$

$$F_{(i+1)\to i} = D_l[c_l(t,z_{i+1}) - c_l(t,z_i)]/\delta z \qquad (B2)$$

and the net change in concentration during a time step δt:

$$\delta c_i = (F_{(i-1)\to i} + F_{(i+1)\to i}) \, \delta t/\delta z \qquad (B3)$$

at the boundaries:

$$F_{(i-1)\to i} = 0 \qquad \text{for} \quad i = 1 \qquad (B4)$$

and

$$F_{(i+1)\to i} = 0 \qquad \text{for} \quad i = n \qquad (B5)$$

where n is the liquid layer adjacent to the interface and, therefore, with $N - n$ layers in the solid phase.

If the liquid–solid interface moves with velocity v, after the time $\Delta\tau = \delta z/v$, the number of liquid layers will be reduced to $n - 1$, taking care that $\Delta\tau$ is an integer multiple of δt. By solving the diffusion equation this way it is easy to include boundary condition (11b) for the segregation. An extra flux added between the layers n and $n + 1$:

$$F_{\text{seg}} = (1 - k')vc_l(t,z_n) \qquad (B6)$$

will be calculated, and for the last liquid layer we will have $F_{(n+1)\to n} = F_{\text{seg}}$ instead of Eq. (B2), while for the first solid layer we will have

$$F_{n\to(n+1)} = -F_{\text{seg}} \qquad (B7)$$

so

$$c_l(t + \delta t,z_n) = F_{(n-1)\to n} \frac{\delta t}{\delta z} + F_{\text{seg}} \frac{\delta t}{\delta z} + c_l(t,z_n) \qquad (B8)$$

and

$$c(t + \delta t,z_{n+1}) = -F_{\text{seg}} \frac{\delta t}{\delta z} + c(t,z_{n+1}) \qquad (B9)$$

Equations (B6)–(B9) are valid in the case of both $k' < 1$ and $k' > 1$; they are derived simply by a finite-step transcription of the boundary Eq. (11b) where the term $D_l\,(\partial c_l/\partial z)$ is interpreted as particle flux following Fick's law. The spatial and time steps δt and δz cannot be chosen arbitrarily. For the numerical integration of the differential Eq. (9) the standard stability relation has been demonstrated (Lance, 1960):

$$\delta t \lesssim \delta z^2/2D_l \tag{B10}$$

However, Eqs. (11) and (B6) impose other restrictions:

$$v(1 - k')\,\delta t/\delta z \ll 1 \tag{B11}$$

which arises from considering that particle flux due to segregation must be a small perturbation in the concentration of the last liquid layer; otherwise oscillations are created during numerical calculation. The condition

$$\delta t \gtrsim \delta z^2/2D_l \tag{B12}$$

arises from considering that particles passed through the liquid–solid interface are uniformly distributed inside the spatial step δz. By coupling Eqs. (B10)–(B12):

$$\delta z \ll D_l/(1 - k')v \qquad \text{and} \qquad \delta t = \delta z^2/2D_l \tag{B13}$$

Experience shows that a value for δz a factor of 10 lower than $D_l/(1 - k')v$ is small enough to prevent oscillations in the solution.

List of Symbols

C_l	concentration of impurities in liquid (atoms/cm³)
C_s	concentration of impurities in solid (atoms/cm³)
C_p	specific heat at constant pressure (J/g K)
D	heat diffusion coefficient (cm²/sec)
D_l	diffusion coefficient in the liquid (cm²/sec)
ΔH_m	enthalpy of melting per unit mass (J/g)
κ	thermal conductivity (W cm⁻¹ K⁻¹)
k_0	equilibrium segregation coefficient
k'	interfacial segregation coefficient
I	power density (W cm⁻²)
P	power (J/sec)
R	reflectivity
T	temperature (K)
T_m	melting temperature (K)
t	time (sec)
v	velocity (cm/sec)

z	depth inside the sample (cm)
α	absorption coefficient (cm^{-1})
λ	wavelength (cm)
ρ	mass density (g/cm^3)
τ	pulse duration
E_{th}	melt threshold

References

Anderson, C. L., Celler, G. K., and Rozgonyi, G. A., eds. (1980). "Laser and Electron Beam Processing of Electronic Materials." Electrochem. Soc., Princeton, New Jersey.
Auston, D. H., Golovchenko, J. A., Smith, P. R., Surko, C. M., and Venkatesan, T. N. C. (1978). *Appl. Phys. Lett.* **33**, 538.
Baeri, P., Campisano, S. U., Foti, G., and Rimini, E. (1979). *J. Appl. Phys.* **50**, 788.
Baeri, P., Foti, G., Poate, J. M., and Cullis, A. G. (1980a). *Phys. Rev. Lett.* **45**, 2036.
Baeri, P., Campisano, S. U., Grimaldi, M. G., and Rimini, E. (1980b). *In* "Laser and Electron Beam Processing of Materials" (C. W. White and P. S. Peercy, eds.), p. 130. Academic Press, New York.
Baeri, P., Poate, J. M., Campisano, S. U., Foti, G., Rimini, E., and Cullis, A. G. (1980c). *Appl. Phys. Lett.* **37**, 912.
Baeri, P., Foti, G., Poate, J. M., and Cullis, A. G. (1981a). *In* "Laser and Electron Beam Solid Interactions and Materials Processing" (J. F. Gibbons, L. D. Hess, and T. W. Sigmon, eds.), p. 39. North-Holland Publ., New York.
Baeri, P., Foti, G., Poate, J. M., Campisano, S. U., Rimini, E., and Cullis, A. G. (1981b). *In* "Laser and Electron Beam Solid Interactions and Materials Processing" (J. F. Gibbons, L. D. Hess, and T. W. Sigmon, eds.), p. 67. North-Holland Publ., New York.
Baeri, P., Foti, G., Poate, J. M., Campisano, S. U., and Cullis, A. G. (1981c). *Appl. Phys. Lett.* **38**, 800.
Bagley, B. C., and Chen, M. S. (1979). *In* "Laser–Solid Interactions and Laser Processing" (S. D. Ferris, H. J. Leamy, and J. M. Poate, eds.), p. 97. Am. Inst. Phys., New York.
Bean, J. C., Leamy, H. J., Poate, J. M., Rozgonyi, G. A., van der Ziel, J. P., Williams, J. S., and Celler, G. K. (1979). *J. Appl. Phys.* **50**, 881.
Bloembergen, N. (1979). *In* "Laser–Solid Interactions and Laser Processing" (S. D. Ferris, H. J. Leamy, and J. M. Poate, eds.), p. 1. Am. Inst. Phys., New York.
Campisano, S. U., Baeri, P., Grimaldi, M. G., Foti, G., and Rimini, E. (1980). *J. Appl. Phys.* **51**, 3968.
Carslaw, H. S., and Jaeger, J. C. (1959). "Conduction of Heat in Solids." Oxford Univ. Press, London and New York.
Chen, H. S., and Turnbull, D. (1969). *J. Appl. Phys.* **40**, 4214.
Ferris, S. D., Leamy, H. J., and Poate, J. M., eds. (1979). "Laser–Solid Interactions and Laser Processing." Am. Inst. Phys., New York.
Gibbons, J. F., Hess, L. D., and Sigmon, T. W., eds. (1981). "Laser and Electron Beam Solid Interactions and Materials Processing." North-Holland Publ., New York.
Goldsmith, A. (1961). "Handbook of Thermophysical Properties of Solid Materials." Macmillan, New York.
Kokorowski, S. A., Olson, G. L., and Hess, L. D. (1981). *In* "Laser and Electron Beam Solid Interactions and Materials Processing" (J. F. Gibbons, L. D. Hess, and T. W. Sigmon, eds.), p. 139. North-Holland Publ., New York.
Lance, G. N. (1960). "Numerical Methods for High Speed Computers." Iliffe, London.

Lax, M. (1977). *J. Appl. Phys.* **48,** 3919.
Lax, M. (1978). *Appl. Phys. Lett.* **33,** 786.
Ready, J. F. (1971). "Effects of High Power Laser Radiation." Academic Press, New York.
Rimini, E., ed. (1978). *Proc. Laser Eff. Ion Implanted Semicond., Univ. Catania.*
Spaepen, F., and Turnbull, D. (1979). *In* "Laser–Solid Interactions and Laser Processing" (S. D. Ferris, H. J. Leamy, and J. M. Poate, eds.), p. 73. Am. Inst. Phys., New York.
Sze, S. M. (1969). "Physics of Semiconductor Devices." Wiley, New York.
Tiller, W. A., Jackson, K. A., Rutter, J. W., and Chalmers, B. (1953). *Acta Metall.* **1,** 428.
Toulokian, Y. S. (1967). "Thermophysical Properties of High Temperature Solid Materials." Macmillan, New York.
von Allmen, M., Luthy, W., Siregar, M. T., Affolter, K., and Nicolet, M. A. (1979). *In* "Laser–Solid Interactions and Laser Processing" (S. D. Ferris, H. J. Leamy, and J. M. Poate, eds.), p. 43. Am. Inst. Phys., New York.
White, C. W., and Peercy, P. S., eds. (1980). "Laser and Electron Beam Processing of Materials." Academic Press, New York.
White, C. W., Wilson, S. R., Appleton, B. R., and Young, F. W. (1980). *J. Appl. Phys.* **51,** 738.
Williams, J. S., Brown, W. L., Leamy, H. J., Poate, J. M., Rodgers, J. W., Rosseau, D., Rozgonyi, G. A., Shelnutt, J. A., and Sheng, T. T. (1978). *Appl. Phys. Lett.* **33,** 541.
Wood, R. F., Wang, J. C., Giles, G. E., and Kirkpatick, J. R. (1980). *In* "Laser and Electron Beam Processing of Materials" (C. W. White and P. S. Peercy, eds.), p. 37. Academic Press, New York.

Chapter 5

Supersaturated Alloys, Solute Trapping, and Zone Refining*

C. W. WHITE and B. R. APPLETON

Solid State Division
Oak Ridge National Laboratory
Oak Ridge, Tennessee

and

S. R. WILSON

Semiconductor Group
Motorola, Inc.
Phoenix, Arizona

* Research sponsored by the Division of Materials Sciences, U.S. Department of Energy under contract W-7405-eng-26 with Union Carbide Corporation.

111

I. Introduction

In pulsed laser annealing of ion-implanted semiconductors, the rapid deposition of laser energy in the near-surface region leads to melting of the surface region to a depth of several thousand angstroms followed by liquid phase epitaxial regrowth from the underlying substrate at growth velocities calculated to be several meters per second (Wang *et al.*, 1978). If the proper laser annealing conditions are used, the annealed regions are free of any extended defects (Narayan *et al.*, 1978; White *et al.*, 1979a). Chapter 4 (Baeri and Campisano) summarizes the experimental observations and theoretical calculations relating to depth of melting and velocity of recrystallization that can be obtained. At the very rapid growth velocities achieved in pulsed laser annealing, recrystallization of the melted region takes place under conditions that are far from equilibrium at the moving interface. Substitutional impurities can be incorporated into the lattice at concentrations that far exceed equilibrium solubility limits (White *et al.*, 1980a). Values for the (nonequilibrium) interfacial distribution coefficient from the liquid (k') can be determined by comparing model calculations to experimental dopant concentration profiles after laser annealing (White *et al.*, 1980a). These comparisons show that values for k' are much greater than corresponding equilibrium values k_0. The high growth velocity achieved by laser annealing provides one with an opportunity to study high-speed, nonequilibrium crystal growth phenomena under well-controlled experimental conditions.

In this chapter, we discuss studies of high-speed crystal growth achieved by laser annealing of ion-implanted silicon. These studies show that Group III and V impurities can be incorporated by solute trapping into substitutional lattice sites at concentrations that far exceed equilibrium solubility limits. Values for the interfacial distribution coefficient are much greater than equilibrium values because of the high velocity of the liquid–solid interface. Values for k' are functions of both growth velocity and crystal orientation. For each Group III and V dopant there is a maximum concentration C_s^{max} that can be incorporated substitutionally into the lattice, and values for C_s^{max} are functions of growth velocity. Mechanisms that limit substitutional solubilities are discussed. Values obtained for C_s^{max} are compared with recent predictions of thermodynamic limits of solute trapping in silicon. Finally, the behavior exhibited by Group III and V species is contrasted with that exhibited by interstitial species (such as Cu, Fe) where, for low concentrations, complete zone refining to the surface can be achieved.

II. Experimental Approach

Studies of high-speed crystal growth have been carried out for the most part using silicon crystals ion-implanted with various impurities. The use of ion implantation allows one to vary the species, concentration (dose), and depth of the impurity in a reliable and reproducible manner. For the work described here, implantation energies were in the range 30–250 keV, and doses were in the range 10^{14}–10^{17}/cm². Laser annealing was carried out using the Q-switched output of a pulsed ruby laser (15×10^{-9} sec pulse duration time) at energy densities of 1–2 J/cm². These conditions give rise to liquid phase epitaxial regrowth velocities of several meters per second (see Chapter 4). Regrowth velocity can be varied by changing the pulse duration time, the energy density, or the substrate temperature during annealing. The variation in the regrowth velocity achieved using radiation from Q-switched lasers was ~1–6 m/sec.

Crystals were analyzed in the as-implanted and laser-annealed conditions using Rutherford backscattering (RBS) and ion-channeling techniques. These measurements can be used to determine the concentration profile of the dopant, the lattice location of the dopant, the substitutional concentration as a function of depth, and the damage distribution as a function of depth in the crystal. In studies of high-speed crystal growth, knowledge of the lattice location of the dopant and the substitutional concentration as a function of depth is essential because only the substitutional component is in solution in the regrown lattice. Selected crystals were examined subsequently by transmission electron microscopy to determine the nature of the remaining defects (if any) and the microstructure of the near-surface region. These measurements are particularly valuable in the case of very-high-dose implants where constitutional supercooling during regrowth leads to lateral segregation of the dopant and the formation of a well-defined cell structure in the near-surface region.

III. Model Calculations for Dopant Redistribution

In pulsed laser annealing, redistribution of the dopant occurs by liquid phase diffusion (Wang *et al.*, 1978; Baeri *et al.*, 1978a,b; Wood *et al.*, 1981; Baeri and Campisano, Chapter 4, this volume). Several techniques have been used to calculate dopant redistribution in the liquid phase during the time the implanted region remains in the liquid state. The most widely used of these methods is a numerical solution to the mass diffusion equation expressed in finite differences. This approach allows one to account properly for the explicit time dependence of the position of the liquid–solid interface and to calculate solute profiles for systems in which the distribution coefficient is less than unity. Using the finite difference method to calculate solute redistribution as a result of annealing, one must know the as-implanted concentration profile, the time dependence of the position of the liquid–solid interface, the liquid phase diffusivity, and the interfacial distribution coefficient. The as-implanted concentration profile is easily measured by Rutherford backscattering; values exist in the literature for the liquid phase diffusivity (Kodera, 1963), and information on the time dependence of the position of the liquid–solid interface can be obtained from model calculations as indicated in Chapter 4 (Baeri and Campisano). Measurements exist for the interfacial distribution coefficient k_0 at low growth velocities where conditions of local equilibrium can be assumed at the interface (Trumbore, 1960). In pulsed laser annealing, however, cooling rates are so high and regrowth velocities are so great that departures from equilibrium at the interface must be expected. Consequently, by comparing experiments to model calculations, one can treat the interfacial distribution coefficient as a fitting parameter in order to obtain values for k' appropriate to the high-speed growth process.

Interfacial distribution coefficients are defined as

$$k' = C_s/C_L \tag{1}$$

where C_s and C_L are concentrations in solution in the solid and liquid phases at the interface. At low growth velocities, when solidification occurs under conditions of local equilibrium at the interface, the interfacial distribution coefficient is the equilibrium distribution coefficient k_0 defined as (see Fig. 1)

$$k_0 = \left.\frac{C_s}{C_L}\right|_{eq} \tag{2}$$

where C_s and C_L are concentrations in the solid and liquid phases (at a fixed temperature) as determined from the equilibrium phase diagram. In crystal growth, if $k' < 1$, then the solute will accumulate in the liquid at the interface, but the ratio of concentration in the solid and liquid phases at the interface will be k'. Under conditions of equilibrium crystal growth (i.e., $k' = k_0 < 1$), if one starts with a liquid containing a uniform solute concentration n_0 (as illustrated in Fig. 1), the first solid to freeze has a solute composition $k_0 n_0$. As solidification proceeds, solute accumulates in the liquid at the interface until a steady state concentration n_0/k_0 is reached in the liquid at the interface. The rejected dopant will be trans-

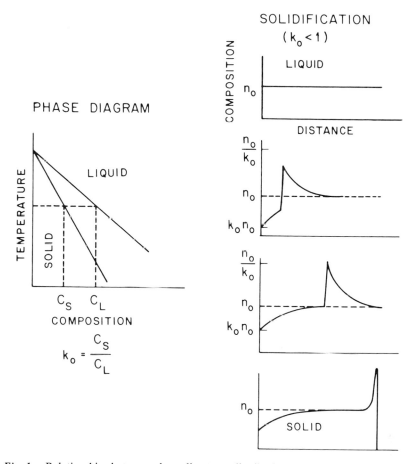

Fig. 1. Relationships between phase diagrams, distribution coefficients k_0, and solidification. The right-hand side shows solute profiles in the liquid and solid at several stages during solidification, assuming $k_0 < 1$.

ported to the surface and will appear as a terminal spike in the near-surface region. Analytic solutions to the case depicted in Fig. 1 (uniform solute concentration in the liquid, constant velocity of regrowth under conditions of local equilibrium, mass transport by liquid phase diffusion only) have been published by Tiller *et al.* (1953), and one need know only n_0, k_0, D, and v to determine the solute profile in the solid. However, if n_0 is not initially uniform, then numerical solution of the mass diffusion equation is required and this is most conveniently accomplished using the method of finite differences.

The method used to calculate dopant distribution during pulsed laser annealing, including the effects of segregation, is based on a finite difference solution to the mass diffusion equation (White *et al.*, 1980a; Wood *et al.*, 1981; Baeri and Campisano, Chapter 4, this volume). In this method the absorbed laser light melts the crystal to a depth dependent on energy density, pulse duration time, and thickness of the amorphous region (Baeri and Campisano, Chapter 4, this volume). As the liquid cools by thermal conduction to the underlying substrate, the melt front begins to recede toward the surface with a velocity v until it begins to encounter diffusing dopant. At this time the liquid layer is divided into segments of equal width Δx_m (typically 25–40 Å). Dopant diffusion in the liquid is calculated by successive numerical solutions to the mass diffusion equation expressed in finite differences:

$$C_m(t + \Delta\tau) = C_m(t) + \frac{D_L \Delta\tau}{(\Delta x)^2} \left[C_{m+1}(t) + C_{m-1}(t) - 2C_m(t) \right] \qquad (3)$$

where D_L is the liquid phase diffusion coefficient and $C_m(t)$ is the solute concentration at position x_m. When the receding melt front begins to encounter diffusing dopant, solidification of the segment Δx_m occurs instantaneously, with dopant concentration $k' C_L'(x_m,t)$ being incorporated into the solid and the remainder $(1 - k') C_L'(x_m,t)$ being rejected into the adjacent liquid layer at x_{m-1}. (C_L' is the dopant concentration in the liquid at the interface prior to solidification). Rejected dopant then diffuses into the remaining liquid, with the melt front remaining stationary for a time $\Delta t = \Delta x/v$, and the resulting profile in the liquid is calculated by n successive solutions, each for a time $\Delta\tau = \Delta t/n$, to the finite difference equation. This procedure results in an exponentially decaying dopant profile in front of the interface. After a time Δt, the melt front then advances instantaneously through the segment Δx_{m-1}, with the partition of dopant between the solid and liquid being given by k'.

Calculations of the type discussed above are continued until the melt front recedes back to within 200 Å of the surface where dopant remaining in the thin surface layer is considered to be segregated at the surface. In this model k' is treated as a fitting parameter, and values for k' are determined by comparing calculations of dopant redistribution to experimental measurements of dopant profiles after laser annealing using least squares analysis. Values for D are taken from the literature (Kodera, 1963), and details on the melt front position as a function of time are taken from calculations described in Chapter 4 (Baeri and Campisano).

This model assumes that the only mechanism for dopant redistribution is liquid phase diffusion in the absence of convection in the liquid. Furthermore, k' is assumed to be independent of dopant concentration. The quality of the agreement between calculated and measured profiles justifies these assumptions. Calculations of the type described above have meaning only when the dopant concentration in the bulk of the crystal is in solution following laser annealing. For this reason it is essential to use ion-channeling analysis to show that the dopant is substitutional or nearly substitutional in the lattice. The calculations also assume a planar liquid–solid interface during regrowth. For Group III and V dopants in silicon, this is valid at low concentrations, but at high concentrations constitutional supercooling during regrowth leads to interfacial instability, lateral segregation of rejected dopant, and the formation of a well-defined cell structure in the near-surface region (White *et al.*, 1980b). For this reason the calculations should be confined to the low-concentration cases, where the interface remains stable during solidification.

IV. Results

A. *Lattice Location for Group III and V Species*

For Group III and V dopants in silicon, Rutherford backscattering and ion-channeling measurements have been used to show that these dopants are substitutional or nearly substitutional in the lattice after pulsed laser annealing (White *et al.*, 1979c, 1980a). The lattice location of the dopant can be determined most easily from detailed angular scans across the major axial directions. Figure 2 shows such a result for the case of ^{75}As in (100) Si following implantation and laser annealing. For these measurements a beam of 2.5-MeV He$^+$ ions was used, and particles scattered from both Si and As atoms in the same depth interval were detected and plotted as a function of tilt angle across the $\langle 110 \rangle$ and $\langle 111 \rangle$ axial directions. In Fig. 2 the yield curve of scattering from As has the same shape and angular width as the yield curve of scattering from Si, demonstrating that the dopant occupies a substitutional site in the recrystallized lattice. The substitutional fraction can be determined from the angular scan results as

$$\text{Fraction substitutional} = 100 \times \frac{1 - \chi_{min}(\text{As})}{1 - \chi_{min}(\text{Si})} \qquad (4)$$

where χ_{min} is the minimum yield defined in the usual manner. For the results shown in Fig. 2, the substitutional fraction is 99%. Corresponding electrical measurements show that all the implanted As is electrically active after laser annealing, thus confirming the channeling results.

Results similar to those in Fig. 2 have been obtained also for the case of implanted Sb (White *et al.*, 1979c), demonstrating that Sb is substitutional

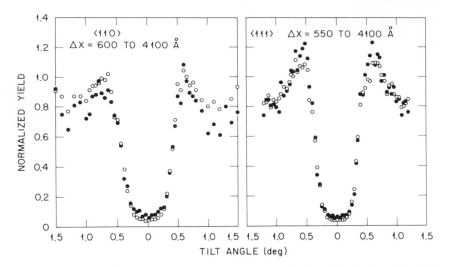

Fig. 2. Angular scans for 2.5-MeV He⁺ ions across the ⟨110⟩ and ⟨111⟩ channels of ⁷⁵As (100 keV, 1.4 × 10¹⁶/cm²)-implanted (100) Si after two-pulse laser annealing. Filled circles (●) refer to scattering from As; unfilled circles (○) refer to scattering from Si; Δx refers to the depth interval for scattering from As and Si. (From White *et al.*, 1979c.)

in the lattice after laser annealing with a similar high degree of substitutionality (99%). For dopants such as Ga and Bi, however, the ion-channeling results show that the dopants are displaced slightly from substitutional lattice sites (White *et al.*, 1980a). This is demonstrated in the angular scan results in Fig. 3 for the case of Bi in Si after laser annealing. As shown in Fig. 3, the yield curves of scattering from Bi are not as wide as those for Si. This implies that at least part of the implanted Bi is displaced slightly from a normal substitutional lattice site. Substitutional fractions obtained using Eq. (3) show Bi to be ~95% substitutional after laser annealing. The angular scan results show that, while Bi is regularly placed in the lattice, it may be displaced slightly from a substitutional site. Similar results have been obtained also for the case of Ga in silicon.

In summary, ion channeling (White *et al.*, 1979c, 1980a) and nuclear reaction analysis results (Swanson *et al.*, 1981) show that Group III and V dopants are regularly placed in the silicon lattice after pulsed laser annealing. Dopants such as As, Sb, B, and P occupy substitutional lattice sites, while Ga, Bi, and probably In are displaced slightly from substitutional lattice sites. In all cases, the substitutional fractions are considerably better than those obtained by thermal annealing, and these very high levels of substitutionality can be achieved even when the dopant concentrations greatly exceed equilibrium solubility limits.

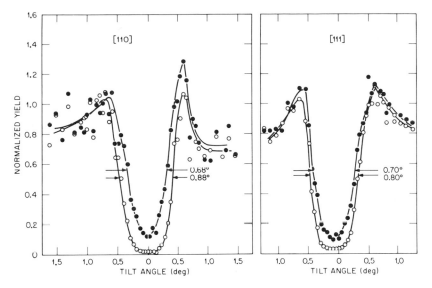

Fig. 3. Angular scans across the [110] and [111] axial directions for ^{209}Bi (250 keV, 1.2×10^{15}/cm²)-implanted (100) Si after laser annealing. ○, Si; ●, Bi. (From White *et al.*, 1980a.)

B. *Interfacial Distribution Coefficients*

The interfacial distribution coefficient is determined from the comparison of model calculations detailed in Section III to experimental measurements of the dopant profile after laser annealing. Figure 4 presents such a comparison for As in silicon (White *et al.*, 1980a). Following laser annealing, ion-channeling results show that As is >95% substitutional in the lattice and is electrically active as determined from Hall effect measurements. This high degree of substitutionality is achieved even though the As concentration in the near-surface region exceeds the equilibrium solubility limit by a factor of ~4. This demonstrates the formation of a supersaturated alloy as a consequence of the high-speed liquid phase epitaxial regrowth process. The solid line in Fig. 4 is a profile calculated using a value for the distribution coefficient of $k' = 1.0$, and the agreement with the experimental profile results (solid circles) is excellent. The value determined for k' is considerably higher than the equilibrium value ($k_0 = 0.3$). The increase in the distribution coefficient relative to the equilibrium value is a consequence of the high regrowth velocity which causes a departure from conditions of local equilibrium at the interface during solidification. The fact that As shows no evidence of segregation at the surface is a further indication that k' must be very close to unity.

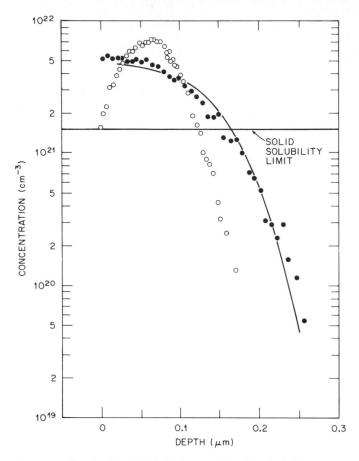

Fig. 4. Dopant profiles for ^{75}As (100 keV, 6.4×10^{16}/cm^2) in (100) Si compared to model calculations. The horizontal line indicates the equilibrium solubility limit, calculated for $k' = 1.0$. Unfilled and filled circles are for, respectively, implanted and laser-annealed profiles. (From White *et al.*, 1980a.)

One other comparison is shown in Fig. 5 for the case of Bi in Si (White *et al.*, 1980a). In this case, as a consequence of laser annealing, approximately 15% of the Bi segregates at the surface, but the concentration remaining in the bulk is >95% substitutional (ion-channeling results), even though this concentration exceeds the equilibrium solubility limit by approximately two orders of magnitude. The solid line in Fig. 5 is a profile calculated using the value $k' = 0.4$ and assuming that the liquid phase diffusivity for Bi in Si is $D_L = 1.5 \times 10^{-4}$. Values for D_L (Bi in Si) have not

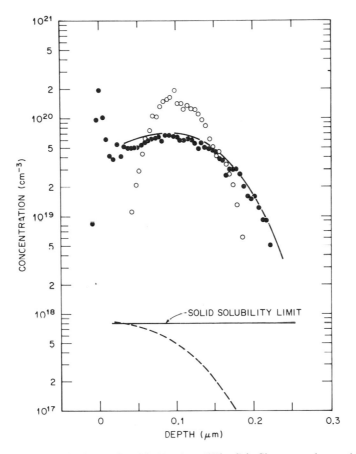

Fig. 5. Dopant profile for [209]Bi (250 keV, 1.2×10^{15}/cm²) in Si compared to model calculations. The horizontal line indicates the equilibrium solubility limit. The solid line profile was calculated assuming $k' = 0.4$. Unfilled and filled circles are for, respectively, implanted and laser-annealed profiles. The dashed line is the profile calculated assuming k' is the equilibrium value $k_0 = 0.0007$. (From White *et al.*, 1980a.)

been reported in the literature, but the value of 1.5×10^{-4} gives the best fit to the experimental results and is in reasonable agreement with the extrapolation of measured liquid phase diffusivities for lower mass impurities in liquid silicon. In Fig. 5, the calculated profile (solid line) is in good agreement with the experimental profiles (filled circles) measured after laser annealing. In contrast, a profile calculated using the equilibrium value for the distribution coefficient of Bi in Si ($k_0 = 7 \times 10^{-4}$) is shown by the dashed curve in Fig. 5. If solidification occurred under conditions of local equilibrium at the interface, very little Bi would remain in the bulk of the crystal and almost all the Bi would be zone-refined to the surface. Clearly, this does not fit the experimental results.

It is interesting to note that the experimental profile results in Figs. 4 and 5 can be fit by a single value for k' over the entire range of concentrations. This indicates that the value for k' is not a strong function of concentration and is determined, to first order, by the regrowth velocity. For the case of Bi in Si, experiments similar to those illustrated in Fig. 5 have been carried out at both higher and lower implanted doses (concentrations). In each case the value determined for k' lies in the range 0.35–0.40, even though the implanted dose was varied by over an order of magnitude. This further reinforces the conclusion that the value for k' is not a strong function of concentration.

By similar methods, values for k' have been determined for a wide variety of Group III and V dopants in silicon at the very high growth velocities achieved by pulsed laser annealing (White *et al.*, 1980a). The values determined for k' during pulsed laser annealing are summarized in Table I and compared with the corresponding equilibrium values k_0. These values for k' were determined at a growth velocity of 4.5 m/sec except for the cases of B, P, and Sb. For these three dopants, a somewhat longer pulse duration time ($\sim60 \times 10^{-9}$ sec) was used for annealing, resulting in a growth velocity of ~2.7 m/sec. Values for the liquid phase diffusivities used to fit the experimental profiles were taken from the literature (Kodera, 1963), except for the case of Bi where a value of $D_L = 1.5 \times 10^{-4}$ cm²/sec was assumed (see previous discussion).

TABLE I

COMPARISON OF DISTRIBUTION COEFFICIENTS
UNDER EQUILIBRIUM (k_0) AND LASER-ANNEALED
(k') REGROWTH CONDITIONS

Dopant	$k_0{}^a$	k'^b
B	0.80	~1.0
P	0.35	~1.0
As	0.30	~1.0
Sb	0.023	0.7
Ga	0.008	0.2
In	0.0004	0.15
Bi	0.0007	0.4

[a] From Trumbore (1960).
[b] Values for k' were determined at a growth velocity of 2.7 m/sec for B, P, and Sb, and at 4.5 m/sec for As, Ga, In, and Bi.

The results presented in Table I show that in every case k' is significantly greater than k_0 by factors that extend up to ~600. The large increase in k' relative to k_0 reflects the nonequilibrium nature of the laser annealing-induced liquid phase epitaxial regrowth process. The departure from conditions of local equilibrium at the interface is brought about by the very high growth velocities (several meters per second) achieved by laser annealing. The results shown in Figs. 4 and 5, and in Table I, demonstrate the potential for detailed fundamental studies on high-speed, nonequilibrium crystal growth processes under well-controlled experimental conditions. The values reported in Table I were the first determination of interfacial distribution coefficients under conditions of high-speed, nonequilibrium crystal growth for any system.

C. *Dependence of Distribution Coefficients on Velocity and Crystal Orientation*

Recent experiments have shown that the interfacial distribution coefficient is a function of both growth velocity and crystal orientation. The velocity dependence is entirely expected because, as the velocity decreases, k' must approach the equilibrium value k_0. The first observations of this expected velocity dependence were for the case of Pt in Si, where it was observed that increasing the growth velocity resulted in more implanted Pt being incorporated into the lattice during laser annealing (Cullis et al., 1980). These results are shown in Fig. 6 for three different growth velocities. Variation in growth velocity was achieved by changing the substrate temperature during laser annealing. For the results shown in Fig. 6, substrate temperatures of 620, 300, and 77 K give rise to growth velocities of 1, 1.8, and 2.5 m/sec (Nd:YAG 1.06-μm laser radiation). At the slowest growth velocity, almost all the Pt is segregated at the surface during laser annealing and very little is observed to be substitutional in the lattice after annealing. In contrast, at the highest growth velocity, measurable amounts of Pt are trapped substitutionally in the lattice and a much smaller amount is segregated at the surface. Although values of k' for Pt in Si were not reported, it is obvious from the results in Fig. 6 that increasing the growth velocity leads to increases in the value for k'. It should be noted, however, that even at the highest growth velocities the Pt in the bulk of the crystal is not 100% substitutional, and electron microscope results showed that Pt underwent lateral segregation as a result of constitutional supercooling to form a well-defined cell structure in the near-surface region (Cullis et al., 1980). If lower implanted doses had been used, the Pt remaining in the bulk of the crystal could have been trapped 100% in substitutional lattice sites.

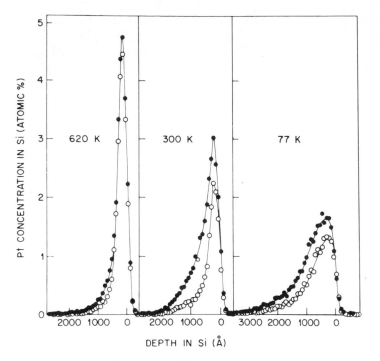

Fig. 6.   Random (●) and aligned (○, ⟨100⟩) 2-MeV He⁺ backscattering spectra for Pt-implanted Si laser-annealed at different substrate temperatures. (From Cullis *et al.*, 1980.)

Similar results are shown in Fig. 7 for the case of Bi in Si (White *et al.*, 1981). Substrate temperatures of 650, 300, and 100 K give rise to regrowth velocities of 1.5, 4.5, and 6.0 m/sec for the laser conditions used for annealing (λ = 6943 Å, 15 × 10⁻⁹ sec, 1.4 J/cm²). At the low growth velocity (1.5 m/sec), almost 55% of the implanted Bi segregates at the surface as a result of laser annealing, while at the highest growth velocity only 5% is segregated at the surface. In each case, the Bi remaining in the bulk of the crystal is ≃95% substitutional in the lattice. The dotted lines in Fig. 7 are calculated profiles using values of k' = 0.1, 0.35, and 0.45 at growth velocities of 1.5, 4.5, and 6.0 m/sec. The agreement between the calculated and experimental profiles in Fig. 7 is excellent, and these results demonstrate that k' and the amount of Bi segregated at the surface are strong functions of growth velocity, as expected. A similar dependence of k' on regrowth velocity has been reported also for the case of In in Si (Baeri *et al.*, 1981a), and similar dependencies should be observable for all group III and V species in silicon. These experiments, if carried out over a wider velocity range, can be expected to provide fundamental

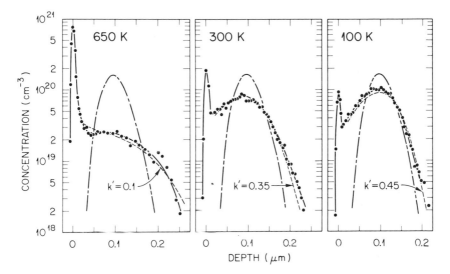

Fig. 7.   Profiles for ^{209}Bi (250 keV, 1.1×10^{15}/cm²)-implanted (100) Si laser-annealed at different substrate temperatures. (— —, As implanted; – ● –, laser annealed. (From White *et al.*, 1981.)

insight into detailed mechanisms of importance to high-speed nonequilibrium crystal growth processes, an area in which little theoretical work has been done to date.

The interfacial distribution coefficient is also a strong function of crystal orientation, at least in certain velocity ranges (Baeri *et al.*, 1981b). This is demonstrated in Fig. 8 for the case of (111) and (100) Si crystals implanted with Bi and laser annealed. Following implantation and laser annealing under identical conditions, more Bi is trapped substitutionally in the lattice of the (111) crystal as compared to the (100) crystal, and there is a factor of 2 difference in the value of k' determined for the two crystal orientations.

Figure 9 shows the velocity dependence of k' for Bi in (100) and (111) Si (Baeri *et al.*, 1981b). For velocities below ~4 m/sec the value for k' in (111) Si is systematically higher than that for (100) Si. For identical laser annealing conditions, the regrowth velocity normal to the surface should be the same, since velocity is determined by heat flow into the underlying bulk. Consequently the anisotropic dependence of k' on growth velocity must be related to differences in detailed mechanisms of crystal growth from the liquid for (100) and (111) crystals. In particular, it has been suggested that a larger interfacial undercooling on the (111) face compared to the (100) face (Baeri *et al.*, 1981b; Jackson, 1981) might explain the differences in dopant incorporation for these two cases.

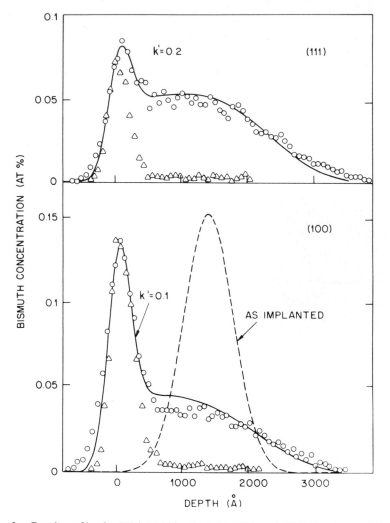

Fig. 8. Depth profiles for 240-keV Bi implanted in (100) and (111) Si and annealed using identical laser treatments (ruby, 2.0 J/cm², 100 nsec). Solid lines are the result of numerical calculations using k' values of 0.2 and 0.1 for the (111) and (100) orientations, respectively. (From Baeri *et al.*, 1981b.)

D.   *Theoretical Treatments of Solute Trapping at High Growth Velocities*

There have been several recent theoretical treatments (Wood, 1980; Jackson *et al.*, 1980; Cahn *et al.*, 1980) of dopant incorporation into Si during high-speed crystal growth. In one approach (Wood, 1980), an expression for the velocity dependence of the distribution coefficient was

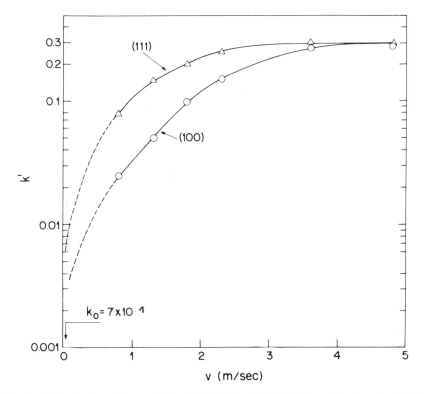

Fig. 9. Velocity dependence of k' for Bi in (100) and (111) Si. (From Baeri *et al.*, 1981b.)

derived by using a velocity-dependent barrier height for escape of dopant atoms from the solid in the reverse rate equations describing the exchange of dopant atoms between the solid and liquid at the interface during solidification. In crystal growth at low velocities impurity atoms exchange many times between the solid and the liquid at the interface before being incorporated permanently into the growing solid. During rapid solidification, new layers of atoms are added to the solid so quickly that dopant atoms have a reduced probability of escaping from the solid being formed. In this sense they cannot diffuse fast enough to stay in the liquid and avoid being incorporated into the growing solid. This is referred to as solute trapping. The model by Wood (1980) accounts for this reduced escape probability through the use of a velocity-dependent barrier height.

Based on a simple model for the velocity dependence of the barrier height (which goes to a limiting value as v approaches infinity), the velocity dependence of the distribution coefficient can be expressed as

$$k' = k_0 \exp\{(-RT \ln k_0)\,[1 - \exp(-v/v_0)]/RT\} \tag{5}$$

In the above expression, $v_0 = D_L/x_0$, where x_0 is the effective width of the interfacial region (taken to be ~ 225 Å by comparison to the experimental results). Based on this model, Fig. 10 shows the predicted velocity dependence of the distribution coefficient for a wide range of dopants in silicon. These predictions are in reasonable agreement with the experimental results in Table I. Thus it appears that the large increases in the distribution coefficients relative to equilibrium values can be accounted for, at least qualitatively, through a velocity-dependent barrier height. The exact functional form for the barrier height cannot be determined until data become available over a much wider velocity range. It is significant to note that the limiting value for k' is unity as v approaches infinity in this model.

In a second approach (Jackson *et al.*, 1980) the assumption was made that during rapid solidification a fraction of the solute atoms in the liquid layer adjacent to the interface would be trapped in the solid being formed.

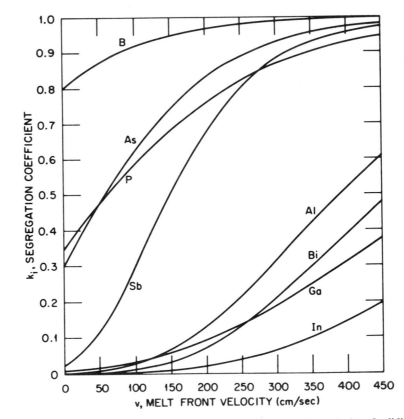

Fig. 10. Predicted dependence of the distribution coefficient on the velocity of solidification for various impurities in silicon. (From Wood, 1980.)

With this assumption in the classical rate equations describing solidification, it was predicted that there would be a maximum dopant concentration that could be incorporated into the solid (this is observed experimentally; see White *et al.*, 1980a). In addition, this treatment predicted that there would be a limiting value for the distribution coefficient that could be less than unity, in contrast to the prediction of Wood (1980). Support for the Jackson *et al.* (1980) model is seen in Fig. 9 where the distribution coefficient for Bi in Si appears to saturate for velocities >3 m/sec. The data in Fig. 7, however, suggest that the value for k' increases somewhat in the range 4.5–6.0 m/sec. It will be necessary to extend experimental measurements to higher velocities in order to determine whether k' saturates at a value less than unity.

E. Maximum Substitutional Solubilities (C_s^{max})

As the implanted dose is increased, for each of the Group III and V species there is a maximum concentration that can be incorporated substitutionally into the Si lattice as a result of pulsed laser annealing (White *et al.*, 1980a; Fogarassy *et al.*, 1980). This is shown in Fig. 11 for the case of Sb in Si, where both the total dopant concentration and the substitutional dopant concentration are plotted as a function of depth after laser annealing. These results were obtained using Rutherford-backscattering–ion-channeling techniques to measure the total dopant concentration and the substitutional concentration as a function of depth. As shown in Fig. 11, up to a concentration of $\sim 2 \times 10^{21}$/cm³ the Sb is almost 100% substitutional in the lattice, but in the near-surface region, where the total concentration is almost a factor of 2 higher, the substitutional component is relatively constant with a value of $\sim 2 \times 10^{21}$/cm³. This value therefore is the maximum substitutional concentration C_s^{max} for Sb that can be incorporated into silicon at this growth velocity (~ 4.5 m/sec). In the near-surface region, the nonsubstitutional Sb is observed to be located in the walls of a well-defined cell structure that arises from interfacial instabilities during regrowth. This is discussed in more detail in Section IV.G.

With similar techniques, values for C_s^{max} have been determined for five Group III and V species in (100) Si at a growth velocity of 4.5 m/sec (White *et al.*, 1980a). These values are listed in Table II and compared to corresponding equilibrium solubility limits C_s^0. [Previously published results for the case of Sb (White *et al.*, 1980a) were obtained at a velocity of 2.7 m/sec.] In Table II, for each dopant C_s^{max} is substantially larger than C_s^0 by factors that range from 4 in the case of As to ~ 500 for the case of Bi. On the equilibrium phase diagram (shown schematically in Fig. 12) each

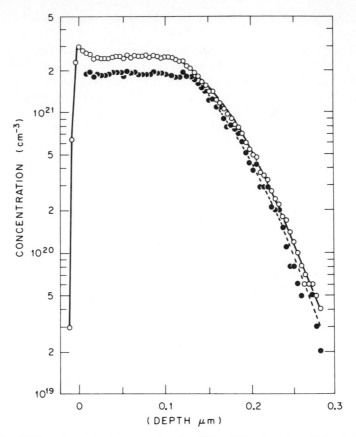

Fig. 11. Total and substitutional profiles for ^{121}Sb (100 keV, 4.5×10^{16}/cm^2) in (100) silicon after laser annealing. $\bigcirc$, Total; $\bullet$, substitutional.

TABLE II

COMPARISON OF EQUILIBRIUM (C_S^0) AND
LASER-ANNEALING-INDUCED ($C_S^{\max}$) SOLUBILITY
LIMITS[a]

Dopant	C_S^0 (cm^{-3})[b]	$C_S^{\max}$ (cm^{-3})[c]
As	1.5×10^{21}	6.0×10^{21}
Sb	7.0×10^{19}	2.0×10^{21}
Bi	8.0×10^{17}	4.0×10^{20}
Ga	4.5×10^{19}	4.5×10^{20}
In	8×10^{17}	1.5×10^{20}

[a] From White *et al.* (1980a).
[b] From Trumbore (1960).
[c] Values were obtained at a velocity of 4.5 m/sec.

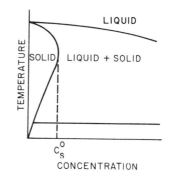

Fig. 12. Phase diagram (schematic) for retrograde alloys.

of these dopants exhibits retrograde solubility in silicon in that the dopant has its maximum solubility C_s^0 at a temperature that is not simply related to a eutectic temperature. As shown by Baker and Cahn (1969), the retrograde maximum concentration cannot be exceeded by solidification from the liquid unless there is a departure from equilibrium at the interface during solidification. In Table II, the large values for C_s^{max} relative to C_s^0 convincingly demonstrate the nonequilibrium nature of the laser-annealing-induced liquid phase epitaxial regrowth process. The departure from conditions of local equilibrium at the interface is brought about by the very rapid regrowth velocities.

Dopant incorporation into the lattice at these high concentrations is a result of solute trapping during solidification. In the simplest terms this means that, if the time required to regrow one monolayer during solidification is significantly shorter than the residence time of the impurity at the interface, then the impurity cannot avoid being incorporated into the growing solid. Theoretical treatments of solute trapping are given in Baker and Cahn (1969), Cahn *et al.* (1980), Wood (1980), and Jackson *et al.* (1980).

It is interesting to note that equilibrium solubility limits can be greatly exceeded also during low-temperature thermal annealing (solid phase epitaxial regrowth) of ion-implanted silicon (Williams and Elliman, 1981; Campisano *et al.*, 1980a,b). Published results indicate that maximum substitutional solubilities obtained after thermal annealing (550°C, 30 min) are only a factor of 2–3 lower than those achieved by pulsed laser annealing. In the thermal-annealing case dopant incorporation appears to be the result of solute trapping at the moving amorphous–crystalline interface. At the temperatures used for annealing (~550°C) the velocity of the amorphous–crystalline interface is only ~10^{-10} m/sec, but the impurity residence time at the interface is very long because this time is inversely proportional to the solid phase diffusivity which is very low. Consequently, even in solid phase epitaxy, a temperature range can be selected such that the impurity residence time at the interface is longer than the monolayer regrowth time, and trapping can occur. If, however, higher temperatures are used for annealing, then precipitation of the dopant concentration in excess of the solubility limit at the annealing temperature will occur.

F. *Measurements of Equilibrium Solubility Limits*

Supersaturated solid solutions of Group III and V species in Si can be readily formed by laser annealing of ion-implanted silicon. These supersatured alloys are infinitely stable at room temperature, but if the crystals subsequently are thermally annealed, precipitation of the dopant concentration in excess of the equilibrium solubility limit at the annealing temperature will occur at a rate determined by the solid phase diffusivity. Subsequent analysis by RBS–ion-channeling techniques can be used to establish experimentally the equilibrium solubility at the annealing temperature. One example is shown in Fig. 13 for the case of Sb in Si (White *et al.*, 1980b). Following implantation and laser annealing, the near-surface region was defect-free and the Sb was measured to be 100% in substitutional lattice sites even though the concentration exceeded the reported equilibrium solubility limit by more than a factor of 2. This crystal was then thermally annealed at 1150°C (a temperature very close to the retrograde temperature) for 30 min. Following thermal processing, RBS–ion-channeling measurements were used to measure the total dopant concentration and the substitutional concentration as a function of depth. The results in Fig. 13 show that, after thermal processing, the maximum substitutional dopant concentration was $8.2 \times 10^{19}/cm^3$, which is taken to be the equilibrium solubility limit for Sb in Si. This result is in good agreement with the value $(7.0 \times 10^{19}/cm^3)$ reported previously for Sb in Si (Trumbore, 1960). Similar results have been obtained for Ga in Si (Bean *et al.*, 1979) and As in Si (White, 1980). In principle, this method could be extended to determine the solidus line on the equilibrium phase diagram.

G. *Mechanisms Limiting Substitutional Solubilities*

The maximum substitutional solubilities that can be achieved by pulsed laser annealing appear to be limited by three mechanisms (White *et al.*, 1981). The first of these, which is related to the materials properties of the implanted region, is lattice strain. This mechanism provides a practical

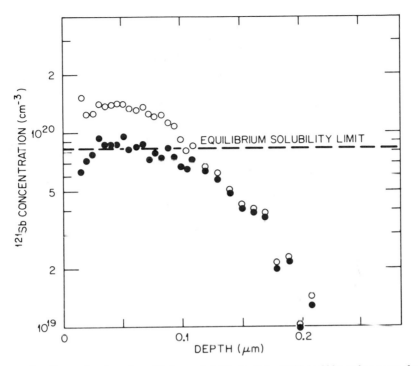

Fig. 13.   Determination of equilibrium solubility limit for ^{121}Sb in Si by subsequent thermal annealing of a metastable solid solution produced by laser annealing. ○, Total; ●, substitutional. (From White *et al.*, 1980b.)

limit to the incorporation of B in silicon by pulsed laser processing. Previous work has shown that, when B is incorporated substitutionally into the silicon lattice during pulsed laser annealing, this causes the lattice to undergo a one-dimensional contraction in the implanted region in a direction normal to the surface (Larson *et al.*, 1978). The lattice contracts because the covalent bonding radius of B is significantly smaller than that of the Si atom it replaces in the lattice. The contraction occurs in one dimension only because of the adherence of the near-surface region to underlying crystal planes. The magnitude of the contraction is proportional to the local boron contraction. Contraction gives rise to strain in the implanted region, and when the strain exceeds the fracture strength of silicon, cracks will develop in the near-surface region. For the case of B in silicon, this occurs when the local boron concentration exceeds ~4 at %. This is shown by the scanning electron microscope (SEM) micrographs in Fig. 14 for the case of B-implanted (35 keV, $6 \times 10^{16}/cm^2$) silicon after laser annealing. Following annealing, cracks (~1-μm wide) are observed

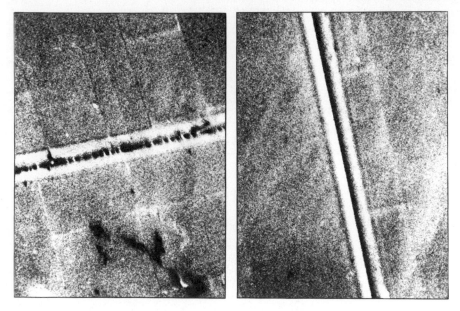

Fig. 14. Cracks in the near surface region of ^{11}B (35 keV, $6 \times 10^{16}/cm^2$)-implanted (100) Si after laser annealing. (From White *et al.*, 1981.)

in the near-surface region. These cracks penetrate to a depth of ~ 1 μm and extend over the entire length of the sample (~ 1 cm). For implanted doses lower than $2.5 \times 10^{16}/cm^2$, the near-surface region is highly strained after laser annealing, but cracks do not develop. If dopants such as Sb, which have a larger covalent radius than Si, are incorporated into the lattice by laser annealing, a one-dimensional expansion instead of a contraction can be produced (Appleton *et al.*, 1979). In order to incorporate more B substitutionally in the silicon lattice by laser annealing, it would be necessary to incorporate such a dopant along with the B. Possible candidates include Ga, In, Bi, and Sb.

A second mechanism that limits substitutional solubility achieved by laser annealing is an interfacial instability that develops during regrowth and leads to lateral segregation of the rejected dopant and the formation of a well-defined cell structure in the near-surface region. Figure 15 shows examples of cell structures formed in the near-surface region as a result of laser annealing for the case of high-dose implants of In, Ga, and Fe in silicon. The interior of each cell in Fig. 15 is an epitaxial column of silicon extending to the surface. Surrounding each column is a thin cell wall containing massive concentrations of the impurity.

Figure 16 shows both concentration profiles and transmission electron

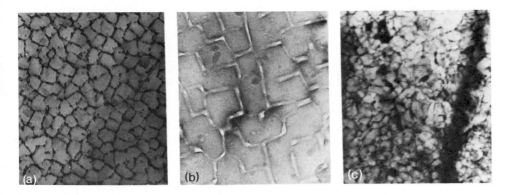

Fig. 15. Cell structure in the near surface region of silicon after laser-annealing high-dose implants of In, Ga, and Fe. (a) ^{115}In (1.3×10^{16}/cm^2); (b) ^{69}Ga (1.2×10^{16}/cm^2); (c) Fe (1.8×10^{16}/cm^2). Results for In and Ga were obtained on ⟨100⟩ Si. Results for Fe were obtained on (111) Si. (From White *et al.*, 1980c.)

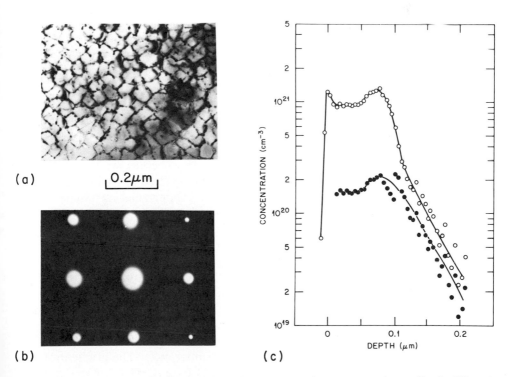

Fig. 16. (a) Microstructure, (b) diffraction pattern, and (c) concentration profiles for ^{115}In (125 keV, 1.3 10^{16}/cm^2) in (100) Si after laser annealing. ○, Total In; ●, substitutional In. (From White *et al.*, 1981.)

microscope (TEM) micrographs for a $\langle 100 \rangle$ Si crystal implanted with ^{115}In (125 keV, 1.3×10^{16}/cm²) and laser-annealed. From the total and substitutional concentration profiles, almost all the In is substitutional, up to a concentration of ~ 1.5–2.0×10^{20}/cm³. However, in the near-surface region where the total concentration extends up to $\sim 10^{21}$/cm³, the substitutional component stays relatively flat at $\sim 1.5 \times 10^{20}$/cm³. The microstructure in Fig. 16 shows the epitaxial columns of silicon (average diameter ~ 450 Å) surrounded by thin cell walls ~ 50 Å thick. The electron diffraction pattern obtained from the near-surface region shows the presence of (weak) extra spots which arise from crystalline In in the cell walls, thus demonstrating nucleation of the second phase during laser annealing. In the concentration profile results, it is the nonsubstitutional In that is located in the cell walls. TEM plan view results (not shown) show that the cell walls penetrate to a depth of ~ 1200 Å, in good agreement with the concentration profile results showing that the nonsubstitutional In in confined to a depth of ~ 1200 Å. The substitutional In in the near surface region is trapped in the interior of the epitaxial columns of silicon. The limiting concentration that can be trapped in the columns depends on the growth velocity and not the implanted dose. If a dose had been used higher than that in Fig. 16, the same limiting substitutional concentration would have been incorporated into the epitaxial Si columns, but the cell walls would have penetrated to a greater depth.

The cell structures depicted in Figs. 15 and 16 arise from lateral segregation of the rejected dopant caused by an interfacial instability that develops during regrowth. Interfacial instabilities occur only when the distribution coefficient is less than unity and when the concentration of the rejected impurity at the interface is large. At lower concentrations, the interface remains planar during regrowth and the rejected impurity is zone refined at the surface (see, for example, Fig. 5). Interfacial instability is caused by constitutional supercooling in front of the liquid–solid interface during regrowth (see, e.g., Jackson, 1975). This phenomenon is represented schematically in Fig. 17. If the distribution coefficient is less than unity, then during solidification rejected dopant accumulates in the liquid at the interface as depicted in Fig. 17a. The concentration gradient of rejected dopant in the liquid leads to a gradient of the freezing temperature of the liquid in front of the interface (Fig. 17a). If the actual thermal gradient in the liquid (G) is less than the gradient of the freezing temperature, then a region in front of the interface will be supercooled, since the actual temperature is less than the liquidus temperature. The gradient of the freezing temperature is given approximately by $\Delta T_f/(D_L/v)$ where ΔT_f is the interfacial undercooling due to rejected dopant, D_L is the liquid phase diffusivity, and v is the solidification velocity.

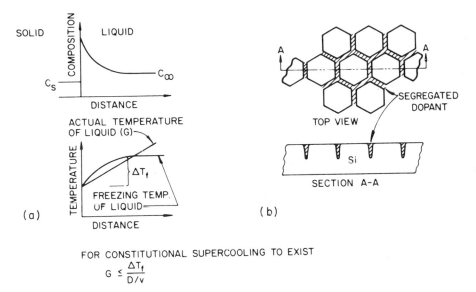

Fig. 17. Constitutional supercooling during solidification (schematic). (a) Conditions necessary for constitutional supercooling; (b) the resulting microstructure.

The condition for the occurrence of constitutional supercooling is that

$$G \leq \Delta T_f/(D/v) \qquad (6)$$

If the inequality is satisfied, then a perturbation on a planar interface can become unstable and grow, leading to lateral segregation and a well-defined cell structure in the near-surface region as illustrated by the microstructure schematic in Fig. 17b. If the inequality is not satisfied, then the interface will remain stable during regrowth. In the above expression, the quantity ΔT_f depends on impurity concentration, thus accounting for the fact that interfacial instability occurs only for high concentrations of rejected impurity at the interface.

Interface instability during solidification has been treated theoretically using a perturbation theory approach (Mullins and Sekerka, 1964). The theory assumes thermodynamic equilibrium in the solid and liquid phases and takes into account constitutional supercooling, surface tension, and latent heat evolution. Surface tension and latent heat produce a stabilizing influence, and the perturbation theory predicts greater stability at high velocities than the simple constitutional supercooling criterion. Although the original theory assumed thermodynamic equilibrium in the solid and the liquid, it has been shown recently (Narayan, 1981; Cullis et al., 1981) that one can make allowance for the large departures from equilibrium during pulsed laser annealing by using in this perturbation theory the values for the interfacial distribution coefficient k' appropriate to this high-speed growth process. Incorporating the predicted velocity dependence of the distribution coefficient, one can predict the concentration at which the interface becomes unstable during regrowth, and the average cell size (Narayan, 1981). Figures 18 and 19 show such predictions for the case of In in Si. Plotted in Fig. 18 is the concentration $C_s(min)$ in the solid

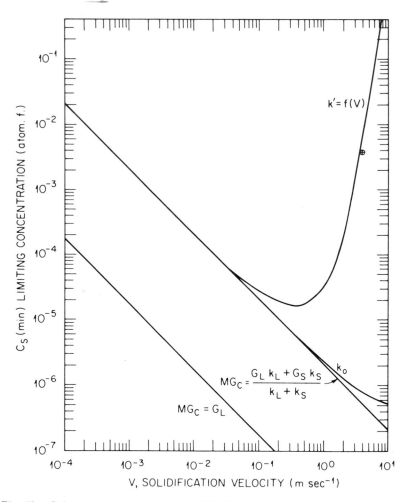

Fig. 18. Solute concentrations in the solid $C'_{s(min)}$ for In in Si above which interface instability occurs as a function of velocity of solidification using $k' = f(v)$, $k' = k_0$, and the original constitutional supercooling criteria. The dot is an experimental measurement. (From Narayan, 1981.)

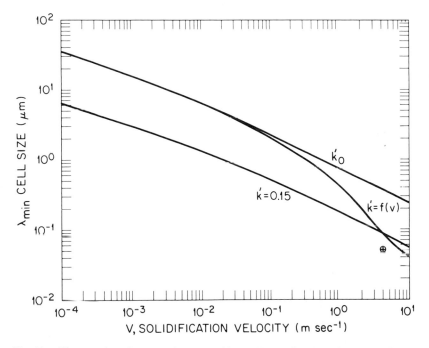

Fig. 19.   The wavelength λ_{min} at the onset of instability during directional solidification of Si containing In using $k' = f(v)$, $k' = k_0$, and $k' = 0.15$. The dot is an experimental measurement. (From Narayan, 1981.)

at which the interface will become unstable during regrowth as a function of growth velocity. The interface will be unstable above the C_s^{min} curve and stable below. The data point in Fig. 18 is the experimentally measured maximum substitutional concentration for In in Si. In Fig. 18, the lower straight line with negative slope is a prediction of $C_s(\mathrm{min})$ using the original constitutional supercooling criterion and the equilibrium distribution coefficient ($MG_c = G_L$). The upper straight line with negative slope is a prediction of $C_s(\mathrm{min})$ using the constitutional supercooling criterion modified to include temperature gradients in both the liquid and solid. The upper two curves in Fig. 18 are predictions of $C_s(\mathrm{min})$ from the Mullins–Sekera theory using $k' = k_0$ and $k' = f(v)$ from the work of Wood (1980). Good agreement with the experimental results is obtained for the prediction of $C_s(\mathrm{min})$, which includes the velocity dependence of the distribution coefficient.

Prediction of cell size as a function of velocity for In and Si are shown in Fig. 19 by the curve labeled $k' = f(v)$ where $f(v)$ is the predicted (Wood, 1980) velocity dependence of the distribution coefficient. The data point is the measured cell size from Fig. 16. (The straight lines of negative slope are predictions of cell size for constant values of $k' = 0.15$ and $k' = k_0$.) The measured cell size is closest to the predicted value when the dependence of k' on velocity is taken into account. The results in Figs. 18 and 19 show that perturbation theory treatment of interface instability can predict with reasonable accuracy the concentration at which interface instability occurs and the size of the resulting cell structure if the proper value for the distribution coefficient is used in the calculations. Values for the distribution coefficient can be determined using the methods discussed previously. The theory also predicts [see the curve labeled $k' = f(v)$ in Fig. 18] that one can increase the maximum substitutional solubilities by increasing the growth velocity. This has been observed experimentally and will be discussed in the next section.

In addition to the limitations on substitutional solubility imposed by lattice strain and by interfacial instability during regrowth (constitutional supercooling), there are predicted thermodynamic limits to solute trapping in silicon (Cahn *et al.*, 1980). Basic ideas underlying these predictions are illustrated schematically in Fig. 20. On a plot of the Gibbs free energy versus composition at fixed temperature, the solidus and liquidus lines intersect at one point, which is the upper limit of the solid composition that can be formed from the liquid at any composition. Plotting the locus of these points at different temperatures on the equilibrium phase diagram defines the T_0 curve, which is the maximum solid composition that can be formed from the liquid at any temperature even at infinite growth velocity. The T_0 curve thus defines the thermodynamic limit to diffusionless solidification. For retrograde systems, thermodynamic arguments can be used to obtain a simple estimate for the maximum concentration C_S^L on the T_0 curve. This maximum concentration on the T_0 curve is the liquidus con-

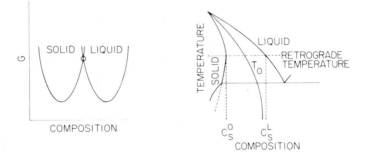

Fig. 20. Schematic representation of the method used to determine the thermodynamic limit of solute trapping.

centration on the equilibrium phase diagram at the retrograde temperature (Cahn *et al.*, 1980).

Predictions of C_S^L for five dopants in silicon (Cahn *et al.*, 1980) are listed in Table III and compared with measurements of maximum substitutional solubilities C_S^{max} obtained at two different growth velocities (White *et al.*, 1981). Laser annealing at temperatures of 300 and 77 K results in regrowth velocities of ~4.5 and ~6.0 m/sec. At either growth velocity, values for C_S^{max} approach predicted thermodynamic limits for solute trapping. Dopants for which k' is very near to unity (As and Sb from Table I) have measured solubilities very close to predicted thermodynamic limits. Measured solubilities for dopants with relatively lower values for k' (Ga, In) are somewhat lower than predictions but are still within an order of magnitude of thermodynamic limits. For Ga, In, and Bi, values for C_S^{max} obtained at a velocity of 6 m/sec are larger by factors of 2–3 than results obtained at a regrowth velocity of 4.5 m/sec. These results are still limited by interfacial instability during regrowth, but they demonstrate that at higher growth velocities the onset of instability can be delayed until higher concentrations accumulate at the interface, as expected from the discussion in the previous section. It is significant to note that a growth velocity of ~6 m/sec the Bi concentration in substitutional lattice sites exceeds the equilibrium solubility limit by more than three orders of magnitude.

For the case of As in Si, Table III shows there is no change in the measured maximum substitutional concentration as the regrowth velocity is increased from 4.5 to 6.0 m/sec. This indicates that the thermodynamic limit for As in Si may have been reached. This is reasonable, since the

TABLE III

Maximum Substitutional Dopant Concentrations (C_S^{max}) Obtained at Growth Velocities of 4.5 and 6.0 m/sec Compared to Equilibrium Solubility Limits (C_S^0) and Predicted Thermodynamic Limits to Solute Trapping (C_S^L)[a,b]

Dopant	C_S^0	C_S^{max} (v = 4.5 m/sec) (cm^{-3})	C_S^{max} (v = 6.0 m/sec) (cm^{-3})	C_S^L (cm^{-3})
As	1.5×10^{21}	6.0×10^{21}	6.0×10^{21}	5×10^{21}
Sb	7.0×10^{19}	2.0×10^{21}	—	3×10^{21}
Ga	4.5×10^{19}	4.5×10^{20}	8.8×10^{20}	6×10^{21}
In	8.0×10^{17}	1.5×10^{20}	2.8×10^{20}	2×10^{21}
Bi	8.0×10^{17}	4.0×10^{20}	1.1×10^{21}	1×10^{21}

[a] From White *et al.* (1981).

[b] Substrate temperatures of 300 and 77 K were used during laser annealing to produce growth velocities of 4.5 and 6.0 m/sec.

determined value for k' is ~ 1.0 (see Table I), which implies that the liquidus and solidus lines on the phase diagram are coincident (i.e., the T_0 limit). The fact that the measured value for C_s^{max} exceeds the predicted thermodynamic limit is probably due to uncertainties on the equilibrium phase diagram from which the predictions were made.

H. Zone Refining of Interstitial Impurities

There are a wide variety of impurities outside Groups III, IV, and V that cannot be incorporated into the Si lattice during nanosecond laser annealing (Baeri *et al.*, 1979; White *et al.*, 1979b). These impurities include Cu, Fe, Zn, Mn, W, Mg, Cr, and Yb. These impurities all segregate toward the surface during pulsed laser annealing, and the degree of surface segregation is a strong function of the concentration of the impurity. This is illustrated in Fig. 21 for the case of Fe implanted in (111) Si at three different doses. At the low dose (i.e., low concentration) one pulse is sufficient to cause complete zone refining of the impurity to the surface. In this case, the amount of impurity remaining in the bulk of the crystal is below the detection limits of standard ion-scattering measurements. At intermediate doses, the first pulse causes significant segregation toward the surface, but significant concentrations remain in the bulk. The second pulse, however, causes all measurable impurity to be segregated to the near-surface region. Finally, at high concentrations, measurable concentrations remain in the bulk even after 10 laser pulses. TEM results (see Fig. 15) show that the impurity remaining in the bulk is localized to the walls of a cell structure that arises as a result of constitutional supercooling during regrowth.

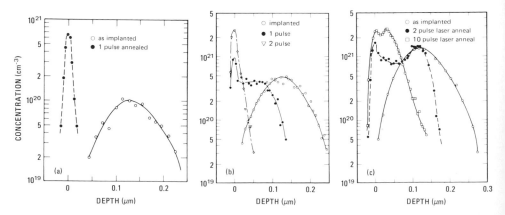

Fig. 21. Profiles for ^{56}Fe (150 keV) in (111) Si. (From White *et al.*, 1980c.)

The Fe remaining in the bulk of the crystal is not located on a regular lattice site. RBS angular scans across the $\langle 110 \rangle$ and $\langle 111 \rangle$ axes are featureless, indicating that no measurable quantities of Fe are left in solution in the lattice after laser annealing. Calculations similar to those described previously can be used only to determine an upper limit for k'. The result is k' (Fe) $< 10^{-2}$, and similar upper limits apply also for the other non-Group III and V species. This value is much lower than those determined for Group III and V impurities (see Table I) and indicates that for these impurities recrystallization takes place under conditions much closer to equilibrium at the interface than for the case of Group III and V impurities.

A behavior similar to that shown in Fig. 21 has been observed for a wide variety of nonsubstitutional impurities (Cu, Zn, Mn, W, Mg, Cr, Yb). At low concentrations these species can be zone-refined at the surface with a single laser pulse. At high concentrations, constitutional supercooling causes the interface to become unstable during regrowth, resulting in lateral segregation and the formation of a cell structure in the near-surface region. The concentration at which the interface becomes unstable is different for each impurity, and studies of this type at different doses can be expected to provide information on the concentrations in the liquid required for instability to develop.

The fact that substitutional species can be efficiently trapped during laser annealing, while nonsubstitutional species cannot, must be related to differences in the residence time at the interface or to differences in the bonding properties of these impurities in the solid. It has been suggested that the impurity solid phase diffusivity plays an important role (Campisano *et al.*, 1980b). The impurity residence time at the interface is given by λ^2 / D_i, where λ is the interface thickness and D_i is the diffusion coefficient in the interfacial region. If the monolayer regrowth time is shorter than the residence time at the interface, then incorporation of the dopant into the solid will be expected. The interface separates two phases (solid and liquid) in which diffusion coefficients may differ by orders of magnitude. All impurities have liquid phase diffusion coefficients of 10^{-5}–10^{-4} cm²/sec. However, solid phase diffusivities D_s for substitutional species are much smaller than solid phase diffusivities for interstitial species. It has been suggested (Campisano *et al.*, 1980b) that the diffusion coefficient in the interfacial region can be approximated by $D_i = (D_s D_L)^{1/2}$. Under this approximation, the diffusion coefficient for interstitial species in the interfacial region will be much higher than substitutional species, and the impurity residence time for interstitial species is (typically) less than the regrowth time for a monolayer. This may account for the fact that Group III and V species can be efficiently incorporated into the solid, while nonsubstitutional species are not. In addition, in the interfacial region bonds must be broken for the impurity to escape from the solid being formed and remain in the liquid. The breaking of bonds requires energies on the order of the activation energy for diffusion in the solid. Activation energies for solid phase diffusion are much higher for substitutional impurities, and this would increase the residence time at the interface. Interstitial species have relatively low activation energies for diffusion in the solid, and they could more easily escape from the solid being formed in the interfacial region and remain in the liquid.

Significantly higher growth velocities will be required in order to trap these interstitial species in the lattice during laser annealing. These velocities can be obtained only by using picosecond laser pulses possibly combined with substrate cooling during irradiation. Calculations (see Chapter 4) indicate that increases in velocity by a factor of 3–4 may be obtained. It will be interesting to see whether these velocity increases are enough to trap the nonsubstitutional species.

V. Conclusions

Laser annealing of ion-implanted silicon has provided very fundamental information on high-speed nonequilibrium crystal growth processes. The level of understanding brought to this relatively new area of materials science is impressive, considering the short period of time during which these studies have been conducted. These advances have been accomplished because one can use two complementary nonequilibrium processing techniques, ion implantation and pulsed laser annealing, in order to carry out experiments under very carefully controlled conditions.

During the rapid liquid phase epitaxial regrowth process, implanted Group III and V species can be incorporated into the lattice at concentrations that exceed equilibrium solubility limits by orders of magnitude. Interfacial distribution coefficients, in many cases as a function of velocity, have been determined for a wide variety of impurities in silicon. Theoretical models have been developed that may explain in a quantitative fashion the solute-trapping mechanisms. The limits of the substitutional solubility that can be achieved by laser annealing have been established, and insight has been gained into the factors that limit substitutional solubility. Measured solubility limits are approaching predicted thermodynamic limits to diffusionless solidification.

The future will see similar experiments carried out at both faster and slower growth velocities. To increase the growth velocity will require the use of picosecond laser pulses. Slower velocities can be achieved by depositing the energy over a longer time period. Interesting questions that need to be investigated more thoroughly involve the saturation value for the distribution coefficients, more conclusive tests of predicted thermodynamic limits of dopant incorporation, orientation effects in solute trapping, and incorporation of interstitial species. Results generated in these experiments should provide a very sound basis for theoretical understanding of high-speed nonequilibrium crystal growth.

References

Appleton, B. R., Larson, B. C., White, C. W., Narayan, J., Wilson, S. R., and Pronko, P. P. (1979). *In* "Laser–Solid Interactions and Laser Processing" (S. D. Ferris, H. J. Leamy, and J. M. Poate, eds.) p. 291. Am. Inst. Phys., New York.

Baeri, P., Campisano, S. U., Foti, G., and Rimini, E. (1978a). *Appl. Phys. Lett.* **33**, 137.

Baeri, P., Campisano, S. U., Foti, G., and Rimini, E. (1978b). *J. Appl. Phys.* **50**, 788.

Baeri, P., Campisano, S. U., Foti, G., and Rimini, E. (1979). *Phys. Rev. Lett.* **41**, 1246.

Baeri, P., Poate, J. M., Campisano, S. U., Foti, G., Rimini, E., and Cullis, A. G. (1981a). *Appl. Phys. Lett.* **37**, 912.

Baeri, P., Foti, G., Poate, J. M., Campisano, S. U., and Cullis, A. G. (1981b). *Appl. Phys. Lett.* **38**, 800.

Baker, J. C., and Cahn, J. W. (1969). *Acta Metall.* **17**, 575.

Bean, J. C., Leamy, H. J., Poate, J. M., Rozgonyi, G. A., Van der Ziel, J., Williams, J. S., and Celler, G. K. (1979). *In* "Laser–Solid Interactions and Laser Processing" (S. D. Ferris, H. J. Leamy, and J. M. Poate, eds.), p. 487. Am. Inst. Phys., New York.

Cahn, J. W., Coriell, S. R., and Boettinger, W. J. (1980). *In* "Laser and Electron Beam Processing of Materials" (C. W. White and P. S. Peercy, eds.), p. 89. Academic Press, New York.

Campisano, S. U., Rimini, E., Baeri, P., and Foti, G. (1980a). *Appl. Phys. Lett.* **37**, 170.

Campisano, S. U., Foti, G., Baeri, P., Grimaldi, M. G., and Rimini, E. (1980b). *Appl. Phys. Lett.* **37**, 719.

Cullis, A. G., Webber, H. C., Poate, J. M., and Simons, A. L. (1980). *Appl. Phys. Lett.* **36**, 320.

Cullis, A. G., Hurle, D. T. J., Webber, H. C., Chew, N. G., Poate, J. M., Baeri, P., and Foti, G. (1981). *Appl. Phys. Lett.* **38**, 642.

Fogarassy, E., Stuck, R., Grob, J. J., Grob, A., and Siffert, P. (1980). *In* "Laser and Electron Beam Processing of Materials" (C. W. White and P. S. Peercy, eds.), p. 117. Academic Press, New York.

Jackson, K. A. (1981). Personal communication.

Jackson, K. A., Gilmer, G. H., and Leamy, H. J. (1980). *In* "Laser and Electron Beam Processing of Materials" (C. W. White and P. S. Peercy, eds.), p. 104. Academic Press, New York.

Jackson, K. F. (1975). *In* "Treatise on Solid State Chemistry" (N. B. Hannay, ed.), Vol. 5, p. 233. Plenum, New York.

Kodera, H. (1963). *Jpn. J. Appl. Phys.* **2**, 212.

Larson, B. C., White, C. W., and Appleton, B. R. (1978). *Appl. Phys. Lett.* **32**, 801.

Mullins, W. W., and Sekerka, R. F. (1964). *J. Appl. Phys.* **35**, 444.

Narayan, J. (1981). *J. Appl. Phys.* **52**, 1289.

Narayan, J., Young, R. T., and White, C. W. (1978). *J. Appl. Phys.* **49**, 3912.

Swanson, M. L., Howe, L. M., Saris, F. W., and Quenneville, A. F. (1981). *In* "Defects in Semiconductors" (J. Narayan and T. Y. Tan, eds.), p. 71. North-Holland Publ., New York.

Tiller, W. A. Jackson, K. A., Rutter, J. W., and Chalmers, B. (1953). *Acta Metall.* **1**, 428.

Trumbore, F. (1960). *Bell Syst. Tech. J.* **39**, 205.

Wang, J. C., Wood, R. F., and Pronko, P. P. (1978). *Appl. Phys. Lett.* **33**, 455.

White, C. W. (1980). Personal communication.

White, C. W., Narayan, J., and Young, R. T. (1979a). *Science (Washington, D.C.)* **204**, 461.

White, C. W., Narayan, J., Appleton, B. R., and Wilson, S. R. (1979b). *J. Appl. Phys.* **50**, 2967.

White, C. W., Pronko, P. P., Wilson, S. R., Appleton, B. R., Narayan, J., and Young, R. T. (1979c). *J. Appl. Phys.* **50**, 3261.

White, C. W., Wilson, S. R., Appleton, B. R., and Young, F. W., Jr. (1980a). *J. Appl. Phys.* **51**, 738.

White, C. W., Wilson, S. R., Appleton, B. R., Young, F. W., Jr., and Narayan, J. (1980b). *In* "Laser and Electron Beam Processing of Materials" (C. W. White and P. S. Peercy, eds.), p. 111. Academic Press, New York.

White, C. W., Wilson, S. R., Appleton, B. R., Young, F. W., Jr., and Narayan, J. (1980c). *In* "Laser and Electron Beam Processing of Materials" (C. W. White and P. S. Peercy, eds.), p. 124. Academic Press, New York.

White, C. W., Appleton, B. R., Stritzker, B., Zehner, D. M., and Wilson, S. R. (1981). "Laser and Electron Beam Solid Interactions and Materials Processing" (J. F. Gibbons, L. D. Hess and T. W. Sigmon, eds.), p. 59. North-Holland Publ., New York.

Williams, J. S., and Elliman, R. G. (1981). *Nucl. Instrum. Methods* **182/183**, 389.

Wood, R. F. (1980). *Appl. Phys. Lett.* **37**, 302.

Wood, R. F., Kirkpatrick, J. R., and Giles, G. E. (1981). *Phys. Rev. B* **23**, 5555.

Chapter 6

Microstructure and Topography

A. G. CULLIS

Royal Signals and Radar Establishment
Malvern, Worcestershire, England

I. Introduction

Over the past few years, the wide-ranging developments in the area of pulsed annealing have offered the prospect of important advances in semiconductor processing. The opportunities being presented to the process engineer are reviewed by Hill (Chapter 13). However, the new annealing techniques also allow the solid state physicist to probe hitherto unexplored regimes of crystal growth under conditions previously unachievable. This will be evident from the topics covered below and from the work presented in other chapters. The time scale during which energy is deposited in a material to be annealed can vary over the tremendous range from <1 nsec to >1 sec. As would be expected, the structure transitions observed are quite diverse and depend upon the specific annealing mechanisms involved.

A wide variety of physical analytical techniques have been employed in the study of annealed layer structures, and perhaps the most widely used method is Rutherford backscattering and channeling of megaelectron volt ions. However, this does not give direct information about individual crystallographic defects that may be present, and, for this purpose, electron microscopy is unrivaled. Indeed, detailed imaging studies have been indispensable in revealing the varied annealing phenomena characteristic of the solid and liquid phase processes that occur. This chapter relies heavily on the results of transmission electron microscope (TEM) investigations in revealing the microstructural properties of annealed layers. In many cases conventional plan view specimens have been adequate to show details of interest. However, the layers often exhibit a defect depth stratification that can be revealed satisfactorily only by use of cross-sectional specimens. Such work has the additional advantage that the final micrographs can be compared directly with the results of ion-backscattering studies (described elsewhere in this volume).

II.   Annealing of Ion-Implanted Semiconductors

Ion implantation is widely exploited in semiconductor processing, and the ability to anneal out the initial lattice damage generally is a crucial factor. Conventional furnace annealing imposes a number of limitations, and, therefore, it is not surprising that potential applications of pulsed processing have been investigated in very great detail.

A.   *Q-Switched Laser Annealing*

1.   CONVENTIONAL IMPLANTED SPECIES IN SILICON

When ions are implanted in a semiconductor, the lattice damage produced depends upon a number of implantation parameters (ion energy, mass, dose, etc.) and can range from an array of isolated point defect clusters to a continuous surface amorphous layer. The nature of this initial damage, in turn, controls the type of defects produced during Q-switched laser processing, particularly in the regime below the threshold for com-

plete annealing. If the initial implanted layer is amorphous, annealing with
Q-switched laser radiation produces a well-defined series of structure
transitions as the energy density of the laser beam is increased. For ruby
laser pulses (694 nm) in the range 20–50 nsec, this is illustrated by TEM
studies of the type shown in Fig. 1 (Tseng *et al.*, 1978). The amorphous
layer is first converted to polycrystals. When single-crystal material appears, generally it contains dislocations near the threshold for its formation, but these can be removed by increasing the radiation energy density
still further. Indeed, the final quality of the best single-crystal annealed
layers is quite remarkable. The freedom from extended crystallographic
defects that can be achieved has been confirmed for a number of im-

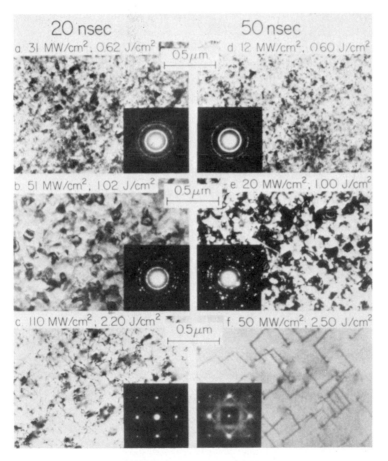

Fig. 1. TEM images and diffraction patterns for Si$^+$ ion-implanted (001) Si samples
irradiated with 20- and 50-nsec ruby laser pulses. (After Tseng *et al.*, 1978.)

planted dopants and, for example, Fig. 2 compares the excellent crystal
quality of fully laser-annealed B-implanted Si with a furnace-annealed
control sample containing dislocation networks (Narayan *et al.*, 1978a).

The precise manner in which structure changes occur during annealing
is perhaps best illustrated by TEM studies on cross-sectional specimens.
A typical sequence for ruby laser (30 nsec)-annealed As$^+$ ion-implanted Si
is shown in Fig. 3 (Cullis *et al.*, 1980a). From this it is clear that, for low
irradiation energy densities, polycrystals first occur in the outer regions of
the initial surface amorphous layer. With increasing energy density, the
polycrystal band progressively penetrates through the amorphous layer
and then coarsens in structure. Next single-crystal material is produced,
although at 0.85 J cm^{-2} it is heavily defective, containing mainly microtwins originating in the region of the original amorphous–single crystal
interface. Increasing the radiation energy density further to 1 J cm^{-2} results in the formation of an irregular interfacial region which contains a
band of point defect clusters. However, the implanted layer is well recrystallized and contains only occasional V-shaped pairs of dislocations.
These extend from the buried point defect cluster band up to the free layer
surface. As mentioned previously, an initial radiation pulse somewhat
higher still in energy density (1.2 J cm^{-2} in Fig. 3) gives a single-crystal
annealed layer free of extended defects.

The sequence of structure transitions described above is totally consistent with the annealing model, outlined by Baeri and Campisano in Chapter 4, based upon transient surface melting. The model is, of course, also
substantiated by a range of other experiments such as observations of

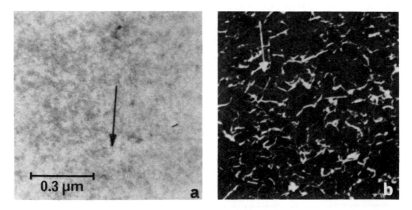

Fig. 2.   TEM images of (001) Si implanted with 35-keV B$^+$ ions (3×10^{15} cm^{-2}). (a) Ruby
laser-annealed at 1.5 J cm^{-2}; (b) furnace-annealed at 1100°C for 30 min. (After Narayan *et
al.*, 1978a.)

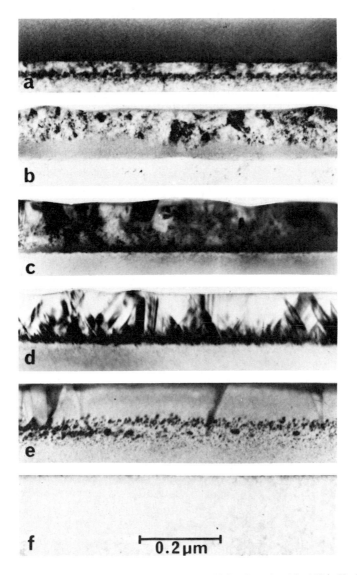

Fig. 3. TEM cross-sectional images of (001) Si implanted with 150-keV As⁺ ions (4 × 10¹⁵ cm⁻²). (a) As-implanted; (b) 0.20 J cm⁻²; (c) 0.35 J cm⁻²; (d) 0.85 J cm⁻²; (e) 1.00 J cm⁻²; (f) 1.20 J cm⁻². (After Cullis *et al.*, 1980a.)

transient surface reflectivity changes (Auston *et al.*, 1978b), measurements of extensive dopant diffusion characteristic of the liquid (Celler *et al.*, 1978), determinations of the velocities of atoms leaving the irradiated surface (Stritzker *et al.*, 1981), and dopant segregation studies of the type described in Section II.A.2. With reference to Fig. 3, it is possible to envisage progressive penetration of the laser-induced melt into the ion damaged layer as the energy density of the incident radiation is increased. First, only the outer portion of the initial amorphous layer melts; since this region is screened from the single crystal matrix by remaining amorphous material, randomly aligned polycrystals are formed on re-solidification (see, e.g., Bertolotti *et al.*, 1979). The final polycrystal–amorphous interface in general appears to be sharp at least down to the scale of about 100 Å, although one study (Gibson and Tsu, 1980) has indicated that solid phase regrowth processes may be significant in this limited region (just beyond the deepest penetration of the original melt). However, it is important to note that, in such samples, most of the unan-nealed lattice damage can retain its initial amorphous character. For example, upon subsequent low-temperature furnace annealing the remaining amorphous Si can be regrown as single-crystal material with the expected recrystallization kinetics (Foti and Cullis, unpublished results).

With a further increase in radiation energy density the thickness of the polycrystal layer increases, although when its inner boundary penetrates to the original amorphous–single-crystal interface, further structural change is delayed over a range of a few tenths of a joule per square centimeter. This phenomenon is entirely consistent with the initial forma-tion of undercooled liquid Si upon melting of the amorphous Si layer (Spaepen and Turnbull, 1979; Bagley and Chen, 1979). As a result, the initially formed molten Si is at too low a temperature to permit further propagation of the melt front into underlying single-crystal matrix ma-terial without additional energy input. Indeed, the reduced melting tem-perature of amorphous Si has been measured (Baeri *et al.*, 1980a) (see Section II.C.1). The annealed layer in Fig. 3d corresponds to the case where the melt (at deepest penetration) has almost achieved the equi-librium melting temperature and has led to epitaxial recrystallization, nevertheless with many planar defects nucleated at the original amor-phous–single-crystal boundary position. (Note also that, if such an un-dercooled melt is quenched sufficiently fast by use, for example, of low energy density picosecond laser pulses (Rozgonyi *et al.*, 1981), it appears possible to reform the amorphous phase on resolidification. Other ex-periments on extremely rapid quenching and the actual amorphization of crystalline Si are described in Section IV.E.)

If sufficient energy is deposited with the laser radiation for the melt just to penetrate only the initial damage boundary, good single-crystal recrystallization occurs, although a relatively small amount of buried disorder remains. The band of point defect clusters (Fig. 3e) contains dislocation loops, and it is thought that, if these fail to close during resolidification, the arms of the defects will extend up to the free surface to give the V-shaped dislocation pairs often seen. However, the nucleation centers are themselves removed if the melt front penetrates still further into the matrix, so that a recrystallized layer free of extended defects is formed (Fig. 3f).

It is clear that, if dislocations remain in a recrystallized layer, they will generally assume characteristic configurations. This is well illustrated by studies on the way in which preexisting dislocations behave when they are only partially consumed in the laser-induced melt. When $a/2$ [110] dislocations remaining after the furnace annealing of an As^+ ion-implanted (001) Si layer are ruby laser-annealed in this way, the segments of dislocation that reform during resolidification do not lie along a [001] direction normal to the surface (Narayan and Young, 1979). As shown in Fig. 4, the re-formed segments at 1, 2, and 3 (originating from the region of deepest melt penetration) lie along or near new inclined ⟨113⟩-type directions. In

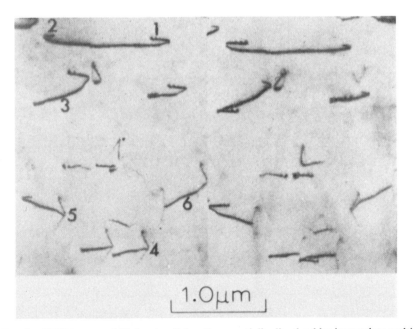

Fig. 4. TEM images of Si showing dislocations partially dissolved by incomplete melting due to ruby laser annealing. Different types of dislocation regrowth are illustrated. (After Narayan and Young, 1979.)

some cases, the initial dislocations split to give two other whole dislocations which also lie along inclined ⟨113⟩-type directions (as at 4, 5, and 6).
It is possible that a reduced line tension along the [113] axis may play a
role in these reorientation effects, although the kinetics of dislocation
growth during resolidification also may be important. Similar laser-
induced dislocation regrowth phenomena with the addition, sometimes, of
stacking fault formation have been observed in P^+ ion-implanted and
furnace-annealed (111) Si by Hofker *et al.* (1979).

It is clearly essential that structural changes occurring during the laser
annealing of an ion-implanted layer be related to corresponding variations
in electrical properties. Curve B in Fig. 5 shows the sheet resistance as a
function of laser radiation energy density for specimens identical to the
As^+ ion-implanted Si in Fig. 3 (Cullis *et al.*, 1980a). As polycrystals begin
to form and grow at low radiation energy densities, there is a rapid fall in
resistance. However, the trend is temporarily arrested while polycrystals
persist, giving an intermediate state (near 0.5 J cm^{-2}) where carrier scattering effects are likely to be important. When single-crystal material
begins to form at higher radiation energy densities, the sheet resistance
once again begins to fall, reaching a low value by about 1.2 J cm^{-2} charac-

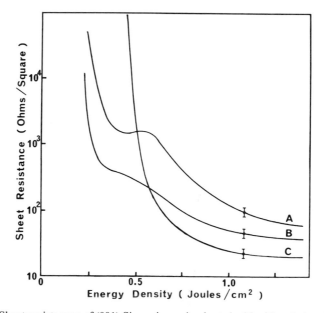

Fig. 5. Sheet resistances of (001) Si specimens implanted with either As^+ or P^+ ions as a
function of laser pulse energy densities used for annealing. A, 2×10^{15} P^+, 100 keV; B,
4×10^{15} As^+, 150 keV; C, 1×10^{16} As^+, 40 keV. (After Cullis *et al.*, 1980a.)

teristic of efficient As dopant activation in excellent single-crystal material. The overall effect is to give a two-stage electrical annealing characteristic. This behavior is even more pronounced for an initially amorphous P$^+$ ion-implanted layer (curve A in Fig. 5). The region of reverse annealing near 0.5 J cm^{-2} may be due to segregation of the dopant to grain boundaries. The general behavior is in some ways reminiscent of the reverse annealing sometimes found during the furnace heat treatment of ion-implanted Si (Bicknell and Allen, 1970)—the underlying mechanisms, however, are different. Figure 5 also illustrates (curve C) the way in which the thickness of the initial amorphous layer controls the radiation energy density required for annealing onset. A decrease in the amorphous layer thickness decreases the absorption of the radiation in the surface region and increases the energy density needed for onset (see Chapter 3 by von Allmen).

At this point it is important to recognize that Q-switched laser annealing is a very nonequilibrium process and, because of the high melt resolidification rate (often >2 msec^{-1}) dopants introduced by high-dose implants are often incorporated at substitutional lattice sites in greatly supersaturated concentrations. While this is most pronounced for low-solubility dopants (see Section II.A.2), conventional dopants such as boron also exhibit the phenomenon to a limited extent. Layers of Si supersaturated with boron in this manner show a strong lattice contraction along the surface normal as a result of the difference in the atomic diameters of the solute and matrix atoms (Larson *et al.*, 1978). Solid solutions of this type betray their metastability upon subsequent furnace annealing when excess boron precipitates as a second phase (Narayan *et al.*, 1978b). Similar precipitation phenomena are observed upon heat treatment of high-dose P$^+$ ion-implanted and laser-annealed Si (Miyao *et al.*, 1980).

The above discussion has focused upon the laser annealing of initially amorphous layers in ion-implanted (001) Si. While the sequence of structure transitions is similar for ion-implanted (111) Si, when the melt front just penetrates through the surface amorphous layer, large numbers of stacking faults can be produced instead of simply the dislocation pairs described earlier (Foti *et al.*, 1978b; Sadana *et al.*, 1980). This is illustrated in Fig. 6 for (111) Si implanted with Si$^+$ ions. Also, somewhat higher radiation energy densities are required to anneal Si in the (111) orientation, although they can be reduced somewhat by the use of low-temperature furnace preanneals (Campisano *et al.*, 1980b).

If the initial ion-implanted layer is not damaged sufficiently to give an amorphous structure, laser annealing at moderate radiation energy densities does not generally yield a polycrystal stage of recrystallization (Foti *et al.*, 1979; McMahon *et al.*, 1980). A typical partial anneal stage is

Fig. 6. TEM image of (111) Si implanted with 50- to 200-keV Si$^+$ ions (10^{16} cm^{-2}) and ruby laser-annealed at 2.5 J cm^{-2}. Note residual stacking faults. (After Foti *et al.*, 1978b.)

shown in Fig. 7 for Si implanted with B$^+$ ions to give initial damage comprising a buried band of point defect clusters. Annealing with a 30-nsec pulse of ruby laser radiation at 1.5 J cm^{-2} melts the Si down to near the center of the band, leaving all the defect clusters in place below this plane. However, the molten (outer) Si layer resolidifies with few extended crystallographic defects present. The only defects visible are occasional dislocation pairs which nucleated at the edge of the remaining portion of the buried damage band. As expected, annealing with substantially higher radiation energy densities can completely dissolve the band of damage, so that the recrystallized layer is free of all extended defects.

In the above discussion, the high quality of laser-annealed Si free of extended defects has been emphasized. However, as will be made clear in the next section, the extremely high resolidification rate toward the end of the annealing cycle can lead to the formation of nonequilibrium solid solutions. In particular, this applies to the incorporation of native point

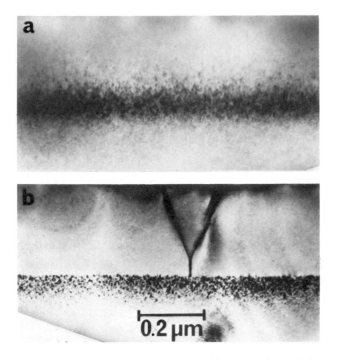

Fig. 7. TEM images of (111) Si implanted with 100-keV B^+ ions (10^{16} cm^{-2}). (a) As-implanted; (b) 1.5 J cm^{-2}. (After McMahon *et al.*, 1980.)

defects in the lattice, and especially high concentrations can be trapped in the quenched, recrystallized surface layer. This effect has been demonstrated (Kimerling and Benton, 1980) in Q-switched laser-annealed Si by the use of deep-level transient capacitance spectroscopy (DLTS). Figure 8 shows the defect state spectrum obtained from n-type Si processed with a 1060-nm radiation pulse of 14 J cm^{-2}. In the spectrum each feature represents the carrier emission process from a defect state to the nearest band. The peak height is proportional to the defect concentration, peaks A and B being the principal features. Experimental measurements show that the laser-induced defects responsible for peak B are donors (activation energy 0.33–0.36 eV) and are essentially confined to the laser-melted Si layer. However, apart from these defects, others may be produced outside the laser-melted region if an ion implant is performed before the laser anneal. This is illustrated in Fig. 9 which shows the photoluminescence spectrum obtained from $^{29}Si^+$ ion-implanted and ruby laser-annealed n-type Si (Skolnick *et al.*, 1981). Peaks W and G are due to defect centers, and peak W is present only after the laser anneal. As can be shown by layer-stripping experiments, the defects corresponding to W (possibly five va-

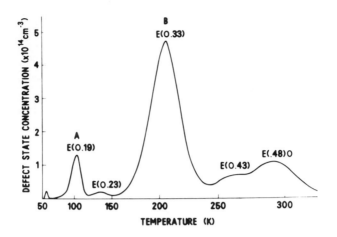

Fig. 8. Capacitance transient defect state spectra of laser-annealed, n-type Czochralski-grown Si. $n = 1.5 \times 10^{16}$ cm^{-3}; $\tau_{t} = 1.8$ msec; 1.06 μm, 14 J cm^{-2}. (After Kimerling and Benton, 1980.)

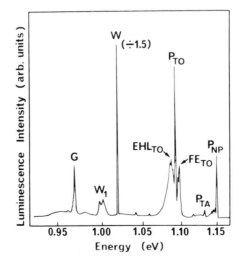

Fig. 9. Photoluminescence spectrum at 4.2 K for Si sample implanted with 80-keV Si$^+$ ions (10^{14} cm^{-2}) and ruby laser-annealed at 2.54 J cm^{-2}. P_{NP}, P_{TA}, and P_{TO} are phosphorus-bound exciton no-phonon and TA and TO phonon replicas. FE_{TO} and EHL_{TO} are the free-exciton and electron–hole liquid TO phonon replicas. The damage-related peaks are W and G; W_1 is a phonon replica of W. (After Skolnick *et al.*, 1981.)

cancy complexes) are concentrated in a buried band beneath the melt front at deepest penetration. Thus, one may conclude that the heat pulse propagating beyond the melted region transforms defects initially present in a channeled ion damage tail (beneath any initial amorphous region) into the observed defect centers (W). In this way, solid phase annealing processes can assume importance even during Q-switched laser annealing work. The point defect centers described are of obvious significance for device fabrication since they may lead to unwanted carrier recombination effects. Centers W (and G), however, can be largely removed by increasing the laser melt depth to consume essentially all the buried ion damage tail.

Although the lattice defects already described can be present within Q-switched laser-annealed layers, perhaps the most clear outward sign that annealing has taken place is given by the appearance of ripples on the sample surface. This is particularly well illustrated in Fig. 10 which shows As$^+$ ion-implanted Si laser-annealed with an overlapped series of pulses from a Nd:YAG laser (Leamy et al., 1978). At low power densities aligned ripples are present. These are thought to be produced by periodic surface melting in the nonuniform radiation field arising from interference between the main laser beam and waves scattered from surface disturbances. As the power density is increased, the melting threshold is exceeded in all main irradiated regions and the periodic ripples are gradually suppressed, remaining longest at the edge of each radiation spot where a range of power densities is present. For the highest power density anneals, the melted regions are relatively smooth, with arclike boundary features exhibiting the appearance of a frozen meniscus.

The concept of wave interference used to explain the periodic ripples described above is not wholly adequate to account for the ripples seen on semiconductor surfaces under all conditions. In some cases, induced surface periodicities have been attributed to nonlinear interaction between simultaneously oscillating axial laser modes, resulting in the production of surface acoustic waves (Maracas et al., 1978). Such an interpretation may assist in accounting for the persistence of wave phenomena often seen when melting is quite deep. However, it is important to note that, if a strongly multimode laser beam is homogenized in a light guide diffuser (Cullis et al., 1979b) to give a beam of uniform intensity and poor local coherence, a small-amplitude surface ripple pattern remains after substantial melting has occurred and is irregular, taking the form shown in Fig. 11. It is immediately clear that this is very similar to the surface structure produced by rapid-pulse electron beam annealing (see Section II.C.1) where the wave nature of the radiation is greatly different. Therefore, it is likely that transient melt oscillations occur in both cases as a result of the general shock wave effects produced, for example, by the momentum

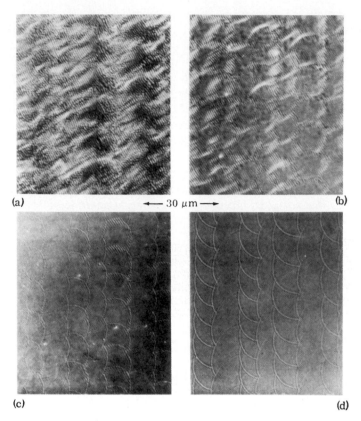

Fig. 10. Optical micrographs (Nomarski interference) of (001) Si implanted with As$^+$ ions $(8 \times 10^{15}$ cm$^{-2})$ and annealed with a scanned Nd:YAG laser beam. Note surface ripples. (After Leamy *et al.*, 1978.)

transferred to the sample from the annealing radiation or by stresses imposed because of the stratified surface heating.

A different and very characteristic type of surface rippling occurs when Si covered with SiO$_2$ is subjected to Q-switched laser annealing (Hill and Godfrey, 1980). As shown in Fig. 12, the ripples take up radial alignment around circular windows, while an isolated annulus of oxide exhibits a greatly reduced ripple amplitude. The surface distortions are thought to occur when underlying Si is transiently molten and the oxide sheet buckles partly because of thermal expansion stresses. However, recent work by Hill has demonstrated that compressive strains remaining in the oxide after growth are also relieved in this way and are particularly important. Isolated small annular regions of oxide avoid buckling by uniform lateral

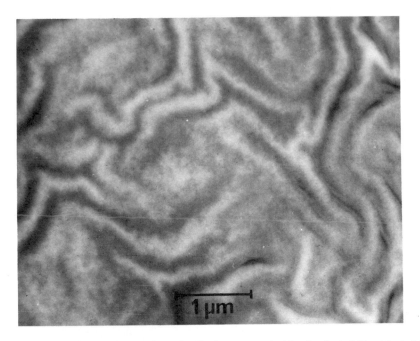

Fig. 11. TEM image showing irregular ripple array on As^+ ion-implanted Si subjected to Q-switched laser annealing with a 30-nsec homogenized multimode pulse from a ruby laser.

expansion. A detailed account of these phenomena, together with the way in which they can be avoided by the device engineer, is given in Chapter 13 by Hill.

2. Low-Solubility Implanted Species in Silicon

Ion implantation is a particularly useful technique for introducing exceptionally high local concentrations of dopants into Si. However, if the impurity element exhibits only a low equilibrium solubility in the Si lattice, annealing of such an implanted layer will lead to precipitation of excess impurity with accompanying second-phase formation. Since Q-switched laser annealing produces transient surface melting, this gives rise to very characteristic impurity segregation phenomena which themselves give important insight into the annealing process.

When an implanted nonconventional impurity is freely soluble in liquid Si, Q-switched laser annealing in the melt regime generally leads to segregation of the impurity out toward the external surface. This has been observed for the heavier Group III and V elements and for certain transi-

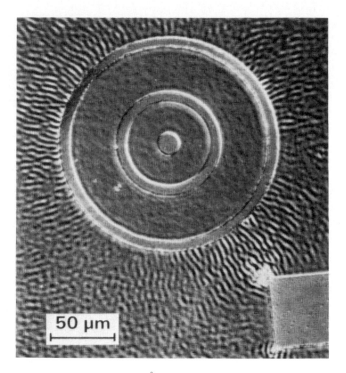

Fig. 12. Optical micrograph of 2000-Å oxide-coated, photoengraved, As$^+$ ion-implanted Si surface after ruby laser annealing at 1.4 J cm^{-2}. (After Hill and Godfrey, 1980.)

tion elements (Baeri *et al.*, 1978; Cullis *et al.*, 1979a; White *et al.*, 1979). The process occurs as a result of the relatively small liquid–solid distribution coefficients exhibited by these elements. This ensures that the elements tend to remain in the last portion of liquid Si to resolidify so that surface segregation takes place. Full details will be found in Chapter 5 by White, Appleton, and Wilson. However, the segregation phenomena are often more complex than has just been outlined, since the impurity can redistribute laterally in the annealed layer to give channels of a precipitate phase, which form the boundaries of small crystalline Si cells (Cullis *et al.*, 1979a). Examples of this are shown in Fig. 13 for Q-switched laser-annealed Fe- and Pt-implanted Si layers. Excess Fe precipitates in the cell walls as the well-defined phase ζ_α-FeSi$_2$, and Fig. 14 demonstrates the way in which the cells can be three-dimensional, penetrating well down into the recrystallized region. The general formation of this cell structure is the result of an instability in the resolidification interface due to the occurrence of constitutional supercooling in the transient melt (Cullis *et*

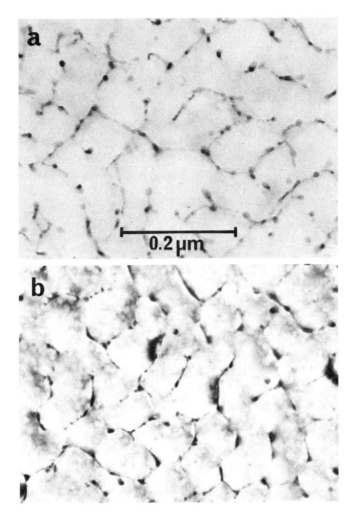

Fig. 13. TEM images of impurity segregation cells in ion-implanted and laser-annealed (001) Si. (a) 150-keV Fe⁺ ions (5 × 10^{15} cm⁻²) ruby laser-annealed at 2.2 J cm⁻². (After Cullis *et al.*, 1980b.) (b) 150-keV Pt⁺ ions (5 × 10^{15} cm⁻²) Nd:YAG laser-annealed at 100 MW cm⁻². (After Cullis *et al.*, 1979a, 1980c.)

al., 1980b,c). The latter phenomenon is well known in the field of crystal growth (Bardsley *et al.*, 1962), and the theory of the conditions under which it occurs (Mullins and Sekerka, 1964) has been tested at conventional growth rates (10^{-5}–10^{-2} cm sec⁻¹). However, the velocity of the recrystallization interface during Q-switched laser annealing is much

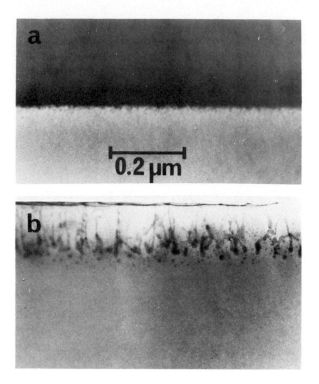

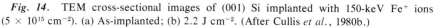

Fig. 14. TEM cross-sectional images of (001) Si implanted with 150-keV Fe^+ ions $(5 \times 10^{15}$ cm^{-2}). (a) As-implanted; (b) 2.2 J cm^{-2}. (After Cullis *et al.*, 1980b.)

greater (often 1–5 msec^{-1}), so that the observation of cells in the present case provides a test for the theory in a new regime of crystal growth. Indeed, interface breakdown conventionally occurs when the destabilizing effects of solute redistribution outweigh the stabilizing effect of the thermal gradient. However, at very high growth rates, the instability wavelength scale becomes very small and the deformed surface has a high curvature. Under these conditions, found during laser processing, the high curvature provides a strongly stabilizing influence through the Gibbs–Thomson effect (Coriell and Sekerka, 1979), so that interface breakdown is inhibited. Hitherto, this prediction of the theory has not been amenable to test, but now calculations (Cullis *et al.*, 1981a) show good agreement with the observed instability wavelengths of 500–1000 Å, with account taken of changes in the distribution coefficient described below.

With melt recrystallization velocities of several meters per second conditions during Q-switched laser annealing are far from equilibrium. As a direct result, kinetically controlled trapping of excess impurity atoms from the melt can occur (see Jackson *et al.*, 1980). This effectively increases the impurity distribution coefficient and leads to the incorporation of metastable high concentrations of dopant at substitutional matrix lattice sites. It has already been noted that this effect can increase the concentration of conventional impurities in Si, but dramatically enhanced solubilities have been observed for nonconventional dopants such as Pt, In, and Ga (Cullis *et al.*, 1979a, 1980c; White *et al.*, 1980). The effect is well illustrated by the laser annealing of Pt-implanted Si. Although the equilibrium solid solubility of Pt in Si has a maximum value of about $10^{17}\,\mathrm{cm}^{-3}$, as is clear from the channeled and random backscattering spectra in Fig. 15, the amount of impurity at substitutional lattice sites in the laser-recrystallized matrix can exceed this by approximately three orders of magnitude. It is particularly important to note that changes in the background substrate temperature can change the prevalence of solute trapping. This occurs because of the corresponding change in the velocity of the recrystallization interface, which is effectively controlled by the Si

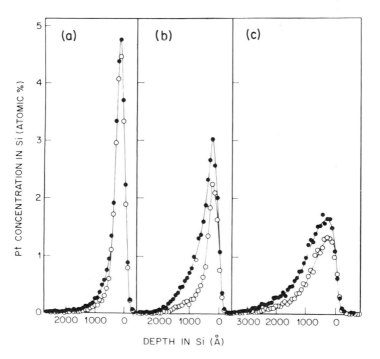

Fig. 15. Random and channeled 2-MeV He⁺ ion backscattering spectra obtained from Pt⁺ ion-implanted and laser-annealed Si layers. Background substrate temperatures were (a) 620, (b) 300, and (c) 77 K. Solid circles, Random; open circles, ⟨100⟩ channeled. (After Cullis *et al.*, 1980c.)

thermal conductivity (Cullis *et al.*, 1980c). Increasing the recrystallization velocity (in the range 1–5 msec^{-1}) increases the solute trapping and hence the effective distribution coefficient of the impurity. In fact, this same change also decreases the degree of constitutional supercooling in the melt, which in turn stabilizes the resolidification interface and can suppress segregation cell formation (Cullis *et al.*, 1981a). If cells are still produced, their size can vary as shown in Fig. 16 where, for In in Si, a change in the resolidification velocity from 4 to 2 msec^{-1} gives a cell size change from ~450 to ~850 Å. The velocity can also be varied by modifying the laser pulse length, as has been demonstrated by observations of the trapping dependence of In in Si (Baeri *et al.*, 1980b).

Implanted low-solubility impurities do not always give a cell structure upon laser annealing. For example, implanted C in Si exhibits only limited solubility ($\gtrsim$0.1 at %) in the transient liquid at the melting point. In this case, ruby laser annealing of an amorphous C-implanted Si layer leads to the segregation of excess C to the surface where a dense accumulation of small β-SiC particles can form (Cullis, 1980; Chiang *et al.*, 1981), probably by direct precipitation in the resolidifying liquid (Fig. 17b). The precipitation process can be minimized (Fig. 17c) by cooling the substrate which, as indicated previously, increases the recrystallization rate. Most significantly, however, marked nonequilibrium trapping of C takes place at substitutional matrix lattice sites (Cullis *et al.*, 1981b), which gives solid solutions containing up to 2×10^{20} substitutional C atoms per cubic centimeter as measured by ir observations of the 607-cm^{-1} local mode absorption peak (Fig. 17e). Such solid solutions are supersaturated with solute

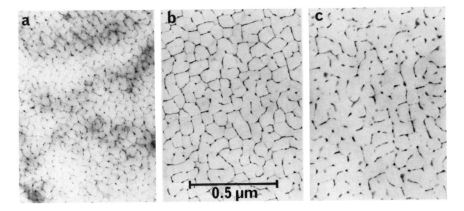

Fig. 16. TEM images of (001) Si implanted with 170-keV In$^+$ ions (3×10^{15} cm^{-2}) and annealed at 1.5 J cm^{-2}. Note dependence of cell size on substrate background temperature. (a) 77 K; (b) 300 K; (c) 410 K. (After Cullis *et al.*, 1981a.)

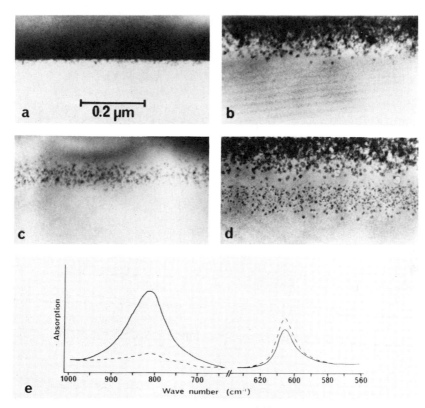

Fig. 17. Carbon-implanted and ruby laser-annealed Si. Cross-sectional TEM images. (a) As-implanted; (b) annealed at 2 J cm⁻² (room temperature); (c) annealed at 2 J cm⁻² (77 K); (d) As in (b) but with a ½-hr furnace anneal at 950°C. The ir absorption spectra in (e) are for specimens (b) (solid curve) and (c) (dashed curve).

concentrations approaching three orders of magnitude more than the equilibrium solubility level, the excess C being precipitated as further β-SiC by heat treatment in a furnace at 1050°C (Fig. 17d).

Impurities such as inert gases are extremely insoluble in both liquid and solid Si. When initially amorphous Ar^+ ion-implanted Si is subjected to ruby laser annealing (Cullis *et al.*, 1980b; Cullis and Webber, 1980; Matteson *et al.*, 1980), as shown in Fig. 18, the resulting surface layer consists of high-quality single-crystal material. This is in contrast to the case of furnace annealing treatment applied to the same implant when the damaged Si recrystallizes to give only polycrystals (Cullis *et al.*, 1978). Also present in the laser-annealed layer is a very high density of bubbles containing precipitated Ar. When such a specimen is examined in cross section (Fig.

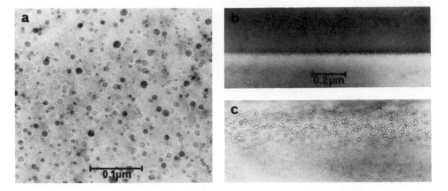

Fig. 18. TEM images of (001) Si implanted with 150-keV Ar$^+$ ions (2×10^{15} cm^{-2}). (a) Plan view, annealed at 2 J cm^{-2}; (b) and (c) cross sections of as-implanted and 2 J cm^{-2}-annealed material, respectively. (After Cullis *et al.*, 1980b.)

18c), it is clear that the bubble distribution has a number density that rises to a maximum in a buried layer, indicating that only limited diffusion of the implanted gas occurs. The final state is consistent with the precipitation of Ar in the transient liquid Si, the relatively immobile bubbles then being trapped near their initial locations upon resolidification (Cullis *et al.*, 1980b).

3. IMPLANTS IN GERMANIUM

Since it is a Group IV element, Ge exhibits many properties similar to those of Si. In particular, Q-switched laser annealing of ion-implanted Ge layers leads to melting (Foti *et al.*, 1978a) and accompanying structure transitions analogous to those in Fig. 3. For example, the ruby laser annealing of an initially amorphous, As$^+$ ion-implanted Ge layer is illustrated in Fig. 19. Here it is clear that with a radiation energy density of only 0.6 J cm^{-2} no extended defects remain and essentially full annealing occurs. An intermediate state of recrystallization showing mostly microtwins with some polycrystals is produced at just 0.2 J cm^{-2}. The relatively low radiation energy densities required to produce these structure changes are a consequence of both the increased radiation absorption coefficient and the decreased melting point of Ge when compared to Si.

4. IMPLANTS IN GROUP III–GROUP V COMPOUND SEMICONDUCTORS

The use of conventional furnace heat treatments to anneal implantation damage in Group III–Group V compound semiconductors is not straight-

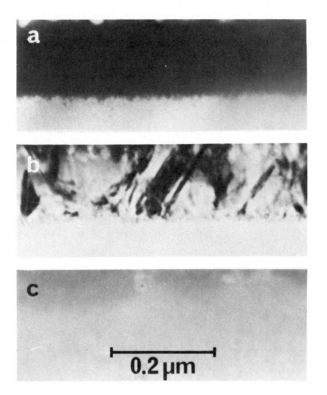

Fig. 19. TEM cross-sectional images of (001) Ge implanted with 150-keV As$^+$ ions (5 × 10^{15} cm^{-2}). (a) As-implanted; (b) and (c) ruby laser-annealed at 0.2 and 0.6 J cm^{-2}, respectively.

forward, since changes in stoichiometry due to preferential loss of the volatile Group V component must be avoided (Eisen and Mayer, 1976). This is generally achieved by sample surface encapsulation. However, the properties of the layer (often SiO$_2$ or Si$_3$N$_4$) used for this purpose can strongly affect the characteristics of the annealed region. Therefore, a potentially attractive feature of pulsed annealing has concerned the possibility that, because the processing time scale is so short, it may not be necessary to use an encapsulating layer.

The Q-switched laser annealing of ion-implanted GaAs has been studied in some detail. It has been found (Campisano *et al.*, 1980a; Sadana *et al.*, 1980) that the general structural changes occurring in an initially amorphous layer are similar to those described earlier for Si. This is illustrated in Fig. 20 which shows the ruby laser annealing of Te-implanted GaAs. With low energy densities polycrystals are formed in the surface region but, when the laser-induced melt penetrates through the initial damaged layer

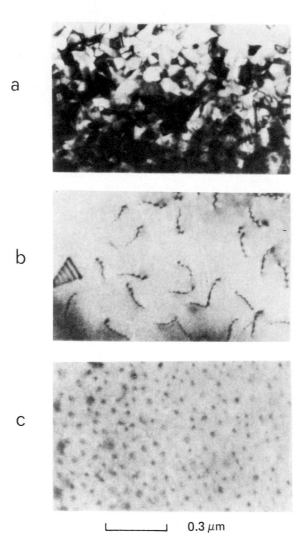

a

b

c

L_____J 0.3 μm

Fig. 20.   TEM images of (001) GaAs implanted with 400-keV Te$^+$ ions (10^{15} cm^{-2}) and ruby laser-annealed at (a) 0.6 J cm^{-2}, (b) 1.0 J cm^{-2}, and (c) 1.4 J cm^{-2}. (After Campisano *et al.*, 1980a.)

at 1 J cm^{-2}, the recrystallization improves dramatically to give single-crystal material with only dislocations and occasional stacking faults. A further increase in the radiation energy density to 1.4 J cm^{-2} leads to the elimination of these defects, although other features (not identified by Campisano *et al.*, 1980a) begin to appear. In this regard it is important to note that some As is lost from the surfaces of annealed samples when higher radiation energy densities are employed (Barnes *et al.*, 1978; Kular *et al.*, 1979), and excess residual Ga can agglomerate on the surfaces as metallic islands. This represents a major difference from the laser annealing of an elemental semiconductor such as Si or Ge, and it is demonstrated in the scanning electron microscope (SEM) images in Fig. 21, which also show that excess Ga can be removed from the GaAs surface by HCl etching treatment. Since the extent of the preferential As loss is strongly dependent on the energy density of the laser radiation (which controls the melt dwell time), surface disproportionation will be aggravated if the radiation is nonuniform, so that some regions will be excessively annealed.

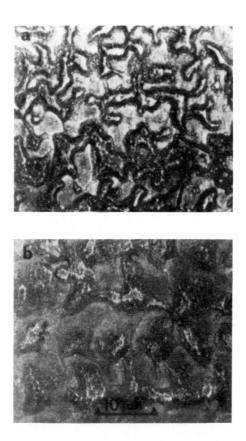

Fig. 21. SEM image of surface of (001) GaAs implanted with Te$^+$ ions and Nd:YAG laser-annealed at 20 MW cm^{-2}. (a) As-annealed; (b) after HCl etching. (After Barnes *et al.*, 1978.)

The laser irradiations just described are generally successful in producing high concentrations of electrical carriers in high-dose-implanted layers, although carrier mobilities are usually depressed below those of furnace-annealed samples. In addition, attempts to activate low-dose-implanted samples electrically have not met with great success. Possible reasons for this situation (including the significance of point defects present in the layers) and the prospects for use of such annealing treatments in device fabrication are described in Chapter 11 by Williams and Chapter 13 by Hill.

Another Group III–Group V compound of particular interest at the present time is InP, a material with an especially high peak electron velocity. Layers implanted with conventional dopants exhibit many of the structure changes seen for GaAs when subjected to Q-switched laser annealing. This is demonstrated in Fig. 22, which shows the amorphous–polycrystal–single-crystal transitions that occur with increasing radiation energy densities (Cullis *et al.*, 1979c). However, the reformed single crystal appears to be of somewhat lower quality than in the case of GaAs, and, for example, the diffraction pattern in Fig. 22c shows weak spot streaking, indicating the presence of small planar defects in the annealed layer. In addition, for InP the transient surface melting can lead to preferential loss of volatile P. Although this is very limited for laser irradiation near the threshold for single-crystal formation, excessive annealing or multiple-pulse treatments can lead to the production of metallic In islands on the sample surface, as illustrated in Fig. 23. Once again, uniformity of the irradiating laser beam is of great importance if this effect is to be suppressed.

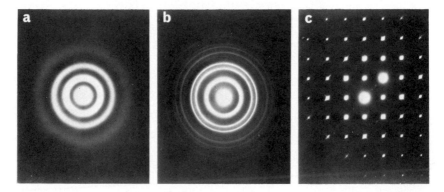

Fig. 22. Transmission electron diffraction patterns from (001) InP implanted with 180-keV S$^+$ ions (6×10^{14} cm^{-2}). (a) As-implanted; (b) 0.2 J cm^{-2}; (c) 0.5 J cm^{-2}. (After Cullis *et al.*, 1979c.)

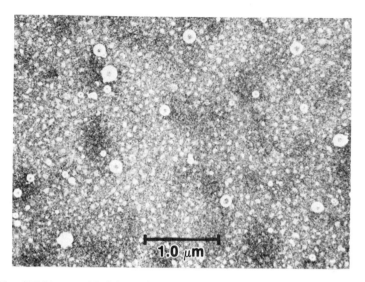

Fig. 23. SEM image of In islands formed on InP subjected to multiple-pulse annealing. (After Cullis *et al.*, 1979c.)

B.   *Scanning Continuous-Wave Laser Annealing*

Processing with a scanning continuous-wave (cw) laser (Gat and Gibbons, 1978) is distinct from the technique discussed in Section II.A, since the radiation power densities achievable at the sample surface are much lower and the local dwell time of the cw beam is on the order of a few milliseconds. While full details of the experimental methods and annealing mechanisms are given by Gibbons and Sigmon in Chapter 10, it is sufficient here to note that the relatively extended heat pulse duration ensures that the dominant annealing mechanism is solid phase regrowth of the irradiated layer (Williams *et al.*, 1978; Auston *et al.*, 1978a). Indeed, the temperature of an irradiated surface has been directly measured (Compaan and Lo, 1980) by observation of the anti-Stokes/Stokes ratio for Raman scattered light. The appearance of typical scanned tracks on the surface of an amorphous ion-implanted Si layer is shown in Fig. 24. The visible contrast is produced by the difference in reflectivity between the amorphous and recrystallized (single-crystal) material. In marked distinction to the Q-switched laser anneal case, the scanned lines in Fig. 24 are mostly flat and featureless since lattice regrowth has occurred in the solid phase. However, because of the Gaussian profile of the scanned laser beam the energy input across each scan line is nonuniform, and Fig. 24

Fig. 24. Optical micrograph showing scanning cw laser-annealed tracks in As^+ ion-implanted Si. Marginal onset of melting along track center is indicated by slight roughening of otherwise smooth surface topography. (Courtesy of Dr. T. W. Sigmon.)

also shows evidence of slight melting (giving surface roughness) in the center of certain tracks.

This method of laser annealing is capable of giving very high-quality regrowth in ion-implanted Si. For initially amorphous ion-implanted layers, TEM studies have shown (Fig. 25) that solid phase regrown material contains few extended lattice defects if the optimum annealing parameters are chosen (Gat *et al.*, 1978b). Wafers are often heated to a few hundred degrees Celsius during the scanning process to achieve adequate control over the annealing, but moderate changes in the laser power or scan speed can still alter the annealed layer structure, slip dislocations being produced if the local temperature is too high (Rozgonyi *et al.*, 1979; Ishida *et al.*, 1980; Mizuta *et al.*, 1980; Uebbing *et al.*, 1980). Since the diffusion coefficient of the As dopant in solid Si is $\sim 10^{-11}$ cm^2 sec^{-1} at the annealing temperature, negligible diffusion (~ 10 Å) takes place during the annealing time (~ 1 msec). Therefore, the final dopant concentration profile shows little change from the as-implanted state. However, metastable high concentrations of As can be introduced at Si substitutional lattice sites with essentially complete electrical activity, deactivation to equilibrium solubility levels taking place during subsequent furnace anneals (Lietoila *et al.*, 1980).

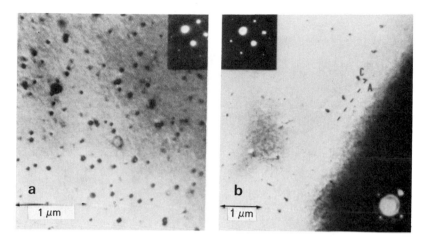

Fig. 25. TEM images of (001) Si implanted with As$^+$ ions. (a) Furnace-annealed at 1000°C for 30 min; (b) scanning cw laser-annealed. (After Gat *et al.*, 1978b.)

When ion implantation does not give an initial surface amorphous layer, excellent recrystallization can still take place. For example, B$^+$ ion implant doses in this category can give very small numbers of extended defects upon scanning laser annealing (Gat *et al.*, 1978c), while conventional furnace annealing gives many dislocation loops and rod defects (Bicknell and Allen, 1970).

Since scanning cw laser annealing is not an equilibrium process, it is to be expected that excess point defects are incorporated into the annealed layers. Their presence can be revealed by DLTS studies and, as shown in Fig. 26, hole emission centers at $E_v + 0.28$ eV are found in material initially implanted with As$^+$ ions (Johnson *et al.*, 1980). However, as also shown, these centers can be removed by furnace annealing at 450°C, although some electron emission centers at $E_c - 0.28$ eV then remain. This observed reduction in defect density upon furnace annealing is also accompanied by a decrease in diode reverse bias leakage current and so may ultimately be of importance for device processing.

C. *Electron Beam Annealing*

The use of electron beam techniques in semiconducting device fabrication is gradually being extended. Already there are important applications in high-resolution lithography and for finished device inspection. Most recently, detailed studies of electron beam annealing have been carried out, and the results are very promising.

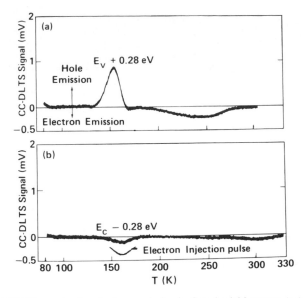

Fig. 26. (a) DLTS spectra for electron emission in Czochralski-grown As$^+$-implanted cw laser-annealed ($T_s = 25°C$) p-type (100) Si and (b) in the same material after an additional 450°C furnace anneal; $P-n$ junction; $e_0 = 347$ sec^{-1}. (After Johnson *et al.*, 1980.)

1. RAPID-PULSE ANNEALING

The first electron beam annealing technique to receive detailed investigation is analogous in many ways to Q-switched laser annealing. It relies upon energy deposition using an electron beam with typical average energy of 10 keV in a single, large-area, short-duration (~100 nsec) pulse (Greenwald *et al.*, 1979). The basic annealing process involves transient surface melting. As described in Chapter 2 by Spaepen and Turnbull and Chapter 3 by von Allmen, since the coupling between the incident beam and the irradiated sample is independent of the physical state of the latter, most of the energy is deposited in the outermost 1 μm of material so that the initial temperature gradient in the first few thousand angstroms is much shallower than for Q-switched laser annealing.

The structure transitions that take place in an ion-implanted sample are conveniently illustrated in Fig. 27, which shows cross-sectional specimen images corresponding to irradiations with two different electron beam energy densities. Melt penetration through the initial amorphous surface layer at 0.55 J cm^{-2} is sufficient to give almost complete liquid phase epitaxial recrystallization. Indeed, slightly higher irradiation energy densities give no residual extended crystallographic defects, although shock

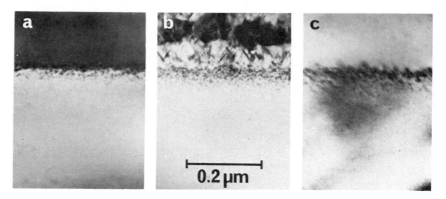

Fig. 27. TEM cross sectional images of (001) Si implanted with 90- to 130-keV P⁺ ions (9 × 10¹⁵ cm⁻²) and pulse electron beam-annealed. (a) As-implanted; (b) 0.5 J cm⁻²; (c) 0.55 J cm⁻². (After Baeri *et al.*, 1980a.)

wave effects accompanying large increases in deposited energy lead to the formation of dislocation networks deep in the crystal. [Stresses in electron beam-annealed Si have been analyzed by Schoen (1980).] However, annealing at just 0.5 J cm⁻² gives the multilevel defect structure shown in Fig. 27b with an outer polycrystal layer and defective epitaxial material immediately below. As described by Baeri *et al.* (1980a), this is consistent with the formation of an initial strongly supercooled melt which recrystallizes with high velocity both from random nucleation sites on the external surface and also from the underlying single-crystal matrix. It should be noted that irradiation at slightly lower energy density levels leads to only partial melting of the initial amorphous layer. Measurement of this melt threshold allows computer modeling to predict a reduced melting temperature for ion-implanted amorphous Si of about 1170 K (Baeri *et al.*, 1980a). Also, modeling of dopant diffusion processes indicates a reduced melting enthalpy for amorphous Si of about 1220 J g⁻¹.

Since annealing in the present case takes place by transient melting, ripples are generally produced on the specimen surface. A typical example of a ripple array is shown in Fig. 28, and it has already been noted that the appearance of these features is similar to that of ripples produced by Q-switched laser annealing using a diffused beam (see Section II.A.1).

When pulsed electron beam annealing is applied to the fabrication of Si electronic device structures, efficient activation of ion-implanted dopants is generally observed. However, dislocation formation can occur in the annealed regions under certain conditions (Leas *et al.*, 1980), and this, together with anomalous dopant redistribution phenomena and pulse nonuniformity problems, requires further study.

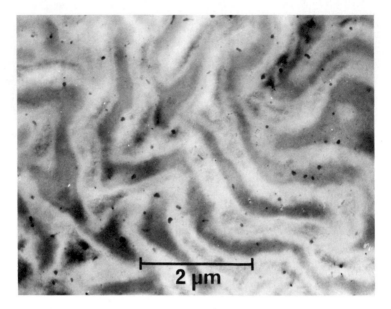

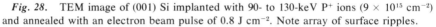

Fig. 28. TEM image of (001) Si implanted with 90- to 130-keV P$^+$ ions (9 × 10^{15} cm^{-2}) and annealed with an electron beam pulse of 0.8 J cm^{-2}. Note array of surface ripples.

Amorphous layers in ion-implanted GaAs also can be structurally recrystallized using the pulsed electron beam technique. Stages of annealing represented by polycrystal material and dislocated single crystal have been identified (Tandon *et al.*, 1979). The electron fluence window available for optimum annealing is quite narrow and the use of too high an electron beam fluence leads to preferential loss of As and to the production of deep damage in the crystal. However, even under optimum annealing conditions, activated dopants generally exhibit low carrier mobilities, probably as a result of the incorporation of point defects into the lattice (Davies *et al.*, 1980b). Similar results have been obtained for electron pulse-annealed InP, and here P loss has been identified as a major problem leading to the formation of electrically active surface defects (Davies *et al.*, 1980a; see also Section II.A.4).

2. SCANNED BEAM ANNEALING

The second and particularly important electron annealing method requires that the sample surface be scanned with a cw electron beam. The apparatus can resemble a SEM, although it should be capable of delivering somewhat higher beam currents. The method uses the thermal pulse delivered by the scanning beam to give solid phase recrystallization (McMahon and Ahmed, 1979a; Regolini *et al.*, 1979a), which also ensures the retention of a microscopically flat sample surface. The annealing can be carried out in two different ways. In one approach, the specimen areas to be annealed are subjected to a single scan with the focused beam, and sufficient energy is deposited in the material to produce local full annealing in ≤ 10 msec. This method is directly analogous to the scanning cw laser technique: it gives excellent recrystallization of amorphous Si layers implanted with conventional dopants and, once again, the formation of metastable, supersaturated solid solutions is possible (Regolini *et al.*, 1979b). The method is attractive, since isolated, selected areas on a semiconductor wafer can be annealed and, for example, the direct writing of ~ 1 μm, discrete crystalline lines in ion-implanted Si has been demonstrated (Ratnakumar *et al.*, 1979).

The second scanning electron beam annealing method also has great potential importance and uses rapid, repeated scans of a sample to raise its overall temperature to a level at which recrystallization can take place in times typically from 10^{-1} to 10 sec (McMahon and Ahmed, 1979b). This

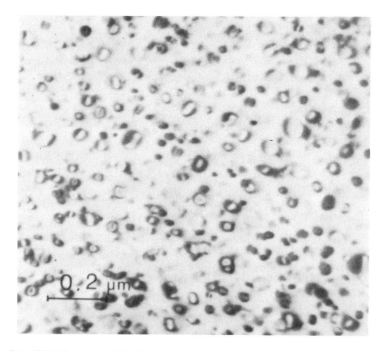

Fig. 29. TEM image of (001) Si implanted with P⁺ ions and annealed with a scanning electron beam. (After Bentini *et al.*, 1980.)

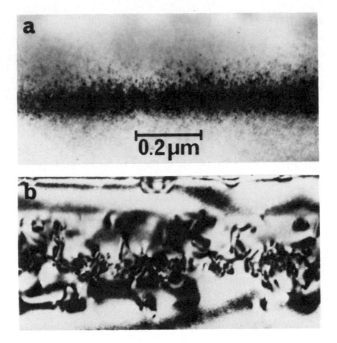

Fig. 30. TEM cross-sectional images of (111) Si implanted with 100-keV B⁺ ions (10^{16} cm^{-2}). (a) As-implanted; (b) multiple-scan electron beam-annealed. (After McMahon *et al.*, 1980.)

multiscanning technique does not require a focused electron beam and effectively provides a "fast furnace" facility. The defect structures in annealed ion-implanted layers can more closely resemble their conventionally annealed counterparts and, for example, an array of dislocation loops formed by a multiscanning anneal of amorphous P⁺ ion-implanted Si (Bentini *et al.*, 1980) is shown in Fig. 29. When the initial ion implantation damage comprises only a buried band of point defect clusters, this can evolve into a tangled network of dislocations (McMahon *et al.*, 1980), as illustrated in Fig. 30 for B⁺ ion-implanted Si. In each case excellent electrical activation of the dopant is measured, demonstrating the potential usefulness of the technique to the device manufacturer.

III. Annealing of Metal Films on Semiconductors

The interaction of metals with semiconductors is of considerable technological importance for the formation of electrical contacts to electronic devices. While conventional processing employs furnace annealing treatments, the new pulsed annealing methods present opportunities either to improve contact structures presently in use or to produce novel metal–semiconductor structures with special properties (see Chapter 12 by von Allmen and Lau).

A. Rapid-Pulse Annealing

The alloying of a metal film to an underlying semiconductor by a very short pulse anneal takes place by surface melting, followed by interdiffusion of the components and resolidification with the possibility of the formation of discrete compound phases. This is illustrated by the work of Poate *et al.* (1978), in which Pt, Pd, and Ni films were reacted with (001) Si using Q-switched Nd:YAG laser pulses. As shown in Fig. 31, a 450-Å Pt film on Si when annealed yields a cell structure as a result of the effects of

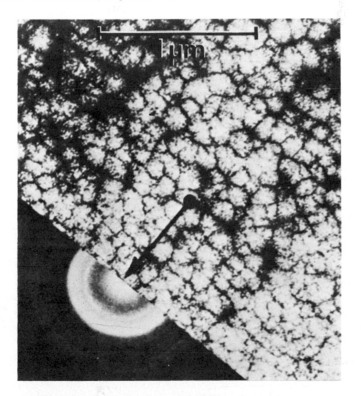

Fig. 31. TEM image and diffraction pattern from Pt film on (001) Si annealed with Nd:YAG laser radiation at 18 MW cm^{-2}. (After Poate *et al.*, 1978.)

melt supercooling (see Section II.A.2). The cell walls within the recrystallized layer in this case contain predominantly PtSi, although other silicides such as Pt_2Si, $Pt_{12}Si_5$, and Pt_3Si (and amorphous phases) can be produced, often in a complex mixture (Wittmer and von Allmen, 1979). Indeed, a metastable phase of approximately $PtSi_2$ stoichiometry has been produced by millisecond pulse irradiations (Conti et al., 1980). Alloyed PtSi layers can yield ohmic or Schottky barrier contacts depending upon the p- or n-type character of the Si, respectively. When Pt is laser-annealed into As^+ ion-implanted p-type Si, an ohmic contact with the $p-n$ junction is formed (Doherty et al., 1980).

The alloying of other metal layers to Si also produces a cellular segregation structure similar to that described above. For example, the result of Nd:YAG laser annealing of Co on Si (van Gurp et al., 1979) is shown in Fig. 32. Here the cell walls are composed of $CoSi_2$, and this is clearly demonstrated in dark-field TEM images. The work by van Gurp et al. (1979) and by Possin et al. (1981) has also shown that convective flow, especially in small-diameter melt pools, can lead to metal segregation on a somewhat larger cellular scale.

Particularly novel results can be obtained when some metal–Si couples are laser irradiated. Rapid melting and solidification of Au–Si thin films leads to the formation of a new amorphous phase (von Allmen et al., 1980a) which is markedly metastable and slowly decomposes even at room temperature. In contrast, when W layers on Si are Q-switched laser-annealed, thin needles of WSi_2 are formed (von Allmen et al., 1980b). However, a spherical cavity is attached to the base of each needle, and it is thought that such cavities form as vapor bubbles in the transiently superheated molten Si.

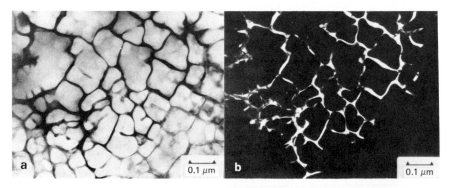

Fig. 32.   TEM images of an initial 250-Å Co film on (001) Si after a Nd:YAG laser anneal at 40 MW cm^{-2}. (a) Bright field; (b) dark field in $CoSi_2$ 200 reflection. (After Stacy et al., 1980.) This figure was originally presented at the Spring 1980 Meeting of The Electrochemical Society, Inc., held in St. Louis, Missouri.

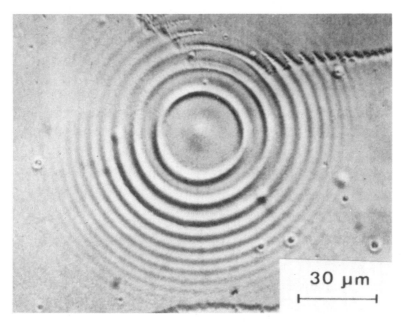

Fig. 33. Optical micrograph (Nomarski interference) of a circular fringe pattern on the surface of Pd-coated Si after Nd:YAG laser irradiation. (After Affolter *et al.*, 1980.)

Q-switched laser processing also can be used to give good electrical contacts with group III–group V compound semiconductors. Barnes *et al.* (1978) demonstrated that low-resistance contact could be made to Te$^+$ ion-implanted and laser-annealed GaAs after stripping off a near-surface layer denuded of dopant. It is also possible to laser-alloy Ge into n-type GaAs to form excellent, uniform ohmic contacts (Badertscher *et al.*, 1980). This technology is likely to be important for device fabrication.

It is also interesting to note that unusual ripple effects can sometimes occur on the surfaces of metal-coated, laser-annealed samples. This is illustrated in Fig. 33 which shows the circular ripple patterns that can be observed on annealed Pd–Si layers (Affolter *et al.*, 1980). It is thought that such patterns occur by diffraction scattering of the laser light by particles suspended in the atmosphere in front of the specimen. The resulting radiation intensity fluctuations on the specimen surface would then give a periodic modulation in the resulting alloying.

B. Scanned Beam Annealing

Scanning cw laser annealing can also be used to induce a reaction between a deposited metal film and its semiconductor substrate. For metals with high melting points the interaction is generally by a solid phase mechanism. In contrast to the Q-switched laser annealing case, large-area, uniform, essentially single-phase silicide layers can be produced. For example, films of Pd on Si yield single-phase Pd_2Si or PdSi depending upon the laser power used for annealing (Shibata *et al.*, 1980). Not all films behave in this way, however, Pt–Si films giving an intimate mixture of phases after the cw laser treatment.

There is also considerable promise in the application of this annealing technique to the fabrication of contacts on Group III–Group V compound semiconductors. Conventional furnace-annealed contacts are generally formed using a mixed alloy layer on the semiconductor: for example, In–Au–Ge on GaAs. However, the furnace heat treatment often leads to complex interdiffusion and phase separation processes which are not well understood. In particular, contact morphology and structural uniformity are usually poor. The scanning cw laser anneal method offers much better control of the heating and cooling cycle (in addition to its selected area annealing capability). Studies of contacts to GaAs have shown (Eckhardt *et al.*, 1980) that laser-annealed contacts possess considerably finer texture and far superior compositional uniformity when compared to their furnace-annealed counterparts.

In addition to the above possibilities for advances in contact processing, it is important to note that a scanning electron beam can be substituted for the scanning laser in many cases (Sigmon *et al.*, 1980). This promises added flexibility for device production and points to the particular advantages of the adoption of pulsed annealing for the best control of metal–Si interactions.

IV.   Annealing of Other Semiconductor Layers

A.   *Deposited Amorphous Films*

Since Q-switched laser annealing takes place by surface melting, in principle it is possible to recrystallize epitaxially a disordered semiconductor surface layer irrespective of its mode of formation, since its initial structural characteristics should be eliminated in the transient liquid. Thus, it has been shown (Bean *et al.*, 1978; Lau *et al.*, 1978) that laser annealing can give excellent recrystallization of deposited amorphous Si layers on single-crystal Si substrates. This is illustrated in Fig. 34 which shows the boundary of an annealed region. On the right-hand side the melt has penetrated completely through the deposited layer and single-crystal

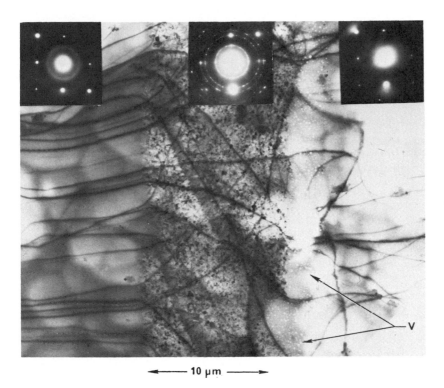

Fig. 34. TEM image of scanning Q-switched Nd:YAG laser-annealed deposited Si film (initially amorphous) on (001) Si. Inset diffraction patterns relate to adjacent areas in the image. (After Bean *et al.*, 1978.)

material has been produced. Partial deposited layer melting results in polycrystal formation (the central region) but, between the two, a narrow band of recrystallized material contains an array of voids. These evolve from free volume initially incorporated into the As-deposited film (Leamy *et al.*, 1980) and may be eliminated by prolonged or repeated surface melting. Once again, it is possible to incorporate metastable high concentrations of dopant elements in the recrystallized layer (Bean *et al.*, 1979), and excellent $p-n$ junctions can be formed (Young *et al.*, 1979). Similar structural recrystallization of deposited Ge layers has also been demonstrated (Golecki *et al.*, 1979; Andrew and Lovato, 1979).

When scanning cw laser annealing is employed, novel film growth phenomena are sometimes observed. An important example is the "explosive recrystallization" found to occur when amorphous Ge films are annealed on amorphous substrates such as fused silica (Fan *et al.*, 1980). Local heating produced by the scanned beam induces the amorphous-to-crystal transformation which is then greatly accelerated by liberation of the corresponding latent heat. Lateral film growth occurs explosively [at ~2 msec^{-1} (Chapman *et al.*, 1980)] in a periodic manner, each burst terminating in a relatively cool region outside the irradiated area. The recrystallized regions consist mainly of large, aligned polycrystals, although amorphous and graded polycrystalline material is also present (see Fig. 35). The explosive nature of the recrystallization, which is initiated in the hottest local regions of the scanned track, is particularly well illustrated in Fig. 36. The bursts of crystal growth give the characteristic overlapping disk structure (Gold *et al.*, 1980), and it is thought that each growth front propagates via a narrow, moving molten zone (Gilmer and Leamy, 1980).

A second novel crystal growth method for which scanning cw laser annealing is important is termed graphoepitaxy. While small, isolated deposited islands can be recrystallized as single crystals (Gibbons *et al.*, 1979), graphoepitaxy involves the recrystallization of large areas of

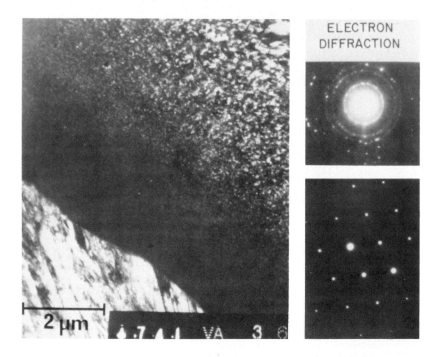

Fig. 35. TEM image of a scanning cw laser-recrystallized deposited Ge film on fused silica, illustrating different microstructural regions. On the right-hand side are transmission diffraction patterns for the fine- and large-grained regions (upper and lower, respectively). (After Fan *et al.*, 1980.)

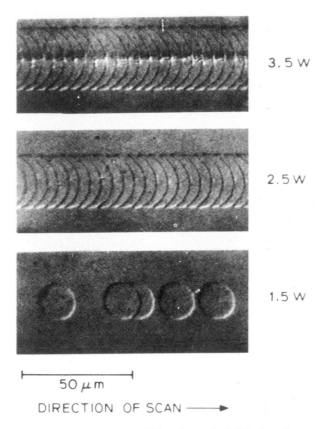

3.5 W

2.5 W

1.5 W

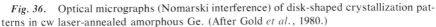

50 μm

DIRECTION OF SCAN ⟶

Fig. 36. Optical micrographs (Nomarski interference) of disk-shaped crystallization patterns in cw laser-annealed amorphous Ge. (After Gold *et al.*, 1980.)

amorphous Si deposited on an amorphous substrate with its surface artificially configured as a fine relief grating (Geis *et al.*, 1979). The dimensions of the grating must be carefully controlled, but annealing can produce sheets of single-crystal material (Fig. 37), the quality improving after repeated laser scans. The ⟨001⟩ directions in the film are generally parallel to the grating lines and perpendicular to the substrate plane, although there can be some orientation rotation from area to area. More recently, it has been found that film misorientations can be suppressed if Si graphoepitaxy is carried out by heating for a few tens of seconds in a strip-heating oven (Geis *et al.*, 1980). Indeed, this method shows particular promise for the production of device-quality Si films on insulating substrates.

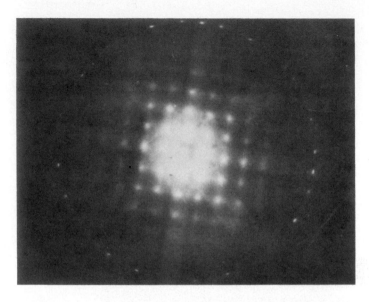

Fig. 37. Transmission electron diffraction pattern from deposited Si film on 3.8-μm silica grating after scanning cw laser anneal. (After Geis *et al.*, 1979.)

B. Deposited Polycrystalline Films

Polycrystalline films are of considerable importance for integrated circuit fabrication, since they are often used for FET gates and to provide interconnections between individual devices. However, dopants usually exhibit relatively low activation in such films, so that interconnect resistance is often higher than is desirable. Therefore, various types of pulsed annealing have been investigated to determine their effect upon the structural and electrical properties of these films. The results indicate that important improvements can be obtained.

When B^+, P^+, or As^+ ion-implanted polycrystalline Si films are subjected to Q-switched laser annealing, the grain size can increase, and essentially complete activation of the dopant can be achieved (Wu and Magee, 1979) with only moderate energy density irradiations. This yields substantial resistivity reductions in comparison to conventionally annealed samples. However, subsequent furnace annealing can increase the resistivity of P-doped layers, possibly because of grain boundary segregation effects (Shibata *et al.*, 1979). When polycrystalline Si is deposited on a single-crystal Si substrate, if Q-switched laser annealing produces a melt that penetrates completely through the layer, the latter can also regrow as single-crystal material (Tamura *et al.*, 1980). The progressive change in layer crystallinity is illustrated in Fig. 38. By use of this method, polycrystalline Si deposited on an oxidized wafer near an oxide window edge can also be converted to single-crystal material by the propagation of orientation information from resolidified Si within the window itself. This repre-

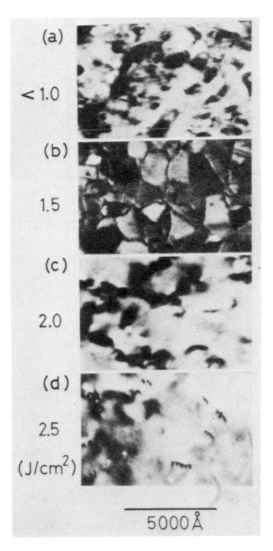

Fig. 38. TEM images showing structural changes in ruby laser-annealed 4000-Å polycrystalline Si film on (001) Si as a function of irradiation energy density. (After Tamura *et al.*, 1980.)

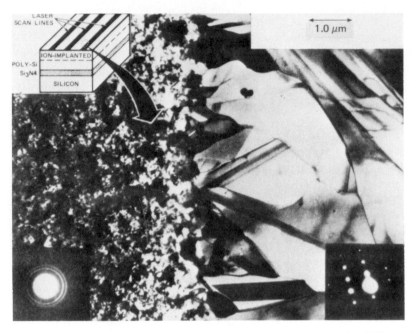

Fig. 39. TEM image of ion-implanted and cw laser-annealed polycrystalline Si film at the boundary of the laser scan line. Insets show diffraction patterns characteristic of each region. (After Gat *et al.*, 1978a.)

sents a further technique by which at least limited areas of single-crystal Si can be grown on an insulating supporting layer.

Scanning cw laser annealing can also be used to recrystallize ion-implanted polycrystalline Si layers. An example of this application is shown in Fig. 39 for B^+ ion-implanted polycrystalline Si deposited on a Si_3N_4 film (Gat *et al.*, 1978a). In this case a very large increase in grain size from ~500 Å to ~25 × 2 μm is observed. The final asymmetry of the grains is due to extension of their long axes in the direction of the laser scan. Dopant activation is also extremely efficient, so that the process could prove to be attractive to the device engineer.

C. *Deposited Films on Sapphire*

Single-crystal Si on sapphire has become an increasingly important material for electronic device fabrication. However, many defects (principally microtwins and stacking faults) are present in as-grown films and originate as a result of mismatch of the lattices at the heterojunction.

Therefore, it is important to devise a process that can reduce the defect density and so improve the performance of devices produced in the layer. Indeed, it has been found (Roulet *et al.*, 1979; Lüthy *et al.*, 1979) that Q-switched laser annealing can enhance the layer crystal quality, as measured by ion-backscattering work, with best results if the transient melt penetrates completely through to the interface. Also, the initial compressive strain state of the Si layer can be substantially modified (Sai-Halasz *et al.*, 1980).

The way in which Q-switched laser annealing can change the defect structure of Si on sapphire is well demonstrated by TEM studies on cross-sectional specimens. The range of structure transitions that occur with increasing ruby laser radiation energy density in As$^+$ ion-implanted material is shown (Hill *et al.*, 1981; Cullis *et al.*, 1981c) in Fig. 40. The initial surface amorphous Si layer recrystallizes at a markedly lower radiation energy density than for homoepitaxial Si, this being due partly to the relatively low thermal conductivity of the sapphire substrate and partly to reflection of radiation at internal boundaries. It is also clear that micro-twins initially present in the Si often reform and penetrate through the laser-recrystallized layer when it is narrow. However, for a radiation energy density of ~0.8 J cm^{-2} the Si melt penetrates more than 3000 Å into the epitaxial layer, with the formation of a greatly increased density of dislocations in the recrystallized material. Many of the dislocations originate at the tips of initial microtwins which terminate within the layer. The melt penetrates completely through the epitaxial layer for irradiation at 1.1 J cm^{-2}. This leads to a transformation of the whole layer structure

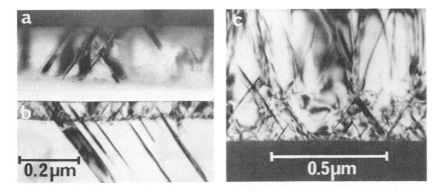

Fig. 40.   TEM cross-sectional images of 0.6 μm (001) Si on sapphire implanted with 40-keV As$^+$ ions (10^{16} cm^{-2}) and ruby laser-annealed. (a) As-implanted; (b) 0.4 J cm^{-2} showing partial recrystallization of implant; (c) 0.8 J cm^{-2} showing newly formed dislocations. (After Cullis *et al.*, 1981c.)

from an initial situation where microtwins dominate to a final condition with a high density of dislocations. The heteroepitaxial interface is the source of many of these defects, undoubtedly because of lattice irregularity at this location.

Important changes in surface morphology also take place as a consequence of Q-switched laser annealing. This is illustrated in Fig. 41, which demonstrates the edge rounding that can be achieved for Si islands on sapphire (Wu and Schnable, 1979). This assists subsequent oxide or metal coverage and leads to device yield improvement.

Scanning cw laser annealing can also be used to modify the structure of Si on sapphire. If carried out in the solid phase regrowth regime, defect densities tend to be increased (Golecki *et al.*, 1980). However, liquid phase processing reduces the defect concentration, although the final Si is autodoped and has a high Al content.

D. Diffused Layers

When dopant elements are diffused into Si by conventional furnace heat treatments, substantial densities of extended lattice defects are also often introduced. These may occur as a result of the lattice strain present if the dopant and host atomic radii are different. In addition, point defect fluxes accompanying the diffusion play a role in extended defect formation, and precipitates of the dopant may occur if the latter is present in excess of local solubility levels. However, Q-switched laser annealing can remove these diffusion-induced imperfections (Young and Narayan, 1978). As shown in Fig. 42, a sample diffused with P from a PH_3 source for 1 hr at 1100°C contains many small dislocation loops and precipitates.

Fig. 41.   SEM images of Si islands. (a) Initial island shape; (b) structure after a ruby laser anneal at 15 MW cm^{-2}. Island edge rounding is evident. (After Wu and Schnable, 1979.)

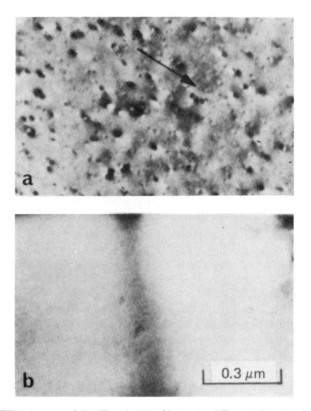

Fig. 42. TEM images of P-diffused (001) Si. (a) As-diffused; (b) ruby laser-annealed. (After Young and Narayan, 1978.)

However, ruby laser annealing at about 1.5 J cm^{-2} removes these defects by dissolution in the transient melt. Indeed, experiments of this type have been used to measure the depth of melting (Narayan, 1979) by determinations of the exact thickness of the layer in which defect removal occurs.

In some cases the need for conventional diffusion processing may itself be eliminated by the advent of laser-based methods. For example, if B is deposited on an Si wafer surface, it can be diffused in by use of Q-switched laser irradiation (Narayan *et al.*, 1978b). This is illustrated in Fig. 43 which shows that after laser annealing few crystallographic defects remain in the diffused region. However, as outlined previously, the Si can be greatly supersaturated with B dopant after this treatment. Thus, as shown in Fig. 43c, subsequent furnace heat treatment can lead to precipitation of the excess impurity.

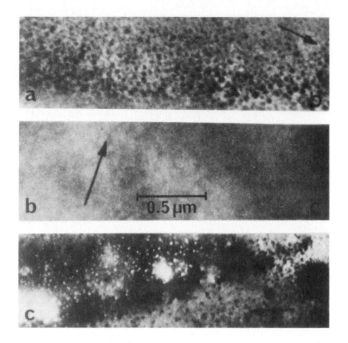

Fig. 43. TEM images of B-diffused (001) Si sample. (a) As-deposited sample, showing B islands on surface; (b) ruby laser-annealed at 1.6 J cm⁻², (c) after subsequent furnace annealing at 900°C for 40 min. (After Narayan *et al.*, 1978b.)

E. As-Grown Single-Crystal Material

In almost all the work described above we have been concerned with the various techniques of pulsed annealing as methods for the removal of lattice damage and the modification of dopant concentrations. However, in Section II.A.1, the way in which point defects could be introduced into a crystal matrix by nanosecond regime laser annealing was described. An extension of these ideas shows that the extremely rapid cooling that can be achieved by use, for example, of controlled picosecond regime laser pulses should lead to an increased final defect density. In practice, it is even possible to produce thin surface amorphous layers in Si by this means (Liu *et al.*, 1979), although the irradiation power density is a very critical parameter since the energy input must lead to the minimum of surface melting to achieve the highest quench rates. Nanosecond regime uv lasers have also been shown to produce similar effects at Si surfaces (Tsu *et al.*, 1979). A particularly detailed study of surface amorphization has been carried out (Cullis *et al.*, 1982a,b) by use of 2.5-nsec pulses of

347-nm radiation. In this work both (001) and (111) Si crystals gave thin (and initially discontinuous) amorphous layers just above the threshold for surface melting (a little below a 0.2 J cm^{-2} at 347 nm—see upper micrographs of Figs. 44a and b). The quench rate was highest under these conditions with calculated 1412°C isotherm peak velocities of $\geqslant$20m sec^{-1}. With increasing radiation energy densities deeper melting occurred, and, at first, a thicker amorphous layer was formed (center micrographs of Figs. 44a and b). However, the quench rate also fell and amorphization ceased

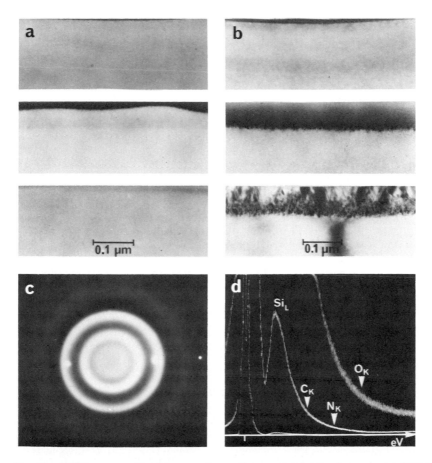

Fig. 44. TEM cross-sectional images of amorphous and defective layers on (a) (001) Si annealed at 0.2 J/cm² (upper), 0.27 J/cm² (center) and 0.4 J/cm² (lower); (b) (111) Si annealed at 0.2 J/cm² (upper), 0.5 J/cm² (center) and 0.6 J/cm² (lower) using 2.5-nsec pulses of 347-nm radiation. (c) Transmission electron diffraction pattern from amorphous Si; (d) transmission electron energy loss spectrum showing absence of light element K-loss peaks (positions indicated).

to occur with 1412°C isotherm peak velocities of $\sim$18 m sec^{-1} (0.3 J cm^{-2}) for (001) Si and $\sim$15 m sec^{-1} (0.55 J cm^{-2}) for (111) Si. With quench rates below these thresholds, while good single crystal was formed on (001) Si, heavily defective (twinned and faulted) Si was produced on (111) Si for 1412°C isotherm velocities down to $\sim$6 m sec^{-1} (illustrated in lower micrographs of Figs. 44a and b). Thus (111) Si shows a greater fundamental tendency for defect formation than (001) Si, this being a probable consequence of the relative difficulty of plane nucleation (leading to increased melt undercooling) and the ease of twin formation on {111} surfaces. These observations, with the orientation-dependent thresholds, appear to be generally consistent with the theory of Jackson (1982) which shows that the recrystallization interface velocity is an orientation-dependent peaked function of melt undercooling (see also Gilmer and Leamy, 1980). The amorphous Si layers which are produced at the higher quench rates show the expected characteristic diffuse scattering bands in transmission electron diffraction patterns (Fig. 44c) signifying the absence of long-range order. They are also substantially free of light-element impurities, as demonstrated by electron energy loss analysis (Fig. 44d) and in agreement with the observations of Liu *et al.* (1981b). Nevertheless, under some conditions, impurities already present (in the ambient and as oxide, etc.) may play a role in stabilizing disordered phases (see, e.g., Liu *et al.*, 1981a). Indeed, this area of high speed crystal growth is currently the subject of vigorous study, and it is not unreasonable to expect many additional exciting and important developments.

Acknowledgments

The author is most grateful for helpful discussions with colleagues over an extended period of time. Particular thanks are due to Dr. J. M. Poate (Bell Laboratories), Prof. G. Foti and Dr. P. Baeri (University of Catania), Dr. D. T. J. Hurle, Mr. H. C. Webber and Mr. N. G. Chew (RSRE).

References

Affolter, K., Lüthy, W., and Wittmer, M. (1980). *Appl. Phys. Lett.* **36**, 559.

Andrew, R., and Lovato, M. (1979). *J. Appl. Phys.* **50**, 1142.

Auston, D. H., Golovchenko, J. A., Smith, P. R., Surko, C. M., and Venkatesan, T. N. C. (1978a). *Appl. Phys. Lett.* **33**, 539.

Auston, D. H., Surko, C. M., Venkatesan, T. N. C., Slusher, R. E., and Golovchenko, J. A. (1978b). *Appl. Phys. Lett.* **33**, 437.

Badertscher, G., Salathe, R. P., and Lüthy, W. (1980). *Electron. Lett.* **16**, 113.

Baeri, P., Campisano, S. U., Foti, G., and Rimini, E. (1978). *Phys. Rev. Lett.* **41**, 1246.

Baeri, P., Foti, G., Poate, J. M., and Cullis, A. G. (1980a). *Phys. Rev. Lett.* **46**, 2036.

Baeri, P., Poate, J. M., Campisano, S. U., Foti, G., Rimini, E., and Cullis, A. G. (1980b). *Appl. Phys. Lett.* **37**, 912.

Bagley, B. G., and Chen, H. S. (1979). *In* "Laser–Solid Interactions and Laser Processing" (S. D. Ferris, H. J. Leamy, and J. M. Poate, eds.), p. 97. Am. Inst. Phys., New York.

Bardsley, W., Boulton, J. S., and Hurle, D. T. J. (1962). *Solid-State Electron.* **5**, 395.

Barnes, P. A., Leamy, H. J., Poate, J. M., Ferris, S. D., Williams, J. S., and Celler, G. K. (1978). *Appl. Phys. Lett.* **33**, 965.

Bean, J. C., Leamy, H. J., Poate, J. M., Rozgonyi, G. A., Sheng, T. T., Williams, J. S., and Celler, G. K. (1978). *Appl. Phys. Lett.* **33**, 227.

Bean, J. C., Leamy, H. J., Poate, J. M., Rozgonyi, G. A. van der Ziel, J. P., Williams, J. S., and Celler, G. K. (1979). *J. Appl. Phys.* **50**, 881.

Bentini, G. G., Galloni, R., and Nipoti, R. (1980). *Appl. Phys. Lett.* **36**, 661.

Bertolotti, M., Vitali, G., Rimini, E., and Foti, G. (1979). *J. Appl. Phys.* **50**, 259.

Bicknell, R. W., and Allen, R. M. (1970). *Radiat. Eff.* **6**, 45.

Campisano, S. U., Foti, G., Rimini, E., Eisen, F. H., Tseng, W. F., Nicolet, M.-A., and Tandon, J. L. (1980a). *J. Appl. Phys.* **51**, 295.

Campisano, S. U., Foti, G., and Servidori, M. (1980b). *Appl. Phys. Lett.* **36**, 279.

Celler, G. K., Poate, J. M., and Kimerling, L. C. (1978). *Appl. Phys. Lett.* **32**, 464.

Chapman, R. L., Fan, J. C. C., Ziegler, H. J., and Gale, R. P. (1980). *Appl. Phys. Lett.* **37**, 292.

Chiang, S. W., Liu, Y. S., and Reihl, R. R. (1981). *In* "Laser and Electron Beam Solid Interactions and Materials Processing" (J. F. Gibbons, L. D. Hess, and T. W. Sigmon, eds.), p. 407. North-Holland Publ., New York.

Compaan, A., and Lo, H. W. (1980). *In* "Laser and Electron Beam Processing of Materials" (C. W. White and P. S. Peercy, eds.), p. 71. Academic Press, New York.

Conti, C. A., Doherty, C. J., Chiu, K. C. R., Sheng, T. T., and Leamy, H. J. (1980). *In* "Laser and Electron Beam Processing of Materials" (C. W. White and P. S. Peercy, eds.), p. 537. Academic Press, New York.

Coriell, S. R., and Sekerka, R. F. (1979). *In* "Rapid Solidification Principles and Technologies II" (R. Mehrabian, B. H. Kear, and M. Cohen, eds.), p. 35. Claitor's Publ. Div., Baton Rouge, Louisiana.

Cullis, A. G. (1980). *Proc.—Annu. Meet., Electron Microsc. Soc. Am.* **38**, 294.

Cullis, A. G., and Webber, H. C. (1980). *In* "Laser and Electron Beam Processing of Electronic Materials" (C. L. Anderson, G. K. Celler, and G. A. Rozgonyi, eds.), p. 220. Electrochem. Soc., Pennington, New Jersey.

Cullis, A. G., Seidel, T. E., and Meek, R. L. (1978). *J. Appl. Phys.* **49**, 5188.

Cullis, A. G., Poate, J. M., and Celler, G. K. (1979a). *In* "Laser–Solid Interactions and Laser Processing" (S. D. Ferris, H. J. Leamy, and J. M. Poate, eds.), p. 311. Am. Inst. Phys., New York.

Cullis, A. G., Webber, H. C., and Bailey, P. (1979b). *J. Phys. E* **12**, 688.

Cullis, A. G., Webber, H. C., and Robertson, D. S. (1979c). *In* "Laser–Solid Interactions and Laser Processing" (S. D. Ferris, H. J. Leamy, and J. M. Poate, eds.), p. 653. Am. Inst. Phys., New York.

Cullis, A. G., Webber, H. C., and Chew, N. G. (1980a). *Appl. Phys. Lett.* **36**, 547.

Cullis, A. G., Webber, H. C., Poate, J. M., and Chew, N. G. (1980b). *J. Microsc. (Oxford)* **118**, 41.

Cullis, A. G., Webber, H. C., Poate, J. M., and Simons, A. L. (1980c). *Appl. Phys. Lett.* **36**, 320.

Cullis, A. G., Hurle, D. T. J., Webber, H. C., Chew, N. G., Poate, J. M., Baeri, P., and Foti, G. (1981a). *Appl. Phys. Lett.* **38**, 642.

Cullis, A. G., Series, R. W., Webber, H. C., and Chew, N. G. (1981b). *In* "Semiconductor Silicon 1981" (H. R. Huff, R. J. Kriegler, and Y. Takeishi, eds.), p. 518. Electrochem. Soc., Pennington, New Jersey.

Cullis, A. G., Webber, H. C., Chew, N. G., Hill, C., and Godfrey, D. J. (1981c). *In* "Microscopy of Semiconducting Materials 1981" (A. G. Cullis and D. C. Joy, eds.), p. 95. Inst. Phys., London.

Cullis, A. G., Webber, H. C., and Chew, N. G. (1982a). *In* "Laser and Electron-Beam Interactions with Materials" (B. R. Appleton and C. K. Celler, eds.), p. 131. North-Holland Publ., New York.

Cullis, A. G., Webber, H. C., Chew, N. G., Poate, J. M., and Baeri, P. (1982b). *Phys. Rev. Lett.* **49**, 219.

Davies, D. E., Kennedy, E. F., Comer, J. J., and Lorenzo, J. P. (1980a). *Appl. Phys. Lett.* **36**, 922.

Davies, D. E., Lorenzo, J. P., and Ryan, T. G. (1980b). *Appl. Phys. Lett.* **37**, 612.

Doherty, C. J., Seidel, T. E., Leamy, H. J., and Celler, G. K. (1980). *J. Appl. Phys.* **51**, 2718.

Eckhardt, G., Anderson, C. L., Colborn, M. N., Hess, L. D., and Jullens, R. A. (1980). *In* "Laser and Electron Beam Processing of Electronic Materials" (C. L. Anderson, G. K. Celler, and G. A. Rozgonyi, eds.), p. 445. Electrochem. Soc., Pennington, New Jersey.

Eisen, F. H., and Mayer, J. W. (1976). *In* "Treatise on Solid State Chemistry" (N. B. Hannay, ed.), Vol. 6, p. 125. Plenum, New York.

Fan, J. C. C., Ziegler, H. J., Gale, R. P., and Chapman, R. L. (1980). *Appl. Phys. Lett.* **36**, 158.

Foti, G., Della Mea, G., Jannitti, E., and Majni, G. (1978a). *Phys. Lett. A* **68A**, 368.

Foti, G., Rimini, E., Tseng, W. F., and Mayer, J. W. (1978b). *Appl. Phys.* **15**, 365.

Foti, G., Campisano, S. U., Baeri, P., Rimini, E., and Tseng, W. F. (1979). *Appl. Phys. Lett.* **35**, 701.

Gat, A., and Gibbons, J. F. (1978). *Appl. Phys. Lett.* **32**, 142.

Gat, A., Gerzberg, L., Gibbons, J. F., Magee, T. J., Peng, J., and Hong, J. D. (1978a). *Appl. Phys. Lett.* **33**, 775.

Gat, A., Gibbons, J. F., Magee, T. J., Peng, J., Deline, V. R., Williams, P., and Evans, C. A., Jr. (1978b). *Appl. Phys. Lett.* **32**, 276.

Gat, A., Gibbons, J. F., Magee, T. J., Peng, J., Williams, P., Deline, V., and Evans, C. A., Jr. (1978c). *Appl. Phys. Lett.* **33**, 389.

Geis, M. W., Flanders, D. C., and Smith, H. I. (1979). *Appl. Phys. Lett.* **35**, 71.

Geis, M. W., Antoniadis, D. A., Silversmith, D. J., Mountain, R. W., and Smith, H. I. (1980). *Appl. Phys. Lett.* **37**, 454.

Gibbons, J. F., Lee, K. F., Magee, T. J., Peng, J., and Ormond, R. (1979). *Appl. Phys. Lett.* **34**, 831.

Gibson, J. M., and Tsu, R. (1980). *Appl. Phys. Lett.* **37**, 197.

Gilmer, G. H., and Leamy, H. J. (1980). *In* "Laser and Electron Beam Processing of Materials" (C. W. White and P. S. Peercy, eds.), p. 227. Academic Press, New York.

Gold, R. B., Gibbons, J. F., Magee, T. J., Peng, J., Ormond, R., Deline, V. R., and Evans, C. A., Jr. (1980). *In* "Laser and Electron Beam Processing of Materials" (C. W. White and P. S. Peercy, eds.), p. 221. Academic Press, New York.

Golecki, I., Kennedy, E. F., Lau, S. S., Mayer, J. W., Tseng, W. F., Eckhardt, R. C., and Wagner, R. J. (1979). *Thin Solid Films* **57**, L13.

Golecki, I., Kinoschita, G., Gat, A., and Paine, B. M. (1980). *Appl. Phys. Lett.* **37**, 919.

Greenwald, A. C., Kirkpatrick, A. R., Little, R. G., and Minnucci, J. A. (1979). *J. Appl. Phys.* **50**, 783.

Hill, C., and Godfrey, D. J. (1980). *J. Phys. (Orsay, Fr.)* **41**(C4), 79.

Hill, C., Godfrey, D. J., Cullis, A. G., Webber, H. C., and Chew, N. G. (1981). Unpublished data.

Hofker, W. K., Oosthoek, D. P., Eggermont, G. E. J., Tamminga, Y., and Stacy, W. T. (1979). *Appl. Phys. Lett.* **34**, 690.

Ishida, K., Okabayashi, H., and Yoshida, M. (1980). *Appl. Phys. Lett.* **37**, 175.

Jackson, K. A. (1982). *Proc. NATO Inst. Surf. Modif. Alloying, Trevi, 1981*.

Jackson, K. A., Gilmer, G. H., and Leamy, H. J. (1980). *In* "Laser and Electron Beam Processing of Materials" (C. W. White and P. S. Peercy, eds.), p. 104. Academic Press, New York.

Johnson, N. M., Bartelink, D. J., Moyer, M. D., Gibbons, J. F., Lietoila, A., Ratnakumar, K. N., and Regolini, J. L. (1980). *In* "Laser and Electron Beam Processing of Materials" (C. W. White and P. S. Peercy, eds.), p. 423. Academic Press, New York.

Kimerling, L. C., and Benton, J. L. (1980). *In* "Laser and Electron Beam Processing of Materials" (C. W. White and P. S. Peercy, eds.), p. 385. Academic Press, New York.

Kular, S. S., Sealy, B. J., Badawi, M. H., Stephens, K. G., Sadana, D. K., and Booker, G. R. (1979). *Electron. Lett.* **15**, 413.

Larson, B. C., White, C. W., and Appleton, B. R. (1978). *Appl. Phys. Lett.* **32**, 801.

Lau, S. A., Tseng, W. F., Nicolet, M.-A., Mayer, J. W., Eckardt, R. C., and Wagner, R. J. (1978). *Appl. Phys. Lett.* **33**, 130.

Leamy, H. J., Rozgonyi, G. A., Sheng, T. T., and Celler, G. K. (1978). *Appl. Phys. Lett.* **32**, 535.

Leamy, H. J., Rozgonyi, G. A., Sheng, T. T., and Celler, G. K. (1980). *In* "Laser and Electron Beam Processing of Electronic Materials" (C. L. Anderson, G. K. Celler, and G. A. Rozgonyi, eds.), p. 333. Electrochem. Soc., Pennington, New Jersey.

Leas, J. M., Smith, P. J., Nagarajan, A., and Leighton, A. (1980). *In* "Laser and Electron Beam Processing of Electronic Materials" (C. L. Anderson, G. K. Celler, and G. A. Rozgonyi, eds.), p. 141. Electrochem. Soc., Pennington, New Jersey.

Lietoila, A., Gibbons, J. F., and Sigmon, T. W. (1980). *Appl. Phys. Lett.* **36**, 765.

Liu, P. L., Yen, R., Bloembergen, N., and Hodgson, R. T. (1979). *Appl. Phys. Lett.* **34**, 864.

Liu, Y. S., Chiang, S. W., and Bacon, F. (1981a). *In* "Laser and Electron Beam Solid Interactions and Materials Processing" (J. F. Gibbons, L. D. Hess, and T. W. Sigmon, eds.), p. 117. North-Holland Publ., New York.

Liu, J. M., Yen, R., Donovan, E. P., Bloembergen, N., and Hodgson, R. T. (1981b). *Appl. Phys. Lett.* **38**, 617.

Lüthy, W., Affolter, K., Weber, H. P., Roulet, M. E., Fallavier, M., Thomas, J. P., and Mackowski, J. (1979). *Appl. Phys. Lett.* **35**, 873.

McMahon, R. A., and Ahmed, H. (1979a). *Electron. Lett.* **15**, 45.

McMahon, R. A., and Ahmed, H. (1979b). *J. Vac. Sci. Technol.* **16**, 1840.

McMahon, R. A., Ahmed, H., and Cullis, A. G. (1980). *Appl. Phys. Lett.* **37**, 1016.

Maracas, G. N., Harris, G. L., Lee, C. A., and McFarlane, R. A. (1978). *Appl. Phys. Lett.* **33**, 453.

Matteson, S., Revesz, P., Farkas, G., Gyulai, J., and Sheng, T. T. (1980). *J. Appl. Phys.* **51**, 2625.

Miyao, M., Itoh, K., Tamura, M., Tamura, H., and Tokuyama, T. (1980). *J. Appl. Phys.* **51**, 4139.

Mizuta, M., Sheng, N. H., Merz, J. L., Lietoila, A., Gold, R. B., and Gibbons, J. F. (1980). *Appl. Phys. Lett.* **37**, 154.

Mullins, W. W., and Sekerka, R. F. (1964). *J. Appl. Phys.* **35**, 444.

Narayan, J. (1979). *Appl. Phys. Lett.* **34**, 312.

Narayan, J., and Young, F. W., Jr. (1979). *Appl. Phys. Lett.* **35**, 330.
Narayan, J., Young, R. T., and White, C. W. (1978a). *J. Appl. Phys.* **49**, 3912.
Narayan, J., Young, R. T., Wood, R. F., and Christie, W. H. (1978b). *Appl. Phys. Lett.* **33**, 338.
Poate, J. M., Leamy, H. J., Sheng, T. T., and Celler, G. K. (1978). *Appl. Phys. Lett.* **33**, 918.
Possin, G. E., Parks, H. G., and Chiang, S. W. (1981). *In* "Laser and Electron Beam Solid Interactions and Materials Processing" (J. F. Gibbons, L. D. Hess, and T. W. Sigmon, eds.), p. 73. North-Holland Publ., New York.
Ratnakumar, K. N., Pease, R. F. W., Bartelink, D. J., and Johnson, N. M. (1979). *J. Vac. Sci. Technol.* **16**, 1843.
Regolini, J. L., Gibbons, J. F., Sigmon, T. W., Pease, R. F. W., Magee, T. J., and Peng, J. (1979a). *Appl. Phys. Lett.* **34**, 410.
Regolini, J. L., Sigmon, T. W., and Gibbons, J. F. (1979b). *Appl. Phys. Lett.* **35**, 114.
Roulet, M. E., Schwob, P., Affolter, K., Lüthy, W., von Allmen, M., Fallavier, M., Mackowski, J. M., Nicolet, M.-A., and Thomas, J. P. (1979). *J. Appl. Phys.* **50**, 5536.
Rozgonyi, G. A., Leamy, H. J., Sheng, T. T., and Celler, G. K. (1979). *In* "Laser–Solid Interactions and Laser Processing" (S. D. Ferris, H. J. Leamy, and J. M. Poate, eds.), p. 457. Am. Inst. Phys., New York.
Rozgonyi, G. A., Baumgart, H., Phillipp, F., Uebbing, R., and Oppolzer, H. (1981). *In* "Microscopy of Semiconducting Materials 1981" (A. G. Cullis and D. C. Joy, eds.), p. 85. Inst. Phys., London.
Sadana, D. K., Wilson, M. C., and Booker, G. R. (1980). *J. Microsc. (Oxford)* **118**, 51.
Sai-Halasz, G. A., Fang, F. F., Sedgwick, T. O., and Segmuller, A. (1980). *Appl. Phys. Lett.* **36**, 419.
Schoen, N. C. (1980). *J. Appl. Phys.* **51**, 4747.
Shibata, T., Iizuka, H., Kohyama, S., and Gibbons, J. F. (1979). *Appl. Phys. Lett.* **35**, 21.
Shibata, T., Gibbons, J. F., and Sigmon, T. W. (1980). *Appl. Phys. Lett.* **36**, 566.
Sigmon, T. W., Regolini, J. L., Gibbons, J. F., Lau, S. S., and Mayer, J. W. (1980). *In* "Laser and Electron Beam Processing of Electronic Materials" (C. L. Anderson, G. K. Celler, and G. A. Rozgonyi, eds.), p. 531. Electrochem. Soc., Pennington, New Jersey.
Skolnick, M. S., Cullis, A. G., and Webber, H. C. (1981). *Appl. Phys. Lett.* **38**, 464.
Spaepen, F., and Turnbull, D. (1979). *In* "Laser–Solid Interactions and Laser Processing" (S. D. Ferris, H. J. Leamy, and J. M. Poate, eds.), p. 73. Am. Inst. Phys., New York.
Stacy, W. T., van Gurp, G. J., Eggermont, G. E. J., Tamminga, Y., and Gijsbers, J. R. M. (1980). *In* "Laser and Electron Beam Processing of Electronic Materials" (C. L. Anderson, G. K. Celler, and G. A. Rozgonyi, eds.), p. 504. Electrochem. Soc., Pennington, New Jersey.
Stritzker, B., Pospieszczyk, A., and Tagle, J. A. (1981). *Phys. Rev. Lett.* **47**, 356.
Tamura, M., Tamura, H., and Tokuyama, T. (1980). *Jpn. J. Appl. Phys.* **19**, L23.
Tandon. J. L., Golecki, I., Nicolet, M.-A., and Sadana, D. K., and Washburn, J. (1979). *Appl. Phys. Lett.* **35**, 867.
Tseng, W. F., Mayer, J. W., Campisano, S. U., Foti, G., and Rimini, E. (1978). *Appl. Phys. Lett.* **32**, 824.
Tsu, R., Hodgson, R. T., Tan, T. Y., and Baglin, J. E. (1979). *Phys. Rev. Lett.* **42**, 1356.
Uebbing, R. H., Wagner, P., Baumgart, H., and Queisser, H. J. (1980). *Appl. Phys. Lett.* **37**, 1078.
van Gurp, G. J., Eggermont, G. E. J., Tamminga, Y., Stacy, W. T., and Gijsbers, J. R. M. (1979). *Appl. Phys. Lett.* **35**, 273.
von Allmen, M., Lau, S. S., Mäenpää, M., and Tsaur, B. Y. (1980a). *Appl. Phys. Lett.* **36**, 205.

von Allmen, M., Lau, S. S., Sheng, T. T., and Wittmer, M. (1980b). *In* "Laser and Electron Beam Processing of Materials" (C. W. White and P. S. Peercy, eds.), p. 524. Academic Press, New York.

White, C. W., Narayan, J., Appleton, B. R., and Wilson, S. R. (1979). *J. Appl. Phys.* **50**, 2967.

White, C. W., Wilson, S. R., Appleton, B. R., and Young, F. W., Jr. (1980). *J. Appl. Phys.* **51**, 738.

Williams, J. S., Brown, W. L., Leamy, H. J., Poate, J. M., Rodgers, J. W., Rousseau, D., Rozgonyi, G. A., Shelnutt, J. A., and Sheng, T. T. (1978). *Appl. Phys. Lett.* **33**, 542.

Wittmer, M., and von Allmen, M. (1979). *J. Appl. Phys.* **50**, 4786.

Wu, C. P., and Magee, C. W. (1979). *Appl. Phys. Lett.* **34**, 737.

Wu, C. P., and Schnable, G. L. (1979). *RCA Rev.* **40**, 339.

Young, R. T., and Narayan, J. (1978). *Appl. Phys. Lett.* **33**, 14.

Young, R. T., Narayan, J., and Wood, R. F. (1979). *Appl. Phys. Lett.* **35**, 447.

Chapter 7

Epitaxy by Pulsed Annealing of Ion-Implanted Silicon

GAETANO FOTI and EMANUELE RIMINI

Istituto di Struttura della Materia
Università di Catania
Catania, Italy

I. Introduction

Ion implantation, as employed in modern semiconductor technology, consists of the introduction of energetic charged particles into a substrate to change the electrical properties of the target. It offers several advantages over classical diffusion including precise control of the dose over the implanted area and good uniformity. In addition, depth and concentration profiles can be varied independently by changing the beam energy and the dose, respectively (Mayer *et al.*, 1970).

A major problem with ion implantation is the creation of damage during the slowing down of the projectile in the target (Lee and Mayer, 1974). Violent collisions of projectiles with the target atoms can displace them

203

from their equilibrium lattice sites. If the knocked-on atom has enough energy, it can create other displacements, giving rise to a collision cascade process. The amount and type of disorder depend on several factors: energy and mass of the projectile, temperature and orientation of the target with respect to the incident beam, and mass of the target atoms. Under suitable conditions the near surface layer of the implanted region can be amorphized. In silicon targets at a temperature lower than or equal to room temperature, implants with heavy ions ($M > 30$ amu) and high fluences ($>10^{14}$ ions/cm²) produce amorphous layers extending to a depth comparable to the projectile's range.

Implants with light ions ($M < 30$) in a Si substrate at room temperature or higher create isolated damage in the form of extended defects. Heavy ions at a low fluence ($<10^{13}$/cm²) produce isolated amorphous regions. At very high fluences ($>10^{16}$/cm²) or with high-dose-rate implants the resulting disorder can take the form of amorphous regions with deep, isolated damage. Partial annealing can occur during such hot implants because of the heating of the substrate. This is only an approximate description of the types of radiation damage, which are schematically shown in Fig. 1a–c.

The lattice damage must be removed to activate electrically the im-

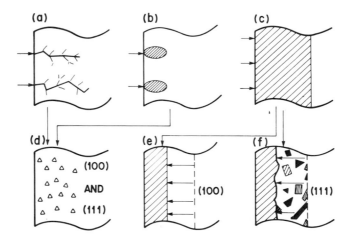

Fig. 1. Schematic of the main types of damage created by ion implantation (a–c) and of the residual disorder after furnace annealing (d–f). (a) Extended defects associated with implants of high-energy light ions, (b) isolated amorphous regions for low-fluence implants of heavy ions, (c) amorphous layer produced by the overlap of isolated damaged regions. The amorphous–single crystal interface is sharp for liquid nitrogen temperature implants. After furnace annealing, extended defects remain in cases (a) and (b) up to 1000°C for ½ hr (d). The amorphous layer regrows epitaxial and free of defects for Si (100) substrates (e) but with twins for Si (111) substrates (f). The epitaxial growth occurs at 550°C in 1 hr.

planted atoms, and the crucial step indeed concerns the annealing process (Pickar, 1975). So far furnaces have been used exclusively for the recovery of crystallinity. In a partially damaged layer the concentration of vacancies available at a temperature higher than 800°C makes defect annihilation possible. For amorphous layers, the system reorders in solid phase epitaxial growth (Crowder, 1971), and defect-free regions result at least for Si ⟨100⟩ substrates (Csepregi *et al.*, 1975). The amorphous–single crystal interface moves with a uniform velocity toward the surface, and growth occurs layer by layer in the temperature range 500–700°C for Si. The process is thermally activated.

On Si ⟨111⟩ substrates the growth velocity at a fixed annealing temperature is not linear with time and is laterally nonuniform. Twins are left in the regrown layer (Csepregi *et al.*, 1976). Dopants dissolved in the amorphous layer influence the regrowth kinetics (Csepregi *et al.*, 1977; Kennedy *et al.*, 1977). Oxygen, carbon, and nitrogen reduce the growth rate, while phosphorus and boron increase it.

Certain types of implantation-induced disorder (defect clusters, dislocation loops, etc.) are rather stable against thermal treatments at temperatures as high as 1100°C (Gyulai and Revesz, 1979).

The structures of postannealed samples are shown schematically in Fig. 1d–f for the damaged structures in Fig. 1a–c. Amorphous layers on Si ⟨100⟩ substrates regrow free of defects, while on ⟨111⟩ substrates twins remain. Microtwins are left in partially damaged Si samples, independent of the crystal substrate.

During furnace annealing, high temperatures are present all over the sample and unimplanted layers are also affected. In certain cases degradation of electrical properties occurs after recovery. Other methods of annealing have been investigated in the past few years to overcome the limitations associated with furnace heating.

Among these, laser and electron beams have been used to supply energy in a short time and over a selected area of the implanted target. In some cases irradiation has been performed by means of a continuous wave (cw) laser (Gat and Gibbons, 1978) or electron beam (McMahon and Ahmed, 1979; Regolini *et al.*, 1979) scanned over the sample. The millisecond dwell time implies annealing processes similar to those occurring in furnace treatment (Gat *et al.*, 1978; Williams *et al.*, 1978). These methods are described in detail in Chapter 10. We consider instead the cases in which energy deposition occurs in very short times on the order of 10–100 nsec.

Pulses of directed energy are suitable for shallow transient heating and are of relevance for near surface treatments. The region in which the energy must be absorbed should be on the order of micrometers. The energy densities involved are on the order of 1 J/cm² for both laser and electron beam single pulses.

The use of a high-power laser pulse to reorder ion-implanted semiconductors was reported for the first time by Russian scientists (Kachurin *et al.*, 1975; Shtyrkov *et al.*, 1975; Khaibullin *et al.*, 1975, 1978; Antonenko *et al.*, 1976). Irradiation of evaporated thin films to induce grain growth or epitaxial regrowth dates back to 1971 (Schwuttke and Howard, 1971). Interest in pulsed annealing increased enormously in 1977, and since then many workshops and international meetings have been dedicated to this subject (Rimini, 1978; Ferris *et al.*, 1979; Anderson *et al.*, 1980; White and Peercy, 1980; Gibbons *et al.*, 1981).

In this chapter we consider in some detail the experiments performed on ion-implanted silicon samples irradiated with a laser or electron beam single pulse. Structural changes from amorphous to crystalline material will be reviewed and described in terms of a liquid transient formation. The energy of the incident beam is absorbed by the matrix and subsequently converted to heat. The surface layer of the irradiated material can be melted, and a defect-free crystalline layer results from liquid epitaxial growth on a single-crystal substrate. Residual defects resulting from irradiation are also described. The melting of amorphous material should require less energy than that required to melt the corresponding amount of crystalline material. In addition the melting of the amorphous layer should occur at a temperature lower than that of a crystalline layer. These phenomena can give rise, under suitable experimental conditions, to regions of considerable undercooling.

II. Phase Transitions Induced by Laser or Electron Beam Pulses

In this section changes in the structure of ion-implanted semiconductors resulting from the deposition of intense energy density pulses in a short time are described. As a starting point we consider the irradiation of amorphous layers created by ion implantation, with the aim of pointing out the threshold characteristics of the transition to single crystal, and a simple explanation in terms of liquid phase epitaxy.

A. *Amorphous-to-Single Crystal Transition*

Amorphous layers having a well-defined thickness and a sharp interface with the single-crystal substrate are created by implanting Si ions at different energies in Si crystals at liquid nitrogen temperature. The thickness of the layer depends on the energy of the impinging ions; for instance, with 250-keV Si ions amorphous layers nearly 5000-Å thick are formed. Multiple energy implants are used to obtain a better degree of amorphization throughout the layer thickness.

The creation of an amorphous layer is not limited, of course, to self-ion implantation. Other impurities can also be used. The self-ion implanted layers are of high purity, and the amorphous–single crystal interface is free of contaminants, in contrast to the usual deposition methods.

In pulsed annealing the samples are irradiated with single shots of laser light or of electrons. The lasers are used in the Q-switched mode, and the pulse duration ranges between 10 and 100 nsec. The energy density involved in the annealing of semiconductor material is on the order of 1 J/cm², which corresponds to a peak power density of 50 MW/cm² for a 20-nsec pulse. Most work has been done using ruby (0.69-μm) and Nd:YAG or Nd:glass (1.06-μm) lasers. The latter are often operated at double frequency (0.53 μm). The wavelength determines the coupling of the irradiated material with the incident light, i.e., the absorbed energy and its depth distribution.

The basic experimental setup is simple for single-shot irradiation in air. The laser beam usually passes through a beam splitter, and a small fraction of the energy is reflected onto a photodetector to measure the light intensity and the waveform. Irradiation with a pulsed electron or ion beam requires a more complicated apparatus and must be performed in vacuum. Nominal electron beam parameters useful for pulsed annealing are 1–30 keV and 10–40 kA, with a pulse width of about 50 nsec (Greenwald *et al.,* 1979). The strong magnetic fields affect the electron trajectories such that they impinge upon the target at oblique angles. The electron energy is then deposited in the first micrometer of the surface layer.

In the study of laser or electron beam irradiation of amorphous layers onto single-crystal substrates it has been found that the transition to single crystal occurs above a threshold energy density. This is illustrated in Fig. 2a and b for pulsed electron beam and ruby laser irradiation, respectively. The amorphous layer created by self-ion implantation in Si (100) substrates was 1900-Å thick.

In many of these experiments, channeling techniques in combination with megaelectron volt He$^+$ backscattering have been used to obtain depth distributions of defects. The yield obtained in an as-implanted sample for a beam incidence parallel to the (100) axis reaches the random level over an energy range corresponding to a surface amorphous layer of 1900 Å. The energy depth conversion for 1.8-MeV He ions in silicon is 21 Å/keV (Chu *et al.,* 1978). The aligned yield recorded after irradiation with a 0.55 J/cm² energy density of the electron beam decreases drastically and

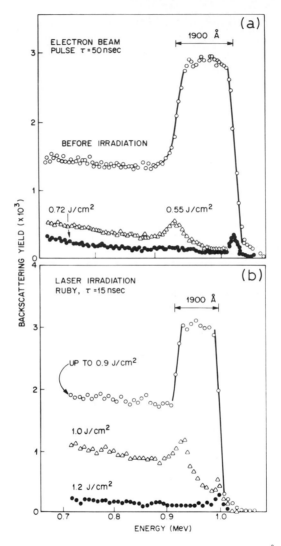

Fig. 2. Channeling spectra showing the recrystallization of a 1900-Å-thick amorphous implanted Si layer on a (100)-oriented Si substrate after irradiation with an electron beam (a) and ruby laser (b) pulse of different energy density. The analysis was performed with 2.0- and 1.8-MeV ^{4}He$^+$ beams, respectively.

at 0.7 J/cm² coincides with that of unimplanted Si single crystals (Fig. 2a). The amorphous layer has become single crystal, with the same orientation as the substrate. For values less than 0.55 J/cm² the yield is still high.

Similar results are found for ruby laser irradiation (Fig. 2b). Up to 0.9

J/cm² the aligned yield does not change. It decreases at 1.0, and finally at 1.2 J/cm² it reaches the unimplanted level. The high yield at 0.55 J/cm² for the electron beam and at 1.0 J/cm² for the ruby laser is caused by residual disorder located mainly at the original crystal–amorphous interface.

The transition to single crystal occurs at a well-defined energy density value and involves the entire disordered layer. This growth mechanism differs then from the usual solid phase regrowth which occurs by a uniform movement of the interface toward the surface during annealing. In pulsed laser irradiation the energy density for the transition to single crystal depends on the thickness of the initial amorphous layer. The channeling analysis reported in Fig. 3 indicates for instance that a 2.2-J/cm² ruby laser pulse of 50-nsec duration causes the transition to single crystal of a 4500-Å-thick amorphous layer.

In many experiments reported in the literature, the change from no recrystallization to complete recrystallization is not abrupt. The height of the aligned yield decreases in steps, and the annealing is not complete. This phenomenon is probably due to the nonuniformity of the laser beam with respect to energy density. The local energy density may change from

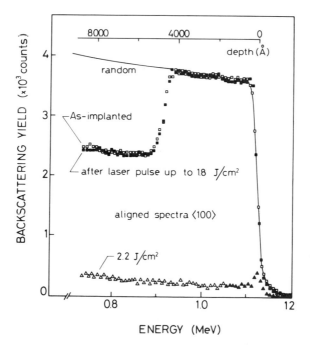

Fig. 3.   Channeling analysis of a 2.0-MeV He⁺ beam showing the effect of 50-nsec ruby laser single-pulse irradiation of a 4500-Å-thick implanted layer. (From Tseng *et al.*, 1978.)

place to place in a random way within the spot. In some regions it is above threshold for recrystallization, and in other contiguous regions it is below threshold.

The radial intensity of the laser pulse follows a Gaussian distribution only under certain experimental conditions, e.g., the TEM_{00} mode of a Nd:YAG laser. Irradiation is usually performed in the multimode condition to use the maximum possible energy output, and the distribution is not uniform. Several methods are now in use to improve beam uniformity. Quite good results are obtained by means of a curved quartz light guide. The beam light suffers a series of reflections at the pipe surface, and at the exit it is homogeneous with respect to the energy density (Cullis *et al.*, 1979). The beam spot size depends on the laser. The mean diameter (FWHM) ranges from a few centimeters to a few micrometers. When the area of the sample to be irradiated is larger than the beam diameter, the laser (or the sample) should be scanned to cover the region of interest.

Experiments on amorphous layers of different thickness in self-ion implanted Si (100) substrates irradiated with a 50-nsec ruby laser pulse are shown in Fig. 4 (Foti *et al.*, 1978a). Thicker amorphous layers require higher energy densities to become single crystals. The transition to single crystal has been followed by the measured ratio $D = (\chi_D - \chi_{min})/(\chi_a - \chi_{min})$, where χ_D, χ_a, and χ_{min} are the normalized aligned yield in the laser-irradiated, as-implanted, and reference unimplanted crystal, respectively. The ratio ranges from 1 when $\chi_D = \chi_a$ to 0 when $\chi_D = \chi_{min}$. The normalized aligned yield has been evaluated just behind the amorphous layer. The energy density required to regrow a 1900-Å amorphous Si layer with a 15-nsec ruby laser pulse (Fig. 2b) is nearly the same as that required to regrow a 1500-Å layer with a 50-nsec duration (Fig. 4). The main parameter is then the energy density required to describe the phase change process (Rimini *et al.*, 1978). Crystallization occurs also for implanted layers in ⟨111⟩-oriented Si substrates after irradiation with a suitable energy density (Foti *et al.*, 1978b). Similar results were found by electron beam irradiation (Kennedy *et al.*, 1979).

The amorphous-to-single crystal transition and the accompanying threshold behavior are shown by other implanted semiconductors such as Ge and GaAs (Fig. 5). The energy density at which reordering occurs also depends on the material; Ge and GaAs amorphous layers require less energy density for the transition to single crystal than that needed by Si samples. This is due mainly to the lower value of the thermal capacity and latent heat of melting.

Silicon amorphous layers deposited on single-crystal Si substrates crystallize after suitable energy density irradiation (Revesz *et al.*, 1978; Hoonhout *et al.*, 1978; Bean *et al.*, 1979). Epitaxial growth is not limited to

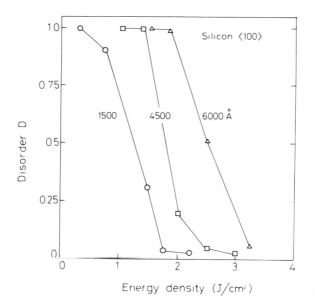

Fig. 4. Residual disorder as measured by channeling, after ruby laser irradiation of amorphous silicon layers of different thickness. (From Foti *et al.*, 1978a.)

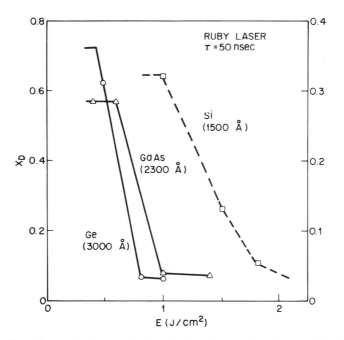

Fig. 5. Transition to single crystal of ion-implanted amorphous layers of Ge, GaAs, and Si after pulsed ruby laser annealing. (From Baeri *et al.*, 1979.)

deposited material of the same type as the single-crystal substrate. Germanium films evaporated onto a silicon wafer followed by electron beam pulsing grew epitaxially into a high-quality monocrystalline surface with minimum diffusion at the Si–Ge interface (Lau *et al.*, 1978). If the deposited film is doped, homo- or heteroepitaxial p–n junctions are formed by pulsed irradiation (Young *et al.*, 1979).

In ion-implanted samples, annealing by pulsed irradiation is usually followed by a redistribution of the implanted impurities to depths as great as 3000–4000 Å (Marquardt *et al.*, 1974; Dvurechensky *et al.*, 1978; Baeri *et al.*, 1978).

B.   Experiments on the Phase Transition

In the previous section channeling data were used to elucidate the characteristics of the amorphous-to-single crystal transition induced by laser or electron beam irradiation. The phase change can be monitored by any other physical property that changes with the lattice order.

Impurities introduced by ion implantation are electrically activated after reordering, and then an abrupt decrease in the sheet resistance should occur at the onset of recrystallization. The data reported in Fig. 6 illustrate this concept. They refer to As- and B-implanted silicon samples irradiated with ruby laser pulses. The threshold energy density for the recovery of electrical activity in the As-implanted Si is lower than that in the B-implanted sample, and it is sharper. For these implant energies the boron distribution and the associated damage distribution extend deeper than that of arsenic. In addition the disorder created during the boron implantation does not result in an amorphous layer. These two considerations explain qualitatively the trend of the sheet resistance relative to the laser energy density.

Changes in the structure of the irradiated layer can also be analyzed by diffraction for both transmission and reflection of high-energy electrons. As an example the patterns obtained by the reflection of electrons (RHEED) are shown in Fig. 7 (Vitali *et al.*, 1977). The electrons impinge at a glazing angle of few degrees from the surface, and they probe the first few hundred angstroms of the sample. The pattern of the as-implanted Si layer is a diffuse halo characteristic of the amorphous structure. After ruby laser irradiation rings appear in the pattern, and with increasing energy density rings change to spots. Rings and spots are associated with polycrystalline and single-crystal structures, respectively.

The polycrystalline layer is produced when the thickness of the melted layer does not penetrate the damaged layer. Solidification occurs before

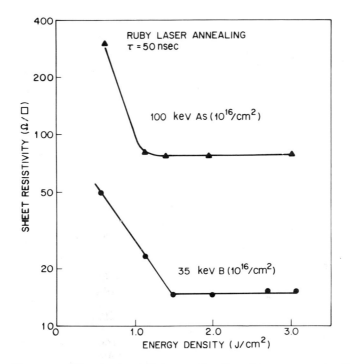

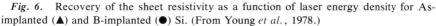

Fig. 6. Recovery of the sheet resistivity as a function of laser energy density for As-implanted (▲) and B-implanted (●) Si. (From Young *et al.*, 1978.)

the liquid wets the single-crystal substrate. The transition to polycrystalline also shows threshold behavior and is related to the amount of energy density needed to melt the surface of the amorphous layer (Foti *et al.*, 1977; Tseng *et al.*, 1978).

Dissolution of precipitates occurs also after pulsed irradiation (Young and Narayan, 1978). High-temperature diffusion of boron or phosporus into silicon leads to the formation of precipitates and small loops. Precipitates result when the dopant concentration exceeds the solid solubility limit, and in transmission electron microscopy they are identified by small clusters with black–white contrast. After Q-switched irradiation, annealed, defect-free regions containing no phosphorus precipitates are found with a thickness increasing with the pulse energy density (see Fig. 8a and b). The increase in the defect-free region thickness is a consequence of the increased melt front penetration at the higher pulse energy density (Narayan, 1979). Melting occurs during irradiation, and dissolution of precipitates can take place under conditions where diffusion coefficients are extremely high. This experiment provides a method for measur-

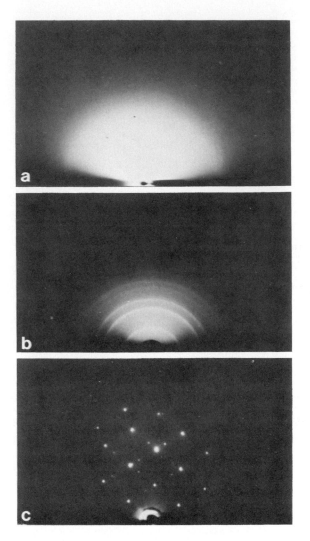

Fig. 7. Reflection high-energy electron diffraction (RHEED) patterns of an as-implanted amorphous Si layer (a) after 1.25-J/cm² (b) and 5.0-J/cm² (c) ruby laser pulses, respectively. The azimuth direction is along the ⟨110⟩ axis. (From Vitali *et al.*, 1977.)

ing the depth at which the melt front penetrates after pulsed irradiation. The solid–liquid interface velocity during solidification prevents agglomeration of dopant in excess of the solid solubility limit.

Noble gas atoms implanted in Si samples can also be used as markers of

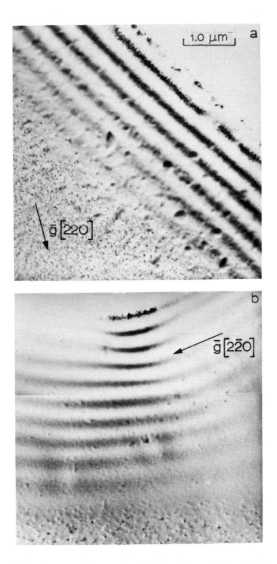

Fig. 8.   Bright-field electron micrograph of a phosphorus-diffused silicon specimen show-ing dissolution of precipitates after ruby laser irradiation. The annealed depth increases with the energy density of the ruby laser single pulse. The thickness of the defect-free regions can be measured by stereomicroscopy or from thickness fringes. (a) 2.0 J/cm², (b) 3.0 J/cm². (From Narayan, 1979.)

the melting depth. These atoms are dissolved in the liquid phase and diffuse out during the subsequent solidification. Loss of atoms located at a certain depth inside the sample implies a liquid layer produced by the irradiation and extending at least to the considered depth.

As an example argon ions implanted at two different energies, 30 and 130 keV, respectively, in Si were used as a marker. After irradiation with a Nd:YAG double-frequency laser pulse the Ar content was measured, and the retained amount is reported in Fig. 9 as a function of the energy density. The decrease in the Ar atom content follows threshold behavior. For the low-energy implant, 0.5 J/cm² is enough to reduce to half the amount of retained Ar, while for the high-energy implant 1.3 J/cm² is required. The transition is not as sharp as for crystallization of the amorphous layer because of the gaussian distribution of the implanted argon atoms around the projected range.

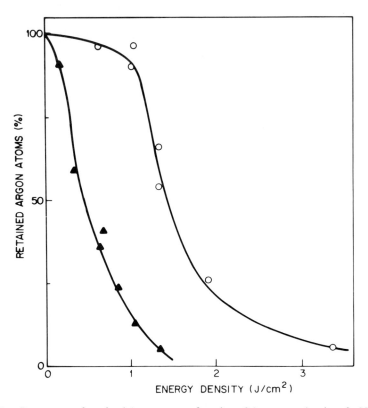

Fig. 9.   Percentage of retained Ar atoms as a function of the energy density of a Nd:YAG frequency-doubled laser pulse. ▲, 30-keV, 10^{16}-atoms/cm² implant; ○, 130-keV, 10^{16}-atoms/cm² implant.

Several other experimental techniques have been used to analyze the amorphous-to-crystal phase change, such as positron annihilation lifetime and angular correlation, electron paramagnetic resonance, Raman scattering (Tsu *et al.*, 1979), and photoluminescence (Nakashima *et al.*, 1979).

III.   Coupling of the Incident Beam with the Matrix

The transition to single crystal of an amorphous layer is described, as previously shown, in terms of liquid phase epitaxy. The liquid formation depends on the energy deposition of the incident beam and on the thermal properties of the irradiated material. The energy deposition is correlated to the coupling of the beam of the matrix (see Chapter 4).

The energy of a pulsed electron or ion beam is transferred to the material mainly through inelastic electronic collisions. In this case the coupling is independent of the structure of the substrate. One can heat a well-defined region inside the sample regardless of the presence of other layers on the specimen surface. In most experiments performed with electrons (in the range 10–100 keV) the beam is not monoenergetic and the incident direction ranges from the normal to the glancing geometry. The energy is then deposited, usually in a 1-μm surface region (Greenwald *et al.*, 1979; Merli and Rosa, 1981). Successful annealing has been reported in implanted Si using pulsed proton beams of 200 keV and 50-ns duration (Hodgson *et al.*, 1980).

The coupling of the laser with a semiconductor is a complex phenomenon because the absorption depends on several processes and is very sensitive to the structure, temperature, doping, etc., of the material (Bloembergen, 1979). A detailed description is given in Chapters 3 and 4, and we now present some experimental results. First, we consider the influence of the temperature of the substrate during Nd:YAG irradiation of thin, amorphized silicon layers. At this wavelength the amorphous layer, which is a few thousand angstroms thick, absorbs only a fraction of the incident light, about 25%. The remainder is absorbed at a much lower rate by the underlying single crystal. In crystalline Si the absorption coefficient increases strongly with the temperature. Small lateral inhomogeneities of the intensity distribution are then amplified. The increase in the absorption coefficient provides a positive feedback that tends to make the heating process inherently unstable (thermal runaway). This deleterious effect is reduced by keeping the sample at an elevated temperature during irradiation. Optical micrographs of samples irradiated at 20 and 250°C with the same Nd:YAG pulse energy densities are shown in Fig. 10a and

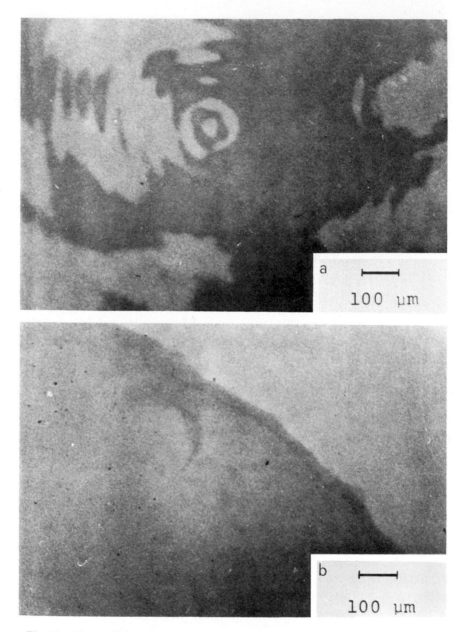

Fig. 10. Nomarski interference micrographs of an implanted surface irradiated with a Nd Q-switched pulse of 1.0 J/cm² at 20°C (a) and at 250°C (b) substrate temperatures. The inhomogeneities disappear at 250°C. (From von Allmen *et al.*, 1979.)

b, respectively. The irradiated area is laterally uniform for the 250°C substrate, and channeling measurements indicate fairly good epitaxial regrowth (von Allmen *et al.*, 1979).

The structure of the irradiated layer determines the absorptivity and thus the threshold energy density value. Layers of silicon implanted with phosphorus ions become electrically active after Nd:YAG irradiation when the implant dose overcomes that required for amorphization (Miyao *et al.*, 1979). The thickness of the amorphous layers created by As implantation in silicon increases with dose. The thicker amorphous layer absorbs more energy in the initial part of the pulse, and the liquid layer therefore forms earlier. More energy is absorbed, thus resulting in a liquid layer that extends deeper into the single-crystal substrate. Impurities at higher doses diffuse deeper inside the sample (Venkatesan *et al.*, 1978).

The annealing to single crystal is not limited to amorphous or heavily damaged layers. Polycrystalline layers can regrow as single crystals on the crystalline matrix after irradiation. The single-crystal transition for a polycrystalline layer requires a density energy value higher than that required for amorphous layers of the same thickness (Fig. 11). For ruby laser irradiation the increase amounts to about 50% for a 4000-Å-thick layer (Vitali *et al.*, 1978). This difference is caused by the lower absorption coefficient of the polycrystalline material with respect to the amorphous material.

High dopant concentrations ($> 10^{20}/cm^3$) increase the absorption coefficients of crystalline silicon because of the large concentration of free carriers. Impurities also play a role in increasing the absorption of an amorphous layer. The overall effect is then a reduction of the threshold energy density for melting. To illustrate this point we consider the irradiation of as-implanted and furnace-annealed As–Si samples with two different laser wavelengths, 1.06 and 0.53 μm, respectively. With the use of the sheet resistance as a probe it has been found that the resistance decreases abruptly for both samples at the same energy density of the 0.53-μm pulse (Fig. 12a). At 1.06-μm irradiation (Fig. 12b) the resistance of the as-implanted samples decreases at an energy density value one-half of that required by the furnace annealed samples. The decrease in the sheet resistance in the furnace-annealed samples is probably associated with the dissolution of As clusters in the liquid silicon and to their incorporation in substitutional lattice sites in excess of the furnace value (Rimini *et al.*, 1981).

The threshold energy density for surface melting of a silicon single crystal and of a silicon crystal covered with 1000-Å-thick amorphous layer should differ by a factor of 2 for irradiation at $\lambda = 0.53$ μm and by a factor of 10 for irradiation at $\lambda = 1.06$ μm. These estimates are based on the

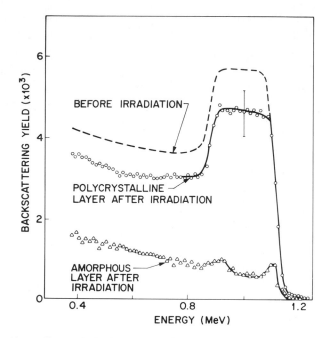

Fig. 11. Channeling analysis showing the effect of the structure on the threshold energy density. A 2.0-J/cm² ruby laser pulse causes the transition to single crystal of a 4500-Å-thick amorphous layer but not of a 4500-Å-thick polycrystalline layer. (From Vitali *et al.*, 1978.)

absorption coefficient of undoped single-crystal and amorphous silicon material. Measurements in the low-intensity regime indicate that the absorption coefficient of Si doped with $4 \times 10^{19}/cm^3$ As is a factor of 100 higher than in undoped specimen for $\lambda = 1.06$ μm.

The trend of the experimental data is in qualitative agreement with the influence of doping on the absorption coefficient. A quantitative comparison requires instead the knowledge of several quantities which are still unknown. The threshold energy density, at which amorphous layers of the same thickness created by different ions become single crystal, provides information on the contribution of impurities to the absorption coefficient. The threshold values for ruby laser irradiation are found to be nearly independent of the implanted species, e.g., Ga, Ge, Sn, As, Sb, and Bi (see Fig. 13) (Hoonhout, 1980). A noticeable dependence of the threshold values on the implanted species is reported instead for irradiation at 1.06 μm (Tamminga *et al.*, 1979). The exact dependence is probably associated with the presence of impurity levels within the forbidden gap.

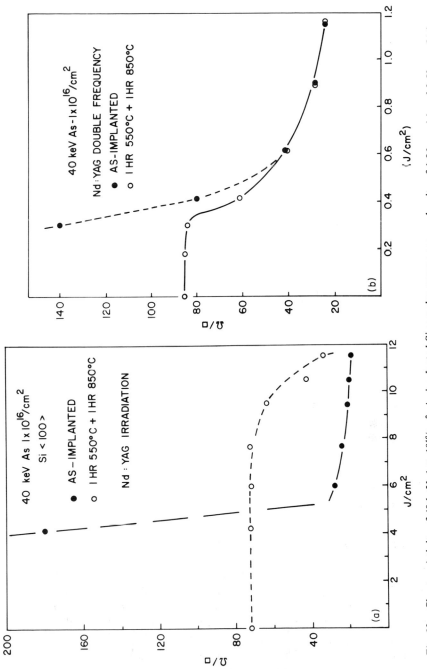

Fig. 12. Sheet resistivity of 40-keV, $1 \times 10^{16}/cm^2$, As-implanted Si samples versus energy density of 1.06-μm (a) and 0.53-μm (b) laser wavelength. The comparison is made between as-implanted and furnace annealed samples. (From Rimini *et al.*, 1981.)

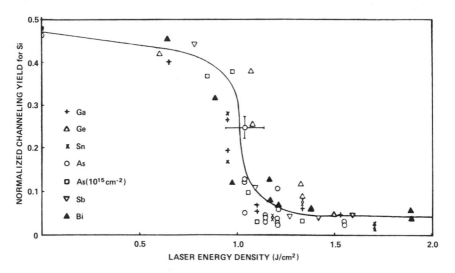

Fig. 13. Normalized aligned yields measured just below the disordered region versus the ruby laser energy density for Si (100) substrates implanted with the indicated elements. (From Hoonhout, 1980.)

IV. Defects in Pulsed Annealing of Disordered Layers

Irradiation of ion-implanted layers can produce a large variety of defects. Their type and final configuration depend on the initial structure of the damaged sample and on the annealing procedure. In the following we will describe the annealing behavior of defect clusters and dislocation networks buried in a single-crystal matrix. The structure of residual defects in irradiated amorphous layers will also be considered, with some emphasis on irradiations performed near the threshold for single-crystal transitions and for different substrate orientations.

A. *Laser Irradiation of Damaged Layers*

Light ion or low-fluence implants produce only partially damaged layers. Self-ion implantation at room temperature shows isolated damaged regions embedded in the single-crystal structure as reported in Fig. 14a (Foti *et al.*, 1979). The TEM image reveals small spots of dark contrast associated with heavily damaged zones 200 Å in size. The (100) diffraction pattern indicates that the disordered regions are surrounded by a single-crystal matrix. This type of damage cannot be removed completely by

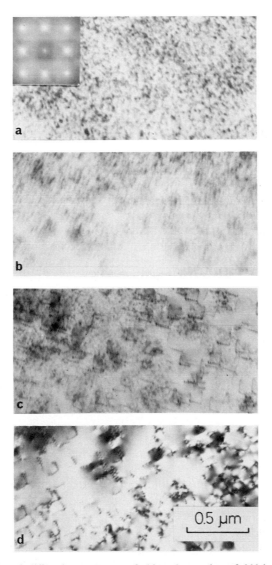

Fig. 14. TEM and diffraction patterns of thinned samples of 200-keV, $10^{15}/cm^2$ Si-implanted Si at room temperature. (a) as-implanted; 50-nsec duration ruby laser pulse at an energy density of 1.5 J/cm² (b), 1.9 J/cm² (c), and 2.4 J/cm² (d). (From Foti *et al.*, 1979.)

furnace annealing, and stable microtwins are observed after a 700°C, 30-min treatment (Gyulai and Revesz, 1979).

The disorder decreases after laser irradiation: at 1.5 J/cm² the structure is still similar to that of the as-implanted sample (Fig. 14b), but with a reduced density of defects. At 1.9 J/cm² a large number of dislocation lines appear near the surface, and black spots are still present in the remaining disordered layer (Fig. 14c). At 2.4 J/cm² the black spots disappear, and dislocations are left in the sample (Fig. 14d). Most of the dislocations are split into two branches, at an angle of $25 \pm 3°$ with the normal to the surface which is very close to the (113) direction. The length of the dislocation lines ranges from 2000 to 4000 Å, and the density is 5×10^8 cm/cm³.

The threshold behavior so clearly evident in the reordering of amorphous layers (see Fig. 13) is absent in these partially damaged layers. The reordering occurs progressively with the energy density of the laser pulse. Laser-induced melting probably initiates in the heavily damaged (amorphous) zones which absorb the light at an higher rate compared with the surrounding crystal. As soon as local melting occurs, the reordering takes place with a decrease in black spot density in the TEM images. No polycrystalline regions or microtwins are detected after laser annealing.

For high-energy-density irradiation (2.4 J/cm²) the molten regions could overlap each other, thus forming a uniform liquid layer. The annealing of the implanted layer is complete except for residual dislocation lines. The mechanism of dislocation growth is still not completely understood; one expects a growth direction normal to the sample surface; instead it aligns close to the (113) direction.

The reordering in these self-ion implanted silicon samples is similar to that of partially damaged layers obtained by light-ion implants. Boron implanted in silicon at room temperature with a fluence of $2.5 \times 10^{16}/cm^2$ and with an energy of 35 keV does not produce an amorphous layer. Irradiation with a ruby laser induces complete reordering for a suitable energy density value. The channeling analysis is reported in Fig. 15, and the results are compared with those of furnace annealing at 900°C for $\frac{1}{2}$ hr (Young *et al.*, 1978). Good recovery of lattice order occurs for 1.5- to 1.7-J/cm² ruby laser pulses.

No damage remains in the form of dislocations, stacking faults, or dislocation loops; but the aligned yield after laser irradiation is slightly higher than that of a virgin undoped single crystal. X-ray diffraction measurements indicate that the silicon lattice undergoes a significant one-dimensional contraction. The atomic radius of boron is lower than that of silicon, and the ions produce a net contraction of the implanted layer. This behavior is not peculiar to boron. Other impurities such as phosphorus

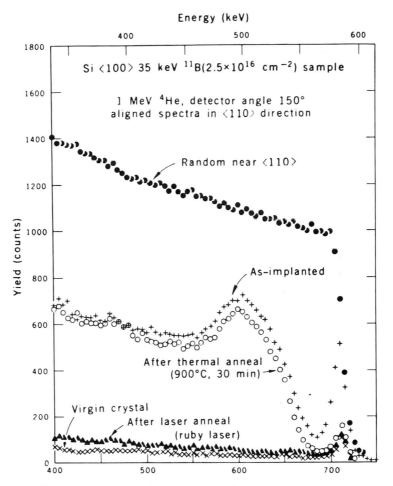

Fig. 15. Channeling analysis of boron-implanted Si crystals in as-implanted, laser-annealed, and thermally annealed samples. (From Young *et al.*, 1978.)

and arsenic produce contraction and expansion, respectively, of the silicon lattice if in excess of the solubility limit.

B. *Growth of Dislocations after Pulsed Melting*

The interaction of extended defects with the liquid layer created by pulsed irradiation has been analyzed in silicon samples using well-defined defect structures (Narayan and Young, 1979). Edge dislocations are left in 100-keV 10^{16}/cm^2 As-implanted silicon samples after thermal annealing at

1000°C for ½ hr. The TEM image before laser irradiation (Fig. 16a) shows that these dislocations lie approximately in (100) planes and extend from the surface to a depth of about 0.6 μm. After 1.5-J/cm² laser pulse irradiation several split dislocations originate at a depth of about 0.4 μm below the surface (Fig. 16b). The dislocations are mixed, with a Burgers vector 29° from a pure screw orientation, and the growth direction is close to the (113) direction. These dislocations are similar to those found in self-ion implanted partially damaged silicon (compare with Fig. 14d). These defects can be removed by increasing the energy density of the laser pulse, i.e., when the original dislocation lines are all inside the thickness of the molten layer.

The growth of dislocations has been also analyzed (Hofker *et al.*, 1979) in (111)-oriented Si substrates. The dislocation network (Fig. 17a) has been introduced by implantation of 50-keV P ions (10^{16}/cm²) and by subsequent thermal annealing at 1100°C for 40 min. The misfit dislocations relieve the stress associated with the high dopant concentration. After irradiation with 1.0-J/cm² laser pulses of Nd : YAG (180 nsec in duration) new extended defects grow from the remaining original dislocations. The new defects are again dislocations of the partial type and include stacking faults. The stacking faults are intrinsic and bounded along their edges with $a/6 \langle 110 \rangle$ stair–rod dislocations. These planar defects are almost uniformly distributed along the equivalent $\{111\}$ planes. The number of dark fringes indicates that they extend to a depth of about 3000 Å (see Fig. 17b).

C. Residual Defects in Irradiated, Ion-Implanted Amorphous Layers

Pulsed irradiation induces epitaxial crystallization of amorphous layers on single-crystal substrates. The epitaxial layer contains extended defects for energy density values just above the threshold for the amorphous-to-single crystal transition. The channeling data of Fig. 2a indicate that residual disorder is left in the 1900-Å-thick amorphous layer after irradiation with an electron beam pulse of 0.55 J/cm². TEM analysis of the same sample shows dislocation pairs distributed throughout the regrown layer (Fig. 18). The Burgers vector and the growth directions are similar to those found in partially damaged self-ion implanted silicon and arsenic-doped silicon (see Figs. 14 and 16). The dislocations are all of the same length and originate from a thin layer located just behind the initial amorphous region. By increasing the energy density the crystallized layer is free of extended defects down to 10 Å in size. This behavior is also found in amorphous layers irradiated with laser pulses. Near the threshold

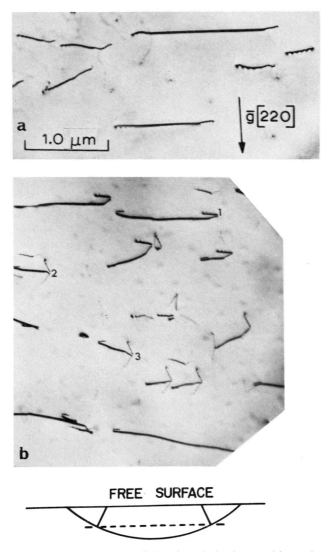

Fig. 16. Growth of edge and screw dislocations during laser melting and solidification of silicon (100). (a) Edge dislocations of $a/2$ [110] Burgers vector before laser irradiation, (b) directions of an edge (at 1) and mixed (at 2 and 3) dislocations after laser irradiation. The inset shows directions of dislocations before and after laser irradiation. (From Narayan and Young, 1979.)

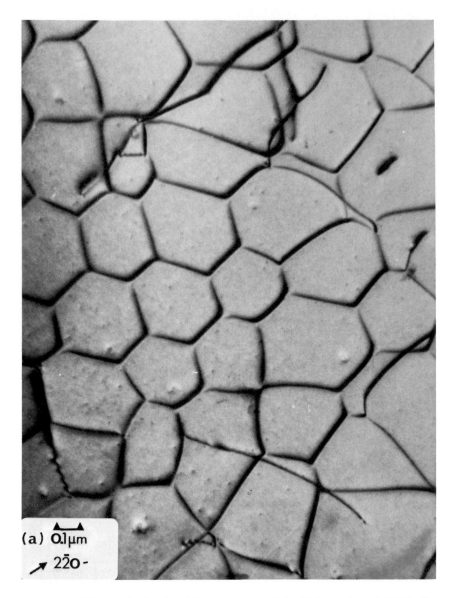

(a) $\overline{0.1}\mu m$

$\nearrow 2\bar{2}0 -$

Fig. 17a. Micrograph showing dislocation network in P$^+$ ion-implanted $\langle 111 \rangle$ silicon after annealing at 1100°C for 40 min. (From Hofker *et al.*, 1979.)

Fig. 17b. Micrograph showing (a) after a subsequent laser pulse irradiation at an energy density not sufficient to completely remove misfit dislocations. (From Hofker *et al.*, 1979.)

Fig. 18. TEM micrograph of a 1900-Å-thick amorphous implanted Si layer after a 0.55-J/cm² electron beam pulse, showing dislocation lines in the irradiated layer.

for the single-crystal transition V-shaped dislocations are always present in the regrown layer on (100) Si.

To investigate the origin and mechanism of formation of these defects, amorphous layers of different thicknesses have been irradiated with ruby laser pulses (50 nsec in duration). Figure 19 compares the TEM micrographs of a 1000- and a 4500-Å-thick amorphous layer after irradiation with 1.5 and 2.5 J/cm², respectively. In the thinner annealed layer the residual defects are small dislocation loops located about 1200 Å below the surface; occasionally the small loops are split into V-shaped dislocations. These dislocations propagate in the 1000-Å regrown layer toward the surface close to the (113) direction (Tsu *et al.,* 1980). In the thicker annealed layer the defect structure is similar, but the small loops are now located at about 4500 Å below the surface. The density (~5 × 10¹⁴/cm³) and the size (~100 Å) of the small loops are comparable in both cases. The length of the V-shaped dislocations increases with the thickness of the amorphous layer, and they originate from the small loops.

The amorphous layer in these samples was obtained by multiple energy implants, but in this case too, the amorphous-to-single crystal interface is not atomically sharp. Damage extends deeper than the thickness of the

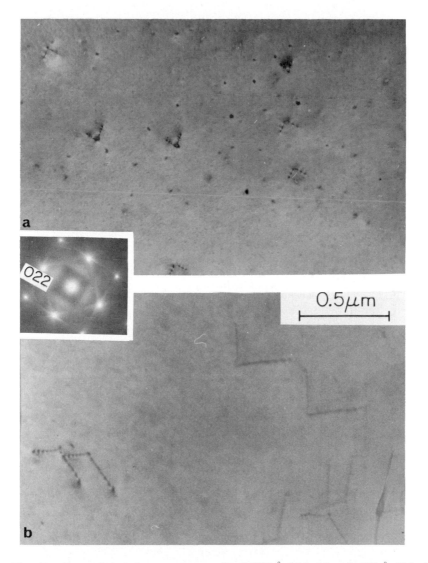

Fig. 19. Transmission electron micrograph of 1000-Å-thick (a) and 4500-Å-thick (b) amorphous implanted Si layers after irradiation with 1.5- and 2.5-J/cm² ruby laser pulses, respectively. The dislocations start from small loops, and their lengths scale with the thickness of the initial amorphous layer.

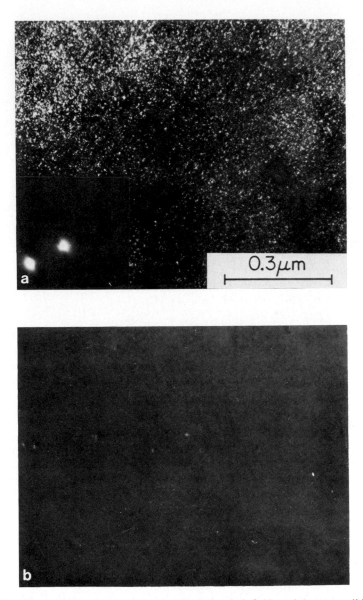

Fig. 20.   Transmission electron micrographs under dark-field weak beam conditions, (a) as-implanted sample after anodic stripping of 450-nm-thick layer, (b) furnace annealing at 450°C for 30 min and anodic stripping of 4500-Å-thick layer. White spots disappear after furnace annealing. (From Campisano *et al.*, 1980.)

amorphous layer and is associated with the tail of the ion distribution. The disorder then decreases from the amorphous level to a low value following approximately the Gaussian distribution of the highest-energy ions. The characteristic transition region is shown in the TEM in Fig. 20a. The initial amorphous layer was removed before analysis by anodic oxidation and stripping. The high-resolution dark-field micrograph reveals the presence of small defect clusters embedded in a crystalline matrix. The defects are removed by low-temperature furnace annealing (Fig. 20b) at 450°C for $\frac{1}{2}$ hr (Campisano *et al.*, 1980).

These defects probably nucleate the long dislocations seen in many irradiated samples. Near threshold the liquid layer extends through the initial amorphous layer and reaches the buried disordered region. These defects act as a seed for the crystallographic imperfections at the liquid–single crystal interface. The propagation toward the surface of the V-shaped dislocations occurs during the rapid solidification of the melt. A schematic illustration of the residual defects in the (100) annealed silicon layer is reported in Fig. 21.

To summarize, the dislocation growth from melting on (100) substrates exhibits the same characteristics independent of the way the crystallographic imperfections are introduced. Defect clusters in partially damaged silicon, disordered regions behind the amorphous layers, and dislocation network all give rise, after pulsed melting, to V-shaped dislocations in the regrown layer. The mechanism responsible for growth along the (113) direction, far from the normal to the sample surface, is not well understood. We speculate that kinetic effects at the liquid–solid interface during

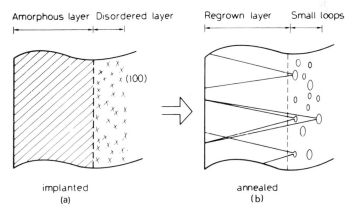

Fig. 21. Schematic illustration of the depth distribution of defects in as-implanted (a) and laser-annealed (b) samples. Extended defects originate from defect clusters behind the amorphous layer.

the very rapid solidification might become relevant, as supported by the dependence of dislocation direction growth on the crystallographic orientation of the substrate.

Amorphous layers on silicon (111) substrates, after pulsed irradiation just above threshold, crystallize and contain stacking faults (Foti *et al.*, 1978b), as shown in Fig. 22. The defects are distributed on the equivalent (111) planes, and they are of intrinsic character. The number of extinction fringes indicates that the faults extend from the surface to a depth of 4100 Å. These defects are similar to those observed after irradiation of a silicon (111) substrate containing a misfit dislocation network (Fig. 17). They also originate from the crystallographic imperfections present in the matrix. The type of resulting extended defect depends then on the orientation of the substrate.

Channeling analysis indicates that the transition to single crystal of an

Fig. 22.   Bright-field micrograph showing intrinsic faults in a 4000-Å-thick amorphous layer on a Si ⟨111⟩ substrate after a 2.5-J/cm² ruby laser pulse of 50-nsec duration. (From Foti *et al.*, 1978b.)

amorphous layer on silicon (111) substrates is not so sharp as for (100)-oriented substrates (see Fig. 23). At 2.5 J/cm² the aligned yield for the (100) sample exhibits a strong reduction compared to the random level, while for the (111) sample the yield is still 12% near the surface and reaches 35% at a depth of 4000 Å. The difference in the yield between the two oriented substrates is due to the large amount of residual stacking faults originating from small clusters just behind the amorphous layer.

Furnace heating at low temperature (450°C, $\frac{1}{2}$ hr) can anneal these clusters (Fig. 20), and then one expects a lower density of stacking faults. To show this behavior, two sets of silicon (111) samples with a 4500-Å-thick

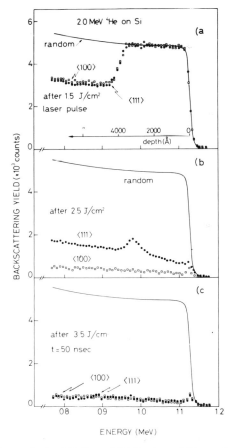

Fig. 23. Channeling analysis of (100)- and (111)-oriented Si samples self-ion implanted at liquid nitrogen temperature and irradiated with a Q-switched ruby laser single pulse of 50-nsec duration. (a) 1.5 J/cm², no melt-through; (b) 2.5 J/cm²; (c) 3.5 J/cm², good epitaxial regrowth to ⟨100⟩ and ⟨111⟩. (From Foti *et al.*, 1978b.)

amorphous layer with and without furnace preannealing were irradiated. The channeling analysis reported in Fig. 24 refers to the preannealed samples. With a 2.0-J/cm², 15-nsec-duration ruby laser pulse, complete crystallization of the 4500-Å-thick amorphous layer occurs. The minimum yield measured at 5000 Å from the surface is shown in the inset in Fig. 24 for samples with and without preannealing. The threshold energy density, measured by channeling, decreases for the preannealed samples down to 2.0 J/cm², and this value agrees very well with that found for the same thickness layers on silicon (100) wafers. The shift in threshold toward a high-energy-density value for the samples without preannealing is due to the fact that the molten layer must extend deeper to include the zone with defect clusters. This result explains the experimental measurements reported in the literature on the crystallization of the amorphous layers on silicon (111) substrates without preannealing (Hoonhout, 1980). The fact

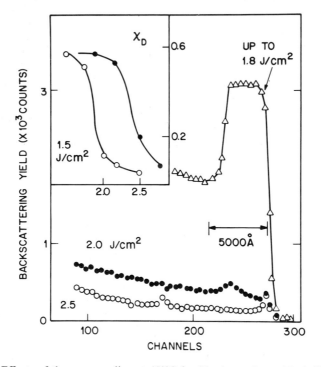

Fig. 24. Effects of the preannealing at 450°C for 30 min on the residual disorder after subsequent irradiation with a ruby laser pulse of 15-nsec duration. The inset reports the normalized aligned yield at a depth of 5000 Å as a function of the laser pulse energy density for samples with (O) and without (●) furnace preannealing. (From Campisano *et al.*, 1980.)

that the liquid formation and the epitaxy growth do not depend on the substrate orientation was demonstrated by measuring the broadening of As atoms implanted in silicon (100) and (111) substrates. The depth profiles are similar within the accuracy of the experiments (i.e., within 5%).

D. *Amorphous-to-Polycrystalline Transition*

For laser irradiation below threshold the amorphous layer becomes polycrystalline. The channeling technique is not able to distinguish between amorphous and randomly oriented crystalline grains. Samples are analyzed by electron transmission and diffraction after back thinning as schematically reported in Fig. 25. Diffraction patterns taken on a 4500-Å-thick amorphous layers indicate the presence of a polycrystalline surface layer with very fine grains (Fig. 26a) for 20-nsec ruby laser irradiation just above 0.4 J/cm². The amorphous-to-polycrystalline transition is also sharp and is characterized by a threshold character. The grain size increases with the energy density of the pulse, as shown in Fig. 26b, c, and d for 0.8, 1.0, and 1.5 J/cm², respectively. At high-energy irradiation the spot patterns of the polycrystalline rings indicate texture growth of large grains. TEM micrographs are shown in Fig. 27a and b for 0.8- and 1.5-J/cm² irradiation, respectively. The average grain sizes are 800 and 1500 Å, respectively. A detailed investigation on the grain size and energy density is summarized in Fig. 28. The relationship between grain dimension and energy density is not linear: At low energy densities the grain size is almost constant and increases suddenly for irradiation above 1.2 J/cm² (Tseng *et al.*, 1978). An in-depth analysis of the grain size by TEM cross

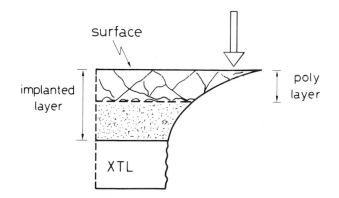

Fig. 25.   Back-thinned implanted silicon sample for TEM analysis (schematic).

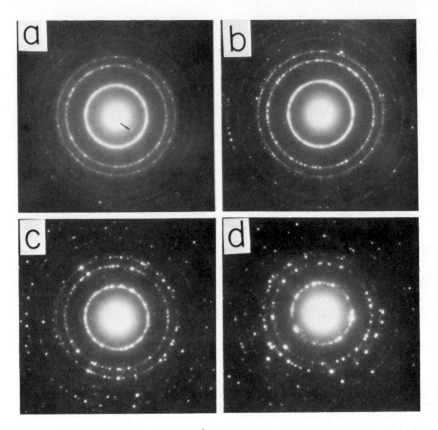

Fig. 26.   Diffraction patterns of a 4500-Å-thick amorphous layer after a 20-nsec ruby laser irradiation at 0.4 J/cm² (a), 0.8 J/cm² (b), 1.0 J/cm² (c), and 1.5 J/cm² (d).

section (Cullis *et al.*, 1980) shows that the size distribution is not uniform. By increasing the energy the observed structures range from very fine grains to large grains which become comparable in size to the thickness of the implanted layer for irradiations just near the threshold for the single-crystal transition.

For 50-nsec ruby laser irradiation the threshold-to-polycrystalline transition in amorphous silicon is about 0.5 J/cm², nearly independent of the layer thickness in the range 1500–5000 Å. The behavior of the amorphous-to-polycrystalline transition appears to need a more detailed description of the liquid formation, including the thermodynamic properties of the amorphous material.

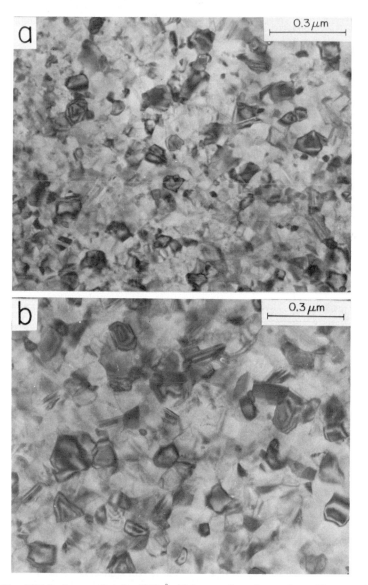

Fig. 27.   TEM micrographs of a 4500-Å-thick amorphous layer after a 20-nsec ruby laser irradiation at 0.8 J/cm² (a) and 1.5 J/cm² (b).

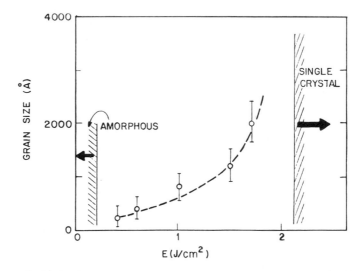

Fig. 28. Grain size in the near surface region measured by TEM versus incident energy density of a 4500-Å-thick implanted amorphous layer on Si (100) irradiated with 20-nsec laser pulses.

V. Amorphous-to-Liquid Transition: Undercooling Effects

In pulsed annealing, the input energy required to melt a given thickness depends on the coupling of the beam with the target and on the thermodynamic properties of the irradiated material. In crystalline samples the thermodynamic parameters as well as their temperature dependence are well known. In amorphous material these parameters will be different from those of the crystalline material. These changes cause new phenomena during irradiation with ultrafast energy pulses. The temperature and the enthalpy of melting of amorphous materials are usually lower than the crystalline values. Calculations for Ge indicate that the difference amounts to 20–30% (Spaepen and Turnbull, 1979; Bagley and Chen, 1979). The free energy of the amorphous state is higher than that of the corresponding solid, as shown schematically in Fig. 29. The free energies of the amorphous and of the liquid are equal at T_α (melting point of the amorphous) lower than T_M (melting point of the crystal).

During slow heating, crystallization of the amorphous layer occurs via a solid state reaction. In silicon, for instance, nucleation and grain growth become reievant at 550°C and crystallization is completed in a few minutes. The use of nanosecond heating techniques has allowed the

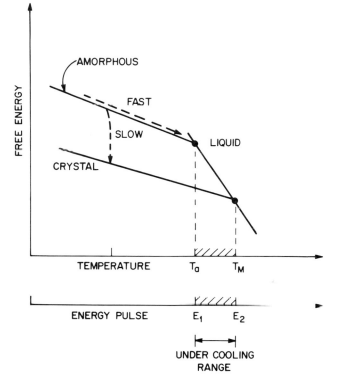

Fig. 29. Free energy of amorphous, crystal, and liquid states. T_α melting point for amorphous and T_M melting point for crystal. E_1–E_2 energy pulse range of undercooling.

amorphous phase to reach the temperature T_α in the absence of any solid state reaction. Experiments performed with electron pulses show that the enthalpy and the melting point of amorphous silicon are lower than the crystalline values by about 30% (Baeri *et al.*, 1980). This implies the existence of a range of pulse energy densities E_1 and E_2 (see Fig. 29) which can melt the amorphous layer but not the crystalline substrate; i.e., the liquid will be undercooled.

For the 15-nsec ruby laser irradiation of a 2000-Å-thick silicon amorphous layer on silicon, E_1 and E_2 are 0.3 and 0.8 J/cm², respectively. At irradiation near E_1 fine grain structure is expected in the annealed layer; by increasing the energy density to E_2 the growth of larger grains occurs up to the single-crystal transition with high-order defects (Tan *et al.*, 1980). Crystallized layers free of defects are obtained at energy density values at least 20% above E_2, because part of the single-crystal substrate

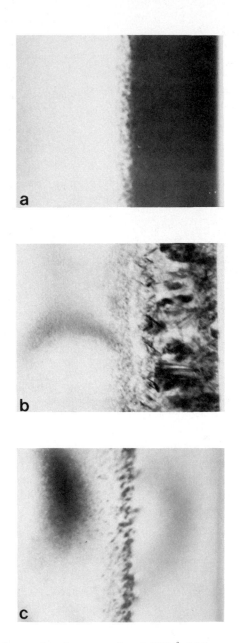

Fig. 30. TEM micrographs of cross-sectioned 1900-Å-thick amorphous layer samples before and after electron beam irradiation (a) as-implanted, (b) 0.5 J/cm², (c) 0.55 J/cm².

must be molten to avoid defect nucleation at the interface. For laser annealing this picture is consistent with the defect structures previously reported.

The TEM micrographs (Baeri *et al.*, 1980) of Fig. 30 refer to a cross section of 1900-Å amorphous silicon layer after electron pulse annealing. A heavily defective layer is observed in the range 0.42–0.5 J/cm² (see Fig. 30b). The structure of the annealed layer is quite complex: Epitaxial regrowth has occurred over a distance of 800 Å, with a large volume fraction of twinned regions; the remaining 1100-Å outer layer consists of polycrystalline material. Some of the grains have a dendritic morphology, indicative of a great speed of crystallization. By increasing the energy pulse above 0.55 J/cm² (see Fig. 30) good epitaxial regrowth with few dislocation lines is observed. The electron annealing behavior can be explained in the following way: In the range 0.42–0.5 J/cm² the amorphous layer melts at a temperature lower than T_M; the undercooled liquid then recrystallizes from both the single-crystal interface and from random nucleation centers at the free surface. These competitive processes produce the observed layered structure. From calculations of transient heating (Baeri *et al.*, 1980) the undercooled layer can exist in the range 0.4–0.55 J/cm² for 2000 Å of amorphous layer, in good agreement with the experimental data where a heavily defective layer is observed in the range 0.42–0.5 J/cm².

References

Anderson, C. L., Celler, G. K., and Rozgonyi, G. A., eds. (1980). "Laser and Electron Beam Processing of Electronic Materials." Electrochem. Soc., Princeton, New Jersey.

Antonenko, A. K., Gerasimenko, N. N., Dvurechensky, A. V., Smirnov, L. S., and Tseitlin, G. M. (1976). *Sov. Phys.—Semicond. (Engl. Transl.)* **10**, 81.

Baeri, P., Campisano, S. U., Foti, G., and Rimini, E. (1978). *Appl. Phys. Lett.* **33**, 137.

Baeri, P., Campisano, S. U., Foti, G., and Rimini, E. (1979). *J. Appl. Phys.* **50**, 788.

Baeri, P., Foti, G., Poate, J. M., and Cullis, A. G. (1980). *Phys. Rev. Lett.* **45**, 2036.

Bagley, B. G., and Chen, H. S. (1979). In "Laser–Solid Interactions and Laser Processing" (S. D. Ferris, H. J. Leamy, and J. M. Poate, eds.), p. 102. Am. Inst. Phys., New York.

Bean, J. C., Leamy, H. J., Poate, J. M., Rozgonyi, G. A., van der Ziel, J., Williams, J. S., and Celler, G. K. (1979). *J. Appl. Phys.* **50**, 881.

Bloembergen, N. (1979). In "Laser–Solid Interactions and Laser Processing" (S. D. Ferris, H. J. Leamy, and J. M. Poate, eds.), p. 1. Am. Inst. Phys., New York.

Campisano, S. U., Foti, G., and Servidori, M. (1980). *Appl. Phys. Lett.* **36**, 279.

Chu, W.-K., Mayer, J. W., and Nicolet, M.-A. (1978). "Backscattering Spectrometry." Academic Press, New York.

Crowder, B. L. (1971). *J. Electrochem. Soc.* **118**, 943.

Csepregi, L., Mayer, J. W., and Sigmon, T. W. (1975). *Phys. Lett. A* **54**, 157.

Csepregi, L., Mayer, J. W., and Sigmon, T. W. (1976). *Appl. Phys. Lett.* **29**, 92.

Csepregi, L., Kennedy, E. F., Gallagher, T. J., Mayer, J. W., and Sigmon, T. W. (1977). *J. Appl. Phys.* **48**, 4234.

Cullis, A. G., Webber, H. C., and Bayley, P. (1979). *J. Phys. C* **12**, 688.
Cullis, A. G., Webber, H. C., Mc Caughan, D. V., and Chew, N. G. (1980). *In* "Laser and Electron Beam Processing of Materials" (C. W. White and P. S. Peercy, eds.), p. 183. Academic Press, New York.
Dvurechensky, A. V., Kachurin, G. A., and Antonenko, A. K. (1978). *Radiat. Eff.* **37**, 179.
Ferris, S. D., Leamy, H. J., and Poate, J. M. eds. (1979). "Laser–Solid Interactions and Laser Processing." Am. Inst. Phys., New York.
Foti, G., Rimini, E., Vitali, G., and Bertolotti, M. (1977). *Appl. Phys.* **14**, 189.
Foti, G., Rimini, E., Bertolotti, M., and Vitali, G. (1978a). *Phys. Lett. A* **65**, 430.
Foti, G., Rimini, E., Tseng, W. F., and Mayer, J. W. (1978b). *Appl. Phys.* **15**, 365.
Foti, G., Campisano, S. U., Baeri, P., Rimini, E., and Tseng, W. F. (1979). *Appl. Phys. Lett.* **35**, 701.
Gat, A., and Gibbons, J. F. (1978). *Appl. Phys. Lett.* **32**, 142.
Gat, A., Gibbons, J. F., Magee, T. J., Peng, J., Williams, P., Deline, V. R., and Evans, C. A., Jr. (1978). *Appl. Phys. Lett.* **33**, 389.
Gibbons, J. F., Hess, L. D., and Sigmon, T. W., eds. (1981). "Laser and Electron Beam Solid Interactions and Materials Processings." North-Holland Publ., New York.
Greenwald, A. C., Kirkpatrick, A. R., Little, R. G., and Minnucci, J. A. (1979). *J. Appl. Phys.* **50**, 783.
Gyulai, J., and Revesz, P. (1979). *Conf. Ser.—Inst. Phys.* No. 46, p. 128.
Hodgson, R. T., Baglin, J. E. E., Pal, R., Neri, J. M., and Hammer, D. A. (1980). *Appl. Phys. Lett.* **37**, 187.
Hofker, W. K., Oosthock, D. P., Eggermont, G. E., Tamminga, Y., and Stacy, W. T. (1979). *Appl. Phys. Lett.* **34**, 690.
Hoonhout, D. (1980). Ph.D. Thesis, University of Amsterdam.
Hoonhout, D., Kerkdigk, C. B., and Saris, F. W. (1978). *Phys. Lett. A.* **66**, 145.
Kachurin, G. A., Pridachin, N. B., and Smirnov, L. S. (1975). *Sov. Phys.—Semicond. (Engl. Transl.)* **9**, 946.
Kennedy, E. F., Csepregi, L., Mayer, J. W., and Sigmon, T. W. (1977). *J. Appl. Phys.* **48**, 4241.
Kennedy, E. F., Lau, S. S., Golecki, I., Mayer, J. W., Tseng, W. F., Minnucci, J. A., and Kirkpatrick, A. R. (1979). *Radiat. Eff. Lett.* **43**, 31.
Khaibullin, I. B., Titov, V. V., Shtyrkov, E. I., Zaripov, M. M., Stashko, V. P., and Kuzmin, K. P. (1975). *Proc. Int. Conf. Ion Implantation, Budapest* (J. Gyulai, T. Lohner, and E. Pastor, eds.), p. 212. Budapest Cent. Res. Inst. Phys., Budapest.
Khaibullin, I. B., Shtyrkov, E. I., Zaripov, M. M., Bayazitov, R. M., and Galyatudinov, M. F. (1978). *Radiat. Eff.* **36**, 225.
Lau, S. S., Tseng, W. F., Nicolet, M.-A., Mayer, J. W., Minnucci, J. A., and Kirkpatrick, A. R. (1978). *Appl. Phys. Lett.* **33**, 235.
Lee, D. H., and Mayer, J. W. (1974). *Proc. IEEE* **62**, 1241.
McMahon, R. A., and Ahmed, H. (1979). *Electron. Lett.* **15**, 45.
Marquardt, C. L., Giuliani, J. F., and Fraser, F. W. (1974). *Radiat. Eff.* **23**, 135.
Mayer, J. W., Eriksson, L., and Davies, J. A. (1970). "Ion Implantation in Semiconductors." Academic Press, New York.
Merli, P. G., and Rosa, R. (1981). *Optik* **58**, 201.
Miyao, M., Tamura, H., Ohyu, K., and Tokayama, T. (1979). *In* "Laser–Solid Interactions and Laser Processing" (S. D. Ferris, H. J. Leamy, and J. M. Poate, eds.), p. 325. Am. Inst. Phys., New York.
Nakashima, H., Shiraki, Y., and Miyao, M. (1979). *J. Appl. Phys.* **50**, 5966.
Narayan, J. (1979). *Appl. Phys. Lett.* **34**, 312.
Narayan, J., and Young, F. W., Jr. (1979). *Appl. Phys. Lett.* **35**, 330.

Pickar, K. A. (1975). *In* "Applied Solid State Science" Vol. 5, p. 151. Academic Press, New York.

Regolini, J. L., Gibbons, J. F., Sigmon, T. W., Pease, R. F. W., Magee, T. J., and Peng, J. (1979). *Appl. Phys. Lett.* **34**, 410.

Revesz, P., Farkas, G., Mezey, G., and Gyulai, J. (1978). *Appl. Phys. Lett.* **33**, 431.

Rimini, E., ed. (1978). *Proc. Laser Eff. Ion Implanted Semicond., Univ. Catania.*

Rimini, E., Baeri, P., and Foti G. (1978). *Phys. Lett. A* **65**, 153.

Rimini, E., Chu, W. K., and Mader, S. R. (1981). *J. Appl. Phys.* **52**, 3696.

Schwuttke, G. H., and Howard, J. K. (1971). U.S. Patent No. 3,585,088.

Shtyrkov, E. I., Khaibullin, I. B., Zaripov, M. M., Galyatudinov, M. F., and Bayasitov, R. M. (1975). *Sov. Phys.—Semicond. (Engl. Transl.)* **9**, 1309.

Spaepen, F., and Turnbull, D. (1979). *In* "Laser–Solid Interactions and Laser Processing" (S. D. Ferris, H. J. Leamy, and J. M. Poate, eds.), p. 73. Am. Inst. Phys., New York.

Tamminga, Y., Eggermont, G. E. J., Hofker, W. K., Hoonhout, D., Garrett, R., and Saris, F. W. (1979). *Phys. Lett. A* **69**, 436.

Tan, T. Y., Tsu, R., and Lankard, J. R. (1980). *In* "Laser and Electron Beam Processing of Materials" (C. W. White and P. S. Peercy, eds.), p. 447. Academic Press, New York.

Tseng, W. F., Mayer, J. W., Campisano, S. U., Foti, G., and Rimini, E. (1978). *Appl. Phys. Lett.* **32**, 824.

Tsu, R., Baglin, J. E., Lasher, G. L., and Tsang, J. (1979). *Appl. Phys. Lett.* **34**, 168.

Tsu, R., Baglin, J. E., Tan, T. Y., and von Gutfeld, R. J. (1980). *In* "Laser and Electron Beam Processing of Electronic Materials" (C. L. Anderson, G. K. Celler, and G. A. Rozgonyi, eds.), p. 382. Electrochem. Soc., Princeton, New Jersey.

Venkatesan, T. N. C., Golovchenko, J. A., Poate, J. M., Cowan, P., and Celler, G. K. (1978). *Appl. Phys. Lett.* **33**, 429.

Vitali, G., Bertolotti, M., Foti, G., and Rimini, E. (1977). *Phys. Lett. A* **63**, 351.

Vitali, G., Bertolotti, M., Foti, G., and Rimini, E. (1978). *Appl. Phys.* **17**, 111.

von Allmen, M., Luthy, W., Thomas, J. P., Fallavier, M., Mackowski, J. M., Kirsh, R., Nicolet, M.-A., and Roulet, M. E. (1979). *Appl. Phys. Lett.* **34**, 82.

White, C. W., and Peercy, P. S., eds. (1980). "Laser and Electron Beam Processing of Materials." Academic Press, New York.

Williams, J. S., Brown, W. L., Leamy, H. J., Poate, J. P., Rodgers, J. W., Rousseau, D., Rozgonyi, G. A., Shelnutt, J. A., and Sheng, T. T. (1978). *Appl. Phys. Lett.* **33**, 542.

Young, R. T., and Narayan, J. (1978). *Appl. Phys. Lett.* **33**, 14.

Young, R. T., White, C. W., Clark, G. J., Narayan, J., Christie, W. H., Murakami, M., King, P. W., and Kramer, S. D. (1978). *Appl. Phys. Lett.* **32**, 139.

Young, R. T., Narayan, J., and Wood, R. F. (1979). *Appl. Phys. Lett.* **35**, 447.

Chapter 8

Epitaxy of Deposited Si

J. M. POATE and J. C. BEAN

Bell Laboratories
Murray Hill, New Jersey

I. Introduction

There is considerable emphasis in this book on the beam annealing of ion-implanted semiconductors. Implantation offers good control over impurity concentration and distribution but is limited by certain practical considerations. Ion beam energies, at present, generally fall in the range 100–500 keV. At such energies common semiconductor impurities penctrate to depths of only 0.1–1.0 μm. Implanted structures are therefore intrinsically shallow. Moreover, ion beam current and wafer throughput constraints limit final impurity concentrations to parts per thousand. Such levels are suitable for electrically active impurities but do not produce the alloy compositions useful in many devices. For these reasons, thick, heavily doped, or alloy layers are generally fabricated by an epitaxial deposition process.

The most common deposition process is chemical vapor deposition (CVD), where semiconductor and dopant species are transported as hydrides or chlorides in the gas phase near atmospheric pressure. The gas stream passes over heated substrate crystals where chemical decomposition and epitaxial deposition occur. The composition of the epitaxial layer is varied by changing the relative flow rates of the gas species. Chemical vapor deposition routinely produces low-cost deposition of doped semiconductor layers. Doping levels can be varied from 10^{13} to 10^{20}/cm^3, deposition rates can exceed 1 μm/min, and layer thicknesses can reach 100 μm. These strengths have made CVD a workhorse of the semiconductor industry. The process, however, is suitable only for the growth of certain materials and does not produce extremely abrupt changes in material composition. These limitations stem from two factors. First, finite gas flow velocities and the presence of stagnant boundary layers make it impossible to change rapidly the composition of the gas at the growing epitaxial surface. Second, the chemical decomposition of the gaseous species requires a rather high substrate temperature (e.g., 950–1200°C for the deposition of crystalline silicon on silicon). At these temperatures solid state impurity diffusion is significant, and the large temperature excursions accentuate grown-in strain when layers of different materials are grown on one another (heteroepitaxy). Further, high-temperature CVD may preclude the overgrowth of materials that, though crystallographically compatible, are thermodynamically unstable in contact with one another.

Most of the limitations of CVD growth can be overcome by the vacuum evaporation process known as molecular beam epitaxy (MBE). This technique is similar to CVD but eliminates both gas stagnation, by growing in vacuum, and high-temperature growth, by the use of monoatomic transport species. Molecular beam epitaxy has produced crystalline epitaxial silicon on silicon at temperatures as low as 450°C. Further, layers have been grown with doping profiles far more abrupt than with CVD, and epitaxial growth has been demonstrated using materials that could not withstand CVD processing temperatures (e.g., silicon–metal silicide–silicon heterostructures). Molecular beam epitaxy is, however, a rather complex technique requiring a large initial capital investment. Further, current wafer throughput is lower than that for CVD, and the growth of device-quality material requires very careful substrate preparation to eliminate the degrading effects of interfacial oxide and carbide contamination.

Important parameters of ion implantation, CVD, and MBE are listed in the first three rows of Table I. Certain of these techniques offer excellent control, rapid growth, thick layers, and cost-effectiveness. At present,

TABLE I

TYPICAL EPITAXIAL PARAMETERS

Technique	Growth rate (μm/min)	Thickness (μm)	Dopant concentration[a]	Substrate T (°C)	$\sqrt{Dt}$ (μm)
Ion implant	Not applicable	0.01–0.3	10	900[b]	0.01
CVD	0.1–1	0.1–100	1	950–1200	0.3
MBE	0.01–0.3	0.0001–10	1	450–900	0
Laser (solid)	10	0.001–1	10	RT–400	0
Laser (liquid)	10^8	0.001–1	1000	RT	0.3

[a] 1 = Solid solubility.
[b] For activation.

however, no single technique offers all these advantages. In the hope of combining these strengths laser processing of deposited layers has been intensively investigated over the past few years. In most experiments, laser epitaxy begins with a CVD- or MBE-like deposition done at so low a temperature that amorphous or polycrystalline growth occurs. The low temperature eliminates diffusion, and one can either exploit the thick, rapid, low-cost deposition of CVD or the good control of MBE. The role of the laser is then to produce the one factor that has been sacrificed, crystallinity. The laser accomplishes this by heating the deposited layer to a temperature where it will crystallize either by solid phase epitaxial growth or by melting and crystalline resolidification. As will be shown, the spatial and temporal localization of the laser irradiation permits crystallization with little or no disturbance of the underlying structure. This has the obvious potential advantage of permitting epitaxy over processed semiconductor structures.

The most active and exciting area of laser epitaxy at present is not the conventional vertical epitaxy but rather lateral epitaxy over amorphous and insulating substrates. The ability to controllably melt and recrystallize Si on SiO_2, Si_3N_4, and fused quartz has opened new dimensions of epitaxy and devices. This field is progressing extremely rapidly, and we shall summarize the methods by which nucleation and growth are controlled.

II. Homoepitaxy

A. *Liquid Phase*

The subject of epitaxial regrowth of amorphous ion-implanted Si layers is discussed in detail in Chapter 7 by Foti and Rimini. What differences do we expect between the epitaxy of amorphous implanted layers and that of deposited layers? There are clear distinctions between the interface and internal structure of deposited and implanted amorphous films. The structures in the near vicinity of the interface can be markedly different because, although the implanted amorphous–crystal interface will be generally exceptionally clean, it will not necessarily be atomically sharp. This point is well illustrated in Chapter 6 by Cullis where there are several transmission electron microscope (TEM) cross-section micrographs of the amorphous–crystal interfaces produced by implantation. The interfaces are not sharp but contain defects propagating into the single-crystal substrates. The defect density will depend upon the conditions of the implantation. For example, the interface may have much better definition for low-temperature implantations. On the other hand the as-deposited interface will have the structure of the surface on which the Si is deposited. In practice this means that the interface morphology can be completely dominated by an interfacial layer of impurities such as the native oxide on Si.

There is ample evidence that the internal structures of deposited amorphous Si films are not homogeneous. For example, evaporated layers can show strong evidence of interconnected void arrays into which atmospheric gases can percolate—we discuss these phenomena in detail in the following section on solid phase epitaxy. No such structures or behavior have been suggested for amorphous implanted films, and it is thought that they are homogeneous and have a higher density than evaporated films.

We first consider the case of liquid phase epitaxy of deposited layers. Several groups (Hoonhout et al., 1978; Lau et al., 1978a; Revesz et al., 1978) have investigated this epitaxy, but we shall discuss our own work (Bean et al., 1978, 1979a,b) as it emphasizes both interface and internal structure effects. Figure 1 shows the Rutherford backscattering and channeling spectra of Ga-doped amorphous Si films before and after laser-induced recrystallization. The films were prepared in an ultrahigh vacuum (UHV) chamber by electron beam evaporation. Before film deposition, the sample surfaces were atomically cleaned by Ar sputtering followed by an 850°C anneal. The Si was then evaporated at typical rates of 5 Å/sec at ambient pressures of 10^{-9} Torr to produce an amorphous layer 3000 Å thick (Fig. 1a). Gallium was coevaporated to produce a uniform doping of 2 at.% Ga in the amorphous Si, as shown in the lower Ga spectrum in Fig. 1b. After the samples were removed from the chamber they were irradiated with a Q-switched Nd:YAG laser ($\lambda = 1.064$ μm) pulsed repetitively at 11.4 kHz with pulse lengths of 124–135 nsec. The laser was

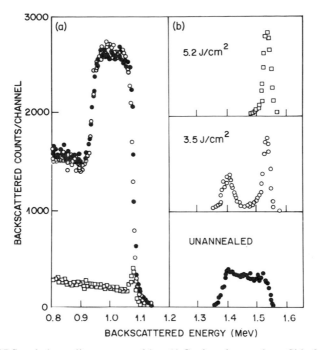

Fig. 1. RBS and channeling spectra of 2 at.% Ga-doped amorphous Si before and after pulsed Nd:YAG irradiation. (From Bean *et al.*, 1979b.)

focused to an $\sim$40-μm optical spot size and scanned across the samples at normal incidence with an 8-μm translation between pulses.

The mechanisms of laser-induced epitaxy are graphically illustrated by the spectra following laser irradiation. At an energy of 3.5 J/cm^2 the melt depth penetrates only about 2500 Å through the amorphous layer. On freezing, most of the Ga within this molten layer is zone refined to the surface. Epitaxy, however, did not result, as the melt did not wet the single-crystal substrate. Irradiation at an energy of 5.2 J/cm^2 causes the melt to penetrate the substrate, and epitaxial growth results, with virtually all the Ga being zone refined to the surface. The total Ga count also dropped by 20%, suggesting evaporation at the sample surface. Scanning electron microscopy analysis in conjunction with energy-dispersive x-ray analysis revealed small Ga droplets on the silicon surface. By etching the sample in HCl and then HF, surface Ga and oxide could be removed.

In order to examine the Ga incorporated within the epitaxial layers, the surface Ga was etched away. Rutherford backscattering and channeling measurements (Fig. 2) showed the Ga to be distributed in an exponential

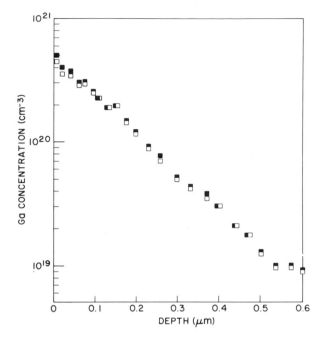

Fig. 2. Laser-annealed α-Si–Ga. Redistribution of Ga in Si layer after Nd:YAG irradiation at 5.2 J/cm^2. ■, Total Ga; □, substitutional Ga. (From Bean *et al.*, 1979b.)

profile with a peak solubility (4.5×10^{20} cm^{-3}) exceeding the maximum solubility by an order of magnitude. This supersaturation was demonstrated by furnace annealing the sample at a temperature of 1250°C which corresponds to the maximum of the retrograde solubility curve. At this temperature the maximum solubility dropped to $\sim 6 \times 10^{19}$ cm^{-3}, along with significant Ga diffusion and surface evaporation. It should be noted that the profile in Fig. 2 does not show the shape typical of zone refining as given by Baeri and Campisano in Chapter 4 and by White, Appleton, and Wilson in Chapter 5 because each spot is melted many times, thus producing an exponential profile with maximum supersaturation at the surface.

One of the scientific advantages of studying the laser epitaxy of amorphous films deposited on Si is that the interface condition, such as impurity content or damage, can be varied. Interface contaminants were studied by growing 15 Å of native oxides on the silicon substrates using the procedure described by Henderson (1972), or etching the surfaces with HF to leave approximately one monolayer of oxide (Feldman *et al.*, 1978). Substrates were also sputtered with 1-kV Ar, but without the subsequent annealing step, so that approximately one monolayer of Ar was incorporated within a narrow damaged surface layer.

In order to evaluate the effect of interface conditions upon recrystallized layer quality, samples were analyzed by Rutherford backscattering and channeling. The ratio (χ_{min}) of the aligned to random yields near the surface is a measure of crystalline quality, and a value of 3.0% was obtained for $\langle 100 \rangle$ channeling in bulk silicon. Typical channeling backscattering spectra are shown in Fig. 3 for an undoped layer deposited on a substrate that was cleaned by argon sputtering followed by a 10-min 850°C thermal anneal. The spectrum of the layer before laser annealing contains an amorphous peak extending from 0.8 to 0.9 MeV. After laser irradiation at a pulse energy of 7.9 J/cm³ the width of the peak is unchanged, indicating that negligible silicon has been lost through evaporation. The near-surface yield (at ~0.87 MeV) has, however, dropped to 38% of the amorphous value, suggesting partial recrystallization within the 1-

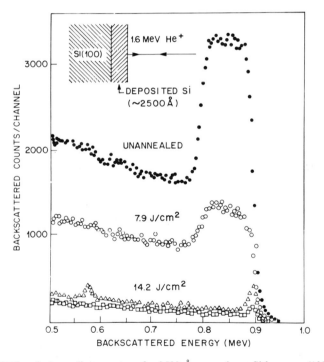

Fig. 3. RBS and channeling spectra of a 2500-Å amorphous Si layer on (100) Si before and after Nd:YAG irradiation. The lowest spectrum corresponds to the 14.2-J/cm² irradiation following a HF etch. (From Bean *et al.*, 1979a.)

mm-diameter region probed by the He^+ backscattering beam. For a 14.2-J/cm² pulse energy, the as-annealed spectrum has a χ_{min} of 3.8% but also contains a somewhat high surface silicon peak at 0.90 MeV and a surface oxygen peak at 0.58 MeV. The χ_{min} value of 3.8% is significantly higher than the 3% measured for perfect (100) silicon and could be explained by partial He^+ dechanneling in a thin amorphous surface oxide. To test this possibility, samples were immersed for 30 sec in HF and reexamined. The 0.58-MeV oxide peak disappeared, and the value of χ_{min} dropped to the bulk figure of 3.0%. By examining very thin unannealed amorphous layers, it was verified that this surface oxide was not present prior to laser annealing. Although the laser-induced oxide concentration was typically only 2–3 × 10¹⁶ oxygen atoms/cm², this quantity was sufficient to cause significant dechanneling. Therefore, χ_{min} values were measured after a short HF immersion. Examination of partially recrystallized or doped layers verified that this etch step did not attack the silicon layers.

The effects of the substrate preparation procedure upon the value of χ_{min} are summarized in Fig. 4. For undoped or lightly doped layers deposited on substrates cleaned by argon sputtering and thermal annealing, crystallization occurs abruptly at a laser energy density of about 10 J/cm². At power densities below 10 J/cm² the molten zone does not penetrate to the crystalline substrate, and as it cools it therefore solidifies into randomly oriented polycrystallites. Above 10 J/cm² the melt penetrates to the substrate, and epitaxial regrowth occurs. Intermediate χ_{min} values, such as that at 38%, are due to nonuniform melt penetration which produces epitaxial recrystallization only near the center of the laser spots. For substrates prepared by HF oxide removal or argon sputtering without thermal anneal, crystallization also occurs at about 10 J/cm², but bulk crystalline perfection is not achieved for pulses of less than 15 J/cm². To recrystallize layers deposited on such surfaces, higher laser energies (and, therefore, greater melt thicknesses) are required to disperse these impurities and produce unfaulted epitaxial silicon. This effect is even more pronounced for layers deposited on substrates retaining ~15 Å of native oxide. For these films, bulk χ_{min} values are attained only with pulse energy densities greater than 20 J/cm².

As we have already discussed, the results of laser-induced epitaxy of deposited layers in the liquid phase regime are quite similar to those for implanted amorphous layers in terms of crystal quality and incorporation of conventional dopants provided the melt penetrates well through any interfacial impurities. There is one area, however, where the results were quite different in that voids were observed (Bean *et al.*, 1978) at the boundary between the recrystallized and amorphous material. Figure 34 in Chapter 6 by Cullis is a transmission electron micrograph of the edge of

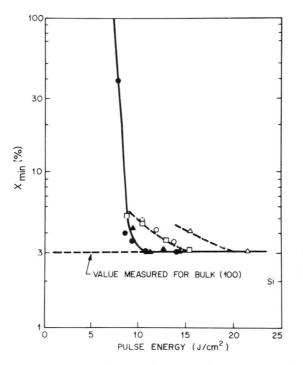

Fig. 4. Channeling minimum yields (χ_{min}) for recrystallized layers as a function of laser power and interface condition. Substrate condition: ●, well ordered, atomically clean; ○, argon sputter damage; □, monolayer oxide; △, 15 Å oxide. (From Bean *et al.*, 1979a.)

such a laser-irradiated region. The amorphous deposit on the left can be identified by the inserted diffraction pattern with its diffuse halo superimposed upon the [100] spot diffraction pattern of the underlying substrate. On the right the laser-processed material consists of defect-free epitaxial silicon. The boundary region between the amorphous and single-crystal regions consists of polycrystallites and voids, labeled "V" in the micrograph. The voids are numerous in number and have small diameters (<5000 Å).

The polycrystalline region to the left of the voids is formed at the boundary of the laser-processed area where the pulse energy is not sufficient to melt through the amorphous Si to the underlying single-crystal Si. The film therefore recrystallizes without the benefit of a seed for epitaxial growth. The polycrystalline transition region is large for the special case of λ = 1.06 μm light because of the larger absorption coefficient of the amorphous layer relative to crystalline material. Once converted to polycrystalline material, the coupling of the light is insufficient to cause remelting. The adjoining scan of laser pulse spots therefore fails to remelt not only the polycrystalline boundary but also an 8-μm-wide region of epitaxially resolidified material. It is within this singly melted region that the voids are observed.

This phenomenon has been investigated further by Leamy et al. (1980a). Amorphous Si films were deposited at $\sim 10^{-6}$ Torr onto chemically cleaned substrates. Epitaxial regrowth was observed at Nd:YAG energies equivalent to those required to recrystallize epitaxially the previous UHV films. Void formation was observed at the boundary of the epitaxially grown material which was melted only once. Experiments were therefore carried out on single shots, and voids were observed over the entire width of the laser-annealed shot. Surface disruption was observed where a void intersected the surface.

The origin and stability of the voids must be associated with a fundamental difference between deposited amorphous Si and amorphous Si produced by conventional implantation. In the next section we show that amorphous deposited material can contain free volume in the form of small, columnar voids which may, indeed, be saturated with atmospheric gases. The cavities in the laser-annealed material are, therefore, probably not true voids but are rather gas-filled bubbles with the gaseous impurity, such as O_2, stabilizing the voids. This contention is supported by the TEM cross-section studies of Cullis et al. (1980), who observed Ar bubbles in Si that had been amorphized by Ar implantation and subsequently recrystallized by laser irradiation. It should be emphasized that void formation is not an inevitable consequence of liquid phase epitaxial regrowth on deposited films. Recently Celler et al. (1981) have melted deposited Si films, with varying density, on Si with Q-switched Nd:YAG irradiation. Increased density of the films was achieved by depositing at high rates on substrates held at 300°C. They observed that the dense films crystallized epitaxially over a wide range of laser energies. In order to recrystallize films epitaxially with 20% lower density, as determined by spectroscopic ellipsometry, higher laser energies were required. Moreover, the less dense films suffered from severe surface pitting. It was believed that coalescence of the excess void volume into microbubbles, stabilized by gaseous contaminants, was responsible for the surface degradation. Indeed, transmission electron microscopy of crystallized samples showed that the less dense samples were filled with a great number of microbubbles, whereas the dense film showed only a few spherical voids. In the next section we present recent measurements of solid phase recrystallization demonstrating that films evaporated at high rates have a much lower interconnected void density.

B. Solid Phase

While many investigations have concentrated on the liquid phase crystallization of deposited layers, relatively few have attacked the problem of
solid phase crystallization. The reason for this paucity of results is simply
due to the fact that most laser crystallization measurements have been
carried out in air and the solid phase crystallization of Si is very sensitive
to entrapped impurities. In this section we first demonstrate some of the
potentialities of solid phase recrystallization of Si and then consider the
problems of impurity entrapment.

Figure 5 (Bean *et al.*, 1979b) shows Rutherford backscattering and
channeling spectra of a 2000-Å Si film deposited on (100) Si. Deposition
and cleaning procedures were the same as those described in the previous
section, except that Ge was codeposited with Si to produce a uniform Ge
doping of approximately 1 at.%. The samples were then irradiated in air
with a continuous-wave (cw) Nd : YAG laser. The laser spot was moved
with dwell times of ~1 msec so that the Si temperature did not exceed
900°C. Good-quality epitaxial crystallization was achieved with a χ_{min} of

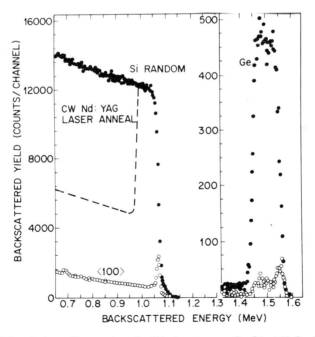

Fig. 5.   RBS and channeling spectra of solid phase regrowth of 1 at.% Ge-doped Si film
annealed with cw Nd:YAG laser. (From Bean *et al.*, 1979b.)

5%, with complete incorporation of Ge at lattice sites. Moreover there was no detectable movement of the Ge, demonstrating that regrowth had occurred in the solid phase. The Ge concentration profile is practically identical to the as-deposited profile. It is conceivable in fact that the Ge boundary is atomically sharp.

The fabrication of such uniformly doped epitaxial layers with extremely sharp interfaces appears to offer interesting device possibilities. The main drawback is that solid phase crystallization, unlike liquid phase crystallization, is extremely sensitive to incorporated impurities. For example, many attempts to produce epitaxial growth in the solid phase resulted only in polycrystalline formation. This point is well illustrated by the data of Hess *et al.* (1980). Films of thickness between 500 and 5000 Å were deposited under UHV conditions at a rate of 1 Å/sec and then Ar laser-annealed in air. The degree of epitaxial regrowth was measured by Rutherford backscattering and channeling. Incomplete recrystallization occurred, with a layer of polycrystalline Si formed at the surface. The thickness of the polycrystalline layer increased monotonically with the starting film thickness, as shown in Fig. 6.

Much work over the past ten years has demonstrated that evaporated Si layers contain interconnected void arrays (see, e.g., Donovan and

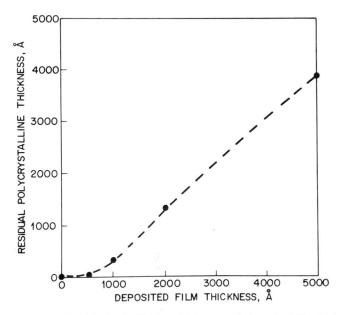

Fig. 6.   Variation of residual poly-Si layer thickness with deposited film thickness for cw Ar annealing in air. (From Hess *et al.*, 1980.)

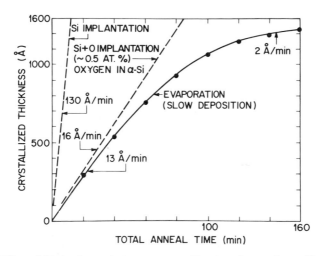

Fig. 7. Effect of fabrication technique on crystallization of amorphous silicon at 562°C. Solid phase recrystallization rates for amorphous layers formed by evaporation and deposition. (From Bean and Poate, 1980a.)

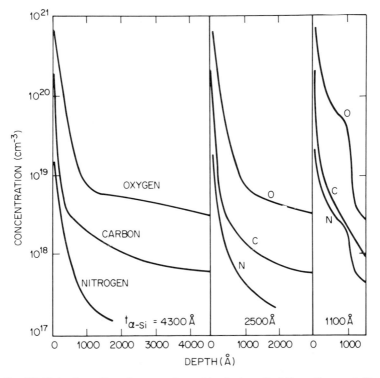

Fig. 8. SIMS depth profiles of gaseous impurities in deposited films after crystallization. (From Foti *et al.*, 1980.)

Heineman, 1971). Recent studies on the furnace annealing of deposited films have shown that the inhibition of regrowth of films exposed to air is due to the incorporation of oxygen by the voids. Figure 7 shows furnace measurements (Bean and Poate, 1980a) of the epitaxial recrystallization of 1900-Å Si films deposited on (100) Si; the films had been exposed to air. Plotted in the same figure are the Caltech data (see Gibbons and Sigmon, Chapter 10, this volume, for the recrystallization of implanted amorphous films). One set was implanted with Si and the other with both Si and O. The regrowth rate of the deposited layer starts off linearly but then slows down appreciably toward the surface. Initially the growth rate is close to that of the Si–O-implanted layer but is an order of magnitude slower than the clean Si implanted layer. Figure 8 shows secondary-ion mass spectroscopy (SIMS) measurements of the O, C, and N concentrations of as-deposited films that had been exposed to air (Foti *et al.*, 1980). For samples thicker than 2500 Å the exponential depth profiles for the O can be explained by gas percolation into randomly connected voids of uniform

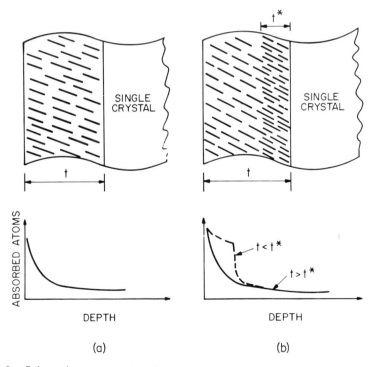

Fig. 9. Schematic representation of two possible distributions of voids. (a) Uniform, (b) nonuniform. Below each is shown the expected distribution of absorbed gases. (From Magee *et al.*, 1981.)

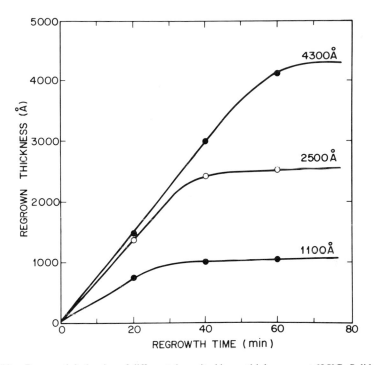

Fig. 10. Regrowth behavior of different deposited layer thicknesses at 625°C. Solid phase recrystallization rates for deposited amorphous Si layers of various thicknesses. (From Foti *et al.*, 1980.)

density. However, the depth profiles for the thinner film suggest that the distribution of voids in the material is not uniform. We assume that the volume or the density of voids increases sharply within a distance t^* of the amorphous–crystal interface. Figure 9 (Magee *et al.*, 1981) shows a schematic of deposited films with uniform and nonuniform depth distributions of voids. In the uniform case (Fig. 9a), the profile of absorbed atoms is independent of the film thickness. For the nonuniform case (Fig. 9b), if the thickness of the film is greater than t^*, the exponential profile of the impurities is almost unchanged. If, however, the thickness is less than t^*, gas diffusion is greatly enhanced. The experimental data in Fig. 8, for example, indicate that t^* is approximately 2500 Å.

Epitaxial recrystallization measurements (Foti *et al.*, 1980) of 1100-, 2500-, and 4300-Å as-deposited films substantiate this observation (Fig. 10). If incorporated O controls the epitaxial regrowth rate, then we would expect the initial regrowth rate for the 2500- and 4300-Å films to be similar. This behavior is experimentally observed after a regrown thickness of

approximately 2000 Å. The initial regrowth rate of the 1100-Å film, however, is approximately a factor of 2 slower than those of the thicker films.

Solid phase epitaxial recrystallization of amorphous Si is therefore
strongly inhibited by the presence of O. This effect is accentuated in
deposited films because of the void network through which gases can
percolate. Implanted layers do not have such networks and are quite
impervious to O percolation. Indeed, Si samples with implanted surfaces
can be stored in air for several years without inhibition of the epitaxial
regrowth rate. The generation of voids in deposited films can be overcome
by depositing films at rates greater than 50 Å/sec (Hess *et al.*, 1980; Hung
et al., 1981; Roth *et al.*, 1981). Auger compositional measurements of such
layers (Hess *et al.*, 1980) show that the O concentrations approach bulk
single-crystal values and that epitaxial regrowth rates are the same as
those of implanted amorphous layers.

The existence of the void network *per se* does not necessarily give rise
to a reduction in the regrowth rates at temperatures above 700°C, as has

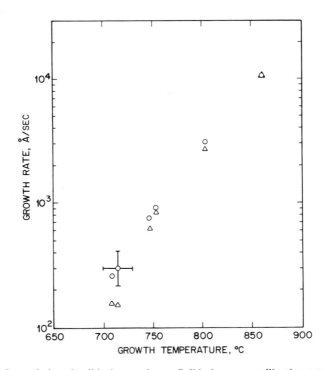

Fig. 11. Laser-induced solid phase epitaxy. Solid phase crystallization rates for amorphous layers formed by ion implantation (○) and UHV evaporation (△), determined from
time-resolved reflectivity measurements. (From Roth *et al.*, 1981.)

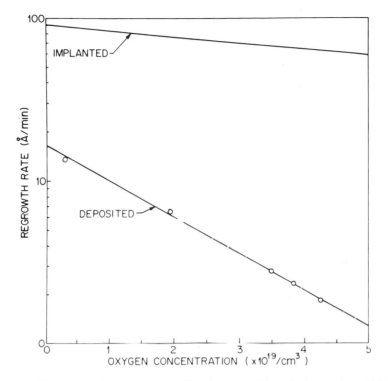

Fig. 12. Comparison of regrowth rates of implanted and deposited amorphous Si films as a function of oxygen concentration for furnace annealing at 550°C. (From Foti *et al.*, 1980.)

been demonstrated in elegant experiments by the Hughes group (Roth *et al.*, 1981). Amorphous Si films were deposited on (100) Si substrates under UHV conditions and at deposition rates of 1 Å/sec. Direct *in situ* measurements under UHV of the solid phase kinetics were obtained by heating with a cw laser and observing the movement of the crystalline–amorphous interface using time-resolved reflectivity techniques (Olson *et al.*, 1980). The regrowth rates of the deposited films were compared directly with those of implanted layers by performing the measurements on samples that contained both a deposited film and an ion-implanted region on the same sample. This sample geometry ensured that measurements were made at the same temperatures.

Figure 11 shows the growth rate for both deposited and implanted films over the temperature range 700–850°C. Allowing for experimental uncertainties, there are no differences in regrowth rates for the two types of amorphous Si for temperatures greater than 750°C. At temperatures in the vicinity of 700°C the regrowth rates of the UHV-deposited films appear substantially slower than those of the implanted films. Similar observations had previously been made by Foti *et al.* (1980) at lower temperatures. The 550°C regrowth rates of deposited Si as a function of oxygen concentration were extrapolated to zero oxygen concentrations. It was observed (Fig. 12) that the deposited films with zero extrapolated oxygen concentrations had regrowth rates approximately a factor of 5 smaller than those of oxygen-free implanted layers. It appears, therefore, that at temperatures below 700°C the void network even without the presence of oxygen contamination influences the kinetics of regrowth.

III. Heteroepitaxy

The ability to melt surface layers raises the possibility of the growth of heteroepitaxial structures without the scrupulous interface cleaning procedures that are necessary in the solid or vapor phase. The first demonstration of heteroepitaxy by pulsed irradiation is due to Lau *et al.* (1978b, 1979a) who grew single-crystal Ge on Si. Amorphous Ge layers 2000–3000 Å thick were evaporated on (100) and (111) substrates at rates of ~30 Å/sec and pressures of 1×10^{-7} Torr. The substrate surfaces were not atomically clean. The samples were irradiated with a 50-nsec electron pulse with a mean energy of 20 keV and a deposited energy of 0.8 J/cm^2.

Figure 13 shows the backscattering spectra of this structure after irradiation. The amorphous Ge film has transformed into a single-crystal film with a minimum yield (χ_{min}) near the surface of 6.5% and increases to 37% near the Ge–Si interface. This increase in the channeling yield is due to defects in the epitaxial layer and is observed, for example, for epitaxial Si layers grown on sapphire ($1\bar{1}02$) substrates. Transmission electron microscopy showed that dislocations were the predominant defects. Similarly, good epitaxy was observed for pulse-irradiated Ge on (111) Si, but the predominant defects were stacking faults. Little interdiffusion is observed at the interface between the Ge and Si. Leamy *et al.* (1980b) have performed similar experiments using pulsed Nd:YAG irradiation. At high energies, however, considerable interdiffusion of the Ge and Si occurs to form epitaxial alloys with a cellular defect microstructure (see von Allmen and Lau, Chapter 12, this volume).

One of the more important heteroepitaxial structures is Si on sapphire (SOS), and there have been numerous attempts to improve the Si crystallinity by pulsed and cw laser annealing. Chemical vapor-deposited Si films on sapphire contain a high density of stacking faults and twins near the sapphire interface (Ham *et al.*, 1977). This large defect density is attri-

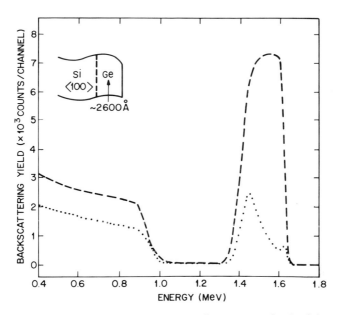

Fig. 13. Rutherford backscattering and channeling spectra of epitaxial regrowth of a 2600-Å Ge film on Si by pulsed 1.5-MeV He⁺ electron beam irradiation. – – –, Random; · · · , ⟨100⟩-aligned. (From Lau *et al.*, 1978b.)

buted to the ≈10% lattice mismatch and the large difference in thermal expansion coefficients. Lau *et al.* (1979b) were able to reduce the defect density by forming a buried amorphous layer by implantation and then furnace annealing at low temperatures (~550°C). Epitaxial regrowth of the amorphous layer starting from the relatively perfect Si surface region leads to a much improved Si crystalline layer.

Several groups (Roulet *et al.*, 1979; Golecki *et al.*, 1980, 1981; Sai-Halasz *et al.*, 1980; Yamada *et al.*, 1980) have shown that surface melting and recrystallization of implanted SOS using cw or pulsed lasers can result in improved crystallinity of the Si over the starting CVD material. In this case, however, liquid phase epitaxy will probably start at the sapphire interface. Golecki *et al.* (1981) have shown that this will result in localized melting of the sapphire and incorporation of Al within the Si. However, Yaron *et al.* (1980) have shown that at certain laser energies the electrical characteristics of SOS devices are not degraded but are improved. They conclude that in their experiments the interface did not melt.

It has recently been demonstrated by Saitoh *et al.* (1980) and Bean and Poate (1980b) that Si–silicide heterostructures can be formed by MBE techniques. In particular, they fabricated a structure of the form Si–CoSi$_2$–Si. The top layer of epitaxial Si was formed by deposition at temperatures ~650°C on the epitaxial CoSi$_2$ which has a cubic structure and lattice match within 1% of those of Si. Ishiwara *et al.* (1981) have deposited amorphous Si on CoSi$_2$ and produced partial epitaxy by melting the Si with pulsed ruby laser irradiation. It is interesting that, although the Si melted, there was little interdiffusion from the CoSi$_2$ into the Si.

IV.   Lateral Si Epitaxy on Amorphous Substrates

A.   *Continuous Amorphous Substrates*

There has been continuing interest in the possibility of producing crystalline Si, or other semiconductors, on insulating or amorphous substrates. For example, Maserjian (1963) described the growth of single-crystal Ge films on sapphire substrates using a scanning electron beam. Films ~4 μm thick could be melted to produce single-crystal grains approximately 1 mm across. Maserjian (1963, p. 478) states:

> By electrically deflecting the electron beam, the molten zone is made to scan the film in a preferred manner. As continuous melting and recrystallization takes place, the adjacent film acts to seed the growth occurring along the trail of the moving molten zone. Very small crystallites grow initially, but by overlapping the path of the moving zone, relatively few large crystallites of preferred orientation survive. This process is developed further by means of a more elaborate scanning pattern, so that a single-crystal is grown over a selected area.

This quotation summarizes the salient points of nucleation and growth on amorphous substrates. In the previous sections describing vertical epitaxy, crystallization was reported to occur from the substrate at velocities of ~1 m/sec without any random nucleation. The situation is very different for the lateral growth of Si on amorphous substrates where the intent is not to melt the substrate.

The potentialities of this technique for Si technology were first demonstrated by the Stanford group (Gat *et al.*, 1978) who showed that the conductivities of polycrystalline Si films could be increased by over a factor of 2 by cw laser scanning. A cw Ar laser with a power of 14 W and a spot size of ~40 μm was scanned at a constant speed of 12.5 cm/sec over 0.6-μm-thick poly-Si that had been deposited by CVD techniques on Si$_3$N$_4$ and SiO$_2$. Figure 39 in Chapter 6 by Cullis is a transmission electron micrograph of such a sample, showing the dramatic increase in grain size from small crystals (~500 Å) to long, narrow crystals (~25 × 2 μm). The surface is considerably smoother following the laser scanning. The growth mechanism is clearly due to the motion of a melt puddle, as has been demonstrated by the diffusion of dopant markers (Gibbons, 1981).

The orientation of the laser-recrystallized Si on SiO_2 on Si_3N_4 has been determined by Kamins *et al.* (1980a). Films of low-pressure chemical vapor-deposited poly-Si were deposited at 625°C to a thickness of 5500 Å. Samples were furnace-annealed at 1100°C, as well as cw Ar laser-recrystallized as described previously. The preferred orientation in the as-deposited films was (110). After furnace annealing the number of (110) grains decreased slightly, but (110) was still the dominant orientation. After laser melting, however, the population of (111), (113), and (331) grains increased significantly.

One of the reasons for the nucleation and growth of several orientations in laser melting is the thermal gradient produced by the circular beam spot. As discussed by Biegelsen *et al.* (1981), a circular shape is just about the worst for controlling nucleation in the trailing, recrystallizing wake of the molten puddle. Figure 14a shows the isotherm in the region of the solidification front for a circular spot. Random nucleation can occur anywhere along the curve, and crystal growth will be driven by the strong thermal gradient perpendicular to the isotherm. Nucleation at points A, for example, will produce crystallites that grow onto the beam path. A preferable spot shape is shown in Fig. 14b. A concave spot shape ensures that competitive crystallites grow out of the beam path, and lateral epitaxy of a single orientation should result. Biegelsen *et al.* (1981) have demonstrated this longitudinal epitaxy for such a spot shape. Figure 15 is

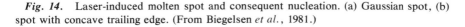

Fig. 14.   Laser-induced molten spot and consequent nucleation. (a) Gaussian spot, (b) spot with concave trailing edge. (From Biegelsen *et al.*, 1981.)

Fig. 15.   TEM photograph of shaped-spot scan on a prethinned 0.5-μm poly-Si on Si$_3$N$_4$ sample. The scan is from left to right, and the ~20 × 100-μm grain has grown from the cool point. (From Biegelsen *et al.*, 1981.)

their TEM photograph of one laser scan from left to right across a prethinned sample of 0.5-μm Poly-Si on Si$_3$N$_4$. The TEM bend contours show that one crystallite quickly dominates and grows out to the extent of the negative radius of curvature.

Large single-crystal grains have been grown on fused quartz substrates (Kamins and Pianetta, 1980; Johnson *et al.*, 1981; Bösch and Lemons, 1982) using scanning cw lasers. Because of the large differences in the thermal coefficients of expansion between Si and quartz, the crystallite films will crack on cooling. Kamins and Pianetta (1980) demonstrated how the cracking could be eliminated by stress relief grooves for defining Si islands. This approach will be discussed in the next section. Figure 16 is a channeling electron micrograph of Bösch and Lemons (1982) of the large-scale grains formed by the cw Ar laser crystallization of polysilicon on quartz. Grains up to 100 μm in length or 40 μm in width have been observed. These grains are considerably larger than those formed on SiO$_2$ on Si. One possible explanation is that the lower thermal conductivity of quartz, as compared to that of SiO$_2$ on Si, ensures that less power is required to melt the Si. Larger melt puddles can therefore be produced on quartz.

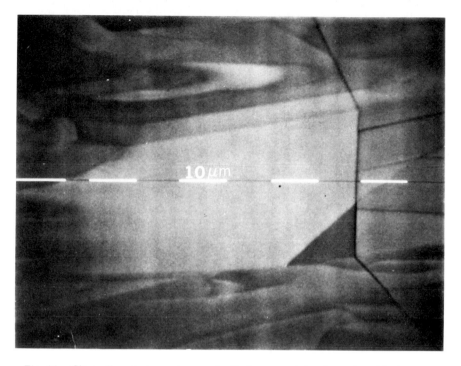

Fig. 16. Channeling electron micrograph of a large crystal grain produced by laser crystallization of a 0.5-μm Si film on fused quartz. The cracks formed on cooling are clearly evident. (From Bösch and Lemons, 1982.)

Much of the stimulus for the work on lateral crystal growth came from the successful attempts to fabricate MOS transitions and integrated circuits directly on beam-recrystallized poly-Si. The first MOSFET devices made on laser-recrystallized poly-Si films were reported by Lee *et al.* (1979). The poly-Si was recrystallized with cw Ar laser parameters similar to those described by Gat *et al.* (1978). The Si was recrystalized on a 1000-Å layer of Si_3N_4. For integrated circuit applications, a thick SiO_2 layer is a better choice for insulating material, and Tasch *et al.* (1979) have successfully fabricated MOSFET devices on laser-annealed poly-Si on SiO_2. Ring oscillators have also been fabricated on laser-recrystallized poly-Si on Si_3N_4 and SiO_2 (Kamins *et al.*, 1980b; Lam *et al.*, 1980a).

The reason that these devices have been fabricated successfully is that they utilize majority-carrier properties and are much less sensitive to the lifetime and grain boundary effects that can dominate the characteristics of bipolar devices (Gibbons, 1981). Moreover the channel lengths for MOS transition are on the order of, or smaller than, the grain sizes in the

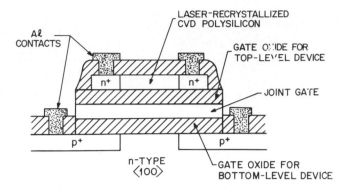

Fig. 17.   Geometry of one-gate-wide CMOS inverter produced by laser crystallization. (From Gibbons and Lee, 1980.) © 1980 IEEE.

laser-recrystallized films. If devices are fabricated on fine-grained poly-Si (Kamins, 1972), the grain boundaries can dominate the electical behavior.

The success of these lateral epitaxial techniques led Gibbons and Lee (1980) to propose a novel device structure in which the bulk Si is used for one device and the bottom of the laser-recrystallized poly-Si film is used for the second device. They have constructed a one-gate-wide CMOS inverter in which the bulk Si is used for the p-channel device and the laser-recrystallized poly-Si film for its n-channel complement. The device structure is shown in Fig. 17. There is a single gate to drive both the n and p-channel devices. This two-level "high-rise" configuration appears to be the first example of a vertical integrated circuit.

B.   Island Growth

Although the recrystallization of continuous Si sheets on amorphous substrates can produce single grains of considerable size, the presence of grain boundaries cannot be completely obviated. Gibbons *et al.* (1979) overcame this problem by prepatterning the poly-Si into islands from 2×20 to 20×160 μm. Argon laser scanning was found to produce single-crystal $\langle 100 \rangle$ material in the 2×20 μm islands, but the larger islands were polycrystalline. Lam *et al.* (1980a) prepatterned poly-Si islands to produce ring oscillators with a 6-μm channel length for the driver transistor. They noted that the laser power window for successful recrystallization was very narrow because of the loss of shape of the poly-Si islands during melting and recrystallization. Surface tension in the molten Si causes the rectangular islands to take on a rounded shape after laser processing. Lam *et al.* (1980b) obviated the rounding effect by use of an oxide retaining wall technique for controlling the shape of the islands. An oxidation process was adopted to produce the oxide retaining wall. In this manner they were able to fabricate MOSFET structures with a higher surface mobility than that of silicon on sapphire.

As discussed previously, the laser recrystallization of Si on quartz substrates can produce very large grains, but crazing or cracking of the films results because of the large differences in thermal expansion. Kamins and Pianetta (1980) prevented cracking by using either stress relief grooves or the oxide retention technique. In this manner they were able to construct MOSFET devices on quartz of reasonably good quality. The sequence of island formation and completed transistor structure is shown in Fig. 18. The ability to fabricate devices on transparent, insulating substrates represents a significant advance in device processing technology.

Biegelsen *et al.* (1981) have investigated in detail the effects of island shape, underlying amorphous layers, and laser spot shape on lateral recrystallization. By tapering the islands, to reduce the probability of competitive nucleation, and using the elliptical beam shape discussed previously, it is possible to grow reproducibly islands of single-crystal Si as large as 20 μm wide on amorphous substrates. They found that encapsulating the whole island with a thin layer ($\leq$200 Å) of Si_3N_4 produced a flat surface on the islands.

C. Seeded and Bridging Epitaxy

A logical step in controlling the nucleation process is to provide a nucleation site at the underlying Si substrate. Tamura *et al.* (1980, 1981) were the first to demonstrate that Si single-crystal growth on SiO_2 could be achieved by seeding from the substrate. They called this technique "bridging epitaxy." In their experiments, (100) wafers were oxidized to form a 2500-Å surface film of SiO_2. Lines 1–5 μm in width were then etched, by standard photolithographic techniques, in the SiO_2, thus exposing the Si substrate. The widths of the SiO_2 lines were also 1–5 μm. Poly-Si films ~3000 Å thick were then deposited on these lined structures. The samples were irradiated with a Q-switched ruby laser with a pulse length of 25 nsec and a beam diameter of 1 cm. In the energy interval of 1–2 J/cm² good lateral epitaxy resulted over the SiO_2 with a maximum width of 4 μm.

At an optimum energy of 1.5 J/cm² Tamura *et al.* observed the formation of (100) single-crystal material on SiO_2. Very few defects were observed in the Si on SiO_2 except for a single, straight grain boundary at the

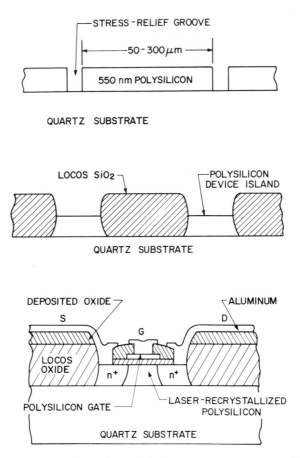

Fig. 18. Cross section of experimental device structures on quartz. (a) Stress relief grooves etched into poly-Si film on quartz before laser recrystallization; (b) poly-Si film defined into device islands by LOCOS oxidation before laser recrystallization; (c) completed transistor structure. The source (S), gate (G), and drain (D) are indicated. (From Kamins and Pianetta, 1980.) © 1980 IEEE.

center. There was, however, a high density of dislocations in the regrown Si on Si. In these experiments the single laser pulse melted a large area, and the direction of solidification depended primarily on the heat flow. Tamura *et al.* propose that solidification starts at the regions over the exposed Si and normal vertical epitaxy results. The solidification of the Si on the SiO_2 comes later because of the lower thermal conductivity of the SiO_2. In this way the lateral crystallization front propagates from *both* oxide edges, thus producing a straight boundary where the two fronts meet. It is intriguing that there is a high density of dislocations in the seeded region but not in the Si on SiO_2. These authors propose that dislocations originating in the seeded region escape to the sample surface along the inclined plane of the SiO_2 edge. The dislocations are not, therefore, propagated during the lateral epitaxy. It is not apparent, however, why there is such a high density of dislocations in the seeded regions. Such high densities are not observed for the vertical epitaxy of deposited amorphous films.

The use of scanning heat sources appears to be a natural extension of this technique. Epitaxy could be initiated over the seeded region and the melt puddle could, consequently, be swept over the oxide in a controllable fashion. Scanning cw Ar lasers have been used by several groups (Lam *et al.*, 1980c; Kamins *et al.*, 1981; Fastow *et al.*, 1981) to produce lateral seeded epitaxy. Lam *et al.* reported that single-crystal growth of 30 μm over the SiO_2 could be achieved by this technique. Much larger growth of 100 μm was achieved by Fastow *et al.* by growth from an Al–Si alloy with a melting temperature of 750°C.

A spectacular increase in lateral epitaxy has recently been achieved by Fan *et al.* (1981). Instead of using a laser for heating they employed a novel form of carbon strip heater. The sample geometry is shown schematically in Fig. 19. The 3.5-μm strips are perpendicular to the (110) direction. Both amorphous and polycrystalline 1-μm films were deposited from a CVD reactor. Most wafers were capped with a SiO_2 layer approximately 2 μm thick. But, apparently, the presence of this capping layer is not essential for achieving epitaxy. The heating geometry consists of a graphite strip heater (Fig. 19) with the Si film facing up. A narrow, movable strip heater is positioned above one end of the sample. The lower graphite strip is resistively heated to 1100–1300°C in about 20 sec, and the amorphous film, for example, is transformed to fine-grained polycrystallites. The upper heater strip is then rapidly heated, thus melting a narrow strip beneath it. This molten zone is moved across the sample by moving the heater at a speed of about 0.5 cm/sec.

Figure 20 shows an optical micrograph of a single-crystal overlayer grown by moving the molten front parallel to the strips (as shown in Fig. 19). Channeling shows an excellent single-crystal film. However, some surface features are apparent midway across the SiO_2. These features are believed to originate from the meeting of the crystallized fronts from adjacent seeded areas. They disappear when the molten front is moved perpendicular to the strips. The photograph shows lateral epitaxy with 50-μm spacing between seeds. Fan *et al.* (1981) report preliminary experiments where single-crystal films were grown by parallel scanning over SiO_2 or Si_3N_4 masks with the strips 500 μm apart on (100) Si substrates and over SiO_2 masks with the strips 50 μm apart on (111) substrates. On

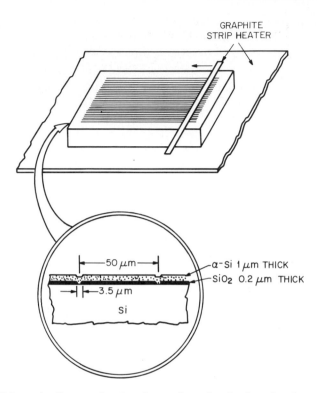

Fig. 19. Schematic diagram showing the configuration for lateral epitaxy by seeded epitaxy using a carbon strip heater. A cross-sectional view of a typical sample is shown. (From Fan *et al.*, 1981.)

the Si_3N_4 masks single-crystal growth was observed in some areas for distances up to 4 mm beyond the strip openings.

D.   Grapho- or Captive Epitaxy

There have been interesting developments in the growth of crystalline Si on amorphous substrates where an artificial surface pattern and SiO_2 cap are used to control orientation in the film. This subject has been pioneered by the MIT group (Geis *et al.*, 1979a,b, 1980), who coined the term "graphoepitaxy" to describe this novel form of crystal growth. Geis *et al.* discovered that, under certain conditions, laser crystallization of amorphous Si on fused silica produced a polycrystalline film with nearly perfect (100) texture and grain sizes of several micrometers. They postulated that the technique could theoretically be extended to produce a

Fig. 20. Optical micrograph of single-crystal film formed by seeded epitaxy using a carbon strip heater. (From Fan *et al.*, 1981.)

uniformly oriented film if a relief structure were etched in the amorphous substrate to contain the orientation of the grains. To realize this concept, square wave gratings were etched into fused silica substrates with periods of 1–4 μm and depths of 1000 Å. Amorphous Si, 0.5 μm thick, was deposited by the CVD process on the grating. Raster scans of an Ar laser operating at 6 W into a spot about 400 μm in diameter produced crystalline films with grain sizes on the order of 100 μm and nearly perfect $\langle 100 \rangle$ orientation.

Recent measurements (Geis et al., 1980) have established that an SiO_2 cap is essential for graphoepitaxy to take place. Graphoepitaxy was initially observed to occur only in air or oxygen atmospheres. Auger measurements showed that the laser scanning in air produced a 1000-Å SiO_2 surface cap on the Si. The role of the laser therefore is not only to provide sufficient heat to recrystallize the Si but also to grow a surface oxide. Better graphoepitaxy has been achieved with a carbon strip heater capable of heating a sample to 1300°C in 10 sec and with Si films capped with 2 μm of SiO_2.

Although the concept of graphoepitaxy has led to interesting developments in lateral epitaxy, it is not clear what the actual processes of crystal growth and alignment are. The fact that both the grating and SiO_2 cap are necessary indicate that several competitive influences are operating.

V.   Explosive Crystallization

Recent experiments on the laser-induced solid phase crystallization of amorphous Ge or Si films on amorphous substrates have produced a novel form of crystallization known as explosive crystallization. Explosive crystallization involves release of the heat of crystallization from a small region by some initiating disturbance such as a pinprick or laser pulse. This release of energy heats the adjacent material which can then crystallize.

The explosive crystallization of amorphous layers of antimony was first studied by Gore (1855). More recently Takamori et al. (1972) and Mineo et al. (1973) reported such recrystallization for 30- to 50-μm films of amorphous Ge. The crystallization was initiated by a localized mechanical shock. The first observations of this phenomenon using laser heating of amorphous Ge films were made by Gold et al. (1980) and Fan et al. (1980). Their studies revealed beautiful periodic structures in the recrystallized materials when the cw laser beams were scanned. Figure 36 in Chapter 6 by Cullis shows the structure obtained by Gold et al. (1980). Similar period structures have been observed in Si films by Auvert et al. (1981) and Lemons and Bösch (1981). Figure 21 from Lemons and Bösch shows the periodic Si structure formed by scanning a 2-μm-diameter 20-kV electron beam at a rate of 2.5 cm/sec. This structure is not due to the simple overlapping of melt puddles, an example of which is shown in Fig. 10 of

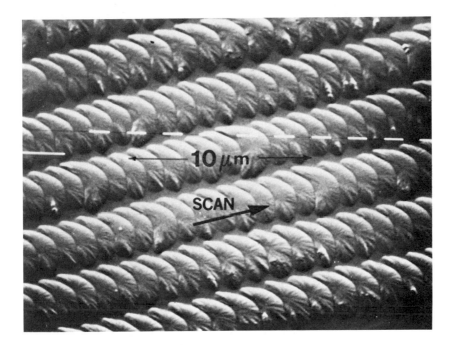

Fig. 21. Scanning electron micrograph of periodic Si structures formed by scanning a 2-μm-diameter electron beam. (From Lemons and Bösch, 1981.)

Chapter 6. The power levels are too low to melt the film, and the spacing of the structure can be independent of the scan speed and substrate temperature. Zeiger *et al.* (1980) have modeled this process in terms of a critical temperature (T_c) where the amorphous material changes phase and releases energy. The crystallization will be self-sustaining when T_c is reached for the surrounding material and, if the substrate bias temperature (T_c) is high enough, the crystallization front can propagate large distances. For the case of localized heating, by laser or electron beam, the crystallization front propagates down the thermal gradient until T_c is reached, thus forming a crescent of crystallized material whose width is determined by the temperature gradient. The process will start again once the beam has heated the interface to T_c. The theory of Zeiger *et al.* gives the period of the crystallization front as a function of T_b/T_c and the heat of crystallization. For $T_b/T_c < 0.5$ the period is nearly constant. Lemons and Bösch (1981) varied T_b from 20 to 500°C and did not observe any variations in the width of the crescent features.

An important feature of the phenomenon is that the recrystallizing interface moves with great velocity. Velocities of 1–5 m/sec have been measured for Ge, and Auvert *et al.* (1981) deduce that for Si the velocities can be as high as 10 m/sec. It is difficult to imagine that such high velocities occur in the solid phase. Gilmer and Leamy (1980) proposed that a thin, supercooled liquid layer must exist at the crystallization front. The layer width is predicted to be ~1% of the film thickness. Indeed, marker experiments (Brown, 1981) have shown that liquid state diffusion of Pb and Sb follows passage of the explosive crystallization front and indicate that the liquid layer is 200–400 Å thick for a 4-μm Ge film.

References

Auvert, G., Bensahel, D., Georges, A., Nguyen, V., Henoc, P., Morin, F., and Coissard, P. (1981). *Appl. Phys. Lett.* **38**, 613.
Bean, J. C., and Poate, J. M. (1980a). *Appl. Phys. Lett.* **35**, 280.
Bean, J. C., and Poate, J. M. (1980b). *Appl. Phys. Lett.* **37**, 643.
Bean, J. C., Leamy, H. J., Poate, J. M., Rozgonyi, G. A., Sheng, T. T., Williams, J. S., and Celler, G. K. (1978). *Appl. Phys. Lett.* **33**, 227.
Bean, J. C., Leamy, H. J., Poate, J. M., Rozgonyi, G. A., van der Ziel, J. P., and Williams, J. S. (1979a). *J. Appl. Phys.* **50**, 885.
Bean, J. C., Leamy, H. J., Poate, J. M., Rozgonyi, G. A., van der Ziel, J., Williams, J. S., and Celler, G. K. (1979b). *In* "Laser–Solid Interactions and Laser Processing" (S. D. Ferris, H. J. Leamy, and J. M. Poate, eds.), p. 489. Am. Inst. Phys., New York.
Biegelsen, D. K., Johnson, N. M., Bartelink, D. J., and Moyer, M. D. (1981). *In* "Laser and Electron Beam Solid Interactions and Materials Processing" (J. F. Gibbons, L. D. Hess, and T. W. Sigmon, eds.), p. 487. North-Holland Publ., Amsterdam.
Bösch, M. A., and Lemons, R. A. (1982). *Appl. Phys. Lett.* **40**, 166.
Brown, W. L. (1981). *In* "Laser and Electron Beam Solid Interactions and Materials Processing" (J. F. Gibbons, L. D. Hess, and T. W. Sigmon, eds.), p. 1. North-Holland Publ., New York.
Celler, G. K., Leamy, H. J., Aspnes, D. E., Doherty, C. J., Sheng, T. T., and Trimble, L. E. (1981). *In* "Laser and Electron Beam Solid Interactions and Materials Processing" (J. F. Gibbons, L. D. Hess, and T. W. Sigmon, eds.), p. 435. North-Holland Publ., New York.
Cullis, A. G., Webber, A. C., Poate, J. M., and Simons, A. L. (1980). *Appl. Phys. Lett.* **36**, 320.
Donovan, T. M., and Heineman, K. (1971). *Phys. Rev. Lett.* **27**, 1974.
Fan, J. C. C., Zeiger, H. J., Yale, R. P., and Chapman, R. C. (1980). *Appl. Phys. Lett.* **36**, 158.
Fan, J. C. C., Geis, M. W., and Tsaur, B.-Y. (1981). *Appl. Phys. Lett.* **38**, 365.
Fastow, R., Leamy, A. J., Celler, G. K., Wong, Y. H., and Doherty, C. J. (1981). *In* "Laser and Electron Beam Solid Interactions and Materials Processing" (J. F. Gibbons, L. D. Hess, and T. W. Sigmon, eds.), p. 495. North-Holland Publ., New York.
Feldman, L. C., Stensgaard, C., Silverman, P. J., and Jackman, T. E. (1978). *In* "Physics of SiO_2 and its Interfaces" (S. T. Pantelides, ed.), p. 344. Pergamon, New York.
Ferris, S. D., Leamy, H. J., and Poate, J. M., eds. (1979). "Laser–Solid Interactions and Laser Processing." Am. Inst. Phys., New York.

Foti, G., Bean, J. C., Poate, J. M., and Magee, C. W. (1980). *Appl. Phys. Lett.* **36**, 840.

Gat, A., Gerzberg, L., Gibbons, J. F., Magee, T. J., Peng, J., and Hong, J. D. (1978). *Appl. Phys. Lett.* **38**, 775.

Geis, M. W., Flanders, D. C., and Smith, H. I. (1979a). *Appl. Phys. Lett.* **35**, 71.

Geis, M. W., Flanders, D. C., Smith, H. I., and Antoniadis, D. A. (1979b). *J. Vac. Sci. Technol.* **16**, 1640.

Geis, M. W., Antoniodas, D. A., Silversmith, D. J., Mountain, R. W., and Smith, H. I. (1980). *Appl. Phys. Lett.* **37**, 454.

Gibbons, J. F. (1981). *In* "Laser and Electron Beam Solid Interactions and Materials Processing" (J. F. Gibbons, L. D. Hess, and T. W. Sigmon, eds.), p. 449. North-Holland Publ., New York.

Gibbons, J. F., and Lee, K. F. (1980). *IEEE Electron Devices Lett.* **EDL-1**, 6.

Gibbons, J. F., Lee, K. F., Magee, T. J., Perg, J., and Ormond, R. (1979). *Appl. Phys. Lett.* **34**, 831.

Gibbons, J. F., Hess, L. D., and Sigmon, T. W., eds. (1981). "Laser and Electron Beam Solid Interactions and Materials Processing." North-Holland Publ., New York.

Gilmer, G. H., and Leamy, H. J. (1980). *In* "Laser and Electron Beam Processing of Materials" (C. W. White and P. S. Peercy, eds.), p. 227. Academic Press, New York.

Gold, R. B., Gibbons, J. F., Magee, T. J., Peng, J., Ormond, R., Deline, V. R., and Evans, C. A. (1980). *In* "Laser and Electron Beam Processing of Materials" (C. W. White and P. S. Peercy, eds.), p. 221. Academic Press, New York.

Golecki, I., Kinoshita, G., Gat, A., and Paine, B. M. (1980). *Appl. Phys. Lett.* **37**, 919.

Golecki, I., Kinoshita, G., and Paine, B. M. (1981). *Nucl. Instrum. Methods* **182**, 675.

Gore, G. (1855). *Philos. Mag.* **9**, 73.

Ham, W. E., Abrahams, M. S., Buiocchi, C. J., and Blane, J. (1977). *J. Electrochem. Soc.* **124**, 634.

Henderson, R. C. (1972). *J. Electrochem. Soc.* **119**, 772.

Hess, L. D., Roth, J. A., Olson, G. L., Dunlap, H. L., von Allmen, M., and Peng, J. (1980). *In* "Laser and Electron Beam Processing of Materials" (C. W. White and P. S. Peercy, eds.), p. 562. Academic Press, New York.

Hoonhout, D., Kernijk, C. B., and Saris, F. W. (1978). *Phys. Lett. A* **66A**, 145.

Hung, L. S., Lau, S. S., von Allmen, M., Mayer, J. W., Ullrich, B. M., Baker, J. E., Williams, P., and Tseng, W. F. (1981). *Appl. Phys. Lett.* **37**, 909.

Ishiwara, H., Saitoh, S., Mitsui, K., and Furukawa, S. (1981). *In* "Laser and Electron Beam Solid Interactions and Materials Processing" (J. F. Gibbons, L. D. Hess, and T. W. Sigmon, eds.), p. 525. North-Holland Publ., New York.

Johnson, N. M., Biegelsen, D. K., and Moyer, M. D. (1981). *In* "Laser and Electron Beam Solid Interactions and Materials Processing" (J. F. Gibbons, L. D. Hess, and T. W. Sigmon, eds.), p. 463. North-Holland Publ., New York.

Kamins, T. I. (1972). *Solid-State Electron.* **15**, 789.

Kamins, T. E., and Pianetta, P. A. (1980). *IEEE Electron Devices Lett.* **EDL-1**, 214.

Kamins, T. I., Lee, F. F., and Gibbons, J. F. (1980a). *Solid-State Electron.* **23**, 1037.

Kamins, T. I., Lee, K. F., Gibbons, J. F., and Saraswat, K. C. (1980b). *IEEE Trans. Electron Devices* **ED-27**, 290.

Kamins, T. I., Cass, T. R., Dell'Oca, C. J., Lee, F. F., Pease, R. F. W., and Gibbons, J. F. (1981). *J. Electrochem. Soc.* **128**, 1151.

Lam, H. W., Tasch, A. F., Holloway, T. C., Lee, K. F., and Gibbons, J. F. (1980a). *IEEE Electron Devices Lett.* **EDL-1**, 99.

Lam, H. W., Tasch, A. F., and Holloway, T. C. (1980b). *IEEE Electron Devices Lett.* **EDL-1**, 206.

Lam, H. W., Pinizzotto, R. F., and Tasch, J. A. F. (1980c). *ECS Ext. Abstr., Hollywood, Fla.* Abstr. No. 481.

Lau, S. S., Tseng, W. F., Nicolet M.-A., Mayer, J. W., Eckardt, R. C., and Wagner, J. S. (1978a). *Appl. Phys. Lett.* **33**, 227.

Lau, S. S., Tseng, W. F., Nicolet, M.-A., Mayer, J. W., Minnucci, J. A., and Kirkpatrick, A. R. (1978b). *Appl. Phys. Lett.* **33**, 235.

Lau, S. S., Tseng, W. F., Golecki, I., Kennedy, E. F., and Mayer, J. W. (1979a). *In* "Laser–Solid Interactions and Laser Processing" (S. D. Ferris, H. J. Leamy, and J. M. Poate, eds.), p. 503. Am. Inst. Phys., New York.

Lau, S. S., Matteson, S., Mayer, J. W., Revesz, P., Gyulai, J., Roth, J., Sigmon, T. W., and Cass, T. (1979b). *Appl. Phys. Lett.* **34**, 76.

Leamy, H. J., Rozgonyi, G. A., Sheng, T. T., and Celler, G. K. (1980a). *ECS Proc.* **80-1**, 333.

Leamy, H. J., Doherty, C. J., Chiu, K. C. R., Poate, J. M., Sheng, T. T., and Celler, G. K. (1980b). *In* "Laser and Electron Beam Processing of Materials" (C. W. White and P. S. Peercy, eds.), p. 581. Academic Press, New York.

Lee, K. F., Gibbons, J. F., Saraswat, K. C., and Kamins, T. I. (1979). *Appl. Phys. Lett.* **35**, 173.

Lemons, R. A., and Bösch, M. A. (1981). *Appl. Phys. Lett.* **39**, 343.

Magee, C. W., Bean, J. C., Foti, G., and Poate, J. M. (1981). *Thin Solid Films* **81**, 1.

Maserjian, J. (1963). *Solid-State Electron.* **6**, 477.

Mineo, A., Matsuda, A., and Kurosu, T. (1973). *Solid State Commun.* **13**, 329.

Olson, G. L., Kokorowski, S. A., McFarlane, R. A., and Hess, L. D. (1980). *Appl. Phys. Lett.* **37**, 1019.

Revesz, P., Farkas, G., Mezey, G., and Gyulai, J. (1978). *Appl. Phys. Lett.* **33**, 431.

Roth, J. A., Olson, G. L., Kokorowski, S. A., and Hess, L. D. (1981). *In* "Laser and Electron Beam Solid Interactions and Materials Processing" (J. F. Gibbons, L. D. Hess, and T. W. Sigmon, eds.), p. 413. North-Holland Publ., New York.

Roulet, M. E., Schwob, P., Affolter, K., Lüthy, W., von Allmen, M., Fallavier, M., Mackowski, J. M., Nicolet, M. A., and Thomas, J. P. (1979). *J. Appl. Phys.* **50**, 5536.

Sai-Halasz, G. A., Fang, F. F., Sedgwick, T. O., and Segmüller, A. (1980). *Appl. Phys. Lett.* **36**, 419.

Saitoh, S., Ishiwara, A., and Furukawa, S. (1980). *Appl. Phys. Lett.* **37**, 203.

Takamori, T., Messier, R., and Roy, R. (1972). *Appl. Phys. Lett.* **20**, 201.

Tamura, M., Tamura, H., and Tokuyama, T. (1980). *Jpn. J. Appl. Phys.* **19**, L23.

Tamura, M., Tamura, H., Miyao, M., and Tokuyama, T. (1981). *Jpn. J. Appl. Phys. Suppl.* **20-1**, 43.

Tasch, A. F., Holloway, T. C., Lee, K. F., and Gibbons, J. F. (1979). *Electron. Lett.* **15**, 435.

White, C. W., and Peercy, P. S., eds. (1980). "Laser and Electron Beam Processing of Materials." Academic Press, New York.

Yamada, M., Hara, S., Yamamoto, K., and Abe, K. (1980). *Jpn. J. Appl. Phys.* **19**, 261.

Yaron, G., Hess, L. D., and Kokorowski, S. A. (1980). *Solid-State Electron.* **23**, 893.

Zeiger, H. J., Fan, J. C. C., Palm, B. J., Gale, R. P., and Chapman, R. L. (1980). *In* "Laser and Electron Beam Processing of Materials" (C. W. White and P. S. Peercy, eds.), p. 234. Academic Press, New York.

Chapter 9

Surface Properties
of Laser-Annealed Semiconductors*

D. M. ZEHNER and C. W. WHITE

Solid State Division
Oak Ridge National Laboratory
Oak Ridge, Tennessee

* Research sponsored by the Division of Materials Sciences, U.S. Department of Energy, under contract W-7405-eng-26 with Union Carbide Corporation.

I. Introduction

The rapid deposition of energy from Q-switched lasers into the near-surface region of semiconductors leads to its melting followed by liquid phase epitaxial regrowth from the substrate at growth velocities calculated to be on the order of meters per second (see Chapter 4). The heating and cooling rates (approaching 10^9 K/sec) that can be achieved by this type of processing are orders of magnitude faster than those achieved by more conventional techniques. It has been shown that, under proper annealing conditions, regions free of extended defects can be formed and substitutional impurities can be incorporated into the lattice far in excess of the equilibrium solubility limits (see Chapter 5). Thus, it may be expected that the surface properties (impurities, structure, electronic energy levels) of laser-annealed semiconductors can be significantly altered by the high heating and cooling rates and rapid velocities of solidification that can be achieved using pulsed lasers.

In this chapter we show that pulsed laser irradiation can be used to produce atomically clean surfaces, metastable surface structures, and surfaces with electronic properties that cannot be achieved by conventional methods. Details concerned with processing of the surface in order to achieve these conditions and the measurement of surface properties are discussed in Section II. In Section III we show that levels of unwanted impurities on semiconductors (O, C, etc.) can be reduced to near the practical detection limits [using Auger electron spectroscopy (AES)]. This is accompanied by the observation of well-defined low-energy electron diffraction (LEED) patterns and evidence of ordered surface structures and is discussed in Section IV. In this section we show also that, for (111)-oriented crystals of Si and Ge, laser irradiation leads to a metastable (1×1) surface structure and that a dynamical LEED analysis of the Si surface suggests that the first and second outermost interlayer spacings exhibit substantial relaxation (in the absence of ordered lateral reconstruction). However, the proposed surface structure is not in agreement with results of photoemission experiments when interpreted using one-electron-band structure calculations. Possible reasons for this discrepancy are cited. Finally, in Section V we show that the combination of ion implantation and laser annealing can be used to produce surfaces with electronic properties that cannot be achieved by conventional methods.

II. Experimental Approach

In most of the laser annealing investigations reported to date, the laser irradiation has been performed in a standard atmospheric environment. If one is concerned with obtaining information about the properties of the surface region (1–20 Å) of laser-annealed semiconductors, the irradiation of the sample and subsequent analysis must take place in an ultrahigh vacuum (UHV) environment ($\leq 10^{-9}$ Torr). This section is concerned with the experimental details specific to such investigations.

A. Sample Preparation

In order to obtain information about the surface properties of laser-annealed crystals, all the measurements to be discussed were made in UHV surface analysis chambers. After bakeout, the background pressure in these chambers was typically less than 2×10^{-10} Torr. The light from a pulsed ruby laser ($\lambda = 6943$ Å) was coupled into the UHV systems through a glass window and irradiated the sample in the vacuum environment. Samples were irradiated using the single-mode (TEM_{00}) output of the ruby laser at energy densities that could be varied between ~0.2 and ~4.0 J/cm². A constant pulse duration time of 15×10^{-9} sec was used in all experiments. The beam diameter was typically between 3.0 and 6.0 mm. Energy densities, which have been corrected for the reflectivity (~10%) of the glass window, were determined by measuring the photon energy delivered through an aperture of known diameter positioned in front of a calorimeter. Implanted samples were prepared in a separate ion implantation facility which was also equipped for making Rutherford backscattering (RBS) measurements. This technique was used to determine the implant profile and characterize the changes in the subsurface region that occurred with laser annealing. An extensive discussion of these results is contained in Chapter 5.

Because of the desire to examine the surface region, the spectroscopic techniques employed used either electrons or photons as the incident probe. In all cases the detected particle was an electron. Because of the short mean free path of electrons with energies between 20 and 1000 eV, only the outermost surface region, ~20 Å, was probed. Several different surface-sensitive spectroscopic techniques were employed at the various surface analysis facilities used in these investigations.

B. *Auger Electron Spectroscopy*

The technique of Auger electron spectroscopy (AES) was used to monitor the levels of both impurities and implanted species in the surface region of the sample. This technique is capable of detecting all elements with $Z \geq 3$. Levels of impurity contamination are quoted in terms of the ratios of the peak-to-peak signals of the impurity Auger transitions to a principal Auger transition of the substrate as measured in the derivative (dN/dE) mode. The energy dependence of the analyzer ($\sim E$), the escape depth of electrons ($\sim E^{1/2}$), and the ionization cross section ($\sim 1/E^2$) have been included in determining these ratios (Morabito, 1975). Although one must be careful in attempting to use this technique to make quantitative measurements, in many situations reasonable estimates of the upper limit of the amount of a particular species present in the surface region can be made.

C. *Low-Energy Electron Diffraction*

Low-energy electron diffraction (LEED) was employed to investigate geometric order in the surface region of the sample. Since the wavelengths for electrons with energies between 20 and 500 eV are 2.7–0.5 Å, it is possible to use the information obtained by observing the wave diffraction of these particles to determine interatomic spacings in the outermost surface layer. By examining the positions of the reflected beams (spot patterns), the symmetry and size of the surface unit cell can be determined. Thus, any changes in these spot patterns with surface modification are a reflection of changes in the geometric arrangement of atoms within the surface region. To obtain detailed information about the positions of the atoms within the unit cell, as well as interlayer spacings, the changes in diffracted intensity as a function of changes in the incident beam energy (I versus V profiles) must be measured for a number of different reflections. These profiles are then compared with the results of model calculations and, by means of a trial-and-error procedure, a model for the geometric structure is obtained.

D. *Photoelectron Spectroscopy*

Photoelectron spectroscopy (PES) was used to obtain information about the electronic properties of the surface region. By using photons in the energy range of 15–150 eV, the photoemitted electron provides information about the surface region as a consequence of its short mean free path. Information about the valence and conduction bands of the solid can be obtained by employing photons with energies, typically less than ~50 eV. With the use of angle-resolved techniques it is possible to map out bands and characterize the symmetry of surface states. Employing photons of higher energies makes it possible to measure the core-level binding energies. By varying the photon energy and thus changing the sampling depth of the emitted electrons, surface core-level shifts can be separated from bulk effects.

III. Preparation of Atomically Clean Surfaces

The production of atomically clean surfaces in UHV is a basic requirement in many areas of technological importance. Basic research directed toward understanding the physical and chemical properties of surfaces requires clean surfaces where the level of unwanted contaminants in the first few monolayers is ≤ 1 at.%. In many device technologies the presence of surface contaminants either during fabrication or during application can contribute greatly to the degradation of device performance. As device dimensions become increasingly smaller, the near-surface region will become even more important and it will be necessary to pay ever-increasing attention to the production of atomically clean surfaces during device processing.

Conventional methods of producing atomically clean surfaces in UHV include physical sputtering, thermal desorption, reactive sputtering, chemical reactions, electron scrubbing, deposition or growth of a film *in situ*, and vacuum cleaving or fracture (Roberts, 1963). In cases where one is dealing with single crystals, it is frequently necessary to anneal the crystal at high temperatures in order to remove damage produced in the surface region by cleaning techniques such as sputtering. One aspect common to many of these techniques is the time required to clean the surface. This time can be measured in hours and sometimes days when cycling between sputtering and annealing is required.

The fact that lasers have been used to generate metal vapors and are capable of raising the temperature in the near-surface region of a crystal to the melting point (see Chapter 4) suggests that they can be used to generate clean surfaces. Although first utilized in 1969 (Bedair and Smith, 1969), only recently have there been a number of investigations concerned with laser irradiation directed toward this objective (Zehner *et al.*, 1980a; McKinley *et al.*, 1980; Rodway *et al.*, 1980; Cowan and Golovchenko, 1980). While a number of different lasers have been used, the results obtained by different investigators are quite similar, especially with respect to Si. Investigations of other elemental and compound semiconductors, as well as metals, are more limited.

A. Silicon

The application of laser irradiation for the purpose of producing atomically clean Si surfaces is demonstrated by the results (Zehner *et al.*, 1980a) shown in Fig. 1. The Auger electron spectrum obtained from an air-exposed Si sample after insertion into a UHV system and following bakeout is shown at the top of Fig. 1. Oxygen [O (510 eV)/Si (91 eV) = 2.3×10^{-1}] and carbon [C (272 eV)/Si (91 eV) = 3.8×10^{-2}] are readily detected on the surface of the as-inserted sample. The features in the region of the Si $L_{2,3}VV$ transition (70–100 eV) are characteristic of Si in silicon dioxide. Measurements made by RBS showed O and C concentrations of 8.1×10^{15} and 3.5×10^{15} atoms/cm², respectively, in the near-surface region. This implies a 20-Å oxide thickness, typical of air-exposed Si. Irradiation with one laser pulse (~2.0 J/cm²) causes a substantial reduction in the levels of O and C present in the surface region. After exposing the same spot to five laser pulses, the Auger electron spectrum shown in Fig. 1 indicates that for the same detection conditions the O and C signals are within the noise level. By increasing the effective sensitivity of the electron detection system, the O and C intensity ratios were determined to be O/Si $\leq 5.5 \times 10^{-4}$ and C/Si $\leq 7.6 \times 10^{-4}$, indicative of surface concentrations of $\leq 0.1\%$ of a monolayer. In addition, the line shape of the Si $L_{2,3}$ VV transition is that expected from a clean Si surface. Although a H transition cannot be detected with AES, PES results (Zehner *et al.*, 1981c), which are discussed later, show the surface region to be free of H.

Consequently, by irradiating the crystal with several (≤ 5) laser pulses, the O and C contaminants have been reduced by factors of 500 and 50, respectively. These reduced contaminant levels, obtained in a processing time of ≤ 1 sec, are comparable to those obtained by repeated sputtering and conventional thermal annealing over a time period of several days. The residual O and C contamination on the surface may arise as a result of adsorption and cracking from the gas phase during the time required to make the Auger measurements.

The effect of pulsed laser irradiation on samples that had been sputtered with Ar⁺ ions (1000 eV, 5 μA, 30 min) was also investigated. Irradiation of the sample with laser pulses of ~2.0 J/cm² produced a similar reduction in the contaminant level Auger signals for O and C as observed for the unsputtered surface. Complete elimination, even under increased sensitiv-

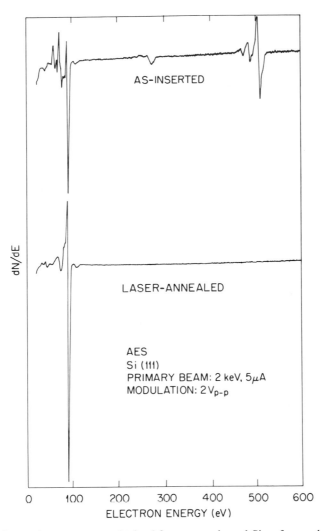

Fig. 1. Auger electron spectra obtained from an uncleaned Si surface and after pulsed laser annealing at ~2.0 J/cm².

ity conditions, of the Ar Auger signal occurred after two or three pulses. In addition, if an atomically clean surface is exposed to O_2 or CO, the surface can be returned to an atomically clean state by irradiation with five laser pulses.

The effect of energy density on the efficiency of removal of impurities was also qualitatively investigated over the range ~0.3–~3.2 J/cm². Two general trends were observed as a function of energy density. First, the higher the energy density, the more extensive the removal of the impurities by the first pulse. Second, at any pulse energy density the larger the number of pulses, the more complete the removal of impurities. At energy densities above ~ 2.0 J/cm^2, little additional change could be observed after five pulses. Visible damage, as observed with optical techniques, occurred at an energy density of 3.2 J/cm^2. At energy densities ≤ 3.0 J/cm^2 there was no visible surface damage. Energy densities greater than 1.0 J/cm^2 were necessary to produce atomically clean surfaces using ruby laser radiation.

Removal of the surface contaminants by laser irradiation was accompanied by a transient increase in background pressure. With laser pulses of ~ 2.0 J/cm^2 and starting with a background pressure of 2×10^{-10} Torr, the first laser pulse on an as-inserted or freshly sputtered sample caused a transient pressure rise into the 5×10^{-8} to 3×10^{-7} Torr range. Subsequent pulses on the same area were accompanied by pressure rises into the 10^{-9}-Torr range, and these continued to drop until the pressure stayed in the 10^{-10}-Torr range by about the fifth pulse.

There are two possibilities as to the ultimate fate of the original oxygen and carbon surface contaminants. The fact that a pronounced pressure rise is observed during the first laser pulse suggests that contaminants are desorbed from the surface during irradiation. It is difficult, however, to quantify these observations and determine the amount of material desorbed from only the crystal surface. Alternatively, since pulsed laser annealing results in the formation of a melted region to a depth of several thousand angstroms, impurities can undergo substantial redistribution by means of liquid phase diffusion during the time the surface region is molten. Redistribution deeper into the sample occurs predominantly for impurities that have a relatively high segregation coefficient from the melt (White *et al.*, 1978; Wang *et al.*, 1978). Impurities that have a low segregation coefficient from the melt have been shown to segregate to the surface during annealing (White *et al.*, 1979). For C and O, the equilibrium segregation coefficients are 0.07 and 0.05, respectively, suggesting that these impurities may be redistributed over a depth interval equivalent to the melt depth (~ 5000 Å for laser energy densities of ~ 2.0 J/cm^2). Complete redistribution over this depth interval would give rise to a surface concentration of $\sim 0.3\%$ of a monolayer for O and $\sim 0.1\%$ of a monolayer for C, both of which are near the detection limits for Auger electron spectroscopy, and these values are consistent with the above measurements. Whether these impurities are desorbed from the surface or redistributed in depth can be determined by experiments designed to measure the total C and O concentrations in the near-surface region before and after irradiation.

Such an investigation has recently been performed in which RBS and resonance nuclear reaction $^{16}O(\alpha,\alpha)^{16}O$ techniques in conjunction with AES have been used to determine the oxygen concentration on the surface and in the near-surface (3000-Å) region both prior to and after pulsed laser annealing (Westendorp *et al.*, 1981). The initial native oxide layer at the surface was observed to have an oxygen concentration of 5.2×10^{15} atoms/cm². After irradiation of a Si (100) sample in UHV with eight pulses at an energy density of ~1.5 J/cm² ($\lambda = 6943$ Å), Auger measurements showed an oxygen concentration at the surface of $\leq 0.3\%$ of a monolayer. The oxygen concentration at a depth of 1100 Å (depth interval 500–1700 Å) was determined to be $\leq 3.1 \times 10^{18}$ atoms/cm³ by nuclear reaction analysis. It was concluded that there was no evidence for oxygen interdiffusion during the laser annealing pulse used for cleaning and that the concentration determined in the bulk was less than the oxygen solubility limit.

Investigations have also been performed to examine the incorporation of oxygen into Si during pulsed laser annealing in an atmospheric environment (Hoh *et al.*, 1980). In these experiments the samples were exposed to an atmospheric pressure of $^{18}O_2$ during the laser irradiation and were subsequently examined using the secondary-ion mass spectroscopy (SIMS) technique. Maximum ^{18}O concentrations of 4×10^{18} and 7×10^{20} atoms/cm³ with depth distributions of 0.6 and 1.4 μm were observed following irradiation with laser pulses with energy densities of 1.4 ($\lambda = 5300$ Å) and 4.7 ($\lambda = 10,600$ Å) J/cm², respectively. Furthermore, it was shown that the incorporation of ^{18}O was blocked when the Si sample surface was covered with SiO_2. Infrared absorption measurements were made to determine the state of the oxygen atoms in the Si samples. From these measurements it was concluded that most of the oxygen atoms were not interstitials but that, in agreement with anomalous etching rates, were possibly included as precipitates.

A possible explanation for the differences in these results may be related to the differences in the dissolution rates of the oxide and oxygen in Si under the time and temperature conditions appropriate to the laser annealing process (Saris, 1981). The dissolution rate of the oxide in Si is known to be about 10^{-5} cm/min (Saris, 1981). Since the thickness of the native oxide layer on Si, irradiated in a UHV environment, is typically on the order of 20 Å, it would take approximately 1 sec to incorporate this layer into the melted region. This is at least six orders of magnitude greater than the time the irradiated region remains above ~600 K. To minimize concern about the possible redistribution of impurities such as O and C, the safest procedure to follow is to sputter the surface mildly to reduce substantially the concentration of these impurities and then laser-anneal the crystal to produce an atomically clean surface region.

B. Germanium

Investigations of the cleaning of Ge samples with pulsed laser irradiation have produced results similar to those obtained with Si samples (Zehner *et al.*, 1980c). The Auger electron spectrum obtained from an air-exposed Ge sample after insertion into a UHV system and following bakeout is shown at the top of Fig. 2. The oxygen and carbon impurity concentrations are typical of those found on all Ge samples investigated. After irradiation with one laser pulse (~ 1.9 J/cm^2), a significant reduction in these levels is observed. By exposing the same spot to five laser pulses, these impurities can be reduced to levels that are within the noise level, as shown in the bottom curve in Fig. 2. With the use of signal-averaging techniques and comparison with the Ge $M_{2,3}M_{4,5}M_{4,5}$ Auger transition, the O and C intensity ratios were determined to be O/Ge $\leq 3.0 \times 10^{-3}$ and

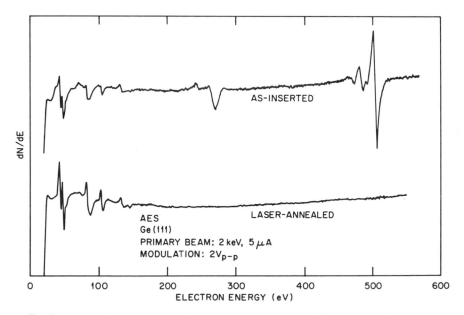

Fig. 2.   Auger electron spectra obtained from an uncleaned Ge surface and after pulsed laser annealing at ~ 1.9 J/cm^2.

C/Ge $\leq 4.2 \times 10^{-3}$, indicative of surface concentrations of $\leq 0.1\%$ of a monolayer.

As observed with Si, sputtering can be used in conjunction with laser irradiation of Ge to produce atomically clean surfaces. This approach permits the elimination of a large fraction of the initial O and C impurities, thus reducing the possibility of diffusion of these constituents into the bulk during the annealing process. The embedded inert gas atoms are easily removed with laser irradiation, and the same level of surface cleanliness is achieved after about five pulses.

The range of energy density that can be used to remove impurities from the surface was found to be ~0.5 to ~2.1 J/cm². A larger number of pulses was needed at low-energy densities to achieve the same level of surface cleanliness. Damage to the surface occurred above an energy density of 2.2 J/cm², and complete removal of surface impurities did not occur below an energy density of ~0.5 J/cm².

C. Group III–V Compounds

The production of clean surfaces on GaAs single crystals was investigated using laser annealing only, as well as sputtering with Ar⁺ ions followed by laser annealing (Zehner *et al.*, 1982a). Examples of Auger electron spectra obtained for these various conditions and typical of all crystal orientations are shown in Fig. 3. The as-inserted samples were found to be covered with large amounts of C and O, and sometimes a trace of Ca could be detected. Similar to that previously observed with Si and Ge, the removal of impurities from the surface region required irradiation with multiple pulses, and the efficiency of cleaning increased with increasing energy density. In order to remove the C and O contaminants to a point where they were within the noise level of the Auger electron spectrum, laser irradiation energy densities in excess of ~0.5 J/cm² were required. An example of a spectrum obtained following multiple-pulse irradiation at ~0.6 J/cm² is shown in Fig. 3. Although C and O are not detectable in this spectrum, it should be noted that there is also a difference in the Ga ($M_{2,3}M_{4,5}M_{4,5}$, 55 eV) and As ($M_{4,5}$ VV, 31 eV) Auger transition intensity ratios and in the Ga Auger transition line shape when compared with that obtained after sputtering, also shown in this figure. The C and O impurities could not be removed completely from the surface region using energy densities of <0.5 J/cm². However, by first sputtering with Ar⁺ ions, it was found that a surface with no impurities present could be produced by irradiation with energy densities between 0.15 and 0.4 J/cm², as illustrated at the bottom of Fig. 3 for the case of 15 pulses at ~0.3

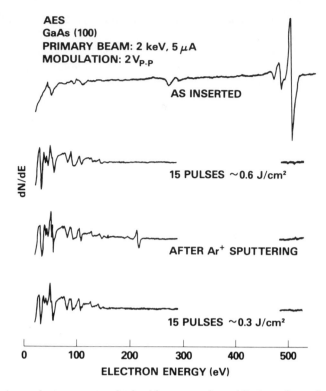

Fig. 3. Auger electron spectra obtained from an uncleaned GaAs surface, after sputtering, and after pulsed laser annealing at ~0.6 and ~0.3 J/cm².

J/cm². At these lower energy densities the As/Ga Auger intensity ratio was observed to be similar to that observed after sputtering. However, the Ga $M_{2,3}M_{4,5}M_{4,5}$ line shape showed the same type of change as that observed at higher energy densities. This change, a reduction in the degree of splitting within the line shape observed after irradiation at all energy densities, is due to the presence of Ga in local regions that are nonstoichiometric. Results with respect to cleaning, similar to those just discussed, have been obtained using glass-bonded (100) GaAs transmission photocathodes (Rodway *et al.*, 1980).

Laser irradiation of cleaved indium phosphide (110) surfaces exposed to air presented problems similar to those observed with GaAs (McKinley *et al.*, 1980). Clean surfaces were produced, and the AES spectra obtained were quite similar to those for clean cleaved surfaces. However, the surfaces were not ordered, and detailed scanning Auger electron studies (with a 1-μm spatial resolution) indicated that, although the overall surface showed a stoichiometry consistent with that of the cleaved face, local regions existed where the deviation from stoichiometry was large.

IV. Surface Structure of Laser-Annealed, Single-Crystal Semiconductor Surfaces

In the previous section it was shown that pulsed laser irradiation could be used to produce atomically clean surfaces in an UHV environment. In this section we discuss the annealing capability of this technique with respect to order in the surface region of a single crystal. The application of laser annealing with regard to the processing of semiconductor materials has been investigated extensively. It has been shown that the technique can be used to anneal completely displacement damage in ion-implanted semiconductors (Narayan *et al.*, 1978). In this application the laser radiation causes the surface region of the crystal to be melted to a depth of several thousand angstroms. The melted layer than regrows from the underlying substrate by means of liquid phase epitaxial regrowth, and the regrown region has the same crystalline perfection as the substrate. It is of interest then to determine if the crystalline order extends to the outermost monolayer subsequent to irradiation in the UHV environment.

The methods commonly used in surface experiments for cleaning or annealing sputter damage involve heating of the sample such that the time for the crystal to return to room temperature from the annealing temperature can range from minutes to hours. In many investigations this becomes the limiting factor with respect to data acquisition rates. Calculations indicate that, for pulse energies of 1–2 J/cm², several tens of microseconds elapse between the irradiation of the sample at room temperature with the laser beam and the return of the sample to ~600 K. Thus this annealing technique provides the capability for doing a number of experiments in which the thermal cycling time is reduced to a minimum. In this section we illustrate this annealing capability by discussing experiments concerned with semiconductor crystals.

A. Silicon

The silicon samples used in these investigations received no cleaning treatment other than a rinse in alcohol prior to insertion into the UHV system. After insertion and following bakeout, an examination with LEED showed that diffraction spots could be observed occasionally, but only at relatively high energies (>250 eV), and AES measurements indicated the presence of both O and C in the surface region. These impurities were removed during the laser irradiation, as discussed in Section III. A comparison of the LEED patterns obtained following laser annealing with those obtained using conventional annealing techniques is shown in Figs. 4 and 5.

1. Si (100) FACE

After irradiation with one laser pulse of ~2.0 J/cm², a well-defined (2 × 1) LEED pattern with moderate background intensity was obtained. Improvement in the quality of the diffraction pattern occurred with subse-

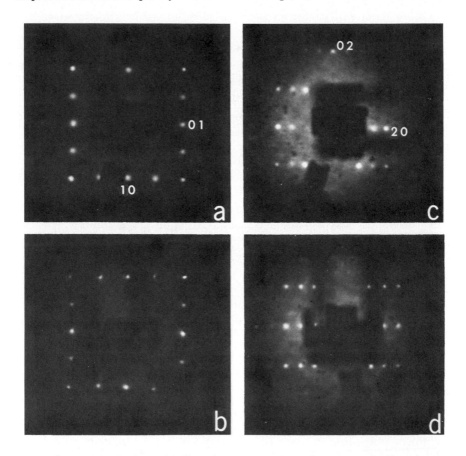

Fig. 4. LEED patterns from clean (a and b) (100) and (c and d) (110) Si surfaces at primary beam energies of (a and b) 49 eV and (c and d) 92 eV. (a and c) Laser annealed, (b and d) thermally annealed.

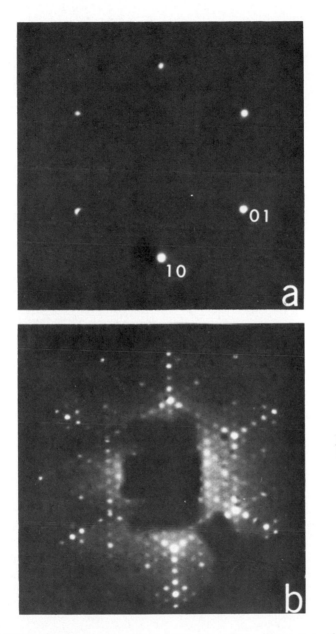

Fig. 5.   LEED patterns from clean (111) Si surfaces at a primary beam energy of (a) 40 eV and (b) 68 eV. (a) Laser annealed, (b) thermally annealed.

quent laser pulses. After five pulses the LEED pattern shown in Fig. 4a
was observed (Zehner *et al.*, 1980b,d). The fact that well-defined LEED
patterns can be obtained indicates that the crystalline order extends to the
outermost monolayers after the liquid phase epitaxial regrowth process.
No detectable change in the LEED patterns was observed with additional
pulses. Similar observations with respect to the quality of LEED patterns
as a function of the number of laser pulses were found for all orientations
investigated. Comparison of the LEED pattern observed after laser annealing with that obtained after using conventional sputter-anneal techniques as shown in Fig. 4b indicates that the same reordered surface
structure is present on the laser-annealed surface. Although quarter-order
LEED beams indicative of a c(4 × 2) structure have been observed occasionally in previous examinations of the (100) surface (Lander and Morrison, 1962), they were not observed in this investigation subsequent to
either laser annealing or conventional thermal treatments. The above results indicate that the atoms in the outermost layers have enough time at a
temperature under the laser annealing conditions used to reorganize into
the reordered arrangement from which the (2 × 1) LEED patterns are
obtained. This is consistent with the proposed surface structure models
for this surface that involve only small displacements of the atoms in the
filled outermost monolayers (Eastman, 1980).

2.   Si (110) FACE

The (1 × 2) LEED pattern obtained from the Si (110) face following
irradiation with five laser pulses of ~2.0 J/cm² is shown in Fig. 4c (Zehner
et al., 1980d). Although in previous investigations of this surface using
conventional annealing techniques, (5 × 1), (4 × 5), and (1 × 2) LEED
patterns were observed (Jona, 1965; Hagstrum and Becker, 1973), only
the (1 × 2) pattern shown in this figure was observed subsequent to laser
annealing. The (1 × 2) LEED pattern obtained from a thermally annealed
surface is shown at the bottom of this figure for purposes of comparison.
While a detailed description of the atomic arrangements in the surface
region of the (1 × 2) structure is lacking at this time, the similarity in the
LEED patterns shows that, as with the (100) surface, the same reordered
arrangement can be obtained following either laser or thermal annealing
procedures.

3.   Si (111) FACE

A sharp, well-defined (1 × 1) LEED pattern, shown at the top of Fig. 5,
was obtained from the Si (111) face subsequent to irradiation with five

laser pulses of ~2.0 J/cm² (Zehner *et al.*, 1980c,d). In contrast to the patterns obtained from the (100) and (110) surfaces, this observation suggests that as a result of laser annealing the normal surface structure (truncation of the bulk) is obtained, and there is no evidence of any ordered lateral reconstruction. Although a (2 × 1) LEED pattern can be obtained from a cleaved Si (111) surface, the (7 × 7) LEED pattern shown at the bottom of Fig. 5 is always observed on a clean thermally annealed crystal surface. Additional investigations showed that the (1 × 1) LEED pattern could be obtained from the (111) surface when the crystal was held at temperatures in the range 100–700 K during laser annealing. The observation of one-seventh-order diffraction spots, indicative of the reconstructed surface, occurred after either annealing the laser-irradiated surface at temperatures greater than ~800 K or holding the crystal at this temperature during the laser annealing process. By heating for a sufficient time (>30 min) at temperatures greater than ~800 K, it was possible to convert the laser-annealed surface to one from which a well-defined (7 × 7) LEED pattern was observed. This indicates that the structure giving rise to the (1 × 1) LEED pattern is metastable, and by combining laser annealing with conventional thermal annealing it is possible to cycle back and forth between the two structures.

Since the initial observation that the (111) surface of Si is reconstructed subsequent to a sputter-anneal treatment, much effort has been expended in determining geometric models appropriate for this atomic structure and in obtaining a variety of experimental data to support these models. Structural models for this reconstructed surface have been divided into two types, smooth and rough (Eastman, 1980). Smooth models are characterized by a complete outermost monolayer with small displacements from bulklike positions. Rough models incorporate either an outermost monolayer or double layer that contains a large number of vacancies. Observations of (1 × 1) LEED patterns suggest that under the combined time and temperature conditions present during the laser annealing process the atoms in the outermost layer are not able to organize, after the regrowth of the molten region, into the ordered geometric arrangements corresponding to the reconstructured surface. However, it is difficult to use the data obtained to data to either eliminate or support any one of the models proposed for the reconstructed surface.

B. Germanium

Investigations of the surface order on Ge single-crystal surfaces produced by laser irradiation yielded results similar to those obtained with Si samples. Figure 6 shows a comparison of the LEED patterns obtained

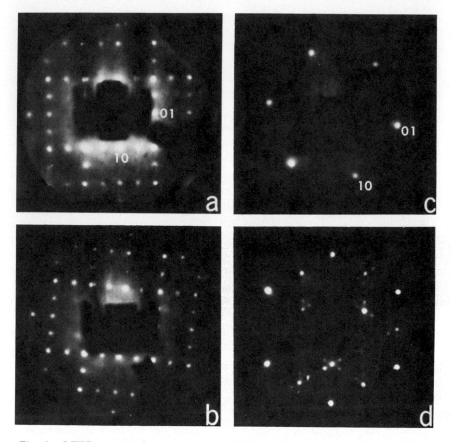

Fig. 6. LEED patterns from clean (a and b) (100) and (c and d) (111) Ge surfaces at primary beam energies of (a and b) 112 eV, (c) 40 eV, and (d) 32 eV. (a and c) Laser annealed, (b and d) thermally annealed.

following laser annealing with those observed after using conventional annealing techniques.

1. Ge (100) FACE

The LEED pattern obtained from the (100) face after irradiation with five pulses at ~1.9 J/cm² is shown in Fig. 6a. This (2 × 1) pattern indicates that, as with the Si (100) surface, a reconstructed surface arrangement is present subsequent to laser annealing. The (2 × 1) LEED pattern obtained from a sample that had been sputtered and thermally annealed is shown in Fig. 6b. There was no evidence of a c(4 × 2) structure following either type of annealing treatment (Lander and Morrison, 1963). The arguments for the observation of the (2 × 1) pattern following laser annealing, which is identical to that obtained with conventional annealing techniques, are similar to those presented for explaining the observations in the Si (100) case.

2. Ge (111) FACE

The sharp (1 × 1) LEED pattern shown in Fig. 6c was obtained from the Ge (111) face subsequent to irradiation with five pulses at ~1.9 J/cm² (Zehner *et al.*, 1980c). The (2 × 8) LEED pattern shown in Fig. 6d, indicative of a reconstructed surface, was obtained from a Ge sample after sputtering with Ar^+ ions with the sample held at a temperature of ~900 K. Thermally heating the laser-annealed Ge sample above ~1100 K resulted in formation of the reconstructed surface, as evidenced by the observation of a (2 × 8) LEED pattern. Heating at higher temperatures improved the quality of the patterns. As with the Si (111) sample, subsequent irradiation of this surface at room temperature with the laser resulted in a (1 × 1) surface structure. The sample temperature range over which the (1 × 1) LEED pattern was observed subsequent to laser irradiation was determined to be between 100 and 1000 K. Although the (111) surfaces of Si and Ge exhibit different LEED patterns subsequent to a sputter-anneal treatment, it has been proposed that the local atomic arrangement within the reconstructed layer is very similar for both surfaces (Himpsel *et al.*, 1981b). The observations of (1 × 1) LEED patterns on (111) surfaces of both Si and Ge after laser annealing may indicate that the local bonding and atomic arrangement are similar on both surfaces, but that formation of long-range order is inhibited during the rapid quenching process.

C. Gallium Arsenide

To investigate the application of laser annealing in producing ordered surface structures on crystal faces of compound semiconductors, especially those in which one of the components is volatile, the low-index faces of GaAs crystals were used. Although LEED patterns were obtained for all orientations investigated, the quality of the surface structure as reflected in these patterns differed significantly. All the results to be discussed were obtained from surfaces that were initially sputtered in order to remove the C and O impurities. As mentioned in Section III, this procedure permitted the use of lower energy densities in order to obtain both clean surfaces and reasonable LEED patterns.

1. GaAs (100) Face

While the most frequently observed LEED patterns obtained from the (100) orientation are (1 × 6) and (2 × 8) (Jona, 1965), depending on thermal treatment or growth conditions, such patterns were never observed subsequent to laser annealing. The LEED pattern obtained from the (100) surface following irradiation at an energy density of ~0.3 J/cm² and shown at the top of Fig. 7 shows that, although there are streaks between the integral-order reflections, it is not possible to determine at what positions these extra reflections are located. Sharp, non-integral-order reflections could not be observed at any incident electron beam energy. This streaking and the high diffuse background intensity are indicative of the presence of a large amount of disorder in the surface region. Although a range of energy densities and a variation in the number of pulses of irradiation were tried, it was not possible to produce a surface from which a better quality LEED pattern could be observed.

2. GaAs (100) Face

The LEED pattern obtained from the (110) surface after irradiation with five pulses at an energy density of ~0.3 J/cm² is also shown in Fig. 7. This (1 × 1) pattern indicates the presence of a surface structure similar to that obtained after cleaving or thermal annealing. Although the twofold rotational symmetry of the surface is apparent and the quality of this pattern is comparable to that previously published (Jona, 1965), the diffuse background intensity may be somewhat higher for the laser-annealed surface when compared to that observed for cleaved surfaces. LEED patterns of this quality were obtained after irradiation with energy densities in the range 0.2–0.4 J/cm².

3. GaAs (111) Face

Both the (111)-A and (111)-B polar faces of GaAs were examined. The (1 × 1) LEED patterns obtained from these two surfaces were similar in quality and indicate that the same degree of surface order is present after laser annealing of either face. These patterns, an example of which is shown at the bottom of Fig. 7, were obtained after irradiation of the sample with five pulses at ~0.3 J/cm². A careful examination of patterns obtained using a range of incident electron beam energies indicates that the diffuse background intensity is higher than that expected for a well-annealed surface and may indicate the presence of some disorder in the

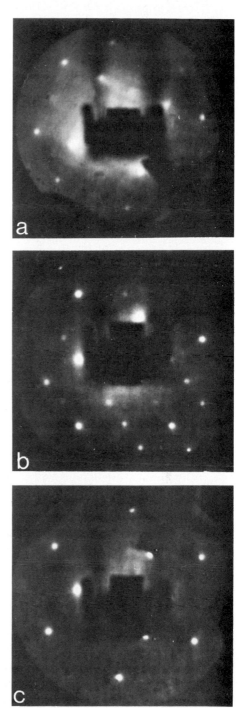

Fig. 7. LEED patterns from clean, laser-annealed (a) (100), (b) (110), and (c) (111) GaAs surfaces at primary beam energies of (a) 113 eV, (b) 123 eV, and (c) 95 eV.

surface region. However, it should be noted that the (2 × 2) and ($\sqrt{19}$ × $\sqrt{19}$) LEED patterns frequently observed from these surfaces following conventional cleaning and annealing techniques (Jona, 1965; Cho and Hayashi, 1971) were never observed in these investigations.

D.   Stepped Surfaces

The electronic properties and chemical activity of a surface can be strongly influenced by the presence of defects such as steps. Although stepped surfaces can be prepared by cleaving or ion etching, the control of the step density and ease of reproducibility have proved difficult using conventional procedures. Thus, it is of interest to determine if well-annealed surfaces with monatomic steps and uniform terrace widths can be produced using laser annealing techniques. In order to investigate this question a Si(111) crystal whose surface was cut at 4.3° from a (111) plane toward the [1$\overline{1}$2] direction was used. For this direction the edge atoms have only two nearest neighbors. The well-defined (1 × 1) LEED pattern obtained from the clean surface and shown in Fig. 8a was observed after irradiating the surface with five pulses at ~2.0 J/cm² (Zehner et al., 1980b). The pattern indicates the existence of a stepped surface which can be indexed [14(111) × ($\overline{1}$1$\overline{2}$)]. The energies at which a given reflection is split or nonsplit give specific information on the step height, and the angular separation between split spots provides information on the terrace widths. With the use of only a kinematic treatment of single scattering from the top layer and methods previously discussed (Henzler, 1970a,b), it is concluded that the surface consists of monatomic steps with an average step height of one double layer (3.14 Å) with terrace widths of ~45 Å as illustrated in Fig. 9. From the results of previous investigations of stepped Si surfaces (Rowe et al., 1975), it has been suggested that strain fields (or other local forces) are responsible for the absence of reconstruction within terraces with widths of less than 40–50 Å. Since fractional-order reflections indicative of reconstruction were not observed for the surface after laser annealing, the local atomic arrangement produced by this annealing procedure may be similar to that obtained by cleaving. The stability of the stepped surface was investigated by subjecting the laser-annealed surface to a series of thermal annealing treatments. As observed for flat (111) surfaces, thermal annealing of the crystal to temperatures greater than ~800 K resulted in a surface from which the (7 × 7) diffraction pattern shown in Fig. 8b was obtained. All indications of the steps were gone (individual spots were no longer split), and the (7 × 7) pattern was obtained over the entire face of the sample in accord with previous

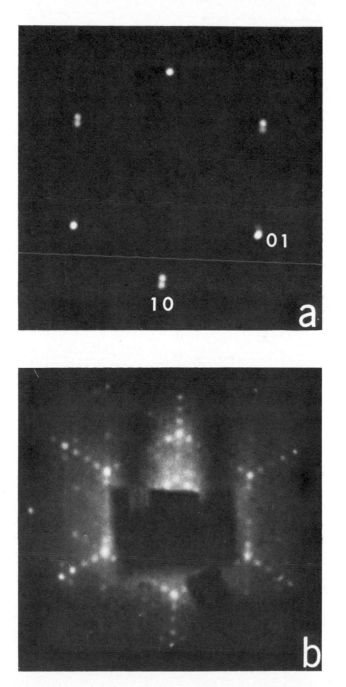

Fig. 8. LEED patterns from clean vicinal (111) Si surfaces at primary beam energies of
(a) 40 eV and (b) 68 eV. (a) Laser annealed, (b) thermally annealed.

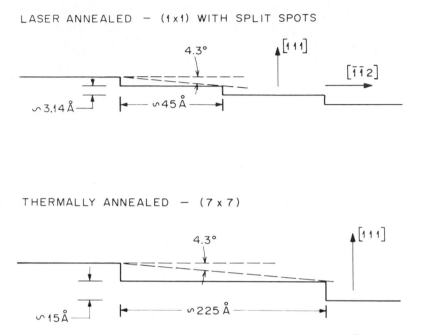

Fig. 9. Schematic illustration of the vicinal surface projected into the ($\overline{1}$10) plane. Top view is for the laser-annealed surface. Bottom view illustrates a possible configuration obtained with thermal annealing.

observations (Olshanetsky and Shkylaev, 1978). The sharpness of the integral-order reflections is consistent with a surface having domains wider than ~200 Å. In order to maintain the average inclination, multilayer steps must be present as illustrated in Fig. 9. After thermal annealing, the stepped surface arrangement could be regenerated by again irradiating the surface with the laser. These observations indicate that it is possible to produce repeatedly a particular step arrangement by initially cutting the crystal to the desired orientation. Recent investigations of vicinal Si (111) surfaces cut along the [11$\overline{2}$] direction have produced results very similar to those discussed above (Chabel *et al.*, 1981b). Steps along this direction contain edge atoms that have three nearest neighbors. Although detailed studies on the angular profiles show the step height to be 3.06 Å in this direction, somewhat less than the double-layer separation, the overall behavior for laser-annealed vicinal surfaces is the same for both types of steps.

E. Si (111)-(1 × 1) Geometric Structure

Following the initial observations that the outermost surface layers of low-index-oriented surfaces of semiconductor single crystals exhibit geometric reconstruction with respect to ideal bulk terminations (Schlier and Farnsworth, 1959), many investigations have concentrated on the examination and determination of both the electronic and geometric properties of these surfaces (Eastman, 1980; Mönch, 1977, 1979). In particular, much of this work has been concerned with the Si (111) surface where (2 × 1) and (7 × 7) reconstructions are observed subsequent to cleaving and thermal annealing, respectively. Although some success has been achieved in determining the atomic positions in the (2 × 1) reconstructed surface layer (Mönch and Auer, 1978), the atomic arrangement within the (7 × 7) reconstruction, because of its increased complexity, is still the subject of much speculation (Eastman, 1980). The observation that (1 × 1) LEED patterns are obtained from the (111) surfaces of Si, Ge, and GaAs after laser irradiation in a UHV environment, coupled with the fact that these surfaces are observed to be atomically clean after this treatment, suggests that they offer the opportunity for investigating clean semiconductor surfaces that exhibit no ordered lateral reconstruction. In particular, it should be possible to obtain information about surface relaxations. To investigate this question it is necessary to measure the intensities of the diffracted electron beams as a function of incident electron energy $(I-V)$ profiles. The experimentally measured profiles must then be compared to results obtained from fully converged dynamical LEED calculations assuming various structural models for the geometric arrangement in the outermost monolayers. A measure of the agreement between the experimental results and the predictions of the model calculations is provided by the R factor (the lower the R-factor value, the better the agreement). Detailed LEED analyses for laser-annealed (111)-(1 × 1) surfaces of Si and Ge have been obtained (Zehner *et al.*, 1981a,b), and the results for the Si surface are discussed below.

A Si (111) surface that had been irradiated with the output of the laser at an energy density of ~2.0 J/cm² was used in these investigations. The intensities of the diffracted beams were measured as a function of electron energy using a Faraday cup operated as a retarding field analyzer. Data were obtained for all of the {10}, {01}, {20}, and {02} beams, and three each of the {11} and {21} beams. Based on observations and conclusions drawn from previous studies, symmetrically equivalent beams were averaged to provide a data base containing six average profiles.

The experimental data base has been compared with the results obtained from fully converged dynamical LEED calculations. A surface region Debye temperature of 550 K, an imaginary component of the optical potential equal to 4.5 eV, and phase shifts obtained from a truncated free-atom potential were employed in these investigations. An initial comparison of calculated profiles with experimental data suggested that d_{12} (bulk value = 0.78 Å) was contracted by approximately 25% from the bulk value. To investigate possible variations in the second interlayer spacing, d_{23} (bulk value = 2.35 Å) calculations were performed in which both d_{12} and d_{23} were varied. The results of the calculations can be conveniently presented using the optimal R factor, for fixed d_{12} and d_{23}, which is the minimum value of R obtained for variation of the real component of the optical potential. The changes in the six-beam optimal R factor with these variations are illustrated by the contour map shown in Fig. 10. Results shown in this figure suggest that d_{12} is contracted by 25.5 ± 2.5%

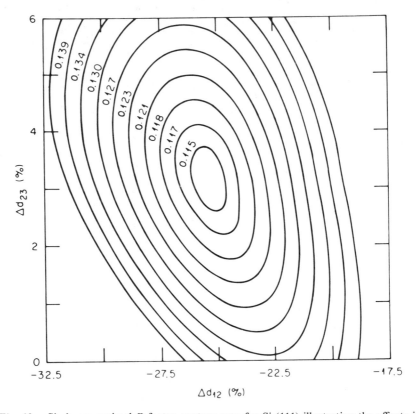

Fig. 10. Six-beam optimal R-factor contour map for Si (111) illustrating the effects in variation of the second interlayer spacing d_{23} with variation in the first interlayer spacing d_{12}. TFA potential, $\theta_D = 550$ K, $V_{oi} = 4.5$ eV.

and d_{23} is expanded 3.2 ± 1.5%. Profiles calculated using these values are shown in Fig. 11, which also contains the corresponding experimental profiles and single-beam reliability factors R determined for each comparison. The six-beam R factor corresponding to Fig. 11 and determined as the minimum in Fig. 10 is 0.115. This value indicates a very good agreement between calculated and experimental profiles in a conventional LEED analysis and suggests that the proposed structural model is highly probable. Furthermore, this R value is significantly lower than any reported value obtained in a LEED analysis of any semiconductor surface. This may be due in part to the proposed model which involves only interplaner spacing changes (relaxation without reconstruction). The changes in interlayer spacings determined from this analysis correspond to nearest-neighbor bond length changes of -0.058 and $+0.075$ Å.

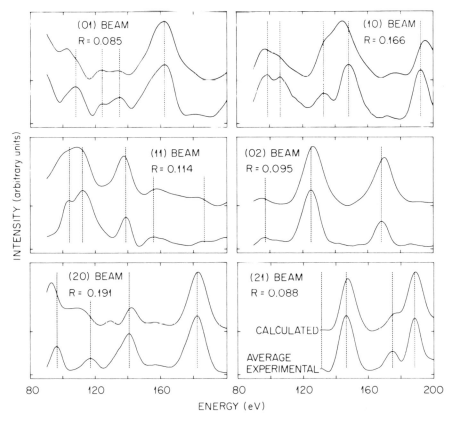

Fig. 11. A comparison of the averaged experimental $I–V$ profiles with calculated results for $\Delta d_{12} = -25.5\%$ and $\Delta d_{23} = 3.2\%$. Other conditions as in Fig. 10.

It should be noted that the interpretation of results obtained in a recent study on the reversible high-temperature ($\sim$1150 K) transition of the (7 × 7) to a (1 × 1) structure has led to the conclusion that a disordered selvedge exists above the transition temperature (Bennett and Webb, 1981a,b). Although the results of the LEED analysis presented above suggest a highly probable structural model for the laser-irradiated surface, only ordered structures could be tested in this analysis. Since the laser-irradiated surface is known to undergo an irreversible transition from (1 × 1) to (7 × 7) at $\sim$700 K, a study on this transition should aid in distinguishing between differing models for each surface structure.

F. Si (111)-(1 × 1) Electronic Structure

An examination of the electronic structure in the surface region of the laser-annealed Si (111) surface is of interest in view of the results of the LEED analysis just discussed. Theoretical one-electron-band calculations (Eastman, 1980) predict that such a surface would be metallic, with a half-filled band of dangling bond states at the Fermi energy E_F, and would be very different from that observed for the annealed Si (111)-(7 × 7) surface as well as for the cleaved Si (111)-(2 × 1) surface. It is assumed that surface states are bandlike in all of these calculations.

Angle-resolved and angle-integrated photoemission studies on both valence band surface states and surface core-level shifts for the laser-annealed Si (111)-(1 × 1) surface prepared in the same manner as was done for the LEED study and for a Si (111)-(7 × 7) surface, which was prepared by thermally annealing the (1 × 1) surface, have been performed (Zehner *et al.*, 1981c). The measurements were made using the display-type spectrometer at the synchrotron radiation source Tantalus I.

In Fig. 12, angle-integrated photoemission spectra are presented for a laser-annealed (1 × 1) surface and for a (7 × 7) surface prepared by thermally annealing the (1 × 1) surface to $\sim$1175 K. For the (1 × 1) surface, the dashed line shows the spectrum obtained after a hydrogen exposure ($\sim$500 langmuirs of activated H) which resulted in about a saturation monolayer coverage of H (same normalization). The difference between the solid curve and the dashed curve within $\sim$3 eV of E_F represents surface state emission. Two predominant surface state features are seen, at -1.8 eV and at -0.85 eV, which are essentially identical for both the (1 × 1) and (7 × 7) surfaces. For the (7 × 7) surface, a third weaker feature is seen at the Fermi level E_F, which is at 0.5 ± 0.05 eV above the valence band maximum E_v. These states at E_F, which correspond to 3% of the total emission intensity from surface states within $\sim$3 eV of E_F,

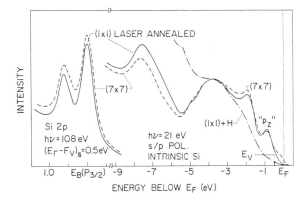

Fig. 12. Angle-integrated photoemission spectra for the valence bands and 2p core levels of laser-annealed Si (111)-(1 × 1) and thermally annealed Si (111)-(7 × 7) surfaces. Two prominent surface-state levels are seen for both; i.e., "p_z-like" levels at -0.85 eV and levels at -1.8 eV.

appear to originate via the (7 × 7) reconstruction from the -0.85-eV states which show a corresponding reduced intensity for the (7 × 7) surface relative to the (1 × 1) surface. The absence of any emission in the gap at E_F for the (1 × 1) surface is not in accord with predictions of one-electron-band calculations for an unreconstructed surface and therefore appears to be inconsistent with the structural model for this surface determined in the LEED analysis. Alternatively, calculations neglecting intrasite correlations may not be adequate to predict surface energy bands.

The angular emission distributions for these surface states were obtained using mixed sp polarization. The surface state at -1.8 eV exhibits a small, sharp feature at the zone center Γ, i.e., normal to the surface, and a more intense emission ring near the hexagonal surface Brillouin zone boundary. For the -0.85-eV surface state a different emission pattern is observed. It consists of a broad, diffuse angular emission distribution centered at Γ, with little emission near the zone boundary. The (7 × 7) surface exhibits essentially identical angular distributions for the corresponding states. The weak surface state feature at E_F, which is seen only for the metallic (7 × 7) surface, exhibits a well-defined hexagonal angular emission distribution that is peaked at the Brillouin zone boundary of a (2 × 2) surface unit cell.

Angle-integrated photoemission spectra of the Si $2p_{3/2}$ core level for the (1 × 1) and (7 × 7) surfaces are shown in Fig. 13 (solid lines) for a photon energy $h\nu = 120$ eV, an energy for which the spectra are surface-sensitive (escape depth 5.4 Å) with about half the emission intensity corresponding

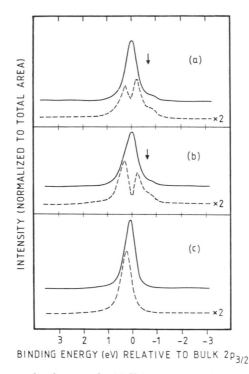

Fig. 13. Si $2p_{3/2}$ core-level spectra for (a) Si (111)-(1 × 1), (b) Si (111)-(7 × 7), and (c) Si (111)-H (1 × 1). The contribution due to the outer double layer of surface atoms is shown by dashed lines. $h\nu = 120$ eV.

to the outer double layer of Si surface atoms. To obtain these spectra the total $2p$ core-level spectra are decomposed into similarly shaped $2p_{1/2}$ and $2p_{3/2}$ contributions, and then a $2p_{3/2}$ spectra for a "bulk" contribution corresponding to the layers below the outer double layer is subtracted. The resulting curves (dashed lines in Fig. 13) show the spectral distributions of surface core levels for the outer two surface layers (one double layer). The bulk line positions were found to be the same (within ±20 meV) for both surfaces; i.e., both have the same band bending at the surface ($E_F - E_v = 0.5 \pm 0.05$ eV).

As seen in Fig. 13, $2p_{3/2}$ surface core-level spectra (dashed curves) for the (1 × 1) and (7 × 7) surfaces are very similar and differ greatly from the spectra for the Si (111)–H (1 × 1). Both show characteristic low-binding-energy peaks, with the (1 × 1) surface showing a peak at −0.8 eV relative to the bulk with an intensity corresponding to ~$(\frac{1}{4} \pm \frac{1}{12})$ of a surface layer of atoms and the (7 × 7) surface showing a peak at −0.7 eV with an intensity corresponding to $\sim(\frac{1}{6} \pm \frac{1}{12})$ of a surface layer of atoms. Both surface spectra also show core levels on the high-binding-energy side of the bulk line at about the same position.

The strong similarity of the valence band surface states and surface core-level spectra for the laser-annealed (1 × 1) surface and thermally annealed (7 × 7) surface indicate that these surfaces have very similar local bonding geometries and differ mainly in long-range order involving geometrical rearrangements that are only a perturbation of the average local bonding geometry. An interesting question then involves why the LEED analysis (Zehner *et al.*, 1981a,b) yields such a good agreement with data using a model (1 × 1) geometry that appears to be different from that needed to describe the surface electronic structure. One possible explanation is that LEED is not particularly sensitive to long-range disorder if it is present on the (1 × 1) surface. Another explanation is that photoemission can rule out the relaxed ordered (1 × 1) geometry only if the surface states are bandlike, as assumed in one-electron-band calculations. However, correlation effects might be very important for these narrow surface levels. Theoretical proposals (Duke and Ford, 1982; Del Sole and Chadi, 1981) have been put forth recently that would make the photoemission data from the (1 × 1) Si surface consistent with the unreconstructed relaxed surface predicted by the LEED analysis. In these models it is assumed that strong correlations dominate the surface state band structure. This idea is, in part, substantiated by experimental results obtained in a study on Si (111)-(2 × 1) (Himpsel *et al.*, 1981a) which show the existence of two dangling bond states. Both theories predict a low-temperature antiferromagnetic ground state and downward dispersion of the dangling bond along Γ-J at low temperature, and these should be tested experimentally.

Two other photoemission studies on laser-annealed Si (111)-(1 × 1) using different annealing conditions (McKinley *et al.*, 1980; Chabal *et al.*, 1981a) have been reported and in agreement with the above results find no occupied states in the gap. In one of these studies (Chabal *et al.*, 1981a), as a consequence of the observation of weak half-order beams and photoemission spectra similar to those obtained from a Si (111)-(2 × 1), it is argued that the laser-annealed surface is buckled with no long-range order but with a short-range (2 × 1) reconstruction. It is suggested that different laser annealing conditions (depth of melt, regrowth velocity) can result in different surface structures. If this is true, then laser annealing with different annealing parameters may aid in determining the driving mechanisms for the reconstructions observed using conventional preparation procedures.

V. Surface Characterization of Ion-Implanted, Laser-Annealed Silicon

A. *Cleaning*

Surfaces of ion-implanted Si crystals can be made atomically clean by employing the procedures previously discussed in Section III. By using multiple-pulse irradiation, the levels of the principal contaminants, O and C, can be reduced to the point where they are not detectable in Auger spectra of the surface region. However, since the degree of diffusion or segregation of the implanted species is known to be a function of the number of laser pulses (see Chapter 5), a more appropriate procedure to follow is initially to sputter the samples. This results in removal of most of the O and C surface contaminants, and the level of cleanliness after one pulse is higher than that obtained without sputtering. Since the depth of material affected during sputtering, ~ 50 Å, is much smaller than the depth of the implanted region, typically ~ 1000 Å, the sputtering process has little or no effect on the subsequent changes in the implant redistribution that occur during laser irradiation. This has been checked by comparing results obtained following laser annealing of implanted crystals that had been sputtered with those that had not. Moreover, since the sputtering species is removed from the surface region during irradiation, as discussed in Section III, it cannot have any effect on the redistribution of the implanted species in the surface region. For most implant conditions, the concentration of implanted species in the surface region is not detectable with AES either after insertion or after sputtering.

B. *Substitutional Implants*

Previous investigations (White *et al.*, 1980b) have shown that both implanted B and As occupy substitutional sites subsequent to laser annealing and that, as a consequence of both the high liquid phase diffusivities and the high values of distribution coefficients, are able to diffuse into the crystal during the regrowth process after irradiation. An example of the redistribution results obtained on a Si (100) sample implanted with ^{75}As (100 keV, 8.3×10^{16}/cm^2) using the RBS technique is shown in Fig. 14. Details of this are discussed in Chapter 5. Although both RBS and secondary-ion mass spectroscopy (SIMS) techniques provide detailed information about the distribution with respect to depth, they provide no information about the concentration in the surface region (≤ 10 Å) and its change with multiple-pulse irradiation. Such information has been ob-

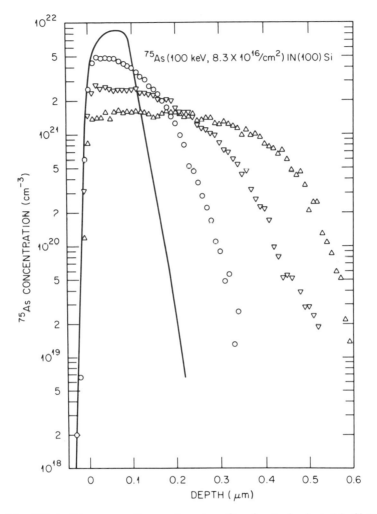

Fig. 14. Effect of laser annealing on dopant profiles for As implanted in Si (100) as determined by RBS. Profile results are shown for As-implanted condition and subsequent to laser annealing at ~2.0 J/cm². —, As-implanted; ○, 1 pulse; ▽, 5 pulses; △, 15 pulses.

tained by performing experiments using the techniques described in Section II.

Auger data were obtained from a Si (100) sample implanted with ^{75}As, the same crystal from which the RBS data shown in Fig. 14 were obtained. With these data, the ratio of the intensity of the As $M_{4,5}VV$ (31 eV) transition to that of the Si $L_{2,3}VV$ (91 eV) transition is shown in Fig. 15,

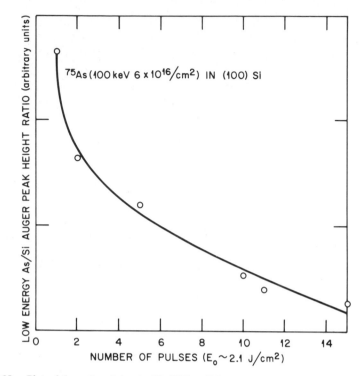

Fig. 15. Plot of the ratio of the As $M_{4,5}VV$ to Si $L_{2,3}VV$ Auger transition intensities as a function of the number of laser pulses.

where it is plotted as a function of the number of pulses. The data show that the relative amount of As in the surface region decreases with an increasing number of pulses, similar to the RBS results obtained for the subsurface region. However, it is difficult to quantify the AES results to the same degree as can be done with the RBS data. Thus, although it can be concluded that a reduction in concentration occurs with multiple pulses, there could still be a possible change in concentration in going from the surface to subsurface region that is a consequence of the surface–vacuum interface. No change is observed in the surface concentration after a large number (10–15) of pulses where RBS results indicate uniform concentration from the subsurface region down to the liquid–solid interface. Similar results have been obtained for a variety of implanted doses of either As or B in Si (100) and (111) crystals.

To examine the effect of the implanted species on surface order, LEED observations have been made also. Shown in Fig. 16 are LEED patterns obtained from the same [75]As-implanted Si (100) crystal. Only a very weak,

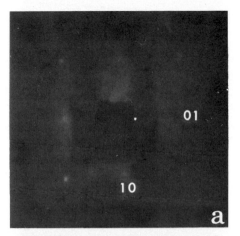

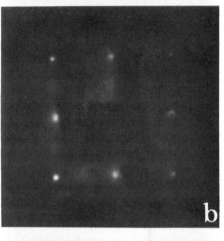

Fig. 16. LEED patterns from an As-implanted Si (100) surface at a primary beam energy of 49 eV. Patterns are shown subsequent to laser annealing at ~2.0 J/cm² for (a) 2, (b) 5, and (c) 10 pulses.

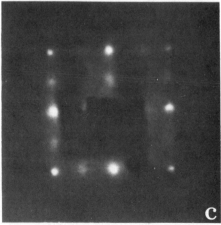

poorly defined LEED pattern was observed after one pulse. Following two pulses of irradiation, the pattern shown at the top of Fig. 16 was obtained. Integral-order beams are observed, as well as weak streaks between them. With additional pulses the streaks begin to coalesce into half-order reflections, indicating the formation of a (2×1) surface structure. They continue to become sharper and more intense with additional pulses, as shown in the figure. However, the pattern observed after ten pulses is not as good as that obtained from a virgin Si (100) crystal and thus indicates the presence of disorder in the surface region. Similar investigations of Si (100) crystals implanted with ^{11}B (35 keV, $1 \times 10^{16}/\text{cm}^2$) show the same type of AES results and much higher-quality LEED patterns after the same number of pulses. This probably reflects the fact that less B can be incorporated into Si, resulting in less disorder in the surface region. However, it is interesting that the (2×1) LEED patterns obtained in both cases show the presence of the reconstructed surface, similar to that obtained from a virgin Si (100) crystal. In contrast to this, although AES results obtained for the (111) crystals implanted with either As or B are similar to those obtained for the (100) surface, (1×1) LEED patterns were observed in all cases. The patterns are of much higher quality after a given number of pulses when compared to those obtained from the (100) surfaces, and they show no evidence of ordered lateral reconstruction.

The observation that laser annealing can be combined with ion- implantation to provide semiconductor surface regions containing novel doping concentrations (supersaturated alloys) suggests that these techniques may be used to alter or tailor the electronic structure in this region. To examine this possibility photoemission techniques, as described in Section IV, have been used to investigate highly degenerate n-type Si (111)-(1×1) surfaces as a function of As concentrations up to $5 \times 10^{21}/\text{cm}^{-3}$ (10 at.%) and degenerate p-type Si (111)-(1×1) surfaces as a function of B concentrations up to $1 \times 10^{21}/\text{cm}^3$ (2 at.%) (Eastman $et\ al.$, 1981). The samples were prepared by ion implanting ^{75}As (100 keV, $7 \times 10^{16}/\text{cm}^2$ and $1 \times 10^{16}\text{cm}^2$) and ^{11}B (35 keV, $2 \times 10^{16}/\text{cm}^2$) and then $in\ situ$ laser annealing. The maximum doping concentrations are about 10 and 3 times the concentrations of electrically active As and B achievable by conventional techniques, respectively.

Angle-integrated photoemission spectra for the valence bands and $2p$ core levels are presented in Figs. 17 and 18 for intrinsic Si (111)-(1×1), degenerate n-type As-doped (4 and 7 at.%) Si (111)-(1×1), and degenerate p-type B-doped (1 at.%) Si (111)-(1×1) surfaces. In Fig. 17, spectra are normalized to constant total emission within 5 eV of E_F, and energies are given relative to the valence band maximum at the surface (E_v^s). The Fermi-level position relative to the valence band maximum at the surface

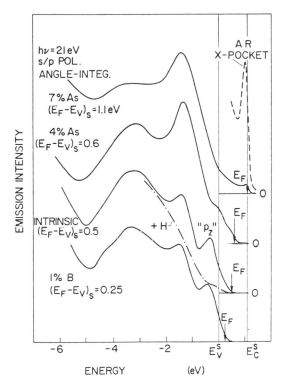

Fig. 17. Photoemission spectra (PDOS) for the valence bands of laser-annealed (111)-(1 × 1) surfaces of intrinsic and highly doped Si. The levels near -0.4 and -1.3 eV are due to surface states. E_v^s, E_c^s, and E_F denote the valence band maximum, conduction band minimum, and Fermi-level positions at the surface.

$(E_F - E_v)_s$ was previously determined to be 0.5 eV for intrinsic Si (111)-(1 × 1) (Zehner *et al.*, 1981c). E_F is seen to shift markedly with doping, i.e., from 0.25 eV above E_v^s for the B-doped sample to the conduction band minimum $E_c = 1.1$ eV for the 7% As-doped sample. For intrinsic Si (111)-(1 × 1) in Fig. 17, the dashed-dotted line shows the effect of an absorbed monolayer of H, which is to remove the two predominant surface state levels at 0.4 and 1.3 eV below E_v^s, as discussed in Section IV. Relative to intrinsic Si, for highly degenerate (1 at.%) B doping, the two states are unaltered and the principal changes are that E_F moves down by 0.25 eV and the surface becomes metallic.

More dramatic effects are seen with As doping. At 4 at.% As doping, the surface states have become significantly altered, while E_F has increased by 0.1 eV relative to the intrinsic Si. That is, the upper "sp_z-like" dangling bond state has become much weaker and shifted upward in

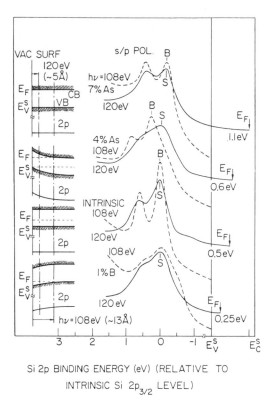

Si 2p BINDING ENERGY (eV) (RELATIVE TO
INTRINSIC Si 2p₃/₂ LEVEL)

Fig. 18. Angle-integrated photoemission spectra for the Si 2p core levels of the four Si surfaces depicted in Fig. 17. The zero binding energy corresponds to the Si $2p_{3/2}$ level for bulk Si (111)-(1 × 1). The schematic diagrams on the left depict band bendings and energy-level positions; the escape depths are $l \sim 5$ Å for $h\nu = 120$ eV and ~13 Å for $h\nu = 108$ eV.

energy by ~0.3 eV, the lower −1.4-eV state has increased significantly in intensity but is unshifted, and the surface has become metallic with new states near E_F. As the doping is further increased from 4 to 7 at.%, E_F rapidly shifts and becomes pinned at the conduction band minimum E_c. Also, the upper sp_z-like surface state continues to diminish in intensity so as to be nearly imperceptible by 7 at.% doping, and the lower surface state becomes extremely intense. The conduction band minima (Δ_{min}) near X become occupied, and emission from these minima is observed as intense elliptical lobes in angle-resolved photoemission spectra (dotted line labeled "AR" in Fig. 17).

The Si 2p core-level spectra shown in Fig. 18 were obtained using $h\nu = 108$ eV (dashed lines) and $h\nu = 120$ eV (solid lines) which provide "bulk-sensitive" and "surface-sensitive" spectra, respectively (Himpsel *et al.*, 1980). S and B denote surface and bulk $2p$ levels. For intrinsic Si and 7 at.% As-doped Si surfaces, the $2p$ levels are the same for both $h\nu = 108$ and $h\nu = 120$ eV; i.e., the surface (S) and bulk (B) $2p_{3/2}$ levels coincide in binding energy. This is indicative of the "flat band" condition at the surface as shown by the schematic diagrams on the left in Fig. 18. (CB and VB denote the conduction band and valence band edges.) A finding of significant interest is that, at an As doping level of 7% or greater, E_F shifts across the entire gap from 0.5 eV for intrinsic (1×1) Si to the conduction band minimum at 1.1 eV. By depositing a thin Au film on this surface it was possible to show via Si $2p$ core-level measurements (not shown) that E_F remained unchanged (within 50 meV). Thus a "zero barrier height" Schottky barrier was formed, although, for electrical purposes, the Au–Si interface is undoubtedly shorted because of the extreme degenerate n-type doping, i.e., short Debye screening length. For 4 at.% As and 1 at.% B doping levels the bands are bent upward and downward in the surface region, respectively. The broadening and smearing of the spectra when compared to that obtained for intrinsic Si reflect the distribution in energy of $2p$ core levels in the sampled region.

C. Interstitial Implants

For Group III–V implants such as As and B, it has been shown that the interfacial distribution coefficient during laser annealing is considerably higher than the equilibrium value (White *et al.*, 1980a) and that it is possible to exceed the equilibrium solubility limit by factors of >500 without causing segregation and/or precipitation to the near-surface region. However, non-Group III–V implants that have been studied exhibit, depending on the implant dose, segregation to the surface as well as the formation of a cell structure subsequent to laser annealing (White *et al.*, 1980c). In order to determine the effects of interstitial implants on surface properties, investigations of the segregation and zone refining of impurities to the surface region following pulsed laser annealing have been performed. To illustrate the results obtained in these studies, we discuss data obtained using Si (111) samples implanted with Fe to doses of 1.13×10^{15} atoms/cm², 6.0×10^{15} atoms/cm², and 1.8×10^{16} atoms/cm² and laser-annealed at ~2.0 J/cm² (Zehner *et al.*, 1982b).

Results obtained using AES showed that, after laser irradiation, Fe could be detected readily in the surface region. For the low-dose case, little increase is observed in the intensity of the Fe Auger signal obtained from a surface irradiated with additional pulses. Although at intermediate doses several pulses (two or three) are sufficient to produce the surface concentration that results in the maximum Fe Auger signal intensity, in the high-dose case at least five pulses are required to produce the same result. These observations are consistent with previous RBS results showing a dependence of the segregation to the surface that is a function of the implant dose and number of pulses used for annealing (White *et al.*, 1980c).

The effect of segregation on surface order was determined by LEED observations. The LEED patterns obtained from each of the implanted samples subsequent to the irradiation with five pulses are shown in Fig. 19. A LEED pattern obtained from a virgin Si (111) crystal for the same incident electron energy is shown at the top of this figure for purposes of comparison. Although (1 × 1) LEED patterns were obtained after one pulse of irradiation on each sample, a higher background intensity was

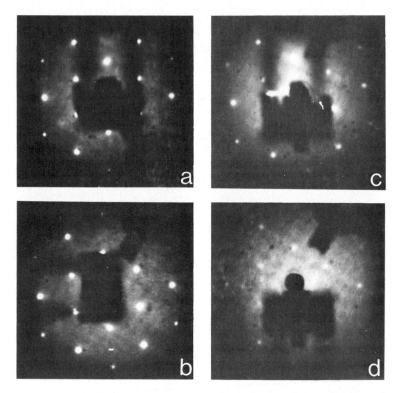

Fig. 19. LEED patterns, at a primary beam energy of 110 eV, from a Si (111) surface and from (111) surfaces of crystals implanted with Fe at various doses. Patterns are shown subsequent to five pulses of laser annealing at ~2.0 J/cm². (a) Unimplanted, (b) $1.13 \times 10^{15}/cm^2$, (c) $6.0 \times 10^{15}/cm^2$, (d) $1.8 \times 10^{16}/cm^2$.

always observed relative to that obtained from the virgin crystal. The effect of increased segregation, at intermediate and high doses, with multiple pulses was to degrade the quality of the LEED patterns. In general, the background intensity increased, as shown in Fig. 19, although the symmetry of the pattern observed was still (1 × 1).

Subsequent examination of the samples with RBS produced the results shown in Fig. 20. Although complete transport of the Fe to the near-surface region has occurred for the low- (after one pulse) and medium- (two or three pulses) dose cases, at higher doses, even after five pulses, substantial quantities of Fe remain in the first 1000 Å of the crystal at an average concentration of 2×10^{21} atoms/cm^3. Furthermore, channeling studies show that Fe in the bulk of the crystal is not in solid solution.

From transmission electron microscopy studies it is known that, in the case of a high-dose Fe-implanted crystal, a well-defined cell structure is observed in the near-surface region subsequent to laser annealing (White *et al.*, 1980c). These results show that subsequent to laser annealing the Fe is not uniformly distributed in the plane of the near-surface region, but instead is highly concentrated in the walls of the cell structure. The interior of each cell is an epitaxial column of silicon extending to the surface (average cell diameter 250 Å). Surrounding each column of silicon is a cell wall, ≤50 Å thick and extending to a depth of ~1000 Å, containing massive quantities of segregated Fe, possibly in the form of Fe silicides. Thus,

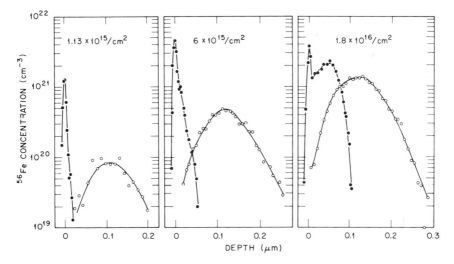

Fig. 20.   Effect of laser annealing on dopant profiles for Fe implanted in Si (111) as determined by RBS. Profile results are shown for as-implanted condition and subsequent to five pulses of laser annealing at ~2.0 J/cm^2. ○, As-implanted; ●, laser annealed.

the (1×1) LEED patterns observed in this case arise from the bulk termination of (111) planes in the columns of silicon at the surface. The absence of any other well-defined diffraction features from the Fe-implanted region shows that no long-range order exists in the termination of the cell walls at the surface. The high background intensities, relative to that obtained from the virgin crystal, observed for all implant conditions indicate the presence of disorder (possibly strain in the region of cell wall boundaries) in the outermost layers, which increases with increasing implant dose. For these samples, as well as those implanted with Cu, sputtering after laser irradiation resulted in the removal of some of the implant from the surface region. However, subsequent irradiation with the laser caused the segregation of large quantities of the implant to the surface region. Furthermore, for samples in which interstitial species such as Cu are present as bulk impurities, laser irradiation can be used to zone-refine these species to the surface region from a depth equivalent to the maximum melt penetration. These impurities can then be removed from the surface with light sputtering, leaving an impurity-free subsurface region (to a depth determined by the melt front penetration). In many device applications involving silicon, Cu and Fe impurities act as very efficient recombination centers and adversely affect minority-carrier lifetime. The above observations show that laser annealing combined with light etching (sputtering) can be used as a rapid purification treatment to produce an impurity-free surface region.

VI.   Conclusions

It has been shown that laser annealing can be used to cause substantial alterations in the properties of the surface region of many semiconductor crystals. The production of atomically clean surfaces in processing times of ≤ 1 sec and the restoration of order to the surface region of a damaged crystal indicate that this technique provides an alternative approach to conventional surface preparation procedures. When combined with ion implantation, laser annealing can be used to tailor both the geometric lattice (interatomic spacings) and electronic structure in the surface region.

In the area of basic research concerned with the physics and chemistry of surfaces, these observations indicate that it may be possible to modulate (clean, adsorb–desorb) surface coverage, study metastable surface structures and effects of well-controlled defects (steps), and investigate alloys that cannot be obtained using conventional crystal growth techniques. For technological applications, the rapid cleaning process may be useful in a number of areas in which surface contamination needs to be removed both swiftly and efficiently. The production of alloys (submicrometer regions) with tailored surface properties should be useful in a number of areas.

These observations indicate that laser annealing of surfaces in UHV has a tremendous potential as a tool for both surface science and practical application. Although it has been tested extensively only on semiconductors, the results obtained suggest that it will be applicable to a wide range of materials. Additional investigations concerned with the surface properties of materials irradiated with the laser as well as the changes that occur by varying the annealing conditions (regrowth velocity, depth of melt) need to be performed in order to characterize and understand this processing technique more completely.

References

Bedair, S. M., and Smith, H. P., Jr. (1969). *J. Appl. Phys.* **40**, 4776.
Bennett, P. A., and Webb, M. B. (1981a). *Surf. Sci.* **104**, 74.
Bennett, P. A., and Webb, M. B. (1981b). *J. Vac. Sci. Technol.* **18**, 847.
Chabal, Y. J., Rowe, J. E., and Zwemer, D. A. (1981a). *Phys. Rev. Lett.* **46**, 600.
Chabal, Y. J., Rowe, R. E., and Christman, S. B. (1981b). *Phys. Rev. B* **24**, 3303.
Cho, A. Y., and Hayashi, I. (1971). *Solid-State Electron.* **14**, 125.
Cowan, P. L., and Golovchenko, J. A. (1980). *J. Vac. Sci. Technol.* **17**, 1197.
Del Sole, R., and Chadi, D. J. (1981). *Phys. Rev. B* **24**, 7431.
Duke, C. B., and Ford, W. K. (1982). *Surf. Sci.* **111**, L685.
Eastman, D. E. (1980). *J. Vac. Sci. Technol.* **17**, 492.
Eastman, D. E., Heimann, P., Himpsel, F. J., Reihl, B., Zehner, D. M., and White, C. W. (1981). *Phys. Rev. B* **24**, 3647.
Hagstrum, H. D., and Becker, G. E. (1973). *Phys. Rev. B* **8**, 1580.
Henzler, M. (1970a). *Surf. Sci.* **19**, 159.
Henzler, M. (1970b). *Surf. Sci.* **22**, 12.
Himpsel, F. J., Heimann, P., Chiang, T.-C., and Eastman, D. E. (1980). *Phys. Rev. Lett.* **45**, 1112.
Himpsel, F. J., Heimann, P., and Eastman, D. E. (1981a). *Phys. Rev. B* **24**, 2003.
Himpsel, F. J., Eastman, D. E., Heimann, P., Reihl, B., White, C. W., and Zehner, D. M. (1981b). *Phys. Rev. B* **24**, 1120.
Hoh K., Koyama, H., and Uda, K. (1980). *Jpn. J. Appl. Phys.* **19**, L375.
Jona, F. (1965). *IBM J. Res. Dev.* **9**, 375.
Lander, J. J., and Morrison, J. (1962). *J. Chem. Phys.* **37**, 729.
Lander, J. J., and Morrison, J. (1963). *J. Appl. Phys.* **34**, 1403.
McKinley, A., Parke, A. W., Hughes, G. J., Fryar, J., and Williams, R. H. (1980). *J. Phys. D.* **13**, 138.
Mönch, W. (1977). *Surf. Sci.* **63**, 79.
Mönch, W. (1979). *Surf. Sci.* **86**, 672.
Mönch, W., and Auer, P. P. (1978). *J. Vac. Sci. Technol.* **15**, 1230.

Morabito, J. M. (1975). *Surf. Sci.* **49**, 318.

Narayan, J., Young, R. T., and White, C. W. (1978). *J. Appl. Phys.* **49**, 3912.

Olshanetsky, B. Z., and Shklyaev, A. A. (1978). *Surf. Sci.* **82**, 445.

Roberts, R. W. (1963). *Br. J. Appl. Phys.* **14**, 537.

Rodway, D. C., Cullis, A. G., and Webber, H. C. (1980). *Appl. Surf. Sci.* **6**, 76.

Rowe, J. E., Christman, S. B., and Ibach, H. (1975). *Phys. Rev. Lett.* **34**, 874.

Saris, F. W. (1981). Personal communication.

Schlier, R. E., and Farnsworth, H. E. (1959). *J. Chem. Phys.* **30**, 917.

Wang, J. C., Wood, R. F., and Pronko, P. P. (1978). *Appl. Phys. Lett.* **33**, 455.

Westendorp, J. F. M., Wang, Z.-L., and Saris, F. W. (1982). *In* "Laser and Electron Beam Interactions with Solids" (B. R. Appleton and G. K. Cellar, eds.), p. 255. North-Holland, New York.

White, C. W., Christie, W. H., Appleton, B. R., Wilson, S. R., Pronko, P. P., and Magee, C. W. (1978). *Appl. Phys. Lett.* **33**, 662.

White, C. W., Narayan, J., Appleton, B. R., and Wilson, S. R. (1979). *J. Appl. Phys.* **50**, 2967.

White, C. W., Wilson, S. R., Appleton, B. R., and Young, F. W., Jr. (1980a). *J. Appl. Phys.* **51**, 738.

White, C. W., Wilson, S. R., Appleton, B. R., Young, F. W., Jr., and Narayan, J. (1980b). *In* "Laser and Electron Beam Processing of Materials" (C. W. White and P. S. Peercy, eds.), p. 111. Academic Press, New York.

White, C. W., Wilson, S. R., Appleton, B. R., and Narayan, J. (1980c). *In* "Laser and Electron Beam Processing of Materials" (C. W. White and P. S. Peercy, eds.), p. 124. Academic Press, New York.

Zehner, D. M., White, C. W., and Ownby, G. W. (1980a). *Appl. Phys. Lett.* **36**, 56.

Zehner, D. M., White, C. W., and Ownby, G. W. (1980b). *Surf. Sci. (Lett).* **92**, L67.

Zehner, D. M., White, C. W., and Ownby, G. W. (1980c). *Appl. Phys. Lett.* **37**, 456.

Zehner, D. M., White, C. W., and Ownby, G. W. (1980d). *In* "Laser and Electron Beam Processing of Materials" (C. W. White and P. S. Peercy, eds.), p. 201. Academic Press, New York.

Zehner, D. M., Noonan, J. R., Davis, H. L., and White, C. W. (1981a). *J. Vac. Sci. Technol.* **18**, 852.

Zehner, D. M., Noonan, J. R., Davis, H. L., White, C. W., and Ownby, G. W. (1981b). *In* "Laser and Electron Beam Solid Interactions and Materials Processing" (J. F. Gibbons, L. D. Hess, and T. W. Sigmon, eds.) p. 111. North-Holland Publ., New York.

Zehner, D. M., White, C. W., Heimann, P., Reihl, B., Himpsel, F. J., and Eastman, D. E. (1981c). *Phys. Rev. B* **24**, 4875.

Zehner, D. M., White, C. W., Appleton, B. R., and Ownby, G. W. (1982a). *In* "Laser and Electron Beam Interactions with Solids" (B. R. Appleton and G. K. Celler, eds.), p. 683. North-Holland, New York.

Zehner, D. M., White, C. W., and Ownby, G. W. (1982b). To be published.

Chapter 10

Solid Phase Regrowth

J. F. GIBBONS and T. W. SIGMON

Stanford Electronics Laboratories
Stanford University
Stanford, California

I. Solid Phase Regrowth and Diffusion Processes

The principal purpose of this chapter is to review the areas of continuous-wave (cw) beam processing of semiconductors that can be adequately understood in terms of solid phase regrowth mechanisms. For this purpose we will begin with a brief review of solid phase regrowth results obtained from low-temperature (400–600°C) furnace annealing experiments and then proceed to discuss parallel results obtained when a scanning cw laser or electron beam is used to promote the regrowth process. In most of this discussion we treat the scanning cw beam system as being equivalent to a high-temperature, short-time-duration annealing furnace with large heating and cooling rates. We shall see that in most cases cw beam processing causes solid phase regrowth and diffusion to occur at rates that can be obtained from a straightforward extrapolation of the low-temperature furnace results.

The physical basis of this parallelism arises directly from the relationship between the dwell time of the beam in a typical scanning beam experiment and the thermal time constants of the materials being processed. The dwell time is typically on the order of 1 msec, while the thermal time constants of the substrates are typically several orders of magnitude less than this, allowing the irradiated surface ample time to come into thermal equilibrium with the beam. Under these conditions it is possible to calculate the spatial temperature profile by assuming steady state heat flow in a moving frame of reference. Such calculations, to be summarized in Section II, provide the theoretical basis for the relationship between the beam processing parameters and solid phase regrowth mechanisms.

A. Concept of Solid Phase Epitaxy

We begin by reviewing the important special case of solid phase *epitaxial* regrowth of silicon. Solid phase epitaxy (SPE) describes a situation where an amorphous layer of a semiconductor recrystallizes epitaxially on a single-crystal substrate at temperatures well below the melting or eutectic point of the relevant material. As suggested in Fig. 1 for silicon, we are concerned with two general forms of solid phase epitaxy. In one of these, Fig. 1a, the amorphous Si layer is in direct contact with the underlying crystalline material. Such a system is characteristic of ion-implanted silicon when the dose and energy of the implanted species are sufficient to produce an amorphous layer (Crowder and Title, 1971). A similar system can also be prepared directly by the evaporation of silicon under appropriate conditions onto an atomically clean single-crystal substrate (Roth and Anderson, 1977).

In Fig. 1b we show the other general system of interest, in which there is a metal (Ottaviani *et al.*, 1972, 1974) or silicide (Canali *et al.*, 1974) layer

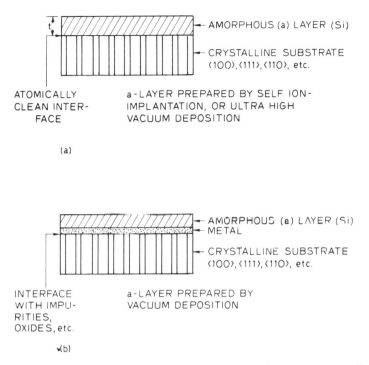

Fig. 1. Systems of interest in solid phase epitaxy. (a) Without transport medium, (b) with transport medium.

interposed between the amorphous and crystalline silicon. In this case the metal is said to serve as a *transport medium*, though in both cases the driving force for the process is the same. The character of this driving force was suggested in Chapter II, Fig. 3 where the free energy of crystalline, amorphous, and liquid silicon was plotted (qualitatively) as a function of temperature. The higher free energy of the amorphous semiconductor as compared to that of the single-crystal material suggests that the free energy of the system can be reduced by recrystallization of the amorphous layer on the underlying substrate.

The difference in free energy can also lead to "explosive" recrystallization which can occur without epitaxial growth. The furnace-activated recrystallization of germanium films on amorphous substrates studied by Takamori *et al.* (1972), Matsuda *et al.* (1973), and more recently by Gold *et al.* (1980), Zeiger *et al.* (1980), and others who used a laser beam to initiate the process is related directly to the reduction in free energy that occurs by this transformation. An essential difference between explosive recrystallization and epitaxy is that randomly ordered crystallites will normally grow when there is no underlying crystalline substrate to provide the necessary template for epitaxy.

B. Solid Phase Epitaxial Regrowth without Transport Media

In this section we review the results of a number of experiments performed to elucidate the parameters that characterize the epitaxial regrowth of amorphous silicon on a single-crystal silicon substrate. Conceptually the simplest system is an amorphous layer deposited directly on a crystalline substrate. However, the presence of surface oxides will normally prevent epitaxial regrowth on the substrate unless the interfacial layer is destroyed prior to the recrystallization process (Sigmon and Tseng, 1978). An experimentally simpler case is afforded by ion implantation of heavy impurities in silicon at relatively low energies, where the damage introduced by the ion implantation can cause the substrate to be driven amorphous from the surface to a depth slightly greater than the projected range of the incident ion. More precisely, we require that the energy deposited in nuclear processes by the primary ion be equal to or greater than 10^{21} keV/cm^3, which leads to the equation

$$\phi \; dE_n/dx \geq 10^{21} \text{ keV/cm}^3 \tag{1}$$

where ϕ is the dose and dE_n/dx is the energy deposition rate in nuclear processes at the plane x (Vook and Stein, 1971). The implantation of heavy elements such as antimony and arsenic in silicon will typically deposit sufficient energy in nuclear processes to satisfy Eq. (1) and therefore produce an amorphous layer from the surface to a depth somewhat greater than the projected range of the ion. A useful alternative view proposed by Morehead and Crowder (1971) is that each ion produces an essentially amorphous damage cluster surrounding its trajectory. The implantation of a sufficient number of ions will then produce complete amorphousness from the surface to a depth somewhat greater than the projected range.

A somewhat different situation exists for the implantation of lighter ions, in which the damage cluster associated with individual implanted projectiles is not itself amorphous and the overlap of damage clusters is required to produce amorphous material (Gibbons, 1972). Under these conditions only a buried amorphous layer may be produced, or the material may be damaged by the implantation but not driven amorphous. The recovery of these materials during annealing is then a good deal more complicated and cannot be conveniently described by a simple solid phase regrowth mechanism (Csepregi et al., 1976).

To circumvent these problems the original studies on solid phase epitaxy (Csepregi *et al.*, 1975) were performed on material that had been driven amorphous by multiple-energy implantation of silicon. Silicon was used so that the regrowth process could be studied free of any potential impurity effects. The basic condition for amorphousness given in Eq. (1) leads to the conclusion that a single-energy silicon implant will normally form an amorphous layer beneath the surface so that regrowth can proceed from both interfaces. Hence to form a continuous amorphous layer from the surface to a depth D, multiple-energy implantation procedures are necessary. It is also essential to prevent substrate heating during the implantation, especially in as much as dislocation networks in the transition region between the amorphous and crystalline materials can occur at elevated temperatures (Seidel *et al.*, 1976; Glowinski *et al.*, 1976). These problems may be avoided by holding the substrate near liquid nitrogen temperature during the implantation and using low beam current densities. With these techniques it is possible to form 0.5-μm-thick, impurity-free amorphous layers in contact with a single-crystal substrate to study the basic epitaxial regrowth process.

The pioneering study on this process (Csepregi *et al.*, 1975) employed channeling effect measurements with megaelectron volt helium ions to determine the thickness of the amorphous layer. The energy spectra of helium particles backscattered from an implantation-amorphized sample after various annealing treatments are shown in Fig. 2. For these experiments the ^{4}He probe was aligned parallel to the $\langle 100 \rangle$ axial rows in the underlying crystal substrate. The yield from the implanted amorphous layer has a leading edge at 1.15 MeV, corresponding to scattering from silicon atoms at the outermost surface into a detector oriented 170° from the incident beam direction. A subsequent reduction in yield is observed for an as-implanted film at 0.94 MeV, corresponding to scattering from the silicon atoms in the amorphous layer at the amorphous–silicon interface. The thickness of the amorphous layer in units of atoms per square centimeter can be determined from the energy width of the random spectrum measured at the half-height. This energy width can be converted to a thickness in angstroms, assuming the amorphous material to have a density equivalent to that of the substrate. This conversion yields a thickness for the amorphous layer of 4300 Å as noted in Fig. 2.

The scattering yield at energies below 0.94 MeV corresponds to scattering from the underlying single-crystal region and is more than an order of magnitude higher than would be obtained from a crystal without an amorphous layer. The reason for this is that the angular divergence of the ^{4}He beam after penetrating the amorphous layer is sufficiently great that few of the helium atoms are still within the critical angle for channeling

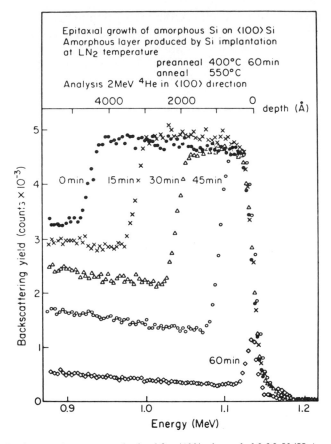

Epitaxial growth of amorphous Si on ⟨100⟩ Si
Amorphous layer produced by Si implantation
at LN₂ temperature
                        preanneal 400°C 60min
                        anneal   550°C
Analysis 2 MeV ⁴He in ⟨100⟩ direction

Fig. 2. Backscattering spectra obtained for ⟨100⟩-channeled 2-MeV ⁴He⁺ on amorphous layers created by self-ion implantation of ⟨100⟩ Si for various anneal times at 550°C.

when they reach the single-crystal material. The yield from the underlying crystal is therefore similar to that which one would obtain from a single-crystal target appropriately misaligned with respect to the ⁴He beam (Lugujjo and Mayer, 1973).

By subjecting the as-implanted sample to an isothermal annealing sequence at a moderate temperature (on the order of 500°C), it is found that the leading edge of the spectra remains fixed but the width of the amorphous layer and the yield of particles from the underlying crystalline material both decrease. These facts imply that the amorphous layer recrystallizes epitaxially on the underlying crystalline substrate. The growth rate can be determined directly from the decrease in the thickness of the amorphous layer following the annealing treatment. For $\langle 100 \rangle$ silicon samples annealed at 550°C, the data in Fig. 2 show that the regrowth is linear in time with a growth rate at 550°C of approximately 90 Å/min. By repeating these experiments at a number of temperatures it is possible to obtain the growth rate versus temperature, with the results shown in Fig. 3. The experimental points are found to scatter closely around a straight line with a slope or activation energy E_a of 2.35 eV for silicon (2.0 eV for germanium; Csepregi *et al.*, 1975, 1977a,b).

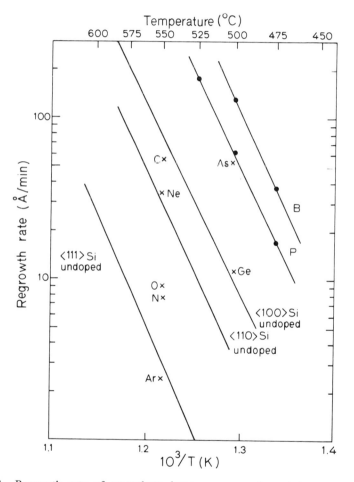

Fig. 3. Regrowth rate of amorphous layers versus reciprocal temperature for Si substrates. The data were obtained from megaelectron volt ⁴He channeling measurements. The activation energy for the process was approximately 2.35 eV. The effect of various impurities is also shown.

1. IMPURITY EFFECTS

The ion implantation process also offers the possibility of incorporating known amounts of impurities into the implanted amorphous layer. The presence of these impurities can have a pronounced influence on the growth rate, as suggested in Fig. 3. The growth rate is reduced appreciably for oxygen concentrations of $10^{20}/cm^3$ and above (Kennedy *et al.*, 1977). In contrast the implantation of phosphorus or boron in an amorphous layer causes an increase in the growth rate (Csepregi *et al.*, 1977a,b). The growth rate versus atomic concentration is shown in Fig. 4

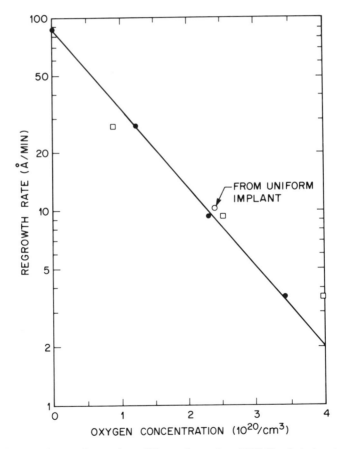

Fig. 4. Regrowth rate of amorphous Si layers formed on ⟨100⟩ Si substrates at 550°C as a function of oxygen concentration.

for oxygen. For an average impurity concentration of approximately 0.5 at.%, the growth rate decreases by roughly an order of magnitude in the presence of oxygen and increases by a factor of 6 for phosphorus. At these concentration levels, implanted nitrogen produces essentially the same effect as implanted oxygen, while implanted arsenic produces an effect similar to that of implanted phosphorus. Further, the growth rate is found to be correlated with the impurity concentration near the amorphous–crystal interface (Csepregi et al., 1977a,b; Kennedy et al., 1977). This suggests that the local impurity concentration near the moving interface controls the growth rate.

One mechanism proposed for the regrowth process (Csepregi et al., 1978) involves a cooperative phenomenon in which one broken bond or vacancy is propagated along the interface, causing a large number of atoms to reorder. Impurities such as oxygen and nitrogen that tend to "scavenge" dangling bonds would then be expected to reduce the growth rate. Similarly, Revesz et al. (1978) found that argon bubbles with an average size of 170 Å were present in argon-implanted samples. The bubbles present after implantation grow in size in the temperature range 500–600°C where recrystallization occurs. Twinning of the regrown layers, as well as retardation of the growth rate, was observed in these experiments. Based on the cooperative phenomenon theory just outlined, the presence of the argon-filled voids is interpreted to act as a sink for broken bonds or vacancies at the growing interface.

2. ORIENTATION EFFECTS

The solid phase epitaxial regrowth of silicon from implanted amorphous layers is also found to depend strongly on the orientation of the underlying substrate (Csepregi et al., 1978). The growth rate for ⟨100⟩ samples is about 3 times that for the ⟨110⟩ direction and about 25 times the *initial* growth rate for the ⟨111⟩ orientation. The growth rate is found to be constant at a given temperature for both the ⟨100⟩ and ⟨110⟩ directions but is nonlinear in anneal time for the ⟨111⟩-oriented samples. We show in Fig. 5 growth rate versus $1/T$ for silicon for a number of different orientations. In all cases the regrowth rate at a given temperature was obtained from isothermal annealing experiments carried out over a range of temperatures and fitted with an activation energy of 2.35 eV. This activation energy provides an excellent fit to the data for all orientations.

In Fig. 6 a calculated regrowth rate obtained from the geometrical consequences of the cooperative phenomenon model discussed earlier is plot-

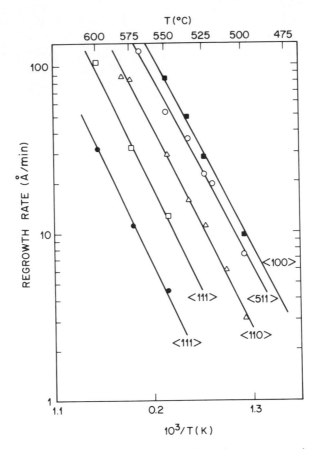

Fig. 5.   Regrowth rate of ^{28}Si-implanted Si amorphous layers versus reciprocal temperature for various Si substrate orientations.

ted with the experimentally determined data. This model requires that two nearest-neighbor atoms at the interface be in crystalline positions for a third to be added correctly. This requirement leads to growth along ⟨111⟩ planes in the diamond lattice and implies no growth whatsoever in the ⟨111⟩ direction. Growth along this direction is then suggested to involve a nucleation phenomenon which is proposed as the basis of nonuniform interfaces and twin formation in the solid phase epitaxial regrowth of implantation-amorphized ⟨111⟩ silicon. This theory explains a number of the features of solid phase regrowth observed experimentally. Other models have also been proposed (Spaepen, 1978), and the cooperative phenomenon model is best regarded as being tentative at the present time.

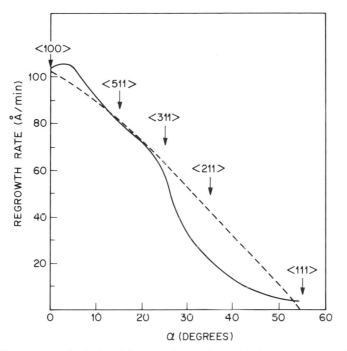

Fig. 6. Comparison of calculated (– – –) and experimental (—) regrowth rate of amorphous Si on crystalline Si as a function of substrate orientation for 550 °C anneals.

3. SOLID PHASE EPITAXY OF DEPOSITED LAYERS

In the previous sections we have indicated that the presence of oxygen will retard the solid phase epitaxial regrowth rate substantially. This effect is perhaps most dramatically demonstrated in experiments in which solid phase epitaxy is attempted on silicon substrates from amorphous layers that have been either evaporated or vapor-deposited. The native oxide that exists between the amorphous film and the crystalline substrate is found to prohibit solid phase epitaxy under normal conditions. On the other hand, solid phase epitaxy of silicon can be obtained if the amorphous layers are electron gun-evaporated onto an atomically clean substrate in a high-vacuum, cryo pumped system. An alternative of potential practical importance has been described by Sigmon and Tseng (1978) in which the native oxide was destroyed *in situ* underneath the amorphous layer by silicon implantation. This technique also produced a thin amorphous layer in the underlying crystalline target so that the solid phase epitaxial process could proceed from a thin amorphized layer in single-crystal material into the deposited film.

Channeling effect measurements combined with transmission electron microscopy were used to examine these films. It was found that recrystallization of a 2900-Å polycrystalline layer could be accomplished after a 12-hr 575 °C anneal. However, the layer was observed to be heavily twinned. Further annealing at 1000 °C for 30 min produced defect-free films. Practical application of these films was demonstrated by fabrication of MOS capacitors on the regrown film. The results of $C-V$ measurements yielded a layer impurity concentration of $N_B \cong 3 \times 10^{16}$ cm^{-3} and a fixed-surface charge density of $Q_{ss} < 2 \times 10^{11}$ cm^{-2}.

C. Solid Phase Epitaxy with Transport Media

The subject of solid phase epitaxy with transport media has been reviewed by Lau and Van der Weg (1978) and can be divided into two categories: eutectic systems where the metal films used as transport media form simple eutectic compositions with the semiconductor, such as the Ge(amorphous)–Al–Ge(crystal) system studied by Ottaviani et al. (1976), and compound-forming systems where the metal film forms a compound with the semiconductor. This system is exemplified by Si(a)–Cr–Si(crystal), where CrSi$_2$ is formed before the solid phase epitaxial process occurs (Lau et al., 1977).

1. EUTECTIC SYSTEMS

The simplest configuration is a eutectic system consisting of a semiconductor substrate onto which a thin metal layer is deposited. As the thin-film composite structure is brought to a temperature well below the eutectic melting temperature, dissolution of the semiconductor in the metal takes place until the solid solubility limit is reached. If the composite structure is then cooled slowly, the dissolved semiconductor atoms will recrystallize epitaxially on the underlying substrate, incorporating metal atoms in the process. By a proper choice of metal layer, the regrown semiconductor can be doped with a simple substitutional impurity. The cleanliness of the substrate–metal interface is clearly critical in this process.

A related configuration is obtained by depositing a layer of amorphous semiconductor onto the metal layer, as in the case of Si(a)–Al–Si(crystal). Upon annealing of this system, the amorphous layer is found to migrate across the aluminum layer and grow epitaxially onto the substrate (Boatright and McCaldin, 1976; Majni and Ottaviani, 1977).

The annealing temperature for these reactions is usually (or can be) well below the eutectic temperature of the system. Of course, epitaxial regrowth can also occur in systems that are heated above the eutectic point, as in the work of Hiraki *et al.* (1972) and Davey *et al.* (1976). Again, the driving force for these reactions originates from the reduction in free energy associated with the amorphous-to-crystalline transition.

The initial stage of epitaxial growth in these systems appears to be laterally nonuniform and characteristic of an island structure. This conclusion is reached by annealing samples to the initial stage of epitaxial regrowth, after which the top layers are removed by etching. Individual islands can then be clearly resolved. Upon further annealing, the individual islands join to form a relatively flat layer. This growth behavior is similar to that observed in Si(a)–Pd–Si(crystal), where the Pd_2Si phase is the transport medium (Liau *et al.*, 1975).

Stacking faults are often found in epitaxial layers grown with a transport medium and are explained tentatively in terms of the island misfit model of epitaxial growth (Booker and Stickler, 1962). In this model, islands are nucleated approximately at random on the substrate so that translation misfit may exist between them. If the relative displacement between two growing islands is close to that allowed by a displacement vector for a stacking fault, it is energetically favorable for the misfit to be accommodated by a stacking fault. Various sources such as oxygen contamination have been suggested as sites for the faulted nuclei. Misfit between islands is also possibly responsible for the presence of twins in epitaxial layers and in regrowth of $\langle 111 \rangle$ implantation-amorphized films as suggested earlier.

2. SILICIDE-FORMING SYSTEMS

As is well known, silicon crystal surfaces often have a thin native oxide layer even after careful chemical cleaning. This oxide layer will normally prevent solid phase epitaxy unless it can be removed prior to film deposition. However, if a layer of metal that has high reactivity with silicon is deposited on the silicon surface, the reactions between the metal and silicon will advance the metal–silicon interface into the substrate. As a result of this reaction or silicide formation, a clean substrate surface for a more uniform nucleation and growth process can be obtained.

The idea of using silicide formation to prepare clean interfaces was first employed for solid phase epitaxy by Canali *et al.* (1974). For these experiments the Pd–Si reaction was used to clean the silicon surface of its native oxide. These authors then observed that transport of silicon could be greatly facilitated and solid phase epitaxy of an overlying evaporated silicon film could be obtained.

The basic experimental configuration for these experiments consisted of

TABLE I

TABULATION OF SILICIDE-FORMING SYSTEMS FOR SPE

SPE system	Transporting silicide	Heat of formation of the transporting silicide (eV/g-atom metal)	Transport temperature at 1000 Å/hr (°C)	Melting point of transporting silicide (°C)	T_{transp}/T_{mel}
Si(a)–Pd–Si(crystal)	Pd$_2$Si	0.45	486	1130	0.47
Si(a)–Pt–Si(crystal)	PtSi	0.68	525	1229	0.53
Si(a)–Ni–Si(crystal)	NiSi	0.89	590	990	0.68
Si(a)–Cr–Si(crystal)	CrSi$_2$	1.26	810	1550	0.59
Si(a)–Fe–Si(crystal)	FeSi$_2$	9.73	660	1212	0.63
Si(a)–Co–Si(crystal)	CoSi$_2$	1.07	700	1326	0.61
Si(a)–Ti–Si(crystal)	TiSi$_2$	1.4	830	1540	0.61
Si(a)–V–Si(crystal)	VSi$_2$	3.18	730	1750	0.50
Si(a)–Rh–Si(crystal)	RhSi	0.70	610	—	—

[a] Average T_{transp}/T_{melt} = 0.58; standard deviation = 0.07.

a $\langle 100 \rangle$ silicon substrate with 1000 Å of Pd and 1 μm of amorphous Si deposited on it. After low-temperature annealing in the vicinity of 300°C, a layer of Pd$_2$Si is formed between the Pd and silicon at both the amorphous and crystalline interfaces. The formation of Pd$_2$Si at both interfaces is believed to be important, since it provides cleaning of each interface. Upon further annealing to 600°C, the amorphous layer thickness is found to decrease and a layer of silicon is observed to grow epitaxially on the substrate. The Pd$_2$Si layer is found from channeling experiments to be displaced toward the surface of the sample. The epitaxial nature of the grown layer is also verified by megaelectron volt ^{4}He channeling measurements, transmission electron microscopy, and other diffraction techniques (Canali *et al.*, 1974; Tseng *et al.*, 1977). Since the discovery of Pd$_2$Si as a transport medium for solid phase epitaxy, a variety of other silicide-forming metals have been used successfully for this process. A list of silicide-forming systems successfully used to date is shown in Table I (Lau, 1978).

D. Summary

In this section we have briefly reviewed the concept of solid phase epitaxial regrowth mechanisms. We have confined the discussion primarily to the low-temperature growth of single-crystal silicon layers. Various mechanisms such as growth through and from a transport medium and growth without a transport medium have been discussed. In the following sections of this chapter it will become clear how these mechanisms can be used to explain many of the scanning cw beam processing phenomena.

II. Temperature Distributions and Solid Phase Reaction Rates Induced by a Scanning cw Laser

To calculate recrystallization rates and a variety of other phenomena related to cw beam annealing, it is necessary to know accurately the temperature distribution produced by the beam in the material being processed. A formalism for calculating the temperature distribution produced by an irradiating beam has been developed for stationary beams by Lax (1977), and for moving circular beams by Cline and Anthony (1977); and calculations based on this formalism have been found to agree well with experimental data. There are also a number of applications in which a ribbon beam, having an elliptical rather than a circular cross section, is preferable, and temperature profiles to be expected for such beams have also been calculated by Nissim *et al.* (1980). In this section we first summarize the temperature distribution theory as developed by Nissim *et al.* (1980) and then apply the results to the calculation of solid phase reaction rates.

A. Solution to the Heat Equation

We first present an expression for the temperature rise induced by a moving elliptical beam in a material assumed to have constant thermal conductivity. Nissim *et al.* (1980) develop this expression by first obtaining an analytical expression for the maximum temperature at the center of the moving beam and then carrying out a numerical integration to obtain all the relevant parameter dependences. Temperature profiles obtained in this case are called linear temperature distributions.

We then present a more refined set of calculations that take into account the temperature dependence of the thermal conductivity. A Kirchoff transformation (Joyce, 1975; Lax, 1978) is used with experimental data on the thermal conductivity of silicon and gallium arsenide to obtain the "true" temperature profiles in these cases.

1. CALCULATION OF THE LINEAR TEMPERATURE

Following Nissim *et al.* (1980) we assume that the laser beam is elliptical in an x,y plane perpendicular to the direction of laser propagation z. The ratio between the major axis r_y and the minor axis r_x of the ellipse is an important parameter in the analysis, defined as $\beta = r_y/r_x$. We assume a Gaussian laser intensity distribution in both directions x and y:

$$I = I_0 \exp(-x^2/2r_x^2) \exp(-y^2/2r_y^2) \tag{2}$$

I_0 can be determined as a function of the power absorbed by the material, assuming an infinite surface:

$$I_0 = P(1 - R)/2\pi r_x r_y \tag{3}$$

where P is the total incident power and R is the reflection coefficient of the irradiated material. Finally, we write the energy absorbed in the solid if the beam is moving with the velocity v in the x direction:

$$Q = \frac{P(1 - R)}{2\pi r_x r_y} \exp\left(-\frac{(x - vt)^2}{2r_x^2} - \frac{y^2}{2r_y^2}\right) f(z) \tag{4}$$

In Eq. (4) $f(z)$ gives the z dependence of the total absorbed energy.

The heat equation to be solved is the diffusion equation in which Q appears as the source term:

$$\frac{\partial T}{\partial t} - D\nabla^2 T = \frac{Q}{C_p} \tag{5}$$

Using an appropriate Green's function G (Carslaw and Jaeger, 1959), the linear temperature rise is found to be

$$\theta = \int_{-\infty}^{t} \int_{-\infty}^{\infty} \int_{-\infty}^{\infty} \int_{-\infty}^{\infty} (Q/C_p)(x',y',z',t')$$
$$\times G(x',y',z',t'/x,y,z,t) \, dx' \, dy' \, dz' \, dt' \tag{6}$$

The integration over the different variables can be done separately. The integrals over x' and y' are Gaussian tabulated integrals.

To perform the integration over z', the function $f(z)$, which expresses the penetration of laser energy into the material with respect to depth, must be specified. Since in most cases of interest the optical absorption depth $\alpha^{-1} \ll r_x, r_y$, $f(z)$ is taken to be a function at the surface of the material. To account for the discontinuity at $z = 0$, the integration is performed over z' from $-\infty$ to $+\infty$, and then twice its value is taken from mirror image considerations.

After carrying out the integration over x, y, z, the linear temperature is found to be

$$\theta = \frac{p}{C_pD} \frac{1}{\sqrt{2}\pi^{3/2}} \int_0^\infty \frac{1}{[(u^2 + 1)(u^2 + \beta^2)]^{1/2}}$$

$$\times \exp\left\{ -\frac{1}{2}\left[\frac{(X + Vu^2)^2}{u^2 + 1} + \frac{Y^2}{u^2 + \beta^2} + \frac{Z^2}{u^2} \right]\right\} du \qquad (7)$$

in which $X = x/r_x$, $Y = y/r_x$, $Z = z/r_x$, $\beta = r_y/r_x$, $p = [P(1 - R)]/r_x$, $V = vr_x/2D$, and $u = (2Dt/r_x^2)^{1/2}$.

The maximum linear temperature at the center of the ellipse ($X = Y = Z = 0$) for a *stationary* beam ($V = 0$) can now be obtained analytically:

$$\theta_{max} = \frac{p}{\beta}\frac{1}{C_pD}\frac{1}{\sqrt{2}\pi^{3/2}} K\left(\frac{\beta^2 - 1}{\beta^2}\right)^{1/2} \qquad (8)$$

where K is the complete elliptical integral of the first kind.

This result can also be used to find the maximum temperature for a circular beam by setting $\beta = r_x/r_y = 1$:

$$\theta_{max}^{circle} = p/C_pD2\sqrt{2\pi} \qquad (9)$$

The analytical integration has now been carried out as far as possible. To analyze the dependence of the actual temperature on beam and substrate parameters, a numerical integration must be performed. To make the results valid for any kind of material, Nissim *et al.* (1980) define the following quantity:

$$\eta = \frac{\theta}{\theta_{max}^{circle}(\beta = 1)} = \frac{2}{\pi}\int_0^\infty \frac{1}{[(u^2 + 1)(u^2 + \beta^2)]^{1/2}}$$

$$\times \exp\left\{ -\frac{1}{2}\left[\frac{(X + Vu^2)^2}{u^2 + 1} + \frac{Y^2}{u^2 + \beta^2} + \frac{Z^2}{u^2} \right]\right\} du$$

$$(10)$$

2. CALCULATED NORMALIZED LINEAR TEMPERATURE CURVES

The numerical integration of Eq. (10) leads to a set of curves that can be chosen to cover a wide range of experimental conditions. A representative set of such curves is presented in Fig. 7 which shows the linearized normal temperature η as a function of the $Y(Y = y/r_x)$ position for different values of the ratio β ranging from 1 to 40. We see that large values of β give a more uniform temperature distribution. Under most experimental conditions, we cannot readily decrease r_x, so the only practical way to obtain high values of β is to increase r_y. A substitution of numbers shows that the power needed to reach typical annealing temperatures at high values of β is very high, in fact, above the limit of currently practical cw Ar or Kr

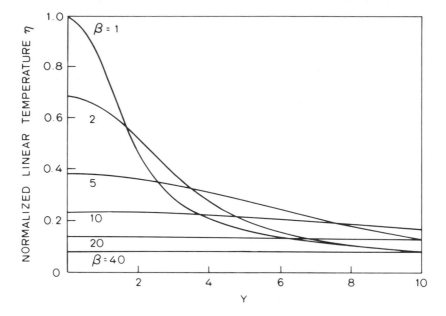

Fig. 7. Linear normalized temperature rise η at $X = Z = V = 0$ as a function of the Y position ($Y = y/r_x$) for different values of β ($\beta = r_y/r_x$) ranging from 1 to 40.

lasers but certainly within the limits of an electron beam. The variations of η in the $X(X = x/r_x)$ and $Z(Z = z/r_x)$ directions are presented in Figs. 8 and 9.

An extended scale in the Z direction has been taken in Fig. 10 (for $r_x = 20\ \mu m$, the full scale represents $1\ \mu m$). This curve shows clearly that the variations near the surface are insignificant and that, for instance, the temperature at the interface of an implanted layer in a semiconductor and underlying substrate is essentially the same as at the surface.

A similar study has been done using beam scanning speed as the parameter for two fixed values of β, namely, $\beta = 1$ and $\beta = 20$. Figures 11 and 12 represent the variation in the Y direction. Again, in Fig. 12 we can see that, if we have enough power for annealing, a high speed at $\beta = 20$ gives a uniform temperature distribution along the Y axis. Similarly, Figs. 13 and 14 represent the variations of η in the X direction. Since the beam is moving in this direction, the nonsymmetrical behavior with respect to the point $X = 0$ appears clearly in these curves. Finally, the variation along Z in Fig. 15 shows again the weak changes in temperature as a function of depth into the material.

To allow the reader to use the curves for any combination of speed and ratio β, the variation of η as a function of speed with β as a parameter is

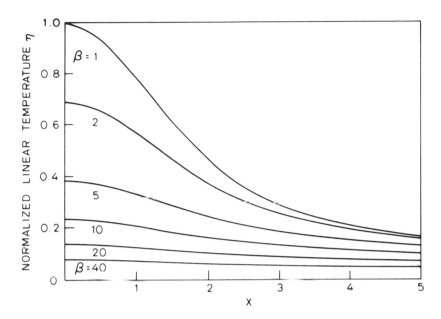

Fig. 8. Linear normalized temperature rise η at $Y = Z = V = 0$ as a function of the X position ($X = x/r_x$) for different values of β ($\beta = r_y/r_x$) ranging from 1 to 40.

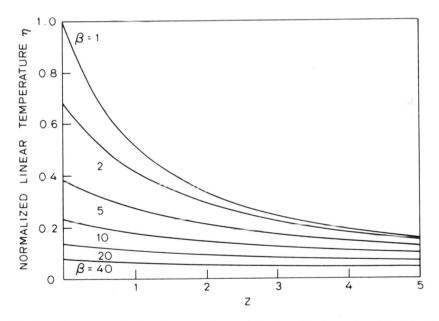

Fig. 9. Linear normalized temperature rise η at $X = Y = V = 0$ as a function of depth $Z(Z = z/r_x)$ for different values of β ($\beta = r_y/r_x$) ranging from 1 to 40.

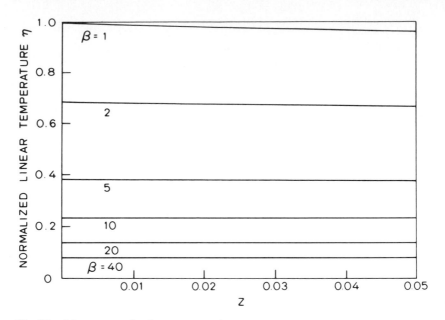

Fig. 10. Linear normalized temperature rise η at $X = Y = V = 0$ as a function of depth close to the surface $Z(Z = r/r_x)$ for different values of β ($\beta = r_y/r_x$) ranging from 1 to 40.

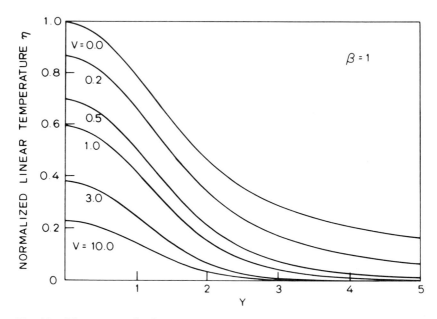

Fig. 11. Linear normalized temperature rise η at $X = Z = 0$ as a function of the Y position ($Y = y/r_x$) for different values of the normalized scan speed $V(V = vr_x/2D)$ ranging from 0 to 10 and for $\beta = 1$ ($\beta = r_y/r_x$).

344

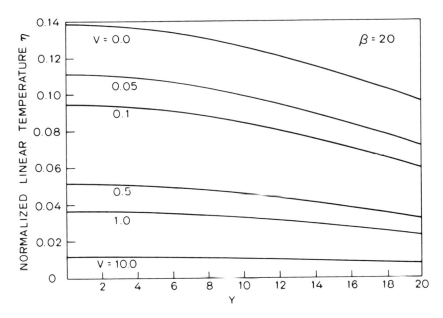

Fig. 12. Linear normalized temperature rise η at $X = Z = 0$ as a function of the Y position ($Y = y/r_x$) for different values of the normalized scan speed V ($V = vr_x/2D$) ranging from 0 to 10 and for $\beta = 20$ ($\beta = r_y/r_x$).

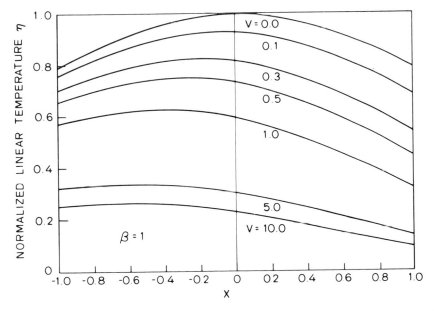

Fig. 13. Linear normalized temperature rise η at $Y = Z = 0$ as a function of the X position ($X = x/r_x$) for different values of the normalized scan speed V ($V = vr_x/2D$) ranging from 0 to 10 and for $\beta = 1$ ($\beta = r_y/r_x$).

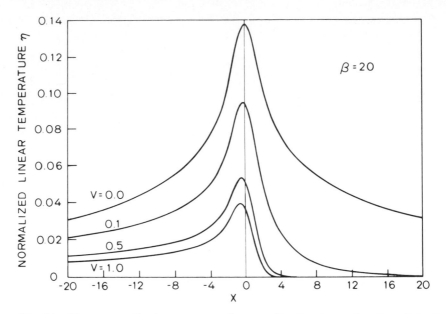

Fig. 14. Linear normalized temperature rise η at $Y = Z = 0$ as a function of the X position ($X = x/r_x$) for different values of the normalized scan speed V ($V = vr_x/2D$) ranging from 0 to 1 and for $\beta = 20$ ($\beta = r_y/r_x$).

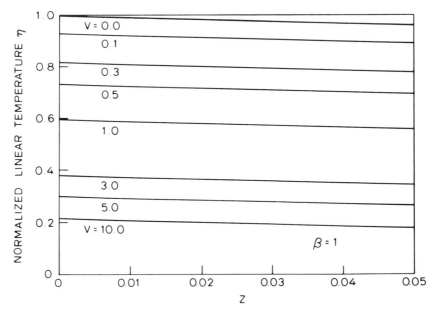

Fig. 15. Linear normalized temperature rise η at $X = Y = 0$ as a function of depth near the surface Z ($Z = z/r_x$) for different values of the normalized scan speed V ($V = vr_x/2D$) ranging from 0 to 10 and for $\beta = 1$ ($\beta = r_y/r_x$).

plotted in Figs. 16 and 17. Figure 17 emphasizes the fact that, for the typical values of speed used in a cw laser system, the solid easily reaches thermal equilibrium.

3. THE TRUE TEMPERATURE RISE IN SILICON AND GALLIUM ARSENIDE

We now employ the Kirchoff transformation (Joyce, 1975) to obtain the true temperature rise induced by the laser beam. We develop these calculations for silicon and gallium arsenide. The temperature-dependent thermal conductivity $\kappa(T)$ has been taken from the literature.

A rational function can be fitted to the experimental data by using the empirical form:

$$\kappa(T) = C_p D(T) = A/(T - B) \tag{11}$$

Using this form the relation between the true and linear temperatures is then

$$T = B + (T_0 - B) \exp(\kappa_0 / A\ \theta) \tag{12}$$

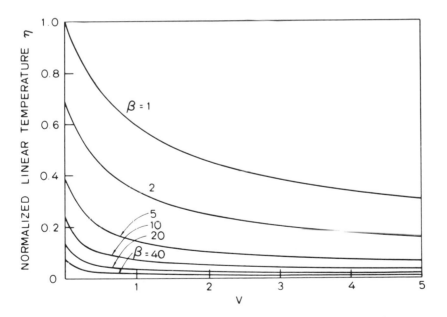

Fig. 16. Linear normalized temperature rise η at $X = Y = Z = 0$ as a function of the normalized scan speed V ($V = vr_x/2D$) for different values of β ($\beta = r_y/r_x$) ranging from 1 to 40.

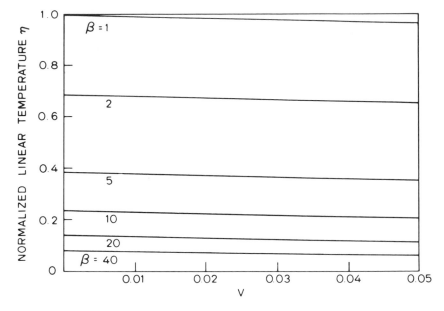

Fig. 17. Extended scale for the linear normalized temperature rise η at $X = Y = Z = 0$ as a function of the normalized scan speed V ($V = vr_x/2D$) for different values of β ($\beta = r_y/r_x$) ranging from 1 to 40.

where T_0 is the temperature of the back surface of the semiconductor and κ_0 is the thermal conductivity at this temperature.

For silicon the experimental data in Ho *et al.* (1974) yield the values $A = 299$ W/cm and $B = 99$ K. A step-by-step numerical integration using published data for $D(T)$ gives a temperature profile essentially identical to that obtained using Eq. (11).

For gallium arsenide, the experimental data of Maycock (1967) result in the values $A = 91$ W/cm and $B = 91$ K. With these values it is possible to obtain the real temperature as a function of $p[p = (1 - R)/P/r_x]$ for both silicon and gallium arsenide. Figures 18 and 19 present these variations (for $\beta = 1$) for silicon and gallium arsenide, respectively.

B. Calculation of Solid Phase Reaction Rates Induced by a Scanning cw Laser

The annealing of ion-implanted silicon and the formation of metal silicides by a scanning cw laser can be interpreted as solid phase reaction processes. As indicated in Section I, the behavior of these systems during low-temperature furnace processing operations can be described by the

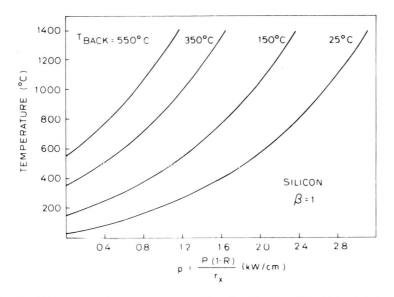

Fig. 18. The true maximum temperature ($X = Y = Z = V = 0$) in Si is plotted versus the normalized power $p\{p = [P(1 - R)]/r_x\}$ for different substrate back-surface temperatures.

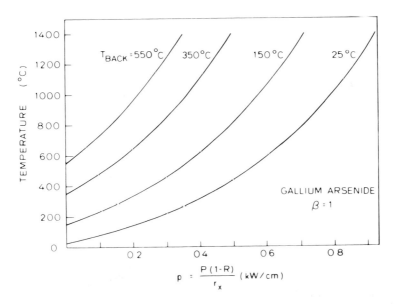

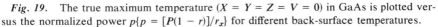

Fig. 19. The true maximum temperature ($X = Y = Z = V = 0$) in GaAs is plotted versus the normalized power $p\{p = [P(1 - r)]/r_x\}$ for different back-surface temperatures.

reacting interface being at a constant temperature throughout the annealing process. This is not true for anneals by a scanning cw laser, however, since the temperature in this case is a rapidly changing function of time. To analyze reaction kinetics in this case, it is necessary to develop a model that accurately represents the transient nature of cw beam annealing. We present such a model in this section, following closely the work of Gold and Gibbons (1980).

1. REACTION RATE CALCULATIONS

For a rate-limited process such as the regrowth of amorphous implanted Si, the reacted-layer (or regrown-layer) thickness resulting from a constant temperature furnace anneal can be expressed as

$$z = R_0 t \, \exp(-E_a/kT) \tag{13}$$

where t and T are the anneal time and temperature, respectively. For diffusion-controlled reactions, such as the formation of certain metal silicides, the z in Eq. (13) is replaced by z^2, but all other features of the following discussion are still applicable.

The transient nature of the temperature rise produced by a scanning cw laser means that the reacted-layer thickness must be expressed by

$$z = R_0 \int_{-\infty}^{\infty} \exp[-E_a/kT(t)] \, dt \tag{14}$$

If the dependence of the temperature on time is known, Eq. (14) can be reduced to Eq. (12) by determination of an effective annealing temperature T_{eff} and an effective annealing time t_{eff}.

Liau et al. (1979) have carried out an appropriate analysis for the case of large-area pulsed laser irradiation, where the temperature is assumed to be uniform across the surface of the wafer and can be calculated as an explicit function of time. The specifics of the analysis described here are quite different because of the nature of the heating by a scanning, focused cw laser. The first step is to calculate the spatial temperature profile induced by the laser, as in Section II.A. Then, based on a knowledge of experimental scan conditions, this spatial profile is converted to a temporal one.

We assume that the laser has a Gaussian intensity distribution, $I = (P/\pi w) \exp(-r^2/w^2)$, where r is the distance from the center of the beam and P is the total absorbed power. As long as the penetration depth of the laser beam is much less than its radius, the linear temperature rise r at the wafer surface is given by

$$\theta(r) = (P/2\pi^{1/2}w\kappa_0) \exp(-r^2/2w^2)I_0(r^2/2w^2) \qquad (15)$$

where $I_0(x)$ is the modified Bessel function of order zero and κ_0 is the thermal conductivity at the wafer back-surface temperature T_0. Although this solution is strictly valid only at the wafer surface, the reacting-layer thickness is much less than the beam radius under typical experimental conditions and consequently the solution is also valid at the reacting interface. To account for the strong temperature dependence of thermal conductivity, the linear temperature rise $\theta(r)$ is then related to the true temperature $T(r)$ through use of the Kirchoff transformation:

$$\theta(r) = \int_{T_0}^{T(r)} [\kappa(T')/\kappa(T_0)] \, dT' \qquad (16)$$

As mentioned in Section II.A.3, an excellent approximation to the thermal conductivity of Si between 300 K and the melting point is given by the expression $\kappa(T) = A/(T - T_k)$, with the constants A and T_k equal to 299 W/cm and 99 K, respectively. Equation (16) thus reduces to

$$T(r) = T_k + (T_0 - T_k) \exp[\theta(r)/(T_0 - T_k)] \qquad (17)$$

Using this calculated spatial temperature profile, Gold and Gibbons (1980) have calculated the reacted-layer thickness expected for two different scan conditions: (a) a single scan through the point of interest, and (b) multiple overlapping scans. The former case is applicable if one wishes to determine the threshold for complete reaction in the center of a scan line, while the latter case is for analyzing uniformly reacted samples, such as those studied by Rutherford backscattering.

2. SINGLE SCAN

We assume that the laser beam is scanning with a velocity v and passes through the point of interest at $t = 0$. Since the temperature rise reaches a constant value in times much shorter than typical laser dwell times, we can treat this problem as a steady state temperature distribution in a moving frame of reference and use the results of Eqs. (15) and (17) with $r = vt$. The subsequent solution of Eq. (3), described in detail by Gold and Gibbons (1980), yields the following expression for the reacted-layer thickness z resulting from a single scan through the point of interest:

$$z \cong R_0 t_{\mathrm{eff}} \exp(- E_\mathrm{a}/kT_{\mathrm{eff}}) \qquad (18)$$

where

$$T_{\mathrm{eff}} = T_{\max} = {}^1T_k + (T_0 - T_k) \exp(P/2\pi^{1/2}wA) \qquad (19)$$

and

$$t_{\text{eff}} = \frac{2w}{v} \left\{ \frac{\pi k T_{\text{max}}^2}{2E_a(T_{\text{max}} - T_k) \ln[(T_{\text{max}} - T_k)/(T_0 - T_k)]} \right\}^{1/2} = \frac{2w}{v} f \quad (20)$$

We see that the effective temperature T_{eff} is equal to the maximum temperature T_{max} and that the effective time t_{eff} is equal to the laser beam dwell time $2w/v$ multiplied by a dwell time reduction factor f. The calculated dependence of f on T_0 and T_{max} is shown in Fig. 20. The activation energies of 1.5 and 2.35 eV correspond, respectively, to typical values for the formation of metal silicides and for the regrowth of ion-implanted Si. It can be seen that the effective time is on the order of one-third the actual beam dwell time.

3. MULTIPLE OVERLAPPING SCANS

In this case the reacted layer produced at a given point is a result of many different laser scans, not all of which pass directly through the point, so we can no longer simply assume that $r = vt$. The problem can be easily solved, however, if we assume that the overlap is sufficient to create a uniformly thick reacted layer. We then need only to calculate the total

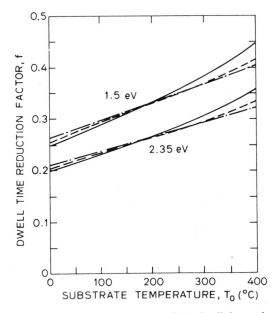

Fig. 20. Calculated dependence on T_0 and T_{max} of the dwell time reduction factor of Eq. (20). —, $T_{\text{max}} = 800°C$; ----, $T_{\text{max}} = 1000°C$; – · – $T_{\text{max}} = 1200°C$.

volume of reacted material produced in a given time by a nonmoving beam and then divide this by the total area being scanned to obtain the average reacted-layer thickness:

$$\langle z \rangle = \frac{1}{A} \int_0^\infty z(r) 2\pi r \, dr$$

$$= \frac{2\pi R_0 t}{A} \int_0^\infty \exp\left[\frac{-E_a}{kT(r)}\right] r \, dr \tag{21a}$$

where t and A represent the total scan time and the total scan area, respectively. This integral can then be solved, again using the results of Eqs. (15) and (17), to yield

$$\langle z \rangle \cong R_0 t_{eff} \exp(-E_a/kT_{eff}) \tag{21b}$$

where

$$T_{eff} = T_{max} = T_k + (T_0 - T_k) \exp(P/2\pi^{1/2} wA) \tag{21c}$$

and

$$t_{eff} = t(2wf)^2/A \tag{21d}$$

The effective temperature for this case is the same as that calculated in the previous section, but the effective time is different, although it still includes the same reduction factor f. An example is given in the next section for the use of Eq. (21) to determine the values of E_a and R_0 for the laser-induced formation of metal silicides.

C. Comparison of Reaction Rate Theory with Experiment

In this section we compare the results predicted by the preceding theory with experimental data obtained on the regrowth of implantation-amorphized silicon and the reaction of thin metal films with silicon substrates for silicides.

1. SILICON REGROWTH UNDER SINGLE-SCAN CONDITIONS

We show in Fig. 21 an example of the use of Eq. (18) to calculate T_{max} and the expected thickness of the regrown layer produced by laser annealing of self-implanted silicon. The assumed dwell time of 4×10^{-4} sec corresponds to a beam radius of 20 μm and a scan speed of 10 cm/sec. The values of E_a and R_0 used for this calculation were obtained from the furnace annealing data of Csepregi *et al.* (1975). It is evident that both T_{max} and z are extremely strong functions of laser power in the region of inter-

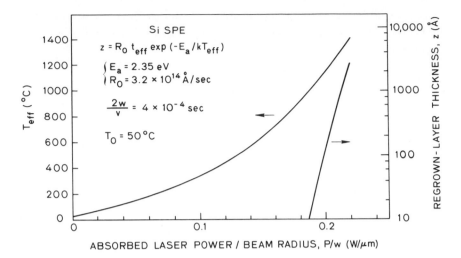

Fig. 21. Maximum temperature and regrown-layer thickness predicted by Eq. (6) for cw laser annealing of self-implanted Si as a function of absorbed laser power $[P = P_{inc}(1 - R)]$ divided by the beam radius.

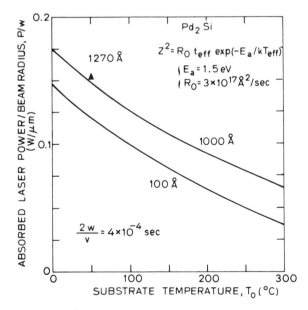

Fig. 22. Calculated combinations of absorbed laser power $[P_{abs} = P_{inc}(1 - R)]$ and substrate temperature required to induce the formation of 100 and 1000 Å of Pd_2Si in a single scan. The effect of the dwell time reduction factor is taken into account. The observation at 1270 Å is from Carslaw and Jaeger (1959).

est. It is also clear that the maximum thickness of silicon that can be regrown in a *single* scan is approximately 3000 Å, since a thickness any greater would require a surface temperature in excess of the melting temperature.

2. SILICIDE FORMATION

As an example of the multiple-scan calculations, we have calculated, using Eqs. (21) the combinations of absorbed laser power and substrate temperature that should induce the growth of Pd_2Si layers with thicknesses of 100 and 1000 Å. These results are shown in Fig. 22 where a laser beam dwell time of 4×10^{-4} sec is again assumed. The extremely strong dependence of thickness on laser power is evident. Also shown are experimental data from Shibata *et al.* (1981a,b), and it can be seen that good agreement between theoretical (1500 Å) and experimental (1270 Å) thicknesses is obtained.

A further comparison of the theory with experimental data will be given in Section III.B.

III. Experimental Analysis of cw Beam-Processed Semiconductor Materials

Extensive research has been performed during the past few years on the application of scanning cw lasers and electron beams for semiconductor processing. The principal results obtained from this work, to be elaborated upon below, are as follows:

(i)   For thin amorphized layers of silicon, the annealing process is a solid phase epitaxial regrowth for which the critical beam parameter has been found to be the beam power divided by the spot size.

(ii) Recrystallization of implant-damaged material is essentially perfect, with no residual damage observed in transmission electron microscopy to a resolution of ~50 Å.

(iii) Little diffusion of the implanted impurities occurs during the annealing of semiconductor silicon, irrespective of whether or not the material has been driven amorphous by implantation.

(iv) The electrical activity of implanted impurities can approach 100% even for impurity concentrations that exceed the solid solubility limit.

(v)   Doped polycrystalline silicon can be annealed to provide both an increase in grain size from ~200 Å to 2×25 μm, with a correspondingly significant reduction in the sheet resistance.

(vi)   Metals commonly used for silicide formation can be reacted by
       scanning laser and electron beams to form silicide compounds of
       large-area uniform dimensions. Although many other interesting
       phenomena can be accomplished with the scanning beam process,
       we will confine ourselves in what follows to a discussion of ion-
       implanted silicon and silicide formation.

A.   Continuous-Wave Beam Annealing of Ion-Implanted Silicon

In this section we consider the basic experimental results that have
been obtained using scanning laser and electron beams to anneal ion-
implanted silicon. The initial investigations of the annealing effects of a
swept cw argon laser on ion-implanted silicon were performed by Gat and
Gibbons (1978). These authors studied the annealing of arsenic implanted
in silicon at concentrations sufficient to create a continuous amorphous
layer from the surface to a depth of approximately 1000 Å. The samples
were typically implanted at 100 keV to a dose of 3×10^{14} As atoms/cm^2
and purposely misoriented to minimize channeling. The implantation was
performed in $\langle 100 \rangle$ silicon substrates. Rutherford backscattering measurements in the channeling mode have shown that this dose is sufficient
to drive the silicon amorphous to a depth of ~ 1000 Å.

A number of experiments were performed to determine the effect of
various annealing powers and scanning speeds on the annealing process.
Evaluation of the annealed layers was performed by Rutherford backscattering, channeling, sheet resistivity measurements, secondary-ion mass
spectroscopy (SIMS), transmission electron microscopy (TEM), and
anodic oxidation and differential van der Pauw measurements. Comparison measurements were made on samples that were thermally annealed at
1000°C for $\frac{1}{2}$ hr.

After laser annealing, the samples were first visually inspected with a
microscope to ascertain that damage to the surface or melting had not
taken place. A rapid indication of the quality of the anneal was then
obtained by measuring the probe-to-probe resistance across an annealed
layer using an ASR Model 100 spreading resistance probe. The results of
this measurement are shown in Fig. 23 for both the thermal and cw laser-
annealed arsenic samples described earlier. It can be seen in Fig. 23 that
both the laser and thermal annealing result in essentially identical probe-
to-probe resistance (~ 70–80 Ω) for both samples. A slight nonuniformity
is observed for the laser-annealed case, which can be explained by insufficient overlapping of individual scan lines during the laser anneal. Further
measurements have shown that this nonuniformity can be eliminated,

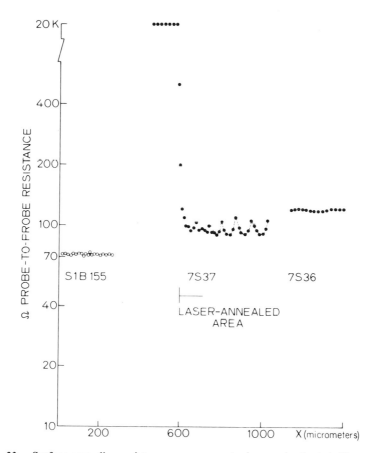

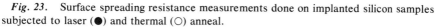

Fig. 23. Surface spreading resistance measurements done on implanted silicon samples subjected to laser (●) and thermal (○) anneal.

with uniform annealing normally requiring that the overlap between individual scan lines be greater than $\sim 30\%$.

A more comprehensive analysis of the annealing (Gat *et al.*, 1978a) is described in Fig. 24, where we show data taken by secondary-ion mass spectrometry on both thermally and laser-annealed samples. The vertical axis (counts) is proportional to the arsenic concentration, while the horizontal axis (sputtering time) is proportional to depth in the material. The upper horizontal axis is obtained by converting sputter time into depth, assuming a uniform sputtering rate. The data show little if any movement of the arsenic profile in the laser-annealed samples. A Pearson IV distribution function is also constructed (using LSS moments; see Gibbons *et al.*, 1975) for the As dose and energy employed. The impurity distribution

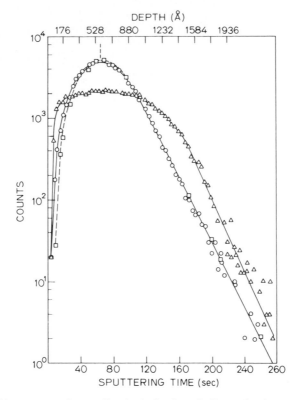

Fig. 24. [75]As concentration profiles in As-implanted silicon after laser (○) and thermal (1000°C, 30 min) (△) anneal. □, As-implanted; – – –, Pearson IV distribution with LSS range statistics.

following laser annealing is seen to fit this theoretical profile very closely. In contrast, significant diffusion of arsenic atoms into the sample is observed for the thermal anneal.

To determine the electrical activity and mobility of the laser-annealed layers, anodic oxidation and differential van der Pauw measurements were performed to determine both the carrier concentration and mobility versus depth in a laser-annealed layer. In Fig. 25 we show results obtained by Gibbons (1979) for laser annealing of the 100-keV $3 \times 10^{14}/cm^2$ arsenic implants. Also shown is the LSS-calculated arsenic distribution expected for this energy and dose of implant. It can be seen in Fig. 25 that an excellent fit between the calculated arsenic concentration profile and the measured carrier concentration profile is obtained over three orders of magnitude. Also, the mobilities measured are essentially equal to those obtained for thermally processed samples. The combination of SIMS and

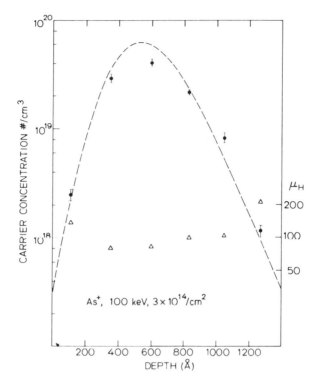

Fig. 25.   Carrier concentration (●) and mobility (△) profiles obtained on As-implanted samples annealed with a scanning Ar cw laser beam. Dashed line is LSS-calculated As distribution.

electrical analysis thus shows that 100% electrical activity has been obtained with no diffusion from the as-implanted profile for the laser-annealed samples.

Structural characterization of both thermally and laser-annealed samples was carried out by Gat *et al.* (1978a) using transmission electron microscopy. The results obtained on both thermally and laser-annealed samples are shown in Fig. 26. It can be seen that the thermally annealed sample (Fig. 26a) shows a single-crystal diffraction pattern, with a bright-field micrograph containing a variety of defect clusters and dislocation loops with dimensions on the order of 200 Å. The laser-annealed sample (Fig. 26b) is seen to be essentially free of any defects observable in a TEM micrograph except near the boundary between the crystallized and amorphous regions (shown as the dashed line in Fig. 26b). These residual defects also disappeared when adjacent scan lines were overlapped by ~30%.

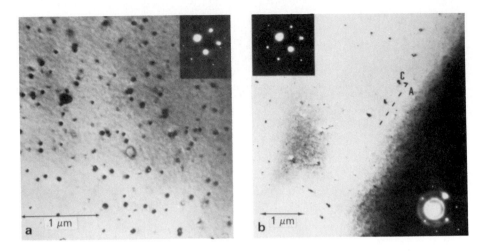

Fig. 26. Electron micrographs of As-implanted silicon subjected to a thermal anneal at 1000°C for 30 min (a) and laser annealing (b). Insets show diffraction patterns typical of their regions.

A set of experiments similar to those described above has been performed on boron-implanted silicon. For these studies boron implantations of $2 \times 10^{15}/cm^2$ at an implantation energy of 35 keV were performed. A krypton laser operated in a multiline mode with an output power of 6 W and a scanning rate of 9.8 cm/sec was utilized to anneal the samples. The sample temperature was held at 178°C during this anneal. Complete experimental details are presented in Gat *et al.* (1978b). The central results have been found to be largely identical to those described for arsenic-implanted silicon. In particular, very high electrical activity (85–95%) was observed, with no diffusion of the implanted species from its as-implanted profile. Recrystallization was also judged to be defect-free by transmission electron microscopy to a resolution of ~20 Å.

Secondary-ion mass spectrometry measurements of the boron profile after implantation, a laser anneal, and a thermal anneal of 1000°C for 30 min are shown in Fig. 27. Here boron ion intensity is plotted versus sputtering time for the described boron implants. It can be seen from these data that the as-implanted and laser-annealed signals are identical, indicating that little if any diffusion of the boron has taken place during the laser anneal. For the thermal anneal (shown as the crosses in Fig. 27) one can observe a spreading out of the profile, as expected from thermal diffusion theory. The electrical activity has been estimated by calculating the sheet resistance for the laser-annealed sample, assuming the impurity profile

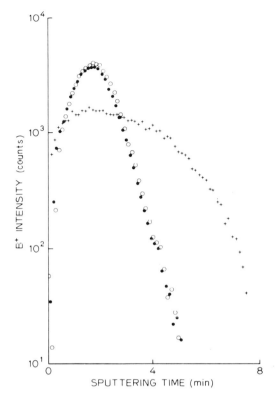

Fig. 27. Impurity profiles with SIMS technique in as-implanted laser-annealed and thermally annealed crystalline Si implanted with 2×10^{15} B$^+$/cm^2 at 35 keV. $\bigcirc$, As implanted; $\bullet$, laser annealed; $+$, thermally annealed (1000°C, 30 min).

measured from SIMS. This calculation suggests that the boron is 85–90% electrically active.

An investigation of deep levels in both scanning electron beam- and laser-annealed implanted silicon has also been carried out by Johnson *et al.* (1979, 1980) using deep-level transient spectroscopy (DLTS). The measurements provided both energy levels and spatial depth profiles for selected defects and were performed in the constant-capacitance mode (CC-DLTS). This technique is particularly appropriate for measuring trap densities that are comparable to shallow dopant concentrations. Measurements have been reported on both arsenic-implanted and annealed *p–n* junction diodes and self-implanted silicon Schottky barriers. In all cases a comparison of laser, electron beam, and furnace annealing was made. A summary of the various dominant traps is listed in Table II for

TABLE II

Summary of Deep Levels Observed in Laser- and Electron Beam-Annealed Si

Material	Laser	Electron beam[a]		Thermal[a]
		SEM	SEBA	
Implanted As$^+$	0.28 eV[b]	0.28[b]	0.28[b]	NO
		0.36[b]	NO	
Si$^+$	0.19 eV[c]	—	—	NO
	0.49 eV[c]	—	—	—
	0.56 eV[c]	—	—	—
	0.28 eV	—	—	—
Unimplanted	—	—	0.19[c]	—
			0.44[c]	—

[a] NO, None observed.
[b] Hole traps.
[c] Electron traps.

the various annealing processes. In general the traps are found to be at concentrations on the order of $10^{13}/cm^3$ or more and remain in this range to depths of more than 1 μm below the annealed region. These defects are not found in furnace-annealed samples and are believed to arise from the propagation of vacancies into the material during both the implantation and the short-duration annealing process. Electron beam-induced current (EBIC) measurements have also been reported on cw laser-annealed, arsenic-implanted silicon by Mizuta *et al.* (1980). These experiments have shown that the laser power window for successful annealing is quite small, with low charge collection efficiency usually resulting unless the window is carefully chosen. To date, the best annealing results have been obtained for highly overlapped laser scans (100-μm beam width with 6-μm scan steps at laser powers ~70–75% of the power required for melting. At higher laser powers, evidence for laser-induced damage has been observed in the EBIC measurements. This evidence also suggests that the damage extends several micrometers below the implanted layer.

B. Silicide Formation

Scanning cw beam processing has also been utilized in other material systems where solid phase processes dominate. One area of considerable interest is the formation of metal silicide compounds from metal layers deposited on silicon substrates. Both scanning cw laser and electron beams have been used to promote these reactions. A complete review of this work has been published elsewhere (Shibata *et al.*, 1981a,b) and will not be presented in detail here. However, we mention briefly for completeness some of the systems in which solid phase reactions have been observed.

The application of scanning cw laser and electron beams in the formation of silicide compounds has led to a process that appears to be superior in many respects to that obtained with pulsed energy beam sources. As is the case for the annealing of ion-implanted amorphous layers, a solid phase process proves to be responsible for the metal–silicon reactions obtained when scanning cw beams are used at sufficiently low power levels. Experiments show that single-phase silicide compounds with little or no silicon precipitation are obtained when scanning cw beams are used to promote metal–silicon reactions. The composition of the film produced (e.g., Pd_2Si or $PdSi$) is determined by the power level employed in the scanning beam. The films are generally found to be homogeneous, with a well-defined silicon–silicide interface.

Although the majority of metal–silicon reactions that have been studied appear to depend on a solid phase interdiffusion mechanism, certain anomalies have been observed. These anomalies are discussed by Shibata *et al.* (1981a,b). For convenience, the following discussion is divided into the reaction of near-noble metals on silicon, including palladium and platinum, and the reaction of refractory metals such as molybdenum, tungsten, and niobium on silicon.

1. NEAR-NOBLE METALS ON SILICON

In Fig. 28 we show backscattering spectra for palladium–silicon samples that were laser-annealed at a normalized power level of $P = 0.71$ or $P = 1.1$. The reference power level is the power required to melt a clean silicon surface. The spectrum for the As-deposited sample is shown by the solid line. The backscattering spectra show that the entire metal layer was reacted with the silicon after the laser irradiation. The average composition of the low- and high-power phases were determined to be Pd_2Si and $PdSi$, respectively. It should be noted that the spectrum of $PdSi$ does not change significantly with laser powers ranging from $P = 0.89$ to $P = 1.4$. These results are quite different from those obtained with pulsed laser annealing, where the average composition of the silicide layer changes continuously with increasing laser power (Poate *et al.*, 1978).

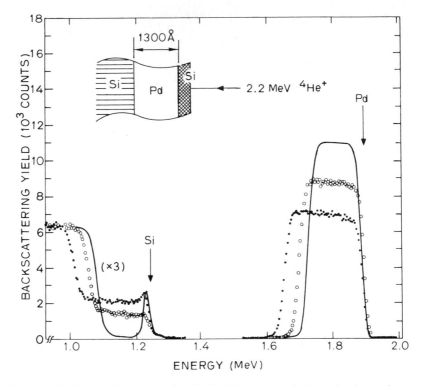

Fig. 28. Backscattering spectra for Si–Pd–Si laser-annealed samples for various normalized laser power levels. —, As deposited; ○, $p = 0.71$ (Pd_2Si); ●, $p = 1.1$ (PdSi).

In Fig. 29 we show the glancing angle x-ray diffraction pattern obtained in a Read x-ray camera from the high-power phase. The diffraction lines from PdSi are indicated by solid circles and the most intense line from Pd_2Si by open circles. It can be concluded that the compound is essentially single-phase PdSi with a trace amount of Pd_2Si. Similar results are obtained for the low-power phase sample; however, here the compound form was mainly Pd_2Si, including trace amounts of PdSi.

The growth mechanisms for the Pd_2Si are well described by a solid phase process. The theoretical development relating laser parameters to this growth process has been presented in Section II. The results of these calculations and the experimental measurements are presented in Fig. 30. In this figure we plot the thickness of the Pd_2Si layer obtained from Rutherford backscattering measurements versus the normalized laser power. Also shown are calculations utilizing the development in Section II for activation energies of 0.93 and 1.5 eV. It can be seen that the activation

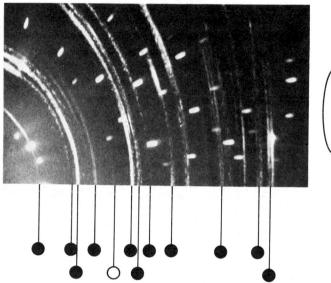

1300 Å Pd/Si

$\left(\begin{array}{c} \text{LASER-ANNEALED} \\ p = 1.1 \\ 1\ \text{SCAN} \end{array} \right)$

Fig. 29. Glancing angle x-ray diffraction pattern from a Read camera on a sample reacted at high power levels ($p = 1.1$). ●, PdSi; ○, Pd₂Si.

energy of 0.93 eV results in an excellent fit to the experimental data using a rate constant of 9.7×10^{14} Å²/sec.

The formation kinetics of the PdSi have not yet been measured. At present it is speculated that the PdSi formation occurs by a nucleation process (similar to that believed to occur for furnace annealing) for the laser reaction process.

A cw laser reaction of platinum–silicon systems has resulted in results somewhat different from those obtained in the furnace annealing case. The average composition of this phase, calculated from backscattering spectra, was determined to be close to Pt_2Si but was found from x-ray diffraction analysis to consist mainly of Pt_3Si and PtSi with trace amounts of $Pt_{12}Si_5$ and Pt_2Si. At higher laser powers the average composition was found to be $\sim PtSi_2$, a phase that does not exist in the equilibrium phase diagram. Again x-ray diffraction patterns identified this phase as a mixture of PtSi, $Pt_{12}Si_5$, and Pt_3Si. Also, there were three definite lines that could not be fit to any data for platinum–silicon compounds listed in the ASTM power diffraction files; these lines were from the Pt_2Si_3 phase recently identified (Tsaur *et al.*, 1980). It is believed that these multiple-phase compounds are the result of a eutectic melting and rapid quenching caused by the scanning laser beam.

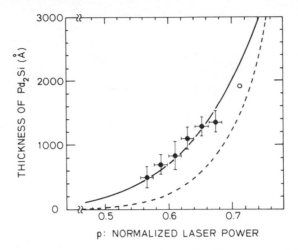

Fig. 30. Thickness of Pd_2Si as a function of normalized laser power. Calculated results are shown for two different activation energies. Solid line: $E_a = 0.93$ eV, $A = 9.7 \times 10^{14}$ Å^2/sec (Cheung *et al.*, 1980); dashed line: $E_a = 1.5$ eV, $A = 3 \times 10^3$ Å^2/sec (Bower *et al.*, 1973). $\bigcirc$, Thickness of fully reacted Pd_2Si film.

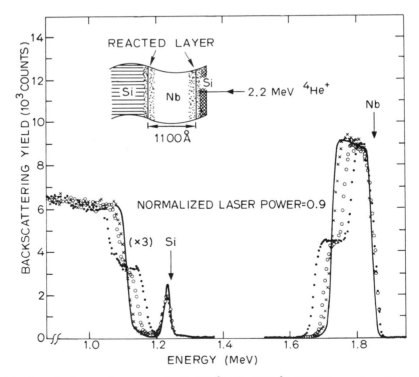

Fig. 31. Backscattering spectra for Si (200 Å)–Nb (1100 Å)–Si ⟨100⟩ samples with multiple laser scans. —, As deposited; ×, 1 scan; $\bigcirc$, 5 scans; ●, 21 scans.

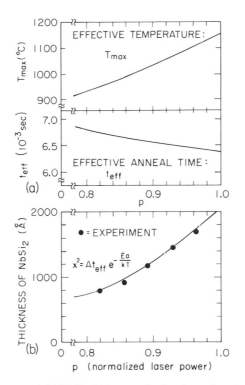

Fig. 32. (a) T_{eff}, t_{eff}, and (b) NbSi$_2$ thickness calculated as a function of normalized laser power. $E_a = 1.47$ eV, $A = 1.08 \times 10^{14}$ Å^2/sec.

In contrast to pulsed laser annealing of refractory metals, where only limited amounts of reaction have been observed, the scanning cw laser reactions of molybdenum, tungsten, and niobium films deposited upon silicon crystals have resulted in the formation of compositionally uniform silicides. We show backscattering spectra in Fig. 31 for the niobium–silicon system. These spectra indicate that formation of the NbSi$_2$ occurs by a solid phase interdiffusion mechanism. Analysis of the experimental data using the theoretical development of Section II allows measurement of the kinetics and activation energy of this solid phase process. The results of the calculations and measurements are shown in Fig. 32.

The reaction of tungsten and molybdenum to form WSi$_2$ and MoSi$_2$ has also been reported for the scanning cw beam process (Shibata *et al.*, 1980). However, at this writing detailed investigations of the kinetics of the formation of the films have not been carried out. It is believed that the process is similar to that observed for the furnace formation case.

IV. Dopant Incorporation and Metastable Phases

In this section we discuss results that have been obtained using scanning cw laser and electron beams both for annealing simple dopants in silicon at concentrations that exceed the solid solubilities and for the formation of superconducting compounds such as Nb_3Si and Nb_3Al.

The ability both to anneal impurities that have been ion-implanted in silicon and to create a high concentration of electrically active impurities is an extremely important aspect of laser processing. Lietoila *et al.* (1979) have shown that it is possible to achieve essentially 100% electrical activity even for implanted impurities that exceed the solid concentrations solubility. However, these impurity concentrations are only metastable, with precipitation occurring when a subsequent anneal is carried out in a furnace at lower temperatures and for longer times.

The formation of metastable phases has also been reported for the case of scanning electron beam processing (Regolini *et al.*, 1979). Certain silicides are of interest, in both superconducting applications that exist in a metastable phase. Rapid quenching of the reacted silicide eutectic has been observed to lead to certain metastable silicide phases (Shibata *et al.*, 1981a,b).

A. *Conventional Dopants in Silicon:* $C < C_{ss}$

The incorporation of conventional dopants in silicon into electrically active sites by scanning beam processing occurs by two processes. The first process is the case where the impurity is incorporated into an amorphous layer on single-crystal material, such as is found with high-dose ion implantation. In this case the impurity is incorporated into the lattice during recrystallization of the amorphous material. This process can be modeled as a solid phase epitaxial process with recrystallization rates obtained by extrapolation from the low-temperature thermal anneal case. Competition between solid phase regrowth and the nucleation of polycrystalline material occurs during the recrystallization process, which leads to a limit on the depth of the amorphous layer that can be recrystallized via scanning laser processing. To date only a few thousand angstroms of amorphous material on silicon have been successfully recrystallized via the scanning process. As indicated above, boron and arsenic are essentially 100% electrically active after the annealing, with essentially no diffusion occurring during the recrystallization. Carrier mobilities are comparable to those observed for single-crystal diffused material.

The second mechanism for incorporating impurities into the silicon crystal via a scanning process is essentially a local diffusion method in which the silicon is brought to a sufficient temperature and maintained there for an adequate time to allow the impurities to become incorporated into the silicon via a standard diffusion process. This process is believed to occur for implants that do not create a heavily damaged or amorphous layer in the material. Again, since the time at which the lattice is at a high temperature is very short, long-range diffusion of these atoms is negligible.

B. Conventional Dopants in Silicon: $C > C_{ss}$

Dopant incorporation for cases in which the impurity concentration is less than its solid solubility at the beam-induced temperature is thus a straightforward extrapolation of results expected in furnace annealing. However, the rapid cooling of the semiconductor material following passage of the annealing beam ($10^6 °C/sec$) allows one to incorporate certain impurities into the silicon lattice at concentrations higher than the thermal equilibrium solid solubility. Lietoila *et al.* (1980) have in fact demonstrated that the cw laser annealing process can completely activate implanted arsenic concentrations in silicon at levels up to 1×10^{21} cm³. The mechanism of incorporation of arsenic into the lattice is believed to be solid phase epitaxy even for $C > C_{ss}$. The process is similar to that described by Blood *et al.* (1979) who showed that indium-implanted silicon could be thermally annealed to achieve In concentrations in excess of the solid solubility in well-annealed material.

In Fig. 33 we show the electron concentration and mobility versus depth for laser-, electron beam-, and thermally annealed samples in which As was implanted to a dose of 10^{16} cm^{-2} at 100 keV. The substrate temperature was maintained at 0°C during the implant to ensure creation of a uniform amorphous layer. Also shown in Fig. 33 is the Pearson IV distribution calculated for this implant. As expected for both the cw electron beam- and laser-annealed layers, the measured impurity concentration (as inferred from the electron concentration) closely matches the calculated one (solid line). An interesting aspect of these data is that the maximum electron concentration exceeds 10^{21}/cm³ for both the electron beam- and laser-annealed samples. This concentration is significantly higher than that measured on thermally annealed samples where identically prepared samples were annealed at 550°C for 30 min (to achieve solid phase epitaxial regrowth of the layer) followed by a 2-min anneal at 1000°C (to minimize diffusion effects) to activate the carriers. For this case a peak elec-

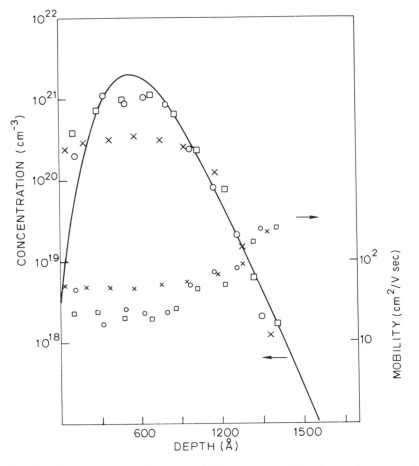

Fig. 33. Electron concentration and mobility versus depth for electron beam-annealed (○), laser-annealed (□), and thermally annealed (×) (550°C, 30 min plus 1000°C, 2 min) samples. —, Pearson IV distribution (after implant).

tron concentration of only $4 \times 10^{20}/cm^3$ was observed. Longer anneal times at 1000°C result only in broadening of the arsenic profile, with the peak electron concentration remaining at or below $10^{20}/cm^3$. Also shown in Fig. 33 is the measured electron mobility as a function of depth in this annealed layer. It can be seen that excellent agreement is observed between this mobility and that calculated from Irwin using the measured electron concentration at all points (except within 300 Å of the surface).

Further work by Lietoila *et al.* (1980) has shown that active concentrations of arsenic in excess of $3 \times 10^{20}/cm^3$ are thermally unstable and relax to about $3 \times 10^{20}/cm^3$ or less during thermal equilibrium annealing at temperatures of 1000°C. It was also found that the value to which the active concentration relaxed depended only upon the annealing temperature. This phenomenon is best understood by referring to Fig. 34 where we plot electron concentration versus depth for several processing conditions. The electron concentrations remaining after various laser-plus-thermal annealing sequences were measured by the differential van der Pauw method. These measurements also give the mobility as a function of the depth. It can be seen that the as-annealed (laser) concentration of arsenic is greater than $10^{21}/cm^3$. However, annealing at temperatures of 1100°C for 1 min causes the concentration of electrically active arsenic to relax to $3 \times 10^{20}/cm^3$. Annealing at lower temperatures produces a further reduction in the peak concentration, each time without significant diffusion of the impurities from regions where $C < C_{ss}$. These data then permit measurement of the solid solubility of arsenic in silicon at various temperatures, with the results shown in Fig. 35.

The kinetics of the relaxation process have been determined over a temperature range of 350–400°C by measuring the sheet conductance as a function of time. In Fig. 36 we show an Arrhenius plot of the time required to decrease the sheet carrier concentration by 7% in the laser-annealed

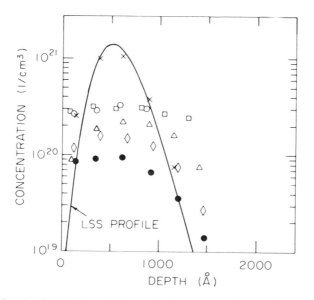

Fig. 34. Results from differential Van der Pauw measurements for samples that were laser-annealed and then subjected to various thermal treatments. ×, Laser only; ○, 1100°C, 1 min; □, 1000°C, 6 min; △, 904°C, 16 min; ◇, 808°C, 8 hr; ●, 707°C, 76 hr.

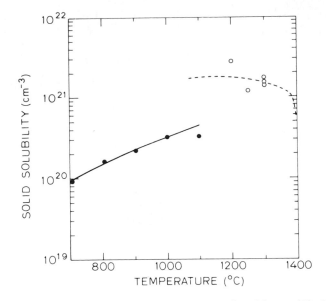

Fig. 35. The solubility of As in Si as an active, noncomplexed dopant (●). ○, Data from Trumbore (1960) for the equilibrium solubility of As in Si.

samples. The activation energy for this process was found to be 2.0 eV, which is close to the energy of vacancy formation in silicon. This measurement suggests a model that was first proposed by Chu and Masters (1979) for the formation of As–V–As complexes in silicon. In their work the samples were thermally annealed after pulsed laser annealing. With the use of angular scanning techniques in the Rutherford backscattering measurement, it was possible to determine the amount of arsenic displaced in the channeling direction after thermal annealing. This measurement suggested a 0.15-Å movement of substitutional arsenic atoms into the channels of the silicon lattice during thermal annealing. Chu and Masters have suggested a possible configuration for this complex, consisting of one doubly charged negative vacancy and two arsenic atoms. This model fits well with the activation energy measured by Lietoila *et al.* (1980).

From these measurements one can conclude that cw laser annealing is capable of activating arsenic in excess of solid solubility to form metastable concentrations. Prolonged thermal annealing causes the metastable active concentration to relax to certain values which depend only upon the temperature. These values are defined as the solubility of arsenic in silicon as an active, noncomplexed dopant. At 900°C, the complexed arsenic atoms appear to reside essentially at substitutional sites. At 500°C the

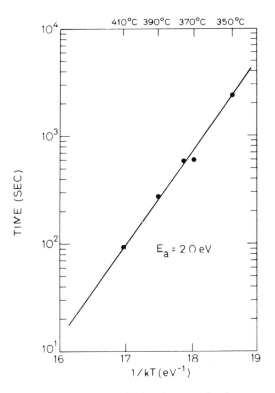

Fig. 36. Arrhenius plot of the time required to decrease the sheet concentration by 7% in the laser-annealed samples.

relaxation is accompanied by a significant increase in the amount of non-substitutional arsenic. As of this writing, experiments on metastable concentrations of dopants other than As in Si have not been performed using the scanning electron or laser annealing process.

C. Superconductors

The formation of superconducting compounds by use of a scanning cw laser has also been investigated. Especially in this case the rapid cooling rates involved in the laser annealing are believed to be responsible for formation of the metastable phases. In particular the successful synthesis of superconducting compounds with the A15 crystal structure, such as Nb_3Ge, Nb_3Al, and Nb_3Si, with scanning lasers has been carried out (Shibata *et al.*, 1981a,b). For these experiments annealing temperatures calculated to be in the range 1500–1900°C are employed at scan rates that lead to dwell times somewhat less than 1 msec. The total annealing time can be increased by employing multiple laser scans, and heating and cooling rates can be controlled by changing the beam scan speed. Improvement in the transition temperature for superconductivity has been observed for niobium, aluminum, and niobium–silicon thin films. Formation of the A15 structure after laser annealing is clearly observed, with weak diffraction lines from other phases visible.

1.   Nb–Al System

Figure 37 shows a typical x-ray diffraction pattern obtained from as-deposited samples of Nb–Al. The principal lines shown were identified as arising from the bcc structure. Some weak lines from the A15 phase were also observed. A variety of laser annealing experiments were performed with calculated laser annealing temperatures T_{max} in the range 1300–1800°C. Figure 37 also shows the x-ray diffraction pattern of a sample of 24.6 at. % Al) that was laser-annealed at 1400°C with 10 laser scan frames. Scan lines were overlapped by ~30% in each frame. Formation of the A15 structure after laser annealing is clearly observed, with some weak diffraction lines from the bcc phase. All laser-annealed Nb_3Al samples examined exhibited similar diffraction patterns, where both the A15 and bcc phases exist. The abundance of the A15 relative to the bcc phase varied substantially with the composition of the films and laser annealing conditions (T_{max} and the number of laser scan frames). One important point is that diffraction lines from the tetragonal Nb_2Al phase, which is the second phase typically obtained at lower temperatures, were not observed. This is particularly important, since the phase boundary between the A15 and Nb_2Al phase is less than 22 at. % Al for temperatures below about 1500°C (Kwo et al., 1980; Lundin and Yamamoto, 1966). This result therefore shows that the cw laser processing succeeded in quenching the high-temperature phase to room temperature without the decomposition into lower-temperature phases that accompanies the conventional thermal quench method.

In Fig. 38 T_c is plotted as a function of the number of laser scan frames. The laser annealing temperature was about 1400°C for all samples. Data are shown for two sets of Nb–Al samples with different compositions (21.0 and 24.6 at. % Al). T_c is seen to increase linearly with the number of scan frames up to 10 frames, after which it remains essentially constant. X-ray diffraction analysis showed that this T_c enhancement was correlated with growth of the A15 phase relative to the bcc phase. Therefore, 10 laser

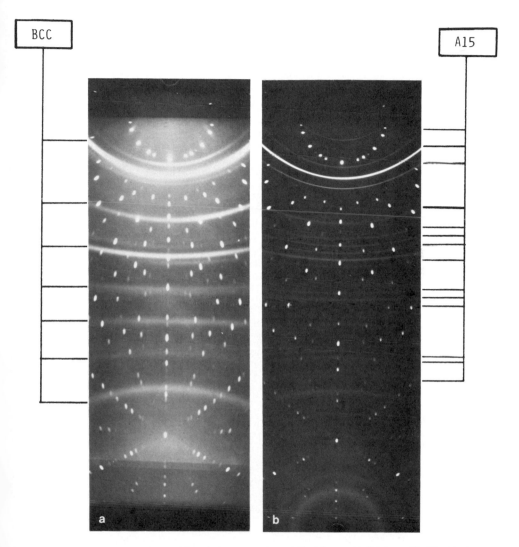

Fig. 37. Read camera diffraction pattern for a Nb–Al sample (24.6 at. % Al). (a) As-deposited, (b) after laser anneal with 10 scans at 1400°C.

scans are probably sufficient to carry the reaction to completion. It is also shown in Fig. 38 that T_c is consistently higher for films with a higher Al composition.

Figure 39 shows the dependence of T_c on the laser annealing temperature for the same two sets of NbAl samples discussed above. Ten laser

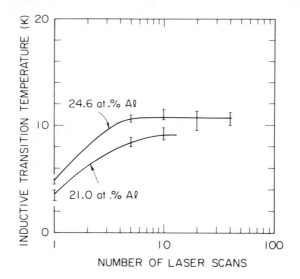

Fig. 38. Superconducting transition temperatures (T_c) as a function of high-temperature laser scans at 1400°C for two sets of Nb–Al samples (of 21.0 and 24.6 at. % Al).

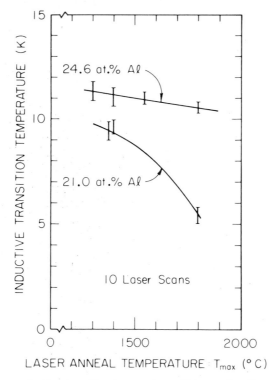

Fig. 39. Superconducting transition temperatures (T_c) as a function of laser annealing temperature for the two sets of Nb–Al samples in Fig. 38.

scans were employed at each temperature. It is seen that a higher annealing temperature reduces T_c, particularly for the low Al composition (21.0
at. % Al) samples. X-ray diffraction analysis showed that the samples
annealed at high temperatures had less A15 phase (and more bcc phase)
than samples annealed at lower temperatures for both Al compositions.
These observations are qualitatively consistent with the equilibrium phase
diagrams of Nb–Al (Lundin and Yamamoto, 1966; Svechinkov *et al.*,
1970), considering the uncertainty in temperature determinations. Briefly,
the compositional range of the bcc solid solution is more Al-rich (≥21 at.
% Al) at high temperatures (≥1800°C). Hence the increased formation of

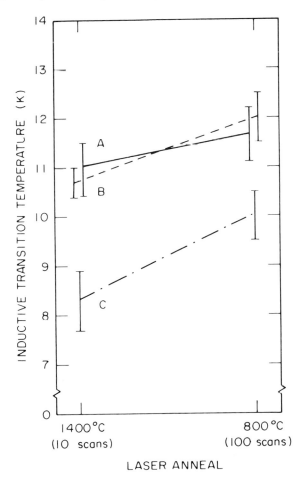

Fig. 40. Enhancement of inductive transition temperature (T_c) by laser annealing at
800°C for 100 scan frames. Also shown are the results at 1400°C.

the bcc phase at high temperatures, particularly for the sample of 21.0 at. % Al, leads to the lower values of T_c.

2.   LOW-TEMPERATURE ANNEALING

From the discussion above one can conclude that the high-temperature state in the phase diagram can be quenched to room temperature by scanning cw laser annealing without a corresponding phase change. However, atomic disorder is usually observed following a rapid quench to room temperature. Low-temperature furnace annealing at 700–800°C can of course improve the ordering, but such annealing usually leads to decomposition of the quenched-in phase to the unwanted low-temperature phases.

As discussed earlier, a single laser beam scan is equivalent to furnace annealing for very short times (approximately millisecond) with very short heating and cooling cycles. By repeated scanning, one can therefore extend the effective annealing time without changing the cooling and heating rates. This is exactly what is required for annealing Nb–Al samples. With such a procedure, three different samples were subjected to 800°C laser annealing for 100 scan frames. Values of T_c before and after this process are shown in Fig. 40. A T_c enhancement of 0.5–1.5 K is observed for all samples. There is no significant change in the amount of the A15 relative to the bcc phase, and second-phase Nb_2Al was not observed in any of these cases. We tentatively attribute this T_c enhancement to the atomic ordering obtained by low-temperature laser annealing.

V.   Future Directions

The processing of materials using scanning cw lasers or electron beams continues to be a fruitful area for materials research. The unique properties offered by this type of thermal processing, i.e., localized heating of the surface, rapid heating and cooling rates controlled in some instances by the scan speed, selective absorption of energy by the use of reflecting or antireflecting coatings, and multiple interference coatings, all offer capabilities unobtainable with standard thermal processing. Exploitation of these unique capabilities will very likely lead to commercial application of cw beam processing techniques.

One area that looks especially promising for future research is the processing of compound semiconductors, typically difficult because of the volatile nature of one or more of the constituent elements of the material.

These materials can be processed by using a capping layer that is transparent to the energy beam. The growth of islands of single-crystal materials by scanning from a seed crystal source through a poly or amorphous material is also an area that has much to offer. Recent results in this field look quite promising both for applications and for bringing about new understanding of microcrystallization techniques. The use of beam processing to assist the diffusion of impurities in materials has been reported. The high-temperature, short-time, beam-assisted diffusion process can produce high impurity concentrations in very shallow depths. An application of such a process may lie in integrated circuit processing for fine-geometry devices. The use of scanning cw beams in the formation of silicides has been discussed, and their application as interconnections and/or gates in Si integrated circuits appears close at hand.

Beam processing techniques for the alteration of metal surfaces, such as corrosion or alloying, represent an area that also seems destined to become important in the near future. The elimination of thick electroplated gold coatings for contacts in electronic systems represents an area where surface alloying of metals by scanning beams can be a cost-effective measure. The ability to create end surface layers with low contact resistivity and good corrosion resistance might be quite easily exploited by energy beam processing.

References

Blood, P., Brown, W. L., and Miller, G. L. (1979). *J. Appl. Phys.* **50**, 173.
Boatright, R. L., and McCaldin, J. O. (1976). *J. Appl. Phys.* **47**, 2260.
Booker, G. R., and Stickler, R. (1962). *J. Appl. Phys.* **33**, 3281.
Bower, R. W., Sigurd, D., and Scott, R. E. (1973). *Solid-State Electron.* **16**, 1461.
Canali, C., Mayer, J. W., Ottaviani, G., Sigurd, D., and Van der Weg, W. F. (1974). *Appl. Phys. Lett.* **25**, 3.
Canali, C., Campisano, S. U., Lau, S. S., Liau, Z. L., and Mayer, J. W. (1977). *Thin Solid Films* **46**, 99.
Carslaw, H. S., and Jaeger, J. C. (1959). "Conduction of Heat in Solids," 2nd ed. Oxford Univ. Press, London and New York.
Cheung, N., Lau, S. S., Nicolet, M.-A., and Mayer, J. W. (1980). In "Thin Film Interfaces and Interactions" (J. E. E. Baglin and J. M. Poate, eds.), p. 494. Electrochem. Soc., Pennington, New Jersey.
Chu, W. K., and Masters, B. J. (1979). In "Laser–Solid Interactions and Laser Processing" (S. D. Ferris, H. J. Leamy, and J. M. Poate, eds.), p. 305. Am. Inst. Phys., New York.
Cline, H. E., and Anthony, T. R. (1977). *J. Appl. Phys.* **48**, 3895.
Crowder, B. L., and Title, R. S. (1971). In "Ion Implantation" (F. H. Eisen and L. T. Chadderton, eds.), p. 87. Gordon & Breach, New York.
Csepregi, L., Mayer, J. W., and Sigmon, T. W. (1975). *Phys. Lett. A* **54A**, 157.
Csepregi, L., Kennedy, E. F., Lau, S. S., Mayer, J. W., and Sigmon, T. W. (1976). *Appl. Phys. Lett.* **29**, 645.

Csepregi, L., Kennedy, E. F., Gallagher, T. J., Mayer, J. W., and Sigmon, T. W. (1977a). *J. Appl. Phys.* **49,** 4234.

Csepregi, L., Kuller, R. P., Mayer, J. W., and Sigmon, T. W. (1977b). *Solid State Commun.* **21,** 1019.

Csepregi, L., Kennedy, E. F., Mayer, J. W., and Sigmon, T. W. (1978). *J. Appl. Phys.* **49,** 3906.

Davey, J. E., Christou, A., and Day, H. M. (1976). *Appl. Phys. Lett.* **28,** 365.

Gat, A., and Gibbons, J. F. (1978). *Appl. Phys. Lett.* **32,** 142.

Gat, A., Gibbons, J. F., Magee, T. J., Peng, J., Deline, V. R., Williams, P., and Evans, C. A. Jr. (1978a). *Appl. Phys. Lett.* **32,** 276.

Gat, A., Gibbons, J. F., Magee, T. J., Peng, J., Williams, P., and Evans, C. A., Jr. (1978b). *Appl. Phys. Lett.* **33,** 389.

Gibbons, J. F. (1972). *Proc. IEEE* **60,** 1062.

Gibbons, J. F. (1979). *Proc. Int. Conf. Solid State Devices, 11th, Tokyo* p. 121.

Gibbons, J. F., Johnson, W. S., and Mylroie, S. W. (1975). "Projected Range Statistics in Semiconductors." Dowden, Hutchinson & Ross, Stroudsburg, Pennsylvania.

Glowinski, L. D., Tu, K. N., and Ho, P. S. (1976). *Appl. Phys. Lett.* **28,** 312.

Gold, R. B., and Gibbons, J. F. (1980). *J. Appl. Phys.* **51** (2), 1256.

Gold, R. B., Gibbons, J. F., Magee, T. J., Peng, J., Ormond, R., Deline, V. R., and Evans, C. A. (1980). *In* "Laser and Electron Beam Processing of Materials" (C. W. White and P. S. Peercy, eds.), p. 211. Academic Press, New York.

Hiraki, A., Lugujjo, E., and Mayer, J. W. (1972). *J. Appl. Phys.* **43,** 3643.

Ho, C. Y., Powell, R. W., and Liley, P. E. (1974). *J. Phys. Chem. Ref. Data, Suppl.* **3,** No. 1, I-588.

Johnson, N. M., Gold, R. B., and Gibbons, J. F. (1979). *Appl. Phys. Lett.* **34,** 704.

Johnson, N. M., Regolini, J. L., Bartelink, D. J., Gibbons, J. F., and Ratnakumar, K. N. (1980). *Appl. Phys. Lett.* **36,** 425.

Joyce, W. B. (1975). *Solid-State Electron.* **18,** 321.

Kennedy, E. F., Csepregi, L., Mayer, J. W., and Sigmon, T. W. (1977). *J. Appl. Phys.* **48,** 4241.

Kwo, J., Hammond, R. H., and Geballe, T. H. (1980). *J. Appl. Phys.* **51,** 1726.

Lau, S. S. (1978). *In* "Thin Film Phenomena—Interfaces and Interactions" (J. E. E. Baglin and J. M. Poate, eds.), p. 269. Electrochem. Soc., Princeton, New Jersey.

Lau, S. S., and Van der Weg, W. F. (1978). *In* "Thin Films—Interdiffusion and Reactions" (J. M. Poate, K. N. Tu, and J. W. Mayer, eds.), p. 433. Wiley, New York.

Lau, S. S., Liau, Z. L., and Nicolet, M.-A. (1977). *Thin Solid Films* **47,** 313.

Lax, M. (1977). *J. Appl. Phys.* **48,** 3919.

Lax, M. (1978). *Appl. Phys. Lett.* **33,** 786.

Liau, Z. L., Campisano, S. U., Canali, C., Lau, S. S., and Mayer, J. W. (1975). *J. Electrochem. Soc.* **122,** 1695.

Liau, Z. L., Tsaur, B. Y., and Mayer, J. W. (1979). *Appl. Phys. Lett.* **34,** 221.

Lietoila, A., Gibbons, J. F., Magee, T. J., Peng, J., and Hong, J. D. (1979). *Appl. Phys. Lett.* **35,** 532.

Lietoila, A., Gibbons, J. F., and Sigmon, T. W. (1980). *Appl. Phys. Lett.* **36,** 765.

Lugujjo, E., and Mayer, J. W. (1973). *Phys. Rev. B* **7,** 1782.

Lundin, C. E., and Yamamoto, A. S. (1966). *Trans. Metall. Soc. M. E.* **236,** 863.

Majni, G., and Ottaviani, G. (1977). *Appl. Phys. Lett.* **31,** 125.

Matsuda, A., Mineo, A., Kurosu, T., and Kikuchi, M. (1973). *Solid State Commun.* **13,** 1165.

Maycock, P. D. (1967). *Solid-State Electron.* **10,** 161.

Mizuta, M., Sheng, N. H., Merz, J. L., Lietoila, A., Gold, R. B., and Gibbons, J. F. (1980). *Appl. Phys. Lett.* **37**, 154.

Morehead, F. F., Jr., and Crowder, B. L. (1971). *In* "Ion Implantation" (F. H. Eisen and L. T. Chadderton, eds.), p. 25. Gordon & Breach, New York.

Nissim, Y. I., Lietoila, A., Gold, R. B., and Gibbons, J. F. (1980). *J. Appl. Phys.* **51**, 274.

Ottaviani, G., Marrello, V., Mayer, J. W., Nicolet, M.-A., and Caywood, J. (1972). *Appl. Phys. Lett.* **20**, 323.

Ottaviani, G., Sigurd, D., Marrello, V., Mayer, J. W., and McCaldin, J. O. (1974). *J. Appl. Phys.* **45**, 1730.

Ottaviani, G., Canali, C., and Majni, G. (1976). *J. Appl. Phys.* **47**, 627.

Poate, J. M., Leamy, H. J., Sheng, T. T., and Celler, G. K. (1978). *Appl. Phys. Lett.* **33**, 918.

Regolini, J. L., Sigmon, T. W., and Gibbons, J. F. (1979). *Appl. Phys. Lett.* **35**, 114.

Revesz, P., Wittmer, M., Roth, J., and Mayer, J. W. (1978). *J. Appl. Phys.* **49**, 5199.

Roth, J. A., and Anderson, C. L. (1977). *Appl. Phys. Lett.* **31**, 689.

Seidel, T. E., Pasteur, G. A., and Tsai, J. C. C. (1976). *Appl. Phys. Lett.* **29**, 648.

Shibata, T., Sigmon, T. W., and Gibbons, J. F. (1980). *In* "Laser and Electron Beam Processing of Electronic Materials" (C. L. Anderson, G. K. Celler, and G. A. Rozgonyi, eds.), p. 520. Electrochem. Soc., Princeton, New Jersey.

Shibata, T., Sigmon, T. W., Regolini, J. L., and Gibbons, J. F. (1981a). *J. Electrochem. Soc.* **128**, 637.

Shibata, T., Gibbons, J. F., Kwo, J., Feldman, R. D., and Geballe, T. H. (1981b). *J. Appl. Phys.* **52**, 1537.

Sigmon, T. W., and Tseng, W. F. (1978). *In* "Thin Film Phenomena—Interfaces and Interactions" (J. E. E. Baglin and J. M. Poate, eds.), p. 99. Electrochem. Soc., Princeton, New Jersey.

Spaepen, F. (1978). *Acta Metall.* **26**, 1167.

Svechinkov, V. N., Pan, V. M., and Latzshev, V. I. (1970). *Metallafigik* **32**, 28.

Takamori, T., Messier, R., and Roy, R. (1972). *Appl. Phys. Lett.* **20**, 201.

Trumbore, F. A. (1960). *Bell Syst. Tech. J.* **39**, 205.

Tsaur, B. Y., Mayer, J. W., and Tu, K. N. (1980). *In* "Thin Film Symposium." (J. E. E. Baglin and J. M. Poate, eds.), p. 264. Electrochem. Soc., Princeton, New Jersey.

Tseng, W., Liau, Z. L., Lau, S. S., Nicolet, M.-A., and Mayer, J. W. (1977). *Thin Solid Films* **46**, 99.

Vook, F. L., and Stein, H. J. (1971). *In* "Ion Implantation" (F. H. Eisen and L. T. Chadderton, eds.), p. 9. Gordon & Breach, New York.

Zeiger, H. J., Fan, J. C. C., Palm, B. J., Gale, R. P., and Chapman, R. L. (1980). *In* "Laser and Electron Beam Processing of Materials" (C. W. White and P. S. Peercy, eds.), p. 234. Academic Press, New York.

Chapter 11

Compound Semiconductors

J. S. WILLIAMS

*Department of Communication and Electronic Engineering
Royal Melbourne Institute of Technology
Melbourne, Victoria, Australia*

I. Introduction

The major thrust of the preceding chapters in this volume has been directed toward transient annealing of elemental semiconductors, and of Si in particular. This reflects the far greater attention given in recent literature to transient annealing of Si compared with compound semiconductors. Nevertheless, a considerable body of data now exists on transient processing of compound semiconductors, although GaAs, a most important compound semiconductor in microelectronic device applications, has been of major interest. In this chapter, laser and electron beam applications in the processing of compound semiconductors are reviewed, with particular emphasis on GaAs.

Transient processing is particularly attractive for GaAs since it offers possible solutions to long-standing difficulties with conventional furnace processing, such as control of decomposition of the near surface, removal of ion implant damage and dopant activation, and formation of reproducible ohmic contacts. Indeed, pioneering Russian work (Kachurin *et al.*, 1976) demonstrated that ion implant damage in GaAs could be successfully removed by pulsed laser annealing. The advantages of laser-induced alloying of films deposited on compound semiconductors were also exploited early, with limited success, in the formation of ohmic contacts (Pounds *et al.*, 1974). Stimulated no doubt by the importance of GaAs as an electronic device material, much of the more recent work has concentrated on an assessment of the electrical properties of transient-annealed GaAs. These studies have essentially confirmed the potential of transient annealing applications to compound semiconductors, although the results to date have not shown the often expected dramatic improvements over conventional furnace processing. In fact, as will become apparent throughout this chapter, many of the problem areas associated with furnace processing of GaAs either remain as difficulties in transient processing or give way to other significant problems associated with rapid heating and cooling effects. Indeed, the lack of detailed studies on the physical processes involved in the transient processing of compound semiconductors, particularly in the removal of implantation-induced damage, has contributed to a much less detailed understanding of transient annealing effects in GaAs than that which has been forthcoming from annealing studies on Si.

Nevertheless, the two basic annealing mechanisms of liquid phase and solid phase regrowth, identified and well characterized for transient processing of Si (Chapters 4, 6, 7, and 10), are applicable to transient annealing studies on compound semiconductors. In this chapter, a distinction is made between scanning continuous-wave (cw) annealing, which has been used to induce structural rearrangements in the solid phase, and pulsed laser and electron beam annealing, which has been generally accepted as inducing liquid phase regrowth in GaAs. However, transient annealing results for GaAs clearly indicate that the distinction between the solid and liquid phase annealing modes is not as clear-cut as has been demonstrated for Si. For example, whereas the kinetics of solid phase epitaxial regrowth are well characterized for Si and indicate the possibility of complete annealing on a millisecond time scale, solid state annealing mechanisms are considerably more complex in GaAs, and complete damage removal in the solid phase is obtained only on transient annealing time scales exceeding about 1 sec (Section II.C). Pulsed annealing (liquid phase) phenomena in GaAs appear to be somewhat similar, in general, to those identified in Si; yet significant differences between GaAs and Si annealing are apparent when details of damage removal, dopant solubility, and impurity redistribution processes are examined (Section III).

This chapter concentrates on the two particular applications of transient annealing that have received the most attention in the literature: (i) the removal of ion implant damage and dopant effects in GaAs, and (ii) the formation of ohmic contacts to compound semiconductors. An attempt has been made to correlate the available physical and electrical data and to present current understanding of the various transient annealing processes. In Section II solid phase annealing of ion-implanted GaAs is discussed, and results of conventional furnace processing are compared with the sparse data available on transient annealing. Section III treats pulsed (liquid phase) annealing in implanted GaAs, a topic that has been of considerable recent interest. Section IV examines transient annealing both in the formation of ohmic contacts to GaAs and in the mixing of deposited layers with GaAs. Finally, Section V reviews the available data on compound semiconductors other than GaAs.

II. Ion-Implanted Layers: Solid Phase Regrowth

A. Ion Implant Damage

Before examining the solid phase regrowth of ion-implanted GaAs, it is important to understand the nature of the implant-induced damage since the subsequent annealing behavior, to a large extent, is determined by it. For example, it has been the usual practice (see Donnelly, 1977, and references therein) to employ elevated temperatures (200–400°C) when implanting dopants into GaAs since such implant conditions appear to produce the optimum electrical characteristics following subsequent higher-temperature furnace annealing. It has been known for some time that elevated temperature implants do not produce an amorphous layer but essentially preserve the crystallinity within the near surface, despite the large number of crystal defects (mainly loops) that are generated (Mazey and Nelson, 1969). In contrast, in GaAs room temperature implants can produce an amorphous layer, but the degree and nature of the implant damage was found in early studies (Mazey and Nelson, 1969; Picraux,

1973) to be most sensitive to the ion dose and dose rate. The difficulties experienced in obtaining satisfactory and reproducible electrical properties from room temperature implants (following subsequent higher-temperature annealing) have been attributed in part to the formation of an amorphous layer (Harris *et al.*, 1972; Eisen, 1975). This behavior is to be contrasted with experiences from ion-implanted silicon, where the removal of amorphous damage by a simple epitaxial process (Lau and Van der Weg, 1978) usually produces complete dopant activation (Dearnaley *et al.*, 1973) and electrical properties far superior to those obtained following elevated temperature implants.

More recent experiments (Williams and Austin, 1980a; Ahmed *et al.*, 1980) have indicated that considerable dynamic annealing can take place during room temperature implantation of GaAs, resulting, in some cases,

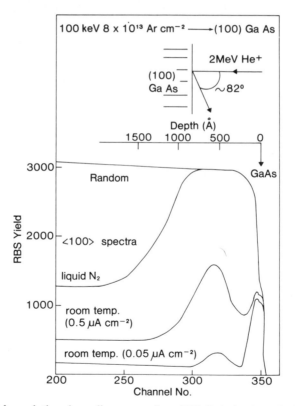

Fig. 1. High-resolution channeling spectra from (100) GaAs implanted with 8×10^{13} Ar^+ cm^{-2} at 100 keV for targets held at room temperature (two dose rates) and liquid nitrogen temperature. (After Williams *et al.*, 1980.) This figure was originally presented at the Fall 1977 Meeting of The Electrochemical Society, Inc., held in Atlanta, Georgia.

in an implant layer containing part amorphous and part crystalline damage. The consequence of these dynamic annealing effects, in terms of the amount of damage that results from implantation, is illustrated in Fig. 1. High-resolution channeling spectra reveal the damage distributions for 8×10^{13} Ar cm^{-2} implanted at 100 keV into GaAs under different implant conditions. At liquid nitrogen implant temperatures, all the displacement damage is frozen in, and a completely amorphous surface layer results. On the other hand, the damage is only partly amorphous for room temperature implants at both dose rates shown. In Figs. 2 and 3, transmission electron microscopy (TEM) micrographs further illustrate the differences in remnant damage that can result from room temperature implantation.

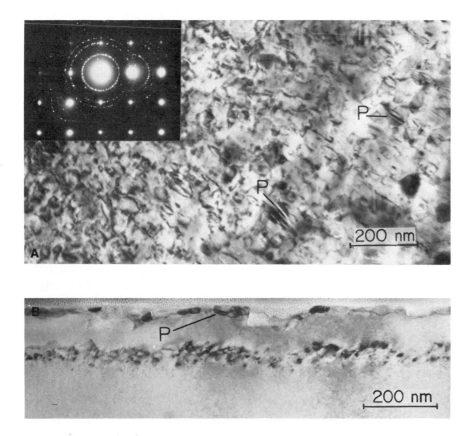

Fig. 2. TEM micrographs of 160-keV Se$^+$-implanted (100) GaAs to a dose of 5×10^{15} ions cm^{-2} at room temperature, indicating that the implanted layer is not amorphous. (a) Plan view with inset diffraction pattern, (b) corresponding cross-section view. "P" denotes polycrystalline regions. (After Fletcher *et al.*, 1981.)

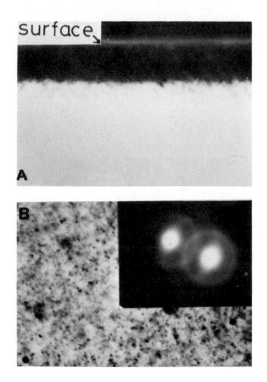

Fig. 3.   TEM micrographs of 150-keV Zn-implanted GaAs to a dose of 1×10^{15} cm^{-2} at room temperature, indicating a completely amorphous near-surface layer. (a) Cross-section view, (b) plan view with diffraction pattern. (After Kular *et al.*, 1980.)

Figure 2 shows an implant layer that is predominantly crystalline following 5×10^{15} Se^{+} cm^{-2} implantation at 160 keV into GaAs, whereas Fig. 3 shows a completely amorphous layer for a 1×10^{15} Zn cm^{-2} implant (at 150 keV). Presumably, the former implant resulted in significant heating during implantation, whereas the latter implant was undertaken at a dose rate just high enough to ensure amorphization.

B. Furnace Annealing

As discussed in Chapters 4 and 10, solid phase annealing of ion-implanted silicon usually results in the epitaxial recrystallization of amorphous zones or layers to produce highly perfect regrown crystals. The regrowth kinetics associated with the amorphous-to-crystalline transition are well characterized for silicon, and the recrystallization process takes place abruptly in the temperature range 500–600°C. In contrast, solid phase annealing of ion-implanted GaAs proceeds via a more complex, multistage process. Early studies (Mazey and Nelson, 1969; Carter *et al.*, 1971) revealed that damage removal of ion implant damage in GaAs occurred over a broad temperature range. The more recent data of Gamo *et al.* (1977), shown in Fig. 4, illustrate this behavior. The Rutherford backscattering (RBS)–channeling spectra for 3×10^{13} Zn cm^{-2} implanted into (100) GaAs at liquid nitrogen temperatures clearly show that annealing of the initially amorphous implant layer takes place over the broad temperature range 200–600°C.

The nature of the solid phase annealing process is further revealed in Fig. 5, from Williams and Harrison (1981). The measured GaAs disorder, obtained from channeling, is plotted in Fig. 5 as a function of the anneal temperature following an amorphizing implant of 8×10^{13} Ar cm^{-2} at 100 keV. These data reveal two apparent annealing stages, the first occurring rather sharply in the temperature range 125–230°C (for 15-min anneals), followed by a second stage in the range 400–600°C. Channeling cannot

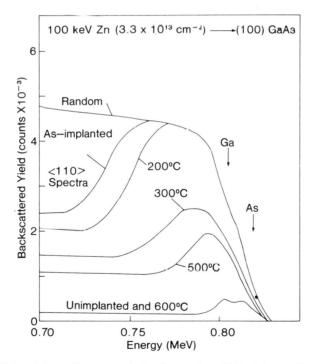

Fig. 4. RBS and channeling spectra indicating isochronal (15 min) annealing of 3.3×10^{13} cm^{-2} 100-keV Zn$^+$ ion implant damage in GaAs. (Adapted from Gamo *et al.*, 1977.)

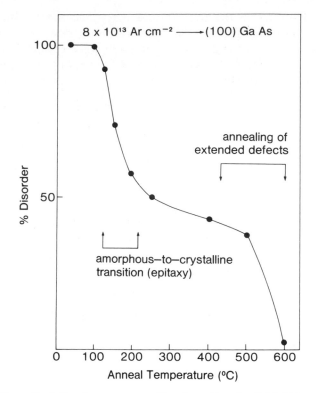

Fig. 5. Normalized disorder, as measured by channeling, for 100-keV 8×10^{13} Ar cm^{-2}- implanted (100) GaAs plotted as a function of furnace annealing temperature (15-min an- neals), indicating two annealing stages. (From Williams and Harrison, 1981.)

unequivocally reveal the nature of the residual disorder but, from obser- vations of characteristic color changes during annealing, Williams and collaborators (Williams and Austin, 1980b; Williams *et al.*, 1980) sug- gested that stage-1 recovery related to the amorphous-to-crystalline tran- sition and that stage 2 related to the annealing of extended (crystalline) defects.

The recent detailed work of Kular *et al.* (1980) correlated channeling, TEM, and electrical measurements from Zn-implanted GaAs furnace- annealed in the temperature range 300–900°C. The series of TEM micro- graphs in Figs. 3 and 6–8 are from this work and clearly reveal the respec- tive anneal stages. The initially 2000-Å-thick amorphous implanted layer (Fig. 3) undergoes a transition to a highly twinned single-crystal layer during a 300°C anneal (Fig. 6). In the temperature range 300–600°C (Fig. 7), the surface layer (twinned crystal) becomes progressively more or-

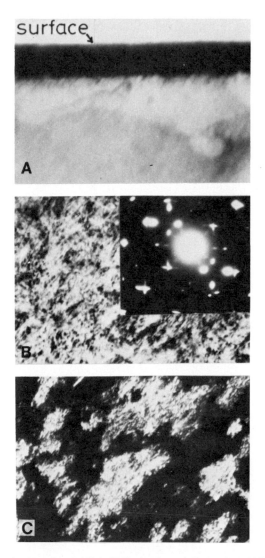

Fig. 6. TEM micrographs of 150-keV Zn-implanted GaAs (1×10^{15} cm^{-2}) annealed at 300°C. (a) Cross section: (b) plan view, bright field (inset, TED pattern); (c) plan view, dark field. (After Kular et al., 1980.)

dered, with the appearance of single dislocation loops. This behavior is essentially in agreement with the channeling measurements in Figs. 4 and 5. In the temperature range up to 900°C (GaAs capped during annealing), the TEM data indicate a third anneal stage (Fig. 8) in which the density of

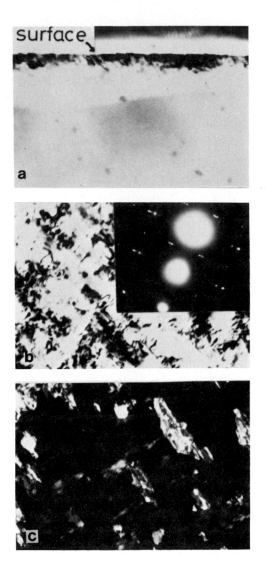

Fig. 7. TEM micrographs of 150-keV Zn-implanted GaAs (1×10^{15} cm^{-2} annealed at 600°C. (a) Cross section; (b) plan view, bright field (inset, TED pattern); (c) plan view, dark field. (After Kular *et al.*, 1980.)

the loops is progressively reduced. Kular *et al.* (1980) comment that channeling is not very sensitive to the damage structure (loops) above 700°C and thus to stage 3, which is in accord with the low damage indicated by channeling in Figs. 4 and 5 after annealing at 600°C.

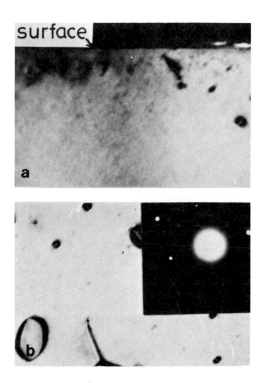

Fig. 8. TEM micrographs of 150-keV Zn-implanted GaAs (1×10^{15} cm^{-2} annealed at 900°C. (a) Cross section; (b) plan view, bright field (inset, TED pattern). (After Kular *et al.*, 1980.)

Thus, there appear to be three important anneal stages associated with the complete removal of ion implant damage in GaAs by solid phase regrowth. It is further significant to note that measurements of electrical activity and mobility for ion-implanted GaAs show continual improvement up to 900°C (Eisen, 1980; Donnelly, 1981; Kular *et al.*, 1980). In view of the anneal process outlined above, the best electrical behavior is at least contingent upon the complete removal of crystal defects that arise during the low-temperature annealing of implanted GaAs. This is not to suggest that complete removal of crystalline extended defects (loops) is sufficient to guarantee good electrical behavior. Other factors may certainly influence the electrical properties, following furnace annealing, such as point defect clusters and material impurities, particularly Cr distributions, and the nature of encapsulants used to protect the surface from decomposition (Eisen, 1980). These aspects are discussed in Section III.E.

Before moving to laser and electron-beam-induced solid phase regrowth in the next section, it is worth briefly examining the factors that appear to influence crystal quality during stage-1 recovery (the amorphous-to-crystalline transition). The work of Williams and Austin (1980b) indicated that good-quality crystalline GaAs (as measured by channeling) could result under carefully controlled implant and annealing conditions. For example, Fig. 9 shows the annealing behavior of Ar-implanted (100) GaAs at two doses sufficient to create an amorphous surface layer for liquid nitrogen implants. For the lower-dose case, almost complete damage removal is obtained following a 180°C anneal, whereas a slightly higher dose does not anneal to good-quality crystal during stage 1 (250°C anneal), and an anneal temperature of 600°C is needed to remove all the defects that are measured by channeling. This suggests a dose dependence in which heavily damaged amorphous layers are harder to regrow via a simple (stage-1) epitaxial process. Williams *et al.* (1980), in attempting to measure the regrowth kinetics, found that the rate of epitaxial growth at temperatures below 200°C was very sensitive to small changes in the implant dose close to the amorphous threshold dose, with the regrowth rate slowing down as the dose increased. More recent, detailed low-temperature annealing studies by Grimaldi *et al.* (1981a,b) have found that the crystal quality following stage-1 recovery depends strongly on the thickness of the amorphous layer generated by ion implantation and less strongly on the ion species, dose, and implant temperature. Single-crystal growth, free of defects as detected by channeling, was achieved only for a very thin amorphous layer (≤ 400 Å). Thus, although these results indicate the possibility of improving the quality of stage-1 regrowth under certain conditions, the most general behavior (particularly for room temperature implants with a part amorphous, part crystalline damage structure) is that anneal temperatures of >600°C are needed to remove remnant crystalline damage.

C. *Continuous-Wave Laser and Electron Beam Annealing*

Applications of scanning cw lasers and electron beams in the annealing of ion implantation damage in GaAs have not received the amount of attention devoted to pulsed annealing. The reasons for this are not clear but may in part arise from practical difficulties in establishing suitable cw annealing conditions and the somewhat limited success and lack of reproducibility generally obtained with respect to electrical properties. However, first reports on cw laser annealing of medium-dose (10^{14} cm^{-2}), Se-implanted GaAs were encouraging (Fan *et al.*, 1979a,b), and the more

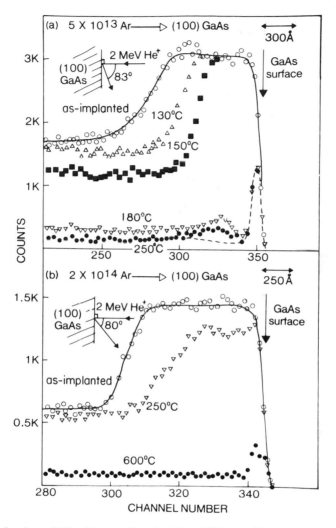

Fig. 9. Isochronal (10 min) annealing of (a) 5×10^{13} Ar cm^{-2}-implanted and (b) 2×10^{14} Ar cm^{-2}-implanted (100) GaAs. The dashed curve in (a) is for unimplanted GaAs. The GaAs was encapsulated with Al metal for the 600°C anneal in (b). (After Williams and Austin, 1980b.)

recent results of Shah *et al.* (1980), using a scanning electron beam to activate low-dose Si^{+}-implanted GaAs, suggest that cw annealing may well constitute a viable method of postimplantation processing of GaAs.

The major difficulties encountered during cw annealing of GaAs are (i) the tendency for the surface to decompose at high temperatures, and

(ii) the severe surface slip and cracking that can result from large thermal gradients present during cw scanning. Minimizing the former effect places an upper limit on both the anneal time and the local surface temperature, whereas avoidance of surface slip places an upper limit on the local heating rate (scan rate and power). As will be illustrated below, the net result is that satisfactory annealing can be achieved only over a narrow power window and, for millisecond annealing, with the bulk substrate held at an elevated temperature (~500°C). Most of the cw annealing studies to date have attempted to optimize the annealing conditions to provide the best electrical characteristics without an attempt to characterize the remnant damage. Consequently, relatively little is known about the solid phase annealing processes that take place during cw annealing. However, considerable insight into the relevant transient annealing processes can be obtained by a careful examination of the various results and the precise annealing conditions used to obtain them.

Williams *et al.* (1980) compared the effectiveness of cw laser and furnace annealing in the removal of amorphous implant damage in GaAs. Typical results are shown by the channeling spectra in Fig. 10 for two doses of Ar^+ implanted into (100) GaAs. Figure 10a shows that both the cw (argon ion) laser and the low-temperature furnace anneal essentially produce an amorphous-to-crystalline recovery (stage-1 annealing) which, in the case of a layer not totally amorphous, results in good crystalline recovery. However, for a higher implant dose, above that required to create a continuous amorphous layer, both low-temperature furnace (250°C) and cw annealing do not produce good-quality crystalline recovery, and furnace anneal temperatures exceeding 600°C are subsequently required to remove remnant crystalline defects. It is interesting to examine the particular cw laser annealing conditions chosen in these experiments. A continuous scan rate of 1 cm sec^{-1} was employed for a focused 150-μm-diameter laser spot, resulting in a laser dwell or anneal time of 15 msec. The laser power chosen was just below the threshold for surface slip, with the target held at room temperature. A range of dwell times (10–100 msec) were investigated at optimum laser power and provided results similar to those shown in Fig. 10.

The significant conclusion from the data in Fig. 10 is that the optimum laser annealing conditions (to preclude slip) for a substrate held at room temperature have facilitated only stage-1 recovery of the ion-implanted damage. This implies that the local surface temperature attained during the cw anneal was not high enough for a sufficiently long time to remove the crystalline defects. Thus, the onset of surface slip emanating from thermal stress appears to preclude the attainment of high surface temperatures under relatively fast laser scan conditions for substrates laser-annealed while held at room temperature.

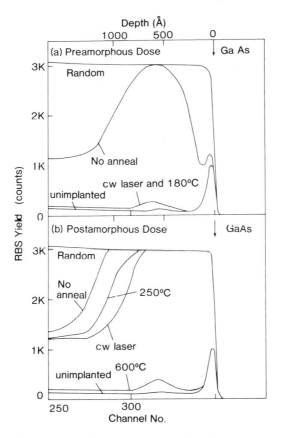

Fig. 10. Channeling spectra indicating the comparison between cw laser and furnace annealing of ion implant damage for (a) 3×10^{13} Ar cm^{-2} and (b) 3×10^{14} Ar cm^{-2} implant doses (100 keV). Laser dwell time, 15 msec. (After Williams *et al.*, 1980.) This figure was originally presented at the Fall 1977 Meeting of The Electrochemical Society, Inc., held in Atlanta, Georgia.

In contrast to the above studies, the work of Fan *et al.* (1979a,b) and Olson *et al.* (1980) utilized elevated temperature substrates (500–580°C). Both these studies used cw Nd:YAG lasers and employed slow scan rates (<0.5 mm sec^{-1}), in addition to the heated substrates, to further reduce local thermal gradients and ensure the attainment of higher surface temperatures before the initiation of surface slip. Significantly, the heated target stage without the additional laser heating was sufficient to remove the amorphous implant damage and significantly reduce the density of crystal defects. The high electrical activation achieved by Fan *et al.* (1979b) and the improved mobilities reported by Olson *et al.* (1980) suggest that the laser annealing conditions provide surface temperatures sufficient to remove residual crystal defects and thus produce electrical characteristics, for medium-dose implants, approaching those attainable from optimum furnace annealing. However, the significant differences obtained for the magnitude of electrical activity by Fan *et al.* (1979b) and Olson *et al.* (1980) may result from subtle differences in anneal conditions (e.g., multiple- or single-laser scanning). Additionally, Olson *et al.* (1980) reported that low-dose *n*-type implants ($<10^{13}$ cm^{-2}) could not be activated under their particular cw annealing conditions.

Surface decomposition and oxidation are significant effects that must also be considered during capless cw annealing of GaAs. Both Olson *et al.* (1980) and Fan *et al.* (1979b) employed forming gas ambients during cw annealing and under these conditions did not observe significant decomposition or oxidation. However, Nissim and Gibbons (1981) reported appreciable surface oxidation for slow cw laser scanning of GaAs in an air ambient. The conditions under which these authors observed annealing and oxide growth are illustrated in Fig. 11, which plots the calculated surface temperature as a function of laser power. Optimum annealing

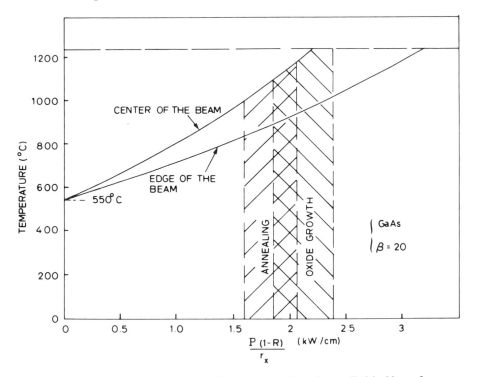

Fig. 11. Temperature induced by cw Ar ion laser on GaAs for a cylindrical lens of aspect ratio $\beta = 20$ plotted as a function of absorbed power $P(1 - R)$ over effective spot radius r_x. (After Nissim and Gibbons, 1981.)

conditions (i.e., best activation of dopants) could not be achieved unless surface temperatures approaching 1200°C were attained, and, under such conditions, a β-Ga_2O_3 layer was grown on the surface during laser scanning. It was found that the formation of oxide prevented surface slip. Furthermore, appreciable activation of low-dose Si^+-implanted (100) GaAs was achieved in the surface region below the oxide, but at the price of low near-surface mobility.

In light of the various results outlined above, it is possible to piece together the annealing behavior of implanted GaAs under a number of cw laser annealing conditions. Figure 12 summarizes the annealing behavior in schematic form. Figure 12a depicts the as-implanted damage structure consisting, for example, of part amorphous, part crystalline disorder. Figures 12b and c illustrate the most probable disorder structure following fast laser scanning, where the local laser dwell time (i.e., hot time) is significantly less than 1 sec. Here, the optimum power (just below the threshold surface damage) does not produce a high enough temperature to remove crystalline defects totally, as in Fig. 12b, whereas higher laser powers result in excessive thermal gradients, hence surface slip, as illustrated in Fig. 12c. Figures 12d and e illustrate slow and/or multiple scanning at lower laser or electron beam power, combined with a heated stage

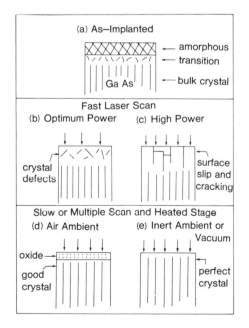

Fig. 12. Schematic illustrating the behavior of cw laser annealing under various conditions.

to minimize thermal gradients. Annealing under such conditions, where the laser dwell time (i.e., local hot time) may be on the order of seconds, and employing an inert or vacuum environment (Fig. 12e) appears to produce optimum results.

In terms of electrical characteristics, the most impressive GaAs annealing results have been obtained by the electron beam scanning technique of Shah *et al.* (1980). These authors utilized a rapid raster scan of a focused electron beam to heat the entire bulk wafer (in vacuum) for a time on the order of 1–5 sec. The more uniform heating obtained using this method

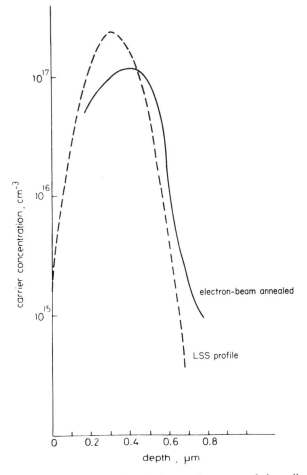

Fig. 13. Carrier concentration profile of electron beam-annealed semiinsulating GaAs implanted with silicon (6×10^{12} cm^{-2}, 360 keV), and LSS range profile. (After Shah *et al.*, 1980.)

allowed surface temperatures close to the melting point to be reached without observable surface slip or other forms of surface degradation. The electrical data for scanned electron beam annealing of low-dose Si^+-implanted GaAs are shown in Fig. 13. It can be observed that high activation (50–60%) of the 6×10^{12} cm^{-2} implant in Cr-doped semiinsulating GaAs has been achieved with little movement of the silicon profile from the expected LSS profile. The measured Hall mobility was 3800 cm^2/V sec which compares favorably with predictions and with the best values obtained via conventional furnace-annealed GaAs using encapsulating layers (see Donnelly, 1977, and references therein). It is interesting that Shah *et al.* (1980) did not observe any significant surface decomposition for 5-sec annealing in vacuum, where the substrate temperatures attained in this time must have exceeded 900°C (to account for the excellent electrical characteristics). Indeed, it appears that the electron beam annealing of Shah *et al.* (1980) was akin to an ultrarapid furnace anneal, whereby the bulk wafer attained high temperatures for short enough times such that no surface decomposition or degradation could occur. The multiple-scan laser annealing method of Fan *et al.* (1979a,b), which also produces high electrical activity with implanted GaAs, may well provide a somewhat similar thermal history.

TABLE I

SOLID STATE ANNEALING: COMPARISON OF THE OBSERVATIONS FOR FURNACE-ANNEALED GaAs WITH THE EXPECTED BEHAVIOR DURING TRANSIENT ANNEALING UNDER VARIOUS CONDITIONS

Furnace temperature		100–300°C	300–600°C	600–950°C
Optimum transient anneals	(i)	Laser ≪ 1-sec dwell		
		Room temperature stage		
	(ii)		Laser ≪ 1-sec dwell	
			Heated stage	
	(iii)		Laser, 1- to 5-sec dwell, heated stage (inert ambient)	
			Electron beam, 1- to 5-sec dwell, raster (vacuum)	
Observed furnace anneal behavior		Amorphous-to-crystalline transition; generally, remnant crystal defects consisting of twins, stacking faults, etc.; poor electrical activation and mobility	Removal of dense defects (twins, poly, stacking faults), leaving loops; channeling suggests a return to good crystal; some activity and improved mobility	Gradual removal of dislocation loops and other remaining defects; best electrical activity and restored mobility to near-bulk value
		(Stage 1)	(Stage 2)	(Stage 3)

Finally, it is interesting to compare the solid phase transient annealing behavior with the better characterized furnace-annealing behavior. For example, in Table I, the observations obtained from TEM, channeling, and electrical measurements are summarized in terms of the three major furnace annealing stages (see Section II.B). The optimum anneal behavior (avoiding surface slip and decomposition) for the various transient annealing situations is correlated with the observed furnace behavior. As shown, the long (1–5 sec) transient dwell time conditions, where precautions have been taken to minimize thermal stress (e.g., heated stage or raster scan for uniform heating), produce results that appear to be similar to those of a high-temperature (950°C) conventional furnace anneal. The major advantage of transient annealing is that the GaAs surface does not need to be protected against decomposition.

III. Ion-Implanted Layers: Liquid Phase Regrowth

A. Introduction

Following the pioneering work of Kachurin *et al.* (1976), pulsed laser annealing of ion-implanted GaAs received considerable attention from several laboratories (Golovchenko and Venkatesan, 1978; Sealy *et al.*, 1978; Campisano *et al.*, 1978; Barnes *et al.*, 1978). These early studies indicated that uncapped wafers could be laser-processed in air at room temperature to provide good crystallinity, impressively high substitutionality of implanted ions, and electrical activities superior to those achievable by conventional furnace annealing. The first reports on pulsed electron beam annealing of GaAs (Tandon and Eisen, 1979) indicated very similar behavior and gave impressive electrical activities for medium-dose implants. However, most of these reports foreshadowed possible problem areas and indicated certain behavior not entirely consistent with the liquid phase regrowth data from pulse-annealed Si.

Some of the significant details revealed in early studies included the following observations. Decomposition of the GaAs surface and significant nonstoichiometry were observed under certain conditions (Barnes *et al.*, 1978), and some authors suggested the use of encapsulants during pulsed annealing (Sealy *et al.*, 1979). Tandon and Eisen (1979) noted that amorphous damage appeared to anneal more satisfactorily than layers that were not amorphized during the implant, and some reports (Campisano *et al.*, 1978; Liu *et al.*, 1979) indicated no significant dopant redistribution during pulsed annealing; these latter observations were apparently in some conflict with the expected liquid phase regrowth processes identified for Si (see Chapters 2 and 5). Furthermore, measured carrier mobilities were anomalously low following pulsed annealing (Sealy *et al.*, 1979; Tandon and Eisen, 1979; Gamo *et al.*, 1979), and difficulty was experienced in activating low-dose *n*-type implants (Liu *et al.*, 1979; Tandon and Eisen, 1979). Some authors (Barnes *et al.*, 1979) observed nonuniform annealing resulting from interference effects when using pulsed Nd:YAG lasers operating at 1.06 μm and suggested that laser wavelength may be an important parameter influencing the annealing of implant damage in GaAs. In this section, all these aspects are further considered in the light of more recent data that have provided considerable insight into pulsed annealing behavior in GaAs.

B. Pulsed Annealing Mechanisms and Damage Removal in GaAs

In contrast to the detailed studies on the annealing mechanism for pulsed laser and electron beam irradiation of Si (Chapters 2, 3, and 5), where liquid phase epitaxy was unequivocally identified as the controlling process, fewer data are available for an appraisal of the detailed processes involved in pulsed annealing of GaAs. However, results clearly point toward liquid phase epitaxy as the operative annealing mechanism. Time-resolved reflectivity measurements during pulsed ruby laser annealing (Venkatesan *et al.*, 1979) suggested the formation of a thin surface layer of the little studied metallic liquid phase of GaAs. This layer was found to last up to several hundred nanoseconds, depending on the annealing pulse energy, in a manner analogous to the process observed for pulsed annealing of Si (Auston *et al.*, 1978).

Less direct evidence for melting phenomena has been obtained by Campisano *et al.* (1980) using correlated channeling and TEM analysis of pulsed ruby laser-annealed GaAs. Typical channeling data are shown in Fig. 14 for GaAs samples initially implanted with 400-keV Te ions to form an amorphous surface layer ~2300 Å. Curve a is representative of the initial implanted sample and samples annealed with pulse energies lying between 0.1 and 0.8 J cm^{-2}. At a laser pulse energy of 0.9 J cm^{-2}, the channeling spectrum (curve b) indicates that part of the amorphous layer has been converted to single-crystal epitaxially on the underlying material. Curve c indicates that laser power levels above 1.0 J cm^{-2} essentially convert the implanted surface layer to near-perfect single-crystal material, as determined by channeling. The TEM measurements of Campisano *et al.* (1980) for samples irradiated at power levels of 0.2–1.4 J cm^{-2} have already been illustrated and discussed in Chapter 6 (see Fig. 18 in Chapter

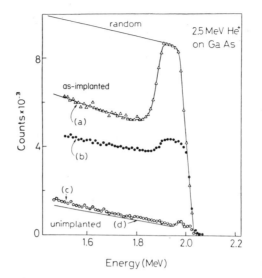

Fig. 14. Backscattering spectra of 2.5-MeV He ions incident in a random direction or in the ⟨100⟩ direction of GaAs samples implanted with 400-keV Te (10^{15} cm^{-2}) after ruby laser irradiation with energy. (a) 0.2–0.8 J cm^{-2}, (b) 0.9 J cm^{-2}, (c) 1.0–1.4 J cm^{-2}. Curve d is obtained from an unimplanted GaAs sample. (After Campisano *et al.*, 1980.)

6). Essentially, polycrystalline material is formed, with the grain size increasing with laser power, for laser energies up to 0.8 J cm^{-2}. At 1.0 J cm^{-2} single-crystal material is formed with stacking faults and dislocations within a surface layer which corresponds closely in thickness to the original amorphous layer. Above 1.0 J cm^{-2}, TEM micrographs indicate near-perfect single-crystal recovery, except for near-surface damage, as determined by stereomicroscopy. A typical microstructure is summarized as a function of laser energy density in Fig. 15. These data clearly indicate a threshold energy density of about 1.0 J cm^{-2} for the transition from polycrystalline to single-crystal recovery. Thus, the work of Campisano *et al.* (1980) is consistent with a process of rapid surface melting and resolidification for pulsed ruby irradiation of implanted GaAs. Polycrystalline recovery results when the melt depth does not penetrate into the underlying single crystal, and liquid phase epitaxy produces near-perfect single-crystal recovery when the melt depth penetrates into the underlying crystal.

The work of Barnes *et al.* (1978) employed pulsed Nd:YAG (1.06 μm) laser irradiation of pulse width 125 nsec to anneal Te-implanted GaAs. The laser conditions differed from the single-zap, large-spot ruby laser annealing of Campisano *et al.* (1980) in that the Nd:YAG laser provided a focused spot of 35-μm diameter which was repetitively pulsed at 11 kHz

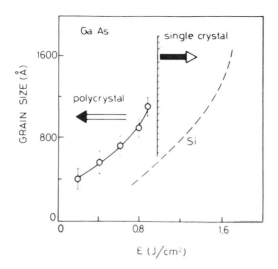

Fig. 15. Average grain size obtained from TEM micrographs of the polycrystalline GaAs layer formed by transient annealing with a ruby pulse of varying energy density. Prior to laser irradiation, the ⟨100⟩ GaAs substrate was implanted with 400-keV Te at room temperature to a dose of 1×10^{15} cm^{-2}. The bars represent the spread in the diameter of the grains. (After Campisano *et al.*, 1980.)

while the spot was translated to provide an 8-μm overlap between adjacent pulses. For 1.06-μm irradiation, the nonimplanted areas of the GaAs were essentially transparent; however, in the amorphous implanted areas, the absorption of laser light was high enough to cause recrystallization of the GaAs. Typical results from the work of Barnes *et al.* (1978) are shown in Fig. 16 for 50-keV Te-implanted GaAs to a dose of 10^{16} ions cm^{-2}. Perfect recrystallization, as measured by channeling, was achieved for an energy density of 2.2 J cm^{-2}. It is interesting that the threshold power level is significantly higher for Nd:YAG irradiation, compared to pulsed ruby laser annealing, since the damaged GaAs layer is still quite transparent to the 1.06-μm irradiation. This partial transparency can result in interference effects, on the implanted front surface between the incident radiation and the radiation reflected from the polished back surface. However, despite the differences between the pulsed ruby and the pulsed Nd:YAG annealing conditions, both processes essentially produce similar-quality annealing. Furthermore, the quality of the recrystallization of a 10^{16} Te cm^{-2} dose, shown in Fig. 16, is far superior to that achieved by conventional furnace annealing (Brawn and Grant, 1976). This result, together with the high Te substitutionality and redistribution, as indicated by the Te portion of the spectra in Fig. 16, suggests that the GaAs recrystallizes by liquid phase epitaxy in pulsed Nd:YAG irradiation.

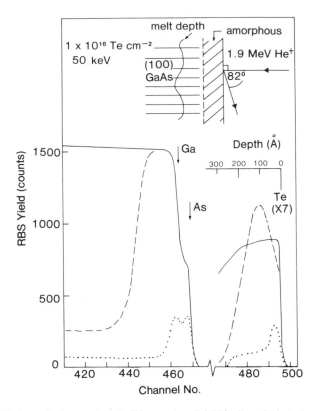

Fig. 16. High-resolution random (solid curve) and ⟨100⟩ aligned (dotted curve) spectra following pulsed Nd:YAG laser annealing of 1×10^{16} Te cm^{-2}-implanted GaAs (50 keV). The dashed curve refers to the aligned spectrum before laser annealing. (From Barnes *et al.*, 1979.)

The work of Tandon and Eisen (1979), Tandon *et al.* (1979), and Golecki *et al.* (1980) on pulsed electron beam annealing of implanted GaAs indicated recrystallization and structural details essentially similar to those described above for pulsed laser annealing. Hence these authors concluded that pulsed electron beam annealing could also be interpreted as occurring via liquid phase regrowth. Certain significant details were revealed in these studies, which are worthy of consideration here. First, above the threshold electron beam energy for single-crystal recovery of the implant layers (0.4 J cm^{-2}), a rather narrow fluence window was observed (between 0.4 and 0.7 J cm^{-2}) within which good recrystallization was observed (Tandon *et al.*, 1979). Above electron beam energies of 0.7 J cm^{-2}, channeling minimum yields were observed to increase to above 0.11 (compared with a virgin GaAs value of 0.04 for 2.4-MeV He$^+$ analysis), and TEM indicated the existence of dislocation lines and other defect patches in the near-surface region. Furthermore, the surfaces of the irradiated samples were found to be Ga-rich. Second, low-dose Se^+ (3×10^{12} cm^{-2}), which produced little damage following implantation (as determined by RBS and TEM), was found to exhibit considerable damage (dense dislocation network and surface patches indicative of Ga-rich regions) following pulsed electron beam annealing in the energy range 0.4– 1.1 J cm^{-2}. It was suggested that this behavior (Tandon et al., 1979) may arise from high thermal stresses produced by rapid temperature changes. Finally, more electron beam-induced damage was produced, at about 0.7 J cm^{-2}, in the high-dose amorphous case than in the low-dose Se implants, and this behavior was explained in terms of expected differences in the thermal properties of amorphous and crystalline GaAs.

These observations, which suggest a narrow energy window for annealing and demonstrate differences between the ability of electron beams to remove amorphous and crystalline damage, are also relevant to laser annealing. For example, Fletcher et al. (1981) observed a similar narrow annealing window for pulsed ruby laser annealing of Se-, Zn-, and Mg-implanted GaAs. The typical implants examined by these authors were not completely amorphous prior to laser annealing, as shown previously in the TEM micrographs in Fig. 2. Following laser annealing, a gradual reduction in implant damage was found as the laser power was increased above 0.2 J cm^{-2}. The progression of the near-defect-free region deeper into the bulk material with increasing laser power (leaving the underlying defect structure unaffected) was consistent with a molten surface layer of increasing thickness. Typical results are shown in the cross-section TEM micrographs (Fletcher et al., 1981) in Figs. 17 and 18. The damage in the implanted layer is clearly shown to decrease with increasing laser power, but the appearance of significant near-surface damage is evident after the high-power irradiation. The detailed TEM study by Fletcher et al. (1981) further revealed that the near-surface damage increased gradually with laser powers above 0.25 J cm^{-2}. This damage was attributed to As evaporation and, significantly, no surface damage was apparent for power levels up to 0.9 J cm^{-2} after the removal of surface Ga by chemical means. These surface damage and decomposition phenomena are discussed more fully in the following section.

C. Surface Decomposition and Damage

The first reports of surface decomposition during pulsed laser annealing of GaAs were made by Barnes et al. (1978). Typical results from this work

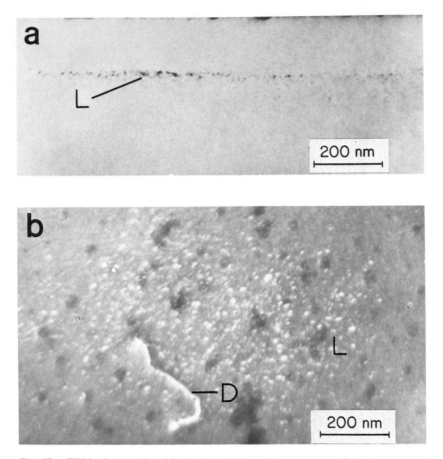

Fig. 17. TEM micrographs of Se-implanted GaAs following pulsed ruby laser annealing at an energy density of 0.36 J cm^{-2}. (a) Cross section, showing remnant defects; (b) plan view micrograph showing loops (L) and dislocations (D). (After Fletcher *et al.*, 1981.)

are illustrated in Fig. 19 which shows channeling spectra for Te-implanted and laser-annealed samples. The hatched area indicates a surface peak containing an excess of approximately 10^{16} Ga cm^{-2} for Nd:YAG laser-annealed GaAs. The surfaces of such annealed samples were analyzed by scanning electron microscopy (SEM), as illustrated in Fig. 20 in Chapter 6. A topographically rough and patchy surface was observed. These irregular surface regions were identified, by x-ray fluorescence measurements within the SEM, as elemental Ga that could be easily removed by etching the surface with warm HCl. Following etching, the Ga surface peak in the channeling spectrum is greatly diminished, as illustrated in

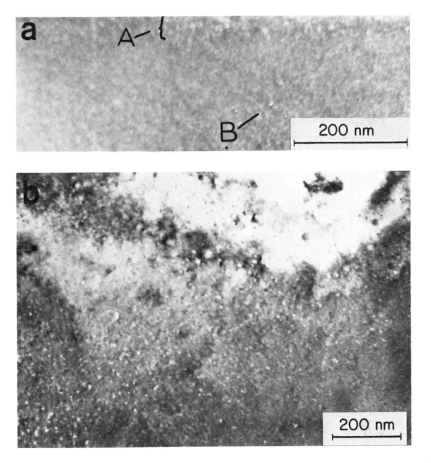

Fig. 18. TEM cross-section micrograph (a) and plan view micrograph (b), indicating surface defects (A) and original implant-induced defects (B) following laser annealing at a high laser power of 0.75 J cm⁻². (After Fletcher *et al.*, 1981.)

Fig. 19, with a corresponding reduction in the channeling minimum yield to a value (4.4%) close to that of virgin (100) GaAs. These results, indicating surface decomposition for Nd:YAG laser irradiation, correlate well with the pulsed ruby anneal behavior described by Fletcher *et al.* (1981), as illustrated by the surface damage in the TEM micrograph in Fig. 18.

A more detailed characterization of decomposition phenomena has been attempted by several authors. For example, Gamo *et al.* (1979, 1980) directly measured the decomposition of the GaAs surface for 30-nsec ruby laser irradiation by placing a quartz plate in close proximity to the laser-irradiated surface and using RBS to measure the amount of Ga and As

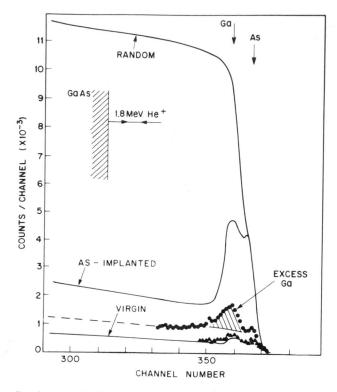

Fig. 19. Random and ⟨100⟩-aligned RBS spectra from Te-implanted (100) GaAs which was Nd:YAG laser-annealed. Spectra taken before (●) and after (▲) etching in warm HCl. (After Barnes *et al.*, 1978.)

deposited on the plate. For 0.63-J cm^{-2} laser energy, 0.5 monolayer was measured, rising to 7 monolayers (7×10^{15} atoms cm^{-2}) for a laser power of 0.83 J cm^{-2} (Gamo *et al.*, 1980). This increase in the magnitude of surface decomposition could be correlated with laser-induced damage, as measured by channeling. The results of Badawi *et al.* (1980), obtained from weight loss measurements following 25-nsec ruby irradiation, also indicated considerable surface evaporation above laser energies of 0.35 J cm^{-2}; however, the excessive loss reported by these authors (exceeding many hundreds of angstroms) is in marked disagreement with other reports (Gamo *et al.*, 1980; Barnes *et al.*, 1978; Venkatesan *et al.*, 1979). Indeed, the work of Venkatesan *et al.* (1979) suggests that decomposition effects can be controlled by optimization of the laser annealing conditions. These authors established that the magnitude of decomposition correlated well with the melt duration. Furthermore, the application of short-pulse-

length lasers (<20 nsec), operating at power levels just sufficient to melt
the ion implant-damaged surface layers, did not produce significant dissociation of the surface. In contrast, other authors (Sealy *et al.*, 1978;
Inada *et al.*, 1979b) suggest that surface decomposition is adequately controlled only when an encapsulating layer such as Si_3N_4 or SiO_2 is used to
protect the surface during pulsed laser irradiation. As we illustrate in
Section III.E, capping can result in better electrical activation following
pulsed annealing. Finally, dissociation effects are not confined to pulsed
laser-induced melting of GaAs but have also been reported by Tandon *et
al.* (1979) for pulsed electron beam annealing.

D. Implant Redistribution and Solubility

In Chapters 4 and 5, the influence of rapid melting and resolidification
processes on impurity redistribution and subsequent solid solubility has
been examined in considerable detail for pulsed energy deposition in Si.
Such phenomena have not been particularly well characterized for pulsed
annealing of GaAs, but the available data indicate that impurity-related
effects are basically similar in Si and GaAs. Early reports noted that
equilibrium solid solubility limits were exceeded following pulsed annealing of implanted GaAs (Golovchenko and Venkatesan, 1978; Barnes *et al.*,
1978). For example, the high-resolution RBS and channeling measurements of Te profiles previously shown in Fig. 16 indicate that >90% of the
Te atoms, for a 10^{16} Te cm^{-2} implant, reside on Ga or As lattice sites
following pulsed annealing. This result implies that the solid solubility is
$>10^{21}$ Te cm^{-3}, which exceeds the equilibrium value and previous values
following furnace annealing (Brawn and Grant, 1976) by more than an
order of magnitude. Furthermore, a comparison of the as-implanted
(dashed curve) and laser-annealed (solid curve) Te profiles in Fig. 16
indicates appreciable redistribution of Te during Nd:YAG laser annealing
at a pulse length of 125 nsec. This latter result is consistent with the
behavior of dopants in Si. However, redistribution of impurities during
pulsed annealing of GaAs has not always been observed, and impurity
redistribution is examined more carefully below.

Whereas certain early reports (Campisano *et al.*, 1978; Liu *et al.*, 1979;
Golovchenko and Venkatesan, 1978) noted insignificant impurity redistribution during pulsed annealing of GaAs, other studies suggested that
appreciable impurity movement could be observed under particular
pulsed annealing conditions (Sealy *et al.*, 1979; Barnes *et al.*, 1979). Furthermore, Sealy *et al.* (1979) reported the diffusion of some impurities
(e.g., Sn, Se, Te) and not others (e.g., Cd, Zn, Si, Ge) for 30-nsec pulsed

ruby laser annealing, whereas Liu *et al.* (1980) demonstrated that the observed redistribution of Si in GaAs was influenced dramatically by implant dose and the laser (pulsed ruby) annealing parameters. These latter observations point to a plausible explanation of the apparent conflict in the literature. If redistribution processes in GaAs are indeed controlled by the extent and duration of the near-surface molten layer (as has been demonstrated to be the case for the annealing of Si in Chapters 4 and 5), then the extent of redistribution would be expected to be quite sensitive to the pulsed annealing parameters on the one hand and the initial disorder structure (which influences the melt threshold) on the other. Since the surface decomposition processes and resultant surface damage become significant at higher laser power levels, many of these pulsed annealing studies on GaAs have concentrated on optimized laser conditions just sufficient to remove the near-surface implant damage. Under such conditions the melt would be essentially confined to the near surface, and little redistribution would be expected.

Examples of the observed redistribution effects are shown in Figs. 20 and 21. Figure 20, from Barnes *et al.* (1979), illustrates that the redistribution of Te-implanted GaAs is enhanced as the laser power (Nd:YAG, 125 nsec) is increased. For example, at a laser power level of 19 MW cm^{-2} (2.5 J cm^{-2} incident energy density) the Te profile had broadened (FWHM) from 200 to 800 Å. In Fig. 21, from Liu *et al.* (1980), SIMS profiles of 1×10^{16} Si cm^{-2} implants in GaAs are given, comparing the redistribution obtained for thermally and pulsed ruby laser-annealed samples. Whereas furnace annealing at 850°C for 20 min resulted in a slight broadening in the tail of the Si distribution, laser annealing at the high power of 2.3 J cm^{-2} produced significant Si redistribution typical of that expected for diffusion in liquid GaAs. However, Liu *et al.* (1980) did not observe any appreciable Si profile broadening for laser energy densities <1 J cm^{-2}, although such laser conditions were sufficient to remove implant damage for doses of $<3 \times 10^{15}$ Si cm^{-2}.

Only recently (Wood *et al.*, 1981) have attempts been made to compare experimentally measured melt depths in GaAs with theory based on heat flow calculations. This study used time-resolved reflectivity to provide temperature–time profiles of the melt, and at low-power levels (0.2 J cm^{-2}) the estimated melt depth (~200 Å) was in agreement with theory. The theoretical calculations are more difficult in GaAs than in Si, since little is known about the thermodynamic parameters for liquid GaAs. Furthermore, Wood *et al.* (1981) measured the redistribution and segregation of impurities Zn, Mg, and Se in GaAs as a function of laser power. The experimental results were in qualitative agreement with theory, and diffusion and segregation coefficients were obtained for impurities in mol-

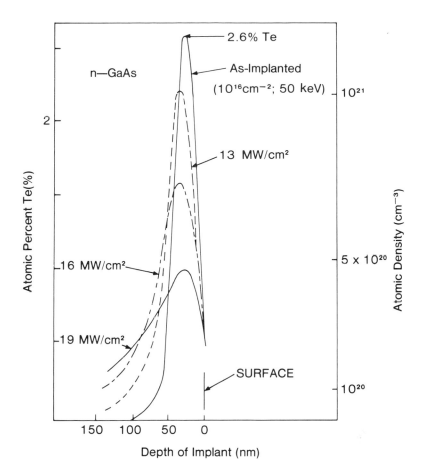

Fig. 20. Te concentration profiles as a function of Nd:YAG laser power density. (From Barnes *et al.*, 1979.)

ten GaAs. These preliminary measurements indicated lower liquid phase diffusivities in GaAs compared with Si, which may account for less significant profile broadening observed following pulsed annealing for impurities in GaAs compared with Si (see Chapter 5).

The SIMS measurements of Wood *et al.* (1981) indicated segregation of Mg at the GaAs surface following pulsed ruby annealing, and these authors suggested that the process was similar to zone migration effects readily observed in pulsed annealing of low-solubility impurities in Si (Chapter 5). Previous studies have also reported zone migration effects in pulse-annealed GaAs (Harrison and Williams, 1980); the observed behavior for pulsed ruby laser (50 nsec) annealing of In-implanted GaAs is

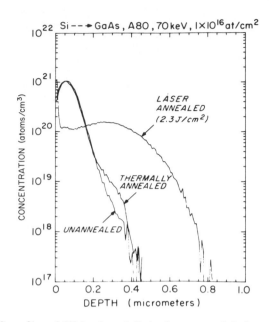

Fig. 21. SIMS profiles of Si⁺-implanted GaAs for unannealed, furnace-annealed, and ruby laser-annealed samples. (After Liu *et al.*, 1980.)

illustrated in Fig. 22. The high-resolution RBS profiles of In distributions clearly show a segregation of In at the GaAs surface following pulsed ruby laser annealing at 0.3 J cm⁻², an energy density sufficient to remove the implant damage. It is interesting to note from Fig. 22 that the In left behind in the bulk GaAs is highly substitutional. This behavior is essentially similar to the observations made for low-solubility dopants in Si (Chapter 5). However, no detailed studies have been undertaken to establish the solid solubility limits that can be achieved by pulsed annealing of GaAs, nor have measurements been made to determine the dependence of solute trapping and segregation processes (e.g., cellular structures) on the velocity of the liquid–solid interface during rapid resolidification (see Chapters 4–6 for such details in Si).

Despite the general lack of data on solid solubility of impurities in GaAs following pulsed annealing, some interesting observations have been made. First, Barnes *et al.* (1978) reported that the measured substitutionality of Te in GaAs as determined by channeling (see Fig. 16), did not correlate with the measured electrical activity. Indeed, less than 20% of the substitutional Te atoms appeared to be electrically active. A suggested explanation for this effect (Barnes *et al.*, 1980) is that short-duration pulsed annealing may quench a significant concentration of impurity

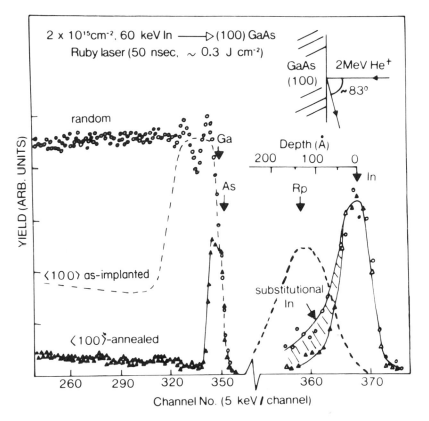

Fig. 22. High-resolution channeling spectra from 60-keV In⁺-implanted GaAs before and after pulsed ruby laser annealing, indicating surface segregation. (After Harrison and Williams, 1980.)

atoms onto the antisite (e.g., Te on a Ga site). Other reports for high-dose implants indicate only 10–30% activation of dopants (Sealy *et al.* 1979; Tandon and Eisen, 1979), and the explanation here may similarly relate to details of the rapid quenching process. However, incomplete activation of dopants is typical of thermally annealed high-dose implants, so that the effect may well have a more basic material-related explanation than one relating to details of the rapid quenching process. Such explanations are discussed more fully in Section III.E.

Before reviewing the electrical measurements on pulsed-annealed GaAs, a further important observation made by Pianetta *et al.* (1980) is worthy of comment. These authors subjected pulsed ruby laser-annealed GaAs to thermal annealing treatments and, for Te-implanted GaAs, measured a substantial drop in the electron concentration as the furnace temperature was increased. Typical data from this work are shown in Fig. 23. Isochronal anneal studies indicate a sharp drop in carrier concentration at 300°C followed by a more gradual drop in the temperature range 600–800°C. At first sight, such data appear to indicate that Te precipitates during the subsequent thermal processing. However, detailed channeling measurements shown in Fig. 24, from Pianetta *et al.* (1981), indicate that no change in Te substitutionality or detailed atom position is observed for annealing up to 460°C. Furthermore, for annealing at 850°C, no large-scale precipitation is observed: a slight narrowing in the angle of the Te channeling dip indicates, at most, only a small displacement of Te atoms from lattice sites. In contrast to measurements in Si, where the electrical activity is found to correlate well with substitutional solid solubility (see Chapter 5), the processes controlling measured electrical activity are more complex in GaAs: possible factors influencing the electrical properties are discussed in the next section.

E. Electrical Characterization

As mentioned earlier, the major thrust of pulsed annealing experiments in GaAs has been directed toward an electrical assessment of ion-implanted samples. Thus, considerable data exist for both n- and p-type dopants in GaAs, and clear trends are now apparent. However, particular

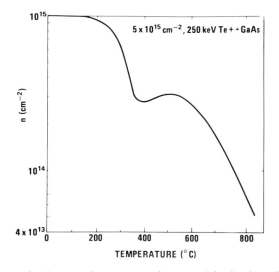

Fig. 23. Changes in sheet carrier concentration caused by isochronal annealing after laser annealing. (After Pianetta *et al.*, 1980.)

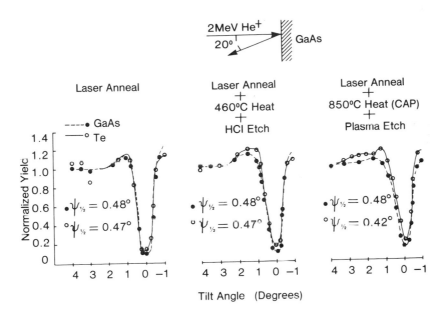

Fig. 24. Angular scans of 2-MeV He⁺ backscattering yield from GaAs and Te along the ⟨100⟩ axis. The samples were laser annealed and heat treated at 460°C and 850°C. (After Pianetta *et al.*, 1981.)

observations are not yet well understood. These include the lack of complete activation of high-dose implants, the inability to activate low-dose *n*-type dopants, and the low carrier mobilities generally observed following both pulsed laser and pulsed electron beam annealing. This section briefly reviews the important electrical characteristics of pulse-annealed GaAs and outlines recent attempts to understand better the processes responsible for the observed electrical behavior.

Figure 25, from Sealy *et al.* (1978), illustrates the typical electrical characteristics achieved from optimized pulse annealing of high-dose *n*-type implants in GaAs. Electron concentrations exceeding 10^{19} cm⁻³ have been obtained for both Te and Se implants using a single ruby laser pulse on uncapped samples. The electron concentration profiles in Fig. 25 extend to depths greater than the range of the implanted ions and, as outlined in Section III.D, this suggests the occurrence of significant diffusion of the implanted species within the molten surface layer during laser annealing at an energy density of 1.3 J cm⁻². The Hall mobilities shown in Fig. 25 are below 10^3 cm²/V sec and much lower than would be expected (1500 cm²/V sec from Sze, 1969) in uncompensated material with electron concentrations exceeding 10^{19} cm⁻³.

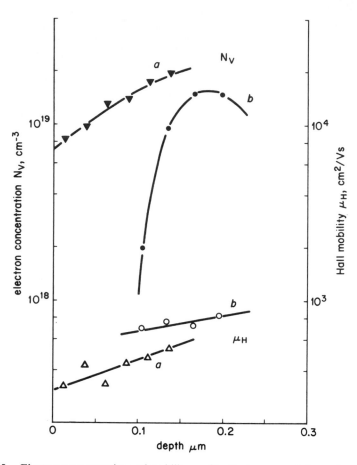

Fig. 25. Electron concentration and mobility profiles for two uncoated samples annealed with a 1.3-J cm⁻² ruby laser pulse. Sample a was implanted at room temperature with 5×10^{15} cm⁻² 100-keV Te ions, and sample b was implanted at 200°C with 1×10^{15} cm⁻² of 200-keV Se ions. (After Sealy *et al.*, 1978.)

In Figure 26, from Liu *et al.* (1980), the improved activation obtained from laser annealing, as compared with conventional furnace annealing, is demonstrated for high-dose Si⁺ implants in GaAs. For Si⁺ implant doses exceeding 5×10^{14} cm⁻², pulsed ruby laser annealing is shown to provide an order-of-magnitude improvement in the sheet electron concentration compared with similar implants subjected to furnace annealing. This behavior has been demonstrated for *n*-type dopants by several groups using pulsed ruby lasers (Tandon and Eisen, 1979; Sealy *et al.*, 1979), pulsed Nd:YAG lasers (Barnes *et al.*, 1978), and pulsed electron beams (Tandon

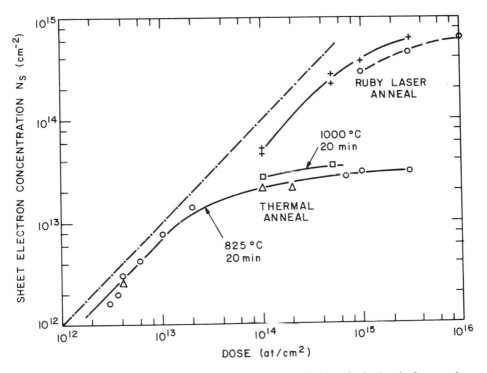

Fig. 26. Sheet electron concentration as a function of Si⁺ dose for both ruby laser and thermally annealed samples. Implant energy: —, 200 keV; – – –, 70 keV. (After Liu *et al.*, 1980.)

and Eisen, 1979; Inada *et al.*, 1979b). Results from Inada *et al.* (1979b) on electron beam annealing of Se-implanted GaAs are shown in Fig. 27. It is interesting to note that the electrical profiles following electron beam annealing are independent of whether the implant was carried out at room temperature, where an amorphous layer would be formed, or at an elevated temperature. In contrast, when furnace annealing is employed, the best electrical activities are obtained for elevated temperature implants (Eisen, 1978). The optimum furnace annealing results are also included in Fig. 27 and show significantly lower activation than the pulse-annealed samples. The widths of the carrier profiles obtained from both pulsed and furnace annealing indicate, rather fortuitously, a similar redistribution of implant into the bulk as compared with the LSS profile (dotted line).

Pulsed annealing of high-dose *p*-type dopants in GaAs results in similarly high hole concentrations, as indicated in Fig. 28. In this example, from the work of Inada *et al.* (1979a), Nd:YAG laser annealing of GaAs implanted with 1×10^{15} Zn⁺ cm⁻² has resulted in a hole concentration

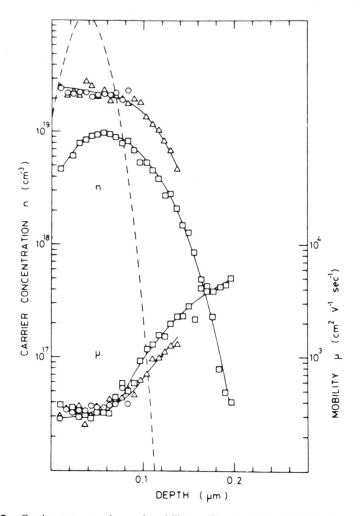

Fig. 27. Carrier concentration and mobility profiles for 100-keV Se⁺-implanted GaAs samples, comparing electron beam-annealed samples (1.2 J cm⁻²) for two implant temperatures with thermally annealed samples (950°C for 60 min). All samples were encapsulated with Si_3N_4 during annealing. Electron beam annealing: O, room temperature implant; △, 400°C implant. Thermal annealing: □, 400°C implant. (After Inada *et al.*, 1979b.)

exceeding 10^{19} cm⁻³. Significant dopant redistribution is evident (in comparison with LSS profiles) for an energy density of 1.5 J cm⁻², but a lower energy density (0.94 J cm⁻²) used to anneal a 1×10^{14} Zn⁺ cm⁻² implant has not produced any marked dopant redistribution. In contrast, thermal annealing at 950°C for 1 min (Inada *et al.*, 1979a) provided lower peak

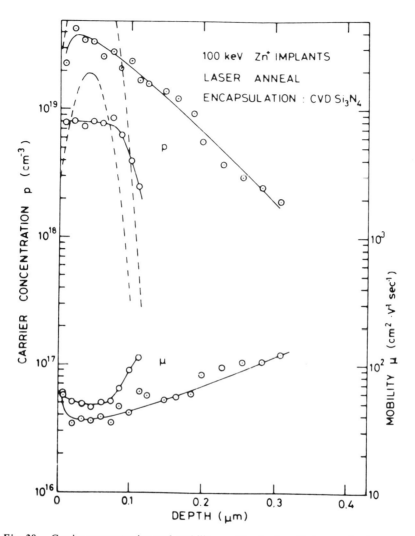

Fig. 28. Carrier concentration and mobility profiles for two GaAs samples implanted at room temperature with a Zn dose of 1×10^{14} (⊙) and 1×10^{15} cm^{-2} (○) and annealed by ruby laser irradiation at a beam density of 0.9 and 1.5 J cm^{-2}, respectively. (After Inada *et al.*, 1979a.)

activity and dramatic Zn redistribution to depths exceeding 1.4 μm. In addition, the mobility values indicated in Fig. 28 are considerably lower than would be expected for uncompensated material having similar hole concentrations (Sze, 1969), and this behavior is consistent with the results obtained for *n*-type dopants. It is further interesting to note that the laser anneals reported in Fig. 28 were carried out using Si_3N_4 as an encapsulating layer to protect the surface against decomposition during pulsed annealing. Several authors report superior electrical activities when encapsulants are used for pulsed annealing of both n- and p-type dopants in GaAs (Badawi *et al.*, 1979; Inada *et al.*, 1979a,b). However, in some cases (Badawi *et al.*, 1979), pulsed annealing through an encapsulating layer can result in the incorporation of capping constituents (e.g., Si) into the substrates, providing unintentional doping effects.

The inability of pulsed annealing to activate low-dose ($<10^{13}$ cm^{-2}) n-type implants in GaAs is a typical observation of many groups using both lasers (Liu *et al.*, 1980; Gamo *et al.*, 1980) and electron beams (Anderson *et al.*, 1980; Golecki *et al.*, 1980). Gamo *et al.* (1980) have established that this behavior is not attributable to the presence or otherwise of an amorphous layer and suggest quenched-in compensating defects as a possible explanation. Golecki *et al.* (1980) examined pulsed annealing, with both lasers and electron beams, of low-dose implants into different substrates, including Cr-doped material, nominally undoped GaAs, and previously amorphized or single-crystal substrates. No activation was measured in any of these substrates, and the authors suggested that the explanation may involve several material parameters such as impurities in the host material (e.g., Cr, O, or C), residual defects, or stoichiometry. Attempts to profile Cr to ascertain if this impurity was responsible for compensation effects (Golecki *et al.*, 1980) did not provide conclusive results.

It is significant that, although the pulsed laser and electron beam results in Figs. 25–28 indicate much improved electrical activity as compared with furnace annealing of high-dose implants, the activity is, at best, only 50% of the implanted dose. This lack of complete activity for high-dose implants and corresponding low electron mobilities are rather general results obtained by many groups, although some disagreement exists regarding the reported levels of activity (ranging from <1 to $\geq 50\%$). Lack of complete activity following furnace annealing has recently been attributed to precipitation of the implant species (Lidow *et al.*, 1980), and measurements of the unprecipitated fraction were found to correlate well with measured electrical activity. It would be convenient to invoke a similar explanation to interpret the laser annealing results, yet the channeling measurements and correlations with electrical activity reported in the previous section (Figs. 16, 23, and 24) suggest that dopants may well be completely soluble following optimum pulsed annealing, but not totally active. Thus, the explanation of incomplete activity in the pulsed annealing experiments may relate to trapping of dopants at antisites, as mentioned previously, or to the presence of quenched-in compensating defects resulting from the rapid resolidification process. In view of the appreciably lower mobility obtained with pulsed annealing, the latter (compensating defect) explanation may be favored. Indeed, support for this conclusion is provided by recent pulsed electron beam annealing studies by Davies *et al.* (1980a,b, 1981) on heavily doped epitaxial GaAs layers, which indicate a reduction in majority-carrier density and a severe curtailment of the carrier mobility. These authors suggest that compensating defects produced by pulsed annealing can account for incomplete activation of high-dose implants, inability to activate low-dose implants, and poor mobility. Furthermore, Davies *et al.* (1980a,b, 1981) suggest that migration of these compensating defects during subsequent furnace processing may account for the loss of activation with moderate heat treatment (see Fig. 23). However, although this explanation is appealing, the nature of the proposed compensating defects is not clear, and more detailed studies are needed to clarify the situation.

IV. GaAs Contacts and Deposited Layers

A. Ohmic Contacts

Possibly the most successful application of transient annealing in terms of GaAs device technology has been in the formation of low-resistance, reproducible ohmic contacts to GaAs. With conventional furnace-annealed contacts, several problems are encountered as a result of the long heating times needed to form the desired alloy from a molten eutectic surface layer and the underlying semiconductor: these include undesirable interdiffusion of some contact and semiconductor constituents, poor wetting between semiconductor and metal layer, formation of high-resistivity intermetallic compounds, and formation of microscopic crystallites giving rise to surface roughness and poor contact edge definition (Eckhardt, 1980). As a result, conventional furnace alloying can lead to nonuniform contacts, regions of high specific contact resistance, and a lack of contact reproducibility and reliability, which are detrimental to device performance. However, transient annealing affords the possibility of short-duration thermal cycling of precisely defined areas on the semiconductor (Barnes *et al.*, 1980) and thus allows a measure of control over the contact formation processes that is not available via furnace annealing. Three approaches for the formation of laser-annealed ohmic contacts are summarized in Fig. 29. These methods and typical results are reviewed below.

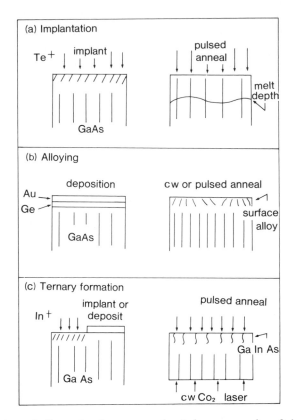

Fig. 29. Schematic illustrating three approaches to laser processing of ohmic contacts to GaAs.

In order to achieve a low-resistance ohmic contact at a metal–high-band-gap semiconductor interface, the free-carrier concentration at the interface must be sufficient to reduce the barrier width to a few tens of angstroms, hence to allow majority carriers to tunnel across the interfacial region (Padovani, 1971). High surface carrier concentrations can, in principle, be produced by ion implantation of a high dose of a suitable dopant followed by an annealing step both to remove lattice damage and to activate the dopant. However, by implantation and furnace annealing is is not possible for n-type GaAs substrates to achieve donor concentrations sufficiently high to allow direct ohmic contact with nonalloyed metal overlays. Pulsed laser (or electron beam) annealing offers a more promising method of activating high-dose n-type implants, as outlined in Section III.E, and the use of this approach for direct contact formation is illustrated in Fig. 29a.

The first report of direct ohmic contacting using a combination of high-dose Te$^+$ implantation and pulsed ruby laser annealing is attributable to Golovchenko and Venkatesan (1978). Subsequent work by Barnes et al. (1978) indicated that the ability to make direct contact to annealed GaAs was most probably a result of free surface Ga and that such contacts exhibited noisy behavior. Details of the Te$^+$ implantation and Nd:YAG annealing results from the work of Barnes et al. (1978, 1979) were discussed in Section III (see Figs. 16, 19, and 20). For contacting applications, when surface Ga was removed by HCl etching (Fig. 19) contact could no longer be made directly to the surface with steel probes, possibly as a result of preferential loss of Te. Consequently, about 50 Å of the surface was removed by rf backsputtering in Ar (2.67 Pa pressure) to provide a high concentration of Te at the surface as confirmed by high-resolution Rutherford backscattering (Barnes et al., 1980). Good electrical contact was then obtained using steel probes and, for measurement of specific contact resistance r_c a transmission line contact was defined using TiPt metallization. Reproducible contacts with $r_c = 2 \times 10^{-5}$ Ω cm^2 were obtained for 10^{16} Te$^+$ implants at 50 keV. More recent studies using short-pulse-length ruby laser irradiation (20–30 nsec) and employing laser powers just above threshold for surface melting have minimized surface decomposition and facilitated direct ohmic contacts to laser-annealed n-GaAs (Pianetta et al., 1980; Liu et al., 1980). The former study employed 5×10^{15} Se$^+$ cm^{-2} (50 keV) implants, and reproducible contacts were formed with $r_c = 5 \times 10^{-6}$ Ω cm^2, as obtained on annealed GaAs using a transmission line contact defined with Al metal. The latter study investigated $1–5 \times 10^{15}$ Si$^+$ cm^{-2} (70 keV) implants, and reproducible contacts with $r_c = 10^{-6}$ Ω cm^2 were achieved as measured from transmission line contacts defined with nonalloyed AuGe–Au metallization. Thus, the combination of implantation and laser annealing for forming reproducible nonalloyed ohmic contacts to GaAs constitutes a viable alternative to furnace annealing in GaAs integrated circuit fabrication.

The second method illustrated in Fig. 29b essentially substitues laser-induced alloying for conventional furnace alloying of deposited metal overlays. The conventional (furnace) technique of forming a high free-carrier concentration at the surface of an n-type material is to deposit a donor-type metal (e.g., Ge) along with Au and to heat the system above the Au–Ge eutectic temperature to dissolve a thin layer of GaAs into which donor metal is incorporated. To facilitate wetting, reduce balling up on resolidification, and improve reproducibility, small amounts of metals (Ni, Pt, Ti, Ag, In, etc.) are often deposited in addition to Au–Ge (Rideout, 1975). The first applications of lasers to form alloyed ohmic contacts were reported by Pounds et al. (1974), as described previously. More recently, Margalit *et al.* (1978) used a Q-switched ruby laser to anneal Ge–Au contacts on GaAs and obtained an r_c value of 7×10^{-5} Ω cm². Optimized laser and electron beam annealing to form alloyed contacts to GaAs with low r_c values has been investigated in considerable detail by several groups (Gold *et al.*, 1979; Eckhardt *et al.*, 1979; Eckhardt, 1980; Tandon *et al.*, 1980; Lee *et al.*, 1980). The pertinent results from these studies and the current understanding of the transient alloying processes are summarized below.

The general behavior is that optimized transient annealing using pulsed (free-running and Q-switched) lasers, pulsed electron beams, and cw lasers results in more reproducible contacts, improved surface morphology, and lower r_c values than can be achieved by conventional furnace processing. The improved surface morphology following laser alloying of the Au–Ge–GaAs system is illustrated in Fig. 30, from Gold *et al.* (1979). The Nomarski interference micrographs illustrate that both a free-running ruby laser (15 J cm^{-2} for 1 msec) and a single scan of an argon laser (2.5 W) result in a near-featureless surface in comparison with the furnace-alloyed surface in Fig. 30a. Gold *et al.* (1979) used Ge as the top layer of the Au–Ge deposition to improve absorption of the incident laser irradiation and to ensure that the laser power was kept below the damage threshold for exposed GaAs. Although the lowest contact resistance values of 2×10^{-6} Ω cm² were obtained, lack of reproducibility may have been attributable to the deposited layer arrangement and the limited supply of Ge.

Eckhardt and co-workers (Eckhardt *et al.*, 1979, 1980; Eckhardt, 1980) have examined the suitability of several laser and electron beam techniques and metallization arrangements for forming alloyed ohmic contacts to GaAs. These authors found that cw lasers produced the most reproducible results and best surface morphology (Eckhardt, 1980). The lowest reproducible values of specific contact resistance approaching 1×10^{-6} Ω cm² were obtained for laser-annealed Au–Ge–In metallization on GaAs (Eckhardt *et al.*, 1980). Depth-profiling of the surface constituents was carried out after deposition and following both optimized laser and furnace annealing. This revealed some interesting features. Laser annealing was found to result in little redistribution of the metallization constituents, whereas furnace annealing produced extensive interdiffusion of constituents, as illustrated for Ge–Au–Ni contacts in Fig. 31, from Eckhardt (1980). The laser-annealed case in Fig. 31b exhibits profiles little different from the as-deposited case except for a reduction in the Ge concentration in the metallization layer. However, the furnace-annealed case in Fig. 31a clearly shows considerable intermixing of all contact constituents. These data suggest that the presence of Ge (donor metal) at the GaAs interface is

Fig. 30. Microscopic appearance of alloyed 115-μm-square Au–Ge contacts to *n*-GaAs. The alloyed metals are Au (200 Å) deposited on the GaAs followed by Ge (100 Å). (a) Thermal anneal, 450°C, 60 sec; (b) ruby laser anneal, 15 J/cm², 1 msec; (c) scanning argon laser anneal, 2.5 W. After Gold *et al.*, 1979.)

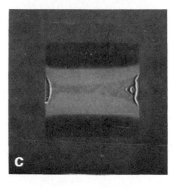

most important for the laser alloying case, where little redistribution takes place.

More recent work by Tandon and co-workers (Tandon *et al.*, 1980) show that intermixing of metals with Ga and/or As is not necessary for good ohmic behavior. A comparison of depth profiles (obtained by Auger electron spectroscopy and sputtering) from furnace and optimized electron

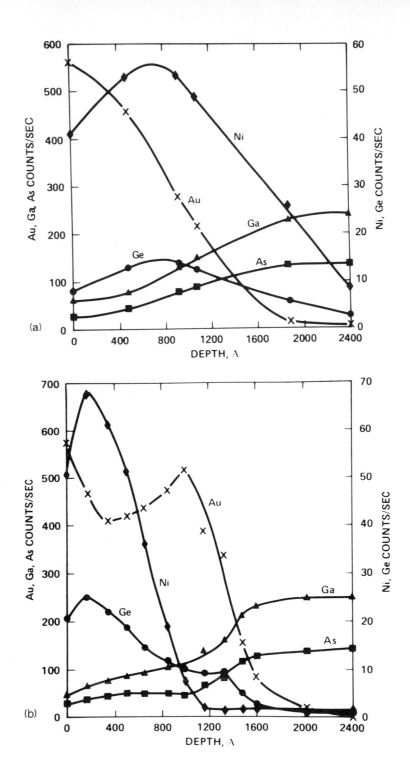

(a)

(b)

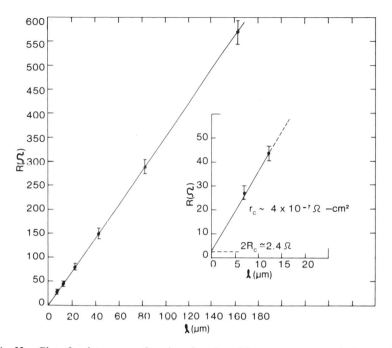

Fig. 32. Plot of resistance as a function of gap length between contact pads for Au–Ge–Pt contacts alloyed with a pulsed electron beam with an energy density of 0.4 J/cm². The inset shows magnified scales near the origin. (After Tandon *et al.*, 1980.)

beam anneals indicated that, for the Ge–Au–Pt system, the heavy doping of a thin GaAs interface layer with a Ge donor played a crucial role in good ohmic behavior. For electron beam annealing this could be achieved without intermixing of metal overlays, but for furnace annealing eutectic melting and subsequent intermixing appeared to be an unavoidable result of the long heating cycle. Impressively low values of specific contact resistance were obtained by Tandon *et al.* (1980) for electron beam annealing (approaching 2×10^{-7} Ω cm²), and typical contact resistance measurements are illustrated in Fig. 32 for pulsed electron beam annealing of a Au–Ge–Pt contact on *n*-GaAs.

Finally, a third method of forming ohmic contacts to GaAs using laser annealing is illustrated in Fig. 29c. This technique involves the implantation or deposition of a specific metal (e.g., In) which, when laser-alloyed

Fig. 31. Depth profiles of constituents of Au–Ge–Ni–Au contact obtained by ion-milled cross section and x-ray (EDX) analysis. (a) Furnace-annealed, (b) laser-annealed. (After Eckhardt, 1980.)

with underlying GaAs, forms a thin layer of a low-band-gap ternary compound (e.g., $Ga_{1-x}In_xAs$) to facilitate direct contact with a metal overlay. The method has only recently been demonstrated by Harrison and Williams (1980) for the $Ga_{1-x}In_xAs$ system. These authors achieved specific contact resistance values below 10^{-4} Ω cm^2 with nonoptimized annealing laser conditions. The most promising laser annealing arrangement for the formation of near-surface ternary compounds appears to involve the use of CO_2 laser light irradiating the GaAs–metal (In) interface from the backside of the GaAs wafer. More detailed studies are necessary to establish whether laser-induced surface ternary formation can constitute a viable approach to the fabrication of ohmic contacts on GaAs.

B. Pulsed Annealing of Deposited Layers

The approaches to ohmic contact formation previously illustrated in Fig. 29b and c provide specific examples of laser-induced diffusion and intermixing of deposited (metal) overlays with underlying GaAs. In the former example, the alloying mechanisms have not been studied in detail. However, for the Au–Ge on GaAs system, intermixing of the (eutectic) metal overlay with underlying GaAs has been assumed to proceed via a rapid melting and resolidification process for pulsed laser (Gold *et al.*, 1979; Barnes *et al.*, 1980), cw laser (Eckhardt *et al.*, 1980) and pulsed electron beam irradiation (Tandon *et al.*, 1980). Indeed, the results for optimized transient annealing of the Au–Ge system suggest that the process proceeds via eutectic melting of the metal overlay and dissolution of a thin GaAs layer during the short-duration melt to produce incorporation of donor metal into GaAs with little long-range interdiffusion of the metals into the underlying GaAs (Lee *et al.*, 1980). Badertscher *et al.* (1980) have recently shown that Ge films on GaAs can be alloyed to produce active donor layers by pulsed laser irradiation. In this case, mixing within a Ge–GaAs molten surface layer was assumed to be the operative alloying mechanism. Both the earlier work of Young *et al.* (1979) and the more recent work of Davies *et al.* (1980a) also demonstrated doping by pulsed diffusion of Mg- and Se-deposited layers, respectively, into GaAs. In both cases, the activity achieved was comparable to that obtained from pulse-annealed ion-implanted layers. In addition, the hole mobility for Mg-doped GaAs (Young *et al.*, 1979) and the electron mobility for Se-doped GaAs (Davies *et al.*, 1980a) were anomalously low and comparable to those obtained from ion-implanted substrates. Interestingly, Davies *et al.* (1980a) employed deposited As_2Se_3 as the diffusion source—the volatile nature and low melting point of As_2Se_3 ensured that the nonalloyed deposit was completely removed from the GaAs surface during a pulsed electron beam anneal.

The mixing of deposited In with underlying GaAs to form a near-surface ternary, as illustrated in Fig. 29c and demonstrated by Harrison and Williams (1980) for pulsed ruby laser irradiation, also appears to proceed via near-surface melting. Recently, Kirkpatrick (1980) suggested pulsed electron beam annealing of deposited Al on GaAs and Hg on CdTe as a means of forming $Ga_{1-x}Al_xAs$ and $Cd_{1-x}Hg_xTe$. These systems have not yet been fabricated, but the possibility exists for the production of multilayer heterostructures on GaAs by deposition and transient annealing techniques.

V. Other Compound Semiconductors

The processing of compound semiconductors other than GaAs by transient annealing methods has not received wide attention in the literature. In this section, a brief account is given of some of the laser applications and typical results that have been obtained in the field.

The early report by Pounds et al. (1974) suggested that pulsed laser annealing may offer some limited advantages for the formation of alloyed ohmic contacts to a range of Group III–V semiconductors including GaAs, GaP, GaSb, InAs, and InSb. However, the measurements of specific contact resistance in this initial study were not as low as those achievable by furnace alloying. The pioneering Russian effort in laser annealing concentrated mainly on Si and GaAs, but Bogatyrev and Kachurin (1977) obtained some interesting early results for laser annealing of InSb. These authors irradiated p-type InSb with a pulsed ruby laser to obtain an n-type surface layer, suggesting that laser-induced damage could be used to produce p–n junctions without the need for a dopant species. More recent work by Parsons et al. (1980) on laser annealing (cw argon ion) of CdSe films provides some insight into type conversion processes which can occur under laser irradiation. In CdSe, increased donor concentration was observed following laser treatment, and subsequent optical absorption studies indicated that this effect resulted from preferential Se loss and build-up of Se vacancies which acted as donors. This result highlights a particular difficulty with laser processing of volatile compound semiconductors, namely, dissociation and preferential loss of one species during laser heating. Indeed, laser-annealing studies on InP have indicated both increased donor levels (Cullis et al., 1979) and measurable phosphorus loss at higher laser powers (Liau et al., 1980a).

Such behavior complicates the interpretation of laser-annealed, ion-implanted compound semiconductors.

The sparse data available on ion-implanted compound semiconductors (other than GaAs) suggest that pulsed laser annealing can recrystallize ion-damaged and amorphous layers. Typical of such behavior is the study of S^+ and Se^+ ion-implanted, laser-annealed InP reported by Cullis *et al.* (1979) and briefly described earlier in Chapter 6. Cullis *et al.* (1979) indicated that the observed amorphous-to-polycrystalline transition at low laser powers and the ultimate amorphous-to-single-crystal transition at higher laser energy densities correlated well with the concept of surface melting. More recent results by Liau *et al.* (1980b) show clearly that Zn redistributes in Zn^+-implanted InP following pulsed Nd:YAG laser annealing in a manner consistent with a melting model. Furthermore, Cullis *et al.* (1979) report appreciable donor activity following laser irradiation for both S^+ and Se^+ implants in InP, but with only partial activation of the implanted dose and with anomalously low mobility. Liau *et al.* (1980a,b) similarly report activation of p-type dopants Zn^+ and Cd^+ in laser-annealed InP. In addition, the channeling results of Bontemps *et al.* (1980) indicate that Pb ion implant damage in InSb is removed by a single ruby laser pulse, and subsequent redistribution of Pb is suggestive of near-surface melting.

Thus, the behavior of ion-implanted InP and InSb under pulsed laser irradiation appears to be consistent with a melting model, and the observed electrical trends for InP are similar to those obtained with GaAs. For cw laser irradiation, the solid phase annealing behavior is less clear, since experiments to date (Fan *et al.*, 1981; Lee *et al.*, 1981) indicate more severe dissociation and surface slip with InP and InSb than such effects in GaAs. However, Fan *et al.* (1981) report annealing of implant damage in InP using a cw Nd:YAG laser operating in the solid phase. In contrast, Lee *et al.* (1981) suggest that liquid phase recrystallization of InSb can take place for cw laser annealing without chemical composition changes normally observed for GaAs under long-lived (≥ 1 μsec) surface melts.

Finally, recent reports of ohmic contact formation with ion-implanted InP using pulsed Nd:YAG lasers (Liau *et al.*, 1980a,b) indicate distinct improvements in specific contact resistance and reproducibility of contacts compared with more conventional furnace alloying methods.

VI. Summary and Conclusions

The results obtained from both furnace annealing and the various transient annealing methods for implanted GaAs are summarized in Table II.

TABLE II

OMPARISON OF THE VARIOUS ANNEALING METHODS FOR ION-IMPLANTED GaAs

	Furnace	Continuous-wave laser/electron beam	Pulsed laser/electron beam
Anneal mechanism	Solid phase	Solid phase	Liquid phase
Optimum anneal conditions	>850°C for best electrical properties	1- to 5-sec dwell with substrate heating (laser) and raster scan (electron beam)	Narrow energy window at energies just sufficient to melt below damage layer
Decomposition	Severe at >600°C; use cap or other preventative measures	Surface oxidation in air; use inert ambient or vacuum	Significant decomposition only at high powers and pulse lengths; surface damage at lower powers
Microstructure	Always some extended defects	Not yet examined	Extended defect-free but often some surface damage; likely quenched-in point defects
Surface topography	Featureless	Featureless (optimum conditions); slip and cracking with excessive thermal gradients	Featureless (optimum conditions); rough surface at high power
Electrical properties	Good activity; always below 10^{19} cm^{-3} for n-type	Limited data but good activity for low- to medium-dose n-type; poor results for nonoptimized anneals; mobilities close to theory, where available	High activity (>10^{19} cm^{-3}) for both n- and p-type; activity decreases on subsequent furnace annealing; cannot activate low-dose n-type; mobilities anomalously low
Implant redistribution	Diffusion broadening a problem with some dopants (e.g., Zn, S) at high temperatures	Few data exist; probably no redistribution	Little redistribution for low power just above threshold; dopant redistribution and "zone refining" can occur in melt
Substitutional solubility (Te GaAs)	Never exceeds 5×10^{19} cm^{-3}	Not measured	Exceeds equilibrium value (>10^{20} cm^{-3}); not good correlation with electrical activity

It is clear that more detailed studies are needed to assess fully the potential of cw annealing and to identify the microstructure, dopant solubility, and redistribution. However, it is apparent that the best results (highest electrical activity and best mobility) are obtained when the cw anneal provides conditions somewhat akin to those of an ultrarapid furnace anneal, with the entire substrate attaining high temperatures ($\geq 950°$) for very short times (1–5 sec).

For pulsed annealing of implanted GaAs, several interesting features are apparent, including the ability to obtain metastable, supersaturated solid solutions of dopants with electrical activities well above those achievable by furnace annealing. In addition, annealing carried out in air can provide extended defect-free recrystallization of damaged layers without any significant surface decomposition. However, despite the apparent ability to activate high doping concentrations, the electrical properties of pulse-annealed GaAs have been poor. For example, mobilities are anomalously low, high dopant activity is substantially reduced on subsequent low-temperature ($<400°C$) furnace processing, and low-dose n-type implants have not been successfully activated. Based on these observations, pulsed annealing has not yet demonstrated immediate potential for the activation of ion-implanted layers in GaAs integrated circuit fabrication. In addition, more detailed investigations are needed to understand fully the nature of quenched-in defect complexes resulting from pulsed annealing.

The recent applications of transient anneal techniques in the formation of ohmic contacts to GaAs have been most successful, since they have provided a degree of reproducibility and contact uniformity not possible with conventional furnace alloying processes. However, in order to provide a better understanding of the appropriate annealing mechanisms, the need exists for more detailed studies on the interdiffusion, alloying, and near-surface doping behavior that occur during both cw and pulsed annealing. Furthermore, the reliability of laser-produced ohmic contacts has not yet been examined in detail, but initial studies by Lee *et al.* (1980) suggest that significant interdiffusion and a corresponding increase in contact resistance can occur, in some cases, during subsequent low-temperature (400°C) furnace processing. Thus, the long-term potential of laser-formed ohmic contacts in device applications cannot as yet be fully assessed.

The few studies that have investigated the transient annealing of compound semiconductors other than GaAs suggest that decomposition effects in more volatile materials may constitute a major problem. However, some success has been obtained in InP and InSb, where ion-implanted dopants have been activated by laser annealing. Transient processing has also shown promise for the formation of reproducible ohmic contacts of low contact resistance to a range of compound semiconductors. Finally, although not yet explored in any detail, transient processing appears to offer interesting possibilities for the formation of near-surface ternary layers and, quite conceivably, for the production of multilayer heterostructures.

References

Ahmed, N. A. G., Carter, G., Christodoulides, C. E., Nobes, M. J., and Titov, A. (1980). *Nucl. Instrum. Methods* **168**, 283.

Anderson, C. L., Dunlap, H. L., Hess, L. D., Olson, G. L., and Vaidyanathan, K. V. (1980). *In* "Laser and Electron Beam Processing of Materials" (C. W. White and P. S. Peercy, eds.), p. 334. Academic Press, New York.

Auston, D. H., Surko, C. M., Venkatesan, T. N. C., Slusher, R. E., and Golovchenko, J. A. (1978). *Appl. Phys. Lett.* **33**, 437.

Badawi, M. H., Akintunde, J. A., Sealy, B. J., and Stephens, K. G. (1979). *Electron. Lett.* **15**, 448.

Badawi, M. H., Sealy, B. J., Kular, S. S., Barrett, N. J., Emerson, N. G., Stephens, K. G., Booker, G. R., and Hockley, M. (1980). *In* "Laser and Electron Beam Processing of Materials" (C. W. White and P. S. Peercy, eds.), p. 354. Academic Press, New York.

Badertscher, G., Salathe, R. P., and Luthy, W. (1980). *Electron. Lett.* **16**, 113.

Barnes, P. A., Leamy, H. J., Poate, J. M., Ferris, S. D., Williams, J. S., and Celler, G. K. (1978). *Appl. Phys. Lett.* **33**, 965.

Barnes, P. A., Leamy, H. J., Poate, J. M., Ferris, S. D., Williams, J. S., and Celler, G. K. (1979). *In* "Laser–Solid Interactions and Laser Processing" (S. D. Ferris, H. J. Leamy, and J. M. Poate, eds.), p. 647. Am. Inst. Phys., New York.

Barnes, P. A., Leamy, H. J., Poate, J. M., and Celler, G. K. (1980). *In* "Laser and Electron Beam Processing of Electronic Materials" (C. L. Anderson, G. K. Celler, and G. A. Rozgonyi, eds.), p. 421. Electrochem. Soc., Pennington, New Jersey.

Bogatyrev, V. A., and Kachurin, J. P. (1977). *Sov. Phys.—Semicond. (Engl. Transl.)* **11**, 56.

Bontemps, A., Campisano, S. U., Foti, G., and Jannitti, E. (1980). *In* "Laser and Electron Beam Processing of Materials" (C. W. White and P. S. Peercy, eds.), p. 379. Academic Press, New York.

Brawn, J. R., and Grant, W. A. (1976). *In* "Application of Ion Beams to Materials" (G. Carter, J. S. Colligon, and W. A. Grant, eds.), p. 59. Inst. Phys., London.

Campisano, S. U., Catalano, I., Foti, G., Rimini, E., Eisen, F., and Nicolet, M.-A. (1978). *Solid-State Electron.* **21**, 485.

Campisano, S. U., Foti, G., Rimini, E., Eisen, F. H., Tseng, W. F., Nicolet, M.-A., and Tandon, J. L. (1980). *J. Appl. Phys.* **51**, 295.

Carter, G., Grant, W. A., Haskell, J. D., and Stephens, G. A. (1971). *In* "Ion Implantation" (L. T. Chadderton and F. H. Eisen, eds.), p. 261. Gordon & Breach, New York.

Cullis, A. G., Webber, H. C., and Roberton, D. S. (1979). *In* "Laser–Solid Interactions and Laser Processing" (S. D. Ferris, H. J. Leamy, and J. M. Poate, eds.), p. 653. Am. Inst. Phys., New York.

Davies, E. D., Ryan, T. G., and Lorenzo, J. P. (1980a). *Appl. Phys. Lett.* **37**, 443.

Davies, E. D., Lorenzo, J. P., and Ryan, T. G. (1980b). *Appl. Phys. Lett.* **37**, 612.

Davies, E. D., Kennedy, E. F., Ryan, T. G., and Lorenzo, J. P. (1981). *In* "Laser and Electron Beam Solid Interactions and Materials Processing" (J. F. Gibbons, L. D. Hess, and T. W. Sigmon, eds.), p. 247. North-Holland Publ., New York.

Dearnaley, G., Freeman, J. H., Nelson, R. S., and Stephen, J. (1973). "Ion Implantation," Chap. 5. North-Holland Publ., New York.

Donnelly, J. P. (1977). *Conf. Ser.—Inst. Phys.* No. 33, p. 167.

Donnelly, J. P. (1981). *Nucl. Instrum. Methods* **182/183**, 553.

Eckhardt, G. (1980). *In* "Laser and Electron Beam Processing of Materials". (C. W. White and P. S. Peercy, eds.), p. 467. Academic Press, New York.

Eckhardt, G., Anderson, C. L., Hess, L. D., and Krumm, C. F. (1979). *In* "Laser–Solid Interactions and Laser Processing" (S. D. Ferris, H. J. Leamy and J. M. Poate, eds.), p. 641. Am. Inst. Phys., New York.

Eckhardt, G., Anderson, C. L., Colborn, M. N., Hess, L. D., and Jullens, R. A. (1980). *In* "Laser and Electron Beam Processing of Electronic Materials" (C. L. Anderson, G. K. Celler, and G. A. Rozgonyi, eds.), p. 445. Electrochem. Soc., Pennington, New Jersey.

Eisen, F. H. (1975). *In* "Ion Implantation in Semiconductors" (S. Namba, ed.), p. 3. Plenum, New York.

Eisen, F. H. (1978). *In* "Ion Beam Modification of Materials" (J. Guylai, T. Lohner, and E. Pasztor, eds.), Vol. 1, p. 147. Hung. Acad. Sci., Budapest.

Eisen, F. H. (1980). *In* "Laser and Electron Beam Processing of Materials" (C. W. White and P. S. Peercy, eds.), p. 309. Academic Press, New York.

Fan, J. C. C., Donnelly, J. P., Bozler, C. O., and Chapman, P. L. (1979a). "GaAs and Related Compounds," p. 472. Inst. Phys., London.

Fan, J. C. C., Chapman, R. L., Donnelly, J. P., Turner, G. W., and Bozler, C. O. (1979b). *Appl. Phys. Lett.* **34**, 780.

Fan, J. C. C., Chapman, R. L., Donnelly, J. P., Turner, G. W., and Bozler, C. O. (1981). *In* "Laser and Electron Beam Solid Interactions and Materials Processing" (J. F. Gibbons, L. D. Hess, and T. W. Sigmon, eds.), p. 261. North-Holland Publ., New York.

Fletcher, J., Narayan, J., and Lownes, D. H. (1981). *In* "Defects in Semiconductors" (J. Narayan and T. Y. Tan, eds.), p. 421. North-Holland, New York.

Gamo, K., Inada, T., Mayer, J. W., Eisen, F. H., and Rhodes, C. G. (1977). *Radiat. Eff.* **33**, 85.

Gamo, K., Katano, F., Yuba, Y., Murakami, K., and Namba, S. (1979). *In* "Laser–Solid Interactions and Laser Processing" (S. D. Ferris, H. J. Leamy, and J. M. Poate, eds.), p. 591. Am. Inst. Phys., New York.

Gamo, K., Yuba, Y., Oraby, A. H., Murakami, K., Namba, S., and Kawasaki, Y. (1980). *In* "Laser and Electron Beam Processing of Materials" (C. W. White and P. S. Peercy, eds.), p. 322. Academic Press, New York.

Gold, R. B., Powell, R. A., and Gibbons, J. F. (1979). *In* "Laser–Solid Interactions and Laser Processing" (S. D. Ferris, H. J. Leamy, and J. M. Poate, eds.), p. 635. Am. Inst. Phys., New York.

Golecki, I., Nicolet, M.-A., Maenpaa, M., Tandon, J. L., Kirkpatrick, C. G., Sadana, D. K., and Washburn, J. (1980). *In* "Laser and Electron Beam Processing of Materials" (C. W. White and P. S. Peercy, eds.), p. 347. Academic Press, New York.

Golovchenko, J. A., and Venkatesan, T. N. C. (1978). *Appl. Phys. Lett.* **32**, 464.

Grimaldi, M. G., Paine, B. M. Maenpaa, M., Nicolet, M.-A., and Sadana, D. K. (1981a). *Appl. Phys. Lett.* **39**, 70.

Grimaldi, M. G., Paine, B. M., Nicolet, M.-A., and Sadana, D. K. (1981b). *J. Appl. Phys.* **52**, 4038.

Harris, J. S., Eisen, F. H., Welch, B., Pashley, R. D., Sigurd, D., and Mayer, J. W. (1972). *Appl. Phys. Lett.* **21**, 601.

Harrison, H. B., and Williams, J. S. (1980). *In* "Laser and Electron Beam Processing of Materials" (C. W. White and P. S. Peercy, eds.), p. 481. Academic Press, New York.

Inada, T., Kato, S., Maeda, Y., and Tokunaga, K. (1979a). *J. Appl. Phys.* **50**, 6000.

Inada, T., Tokunaga, K., and Taka, S. (1979b). *Appl. Phys. Lett.* **35**, 546.

Kachurin, G. A., Pridachin, N. B., and Smirnov, L. S. (1976). *Sov. Phys.—Semicond.* (*Engl. Transl.*) **9**, 946.
Kirkpatrick, A. R. (1980). *In* "Laser and Electron Beam Processing of Electronic Materials" (C. L. Anderson, G. K. Celler, and G. A. Rozgonyi, eds.), p. 108. Electrochem. Soc., Pennington, New Jersey.
Kular, S. S., Sealy, B. J., Stephens, K. G., Sadana, D. K., and Booker, G. R. (1980). *Solid-State Electron.* **23**, 831.
Lau, S. S., and Van der Weg, W. F. (1978). *In* "Thin Films—Interdiffusion and Reactions" (J. M. Poate, K. N. Tu, and J. W. Mayer, eds.), p. 433. Wiley, New York.
Lee, C. P., Tandon, J. L., and Stocker, P. J. (1980). *Electron. Lett.* **16**, 850.
Lee, D. H., Olson, G. L., and Hess, L. D. (1981). *In* "Laser and Electron Beam Solid Interactions and Materials Processing" (J. F. Gibbons, L. D. Hess, and T. W. Sigmon, eds.), p. 281. North-Holland Publ., New York.
Liau, Z. L., De Meo, N. L., Donnelly, J. P., Mull, D. E., Bradbury, R., and Lorenzo, J. P. (1980a). *In* "Laser and Electron Beam Processing of Materials" (C. W. White and P. S. Peercy, eds.), p. 494. Academic Press, New York.
Liau, Z. L., De Meo, N. L., Donnelly, J. P., Hopkins, C. P., Norberg, J. C., Evans, C. A., Jr., and Lorenzo, J. P. (1980b). *In* "Laser and Electron Beam Processing of Materials" (C. W. White and P. S. Peercy, eds.), p. 500. Academic Press, New York.
Lidow, A., Gibbons, J. F., Deline, V. R., and Evans, C. A., Jr. (1980). *J. Appl. Phys.* **51**, 4130.
Liu, S. G., Wu, C. P., and Magee, C. W. (1979). *In* "Laser–Solid Interactions and Laser Processing" (S. D. Ferris, H. J. Leamy, and J. M. Poate, eds.), p. 603. Am. Inst. Phys., New York.
Liu, S. G., Wu, C. P., and Magee, C. W. (1980). *In* "Laser and Electron Beam Processing of Materials" (C. W. White and P. S. Peercy, eds.), p. 341. Academic Press, New York.
Margalit, S., Pekote, D., Pepper, D. M., Lee, C. P., and Yariv, A. (1978). *Appl. Phys. Lett.* **33**, 346.
Mazey, D. J., and Nelson, R. S. (1969). *Radiat. Eff.* **1**, 229.
Nissim, Y. I., and Gibbons, J. F. (1981). *In* "Laser and Electron Beam Solid Interactions and Materials Processing" (J. F. Gibbons, L. D. Hess, and T. W. Sigmon, eds.), p. 275. North-Holland Publ., New York.
Olson, G. L., Anderson, C. L., Dunlap, H. L., Hess, L. D., McFarlane, R. A., and Vaidyanathan, K. V. (1980). *In* "Laser and Electron Beam Processing of Electronic Materials" (C. L. Anderson, G. K. Celler, and G. A. Rozgonyi, eds.), p. 467. Electrochem. Soc., Pennington, New Jersey.
Padovani, F. A. (1971). *In* "Semiconductors and Semimetals" (R. K. Willardson and A. C. Beer, eds.), Vol. 7A, p. 75. Academic Press, New York.
Parsons, R. R., Rostworowski, J. A., Westwood, W. D., and Shepherd, F. R. (1980). *In* "Laser and Electron Beam Processing of Materials" (C. W. White and P. S. Peercy, eds.), p. 367. Academic Press, New York.
Pianetta, P. A., Stolte, C. A., and Hanson, J. L. (1980). *In* "Laser and Electron Beam Processing of Materials" (C. W. White and P. S. Peercy, eds.), p. 328. Academic Press, New York.
Pianetta, P. A., Amano, J., Woolhouse, G., and Stolte, C. A. (1981). *In* "Laser and Electron Beam Solid Interactions and Materials Processing" (J. F. Gibbons, L. D. Hess, and T. W. Sigmon, eds.), p. 239. North-Holland Publ., New York.
Picraux, S. T. (1973). *Radiat. Eff.* **17**, 261.
Pounds, R. S., Saifi, M. A., and Hahn, W. C., Jr. (1974). *Solid-State Electron.* **17**, 245.

Rideout, V. L. (1975). *Solid-State Electron.* **18**, 541.

Sealy, B. J., Kular, S. S., Stephens, K. G., Croft, R., and Palmer, A. (1978). *Electron. Lett.* **14**, 720.

Sealy, B. J., Kular, S. S., Badawi, M. H., and Stephens, K. G. (1979). *In* "Laser–Solid Interactions and Laser Processing" (S. D. Ferris, H. J. Leamy, and J. M. Poate, eds.), p. 610. Am. Inst. Phys., New York.

Shah, N. J., Ahmed, H., Sanders, I. R., and Singleton, J. F. (1980). *Electron. Lett.* **16**, 433.

Sze, S. M. V. (1969). "Physics of Semiconductor Devices," p. 40. Wiley, New York.

Tandon, J. L., and Eisen, F. H. (1979). *In* "Laser–Solid Interactions and Laser Processing" (S. D. Ferris, H. J. Leamy, and J. M. Poate, eds.), p. 616. Am. Inst. Phys., New York.

Tandon, J. L., Golecki, I., Nicolet, M.-A., Sadana, D. K., and Washburn, J. (1979). *Appl. Phys. Lett.* **35**, 867.

Tandon, J. L., Kirkpatrick, C. G., Welch, B. M., and Fleming, P. (1980). *In* "Laser and Electron Beam Processing of Materials" (C. W. White and P. S. Peercy, eds.), p. 494. Academic Press, New York.

Venkatesan, T. N. C., Auston, D. H., Golovchenko, J. A., and Surko, C. M. (1979). *In* "Laser–Solid Interactions and Laser Processing" (S. D. Ferris, H. J. Leamy, and J. M. Poate, eds.), p. 629. Am. Inst. Phys., New York.

Williams, J. S., and Austin, M. W. (1980a). *Nucl. Instrum. Methods* **168**, 307.

Williams, J. S., and Austin, M. M. (1980b). *Appl. Phys. Lett.* **36**, 994.

Williams, J. S., and Harrison, H. B. (1981). *In* "Laser and Electron Beam Solid Interactions and Materials Processing" (J. F. Gibbons, L. D. Hess, and T. W. Sigmon, eds.), p. 209. North-Holland Publ., New York.

Williams, J. S., Austin, M. W., and Harrison, H. B. (1980). *In* "Thin Film Interfaces and Interactions" (J. E. E. Baglin and J. M. Poate, eds.), p. 187. Electrochem. Soc., Pennington, New Jersey.

Wood, R. F., Lowndes, D. H., and Christie, W. H. (1981). *In* "Laser and Electron Beam Solid Interactions and Materials Processing" (J. F. Gibbons, L. D. Hess, and T. W. Sigmon, eds.), p. 231. North-Holland Publ., New York.

Young, R. T., Narayan, J., Westbrook, R. D., and Wood, R. F. (1979). *In* "Laser–Solid Interactions and Laser Processing" (S. D. Ferris, H. J. Leamy, and J. M. Poate, eds.), p. 579. Am. Inst. Phys., New York.

Chapter 12

Silicides and Metastable Phases

MARTIN F. VON ALLMEN

Institute of Applied Physics
University of Bern
Bern, Switzerland

and

S. S. LAU

Department of Electrical Engineering and Computer Sciences
University of California, San Diego
La Jolla, California

I. Introduction

Thin films play a most dominant role in modern technology. Their important uses are exemplified in the science and technology of integrated

439

circuits, solar energy conversion devices, tribology, and other near-surface related technologies. In the area of semiconductors, thin films are used exclusively for Schottky barriers, ohmic contacts, and intercon-nects. The understanding and control of interactions between metal thin films and semiconductors are important issues. In recent years, the in-teractions between metal layers and semiconductors (mostly Si) under steady state (furnace annealing) conditions have been investigated in a systemmatic manner and are subjects of review in the literature (Tu and Mayer, 1978; Muraka, 1980; Mohammadi, 1981). At present, the solid state interactions appear to be relatively well understood. Some of the current research activities are now focused on the atomistic nature of the interfacial region between the metal and the semiconductor (within ~10–20 Å of the interface). With the advent of pulsed laser processing of electronic materials, a new domain of metal–semiconductor interactions has been established. Because of the transient nature of the energy pulse applied to the system, the physical processes known for solid phase in-teractions do not necessarily apply in the case of pulsed laser annealing. One of the most obvious deviations of metal–semiconductor interactions under transient conditions from those under steady state conditions is the metastability of the reaction products. It is the objective of this chapter to discuss reactions in metal–semiconductor systems subject to fast melting and solidification. In our discussion, a brief summary of solid state reac-tions will be given first, followed by an introduction to general concepts of laser-induced melting, mixing, and quenching, and finally experimental results of laser-induced interactions in metal–semiconductor systems are presented.

II. Solid State Reactions—Silicide Formation

A. Phase Sequence

The phase diagrams for metal silicide-forming systems usually show several equilibrium phases, however, not all the equilibrium phases are present as dominant growth phases during solid state reaction in thin-film systems. The first growth phase is difficult to predict and is usually a metal-rich silicide if present in the phase diagram. However, it is some-times possible to use the rule proposed by Walser and Bené (1976) as a guide in predicting the first phase that will grow. This rule states that the first silicide to grow is the highest congruently melting compound next to the lowest melting eutectic composition (for example, Ni_2Si in the Ni–Si system). The Walser–Bené rule is about 80–90% successful in predicting the first growth phase. After the total consumption of metal film to form the first phase, a second equilibrium phase (if present in the phase diagram) will form at higher temperatures and/or for longer annealing times. The second growth phase is also sometimes predictable following the rule proposed by Tsaur *et al.* (1981b). This rule states that the next phase formed is the nearest congruently melting compound richer in the unreacted element; if the compounds between the first phase and the element are all noncongruently melting compounds (such as peritectic or peritectoid phases), the next phase formed will be the one with the smallest temperature difference between the liquidus curve and the peritectic (or peritectoid point). For example, applying this rule to the Ni–Si system, the phase sequence would be (1) Ni_2Si, (2) $NiSi$, (3) $NiSi_2$ for samples with the thickness of Si much larger than that of Ni (bypassing Ni_3Si_2); for samples with the thickness of Ni much larger than that of Si, the phase sequence would be (1) Ni_2Si, (2) Ni_5Si_2, (3) Ni_3Si. These predictions are in agreement with experimental results. Although some of these rules are basically phenomenological observations, it is possible to predict the phase sequence in silicide formation by inspecting the phase diagram.

B. Growth Characteristics

The solid state reactions for silicide formation can be generally divided into two categories: (i) layer-by-layer growth kinetics and (ii) growth that exhibits critical-temperature dependence. For the first category of silicide formation, the reaction takes place over a relatively wide temperature range (usually far below any eutectic or melting temperature of the system), and the growth proceeds in a layer-by-layer manner with relatively sharp interfaces. The growth process can be either diffusion-limited (thickness of silicide proportional to the square root of annealing time) or interfacial reaction-limited (thickness of silicide proportional to annealing time). The layer-by-layer growth characteristics can be clearly demonstrated by the formation of Ni_2Si. Figure 1 shows the Rutherford backscattering spectra of a Ni–Si sample annealed at 250°C for different times (Tu *et al.*, 1975). The growth of the Ni_2Si phase is initiated at the Ni–Si interface at temperatures from 200 to 350°C (only 250°C is shown) for annealing times up to a few hours, with the growth kinetics following a $\sqrt{t}$ dependence.

In the case of linear silicide growth, the silicides usually form at much

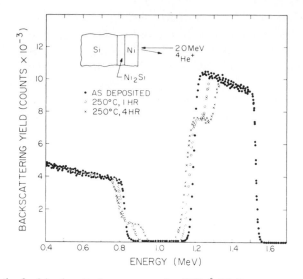

Fig. 1. Rutherford backscattering spectra of a 2000-Å Ni film on Si before and after annealing at 250°C for 1 and 4 hr. (From Tu *et al.*, 1975.)

higher temperatures (~500–700°C) and generally have Si-rich phases. Figure 2 shows the kinetic curves for VSi_2 formation (Krautle *et al.*, 1974). The silicide thickness is seen to increase linearly with time, followed by saturation. The saturation effect is commonly observed in the formation of refractory metal silicides and is suspected to be due to oxygen incursion into the metal film during high-temperature annealing.

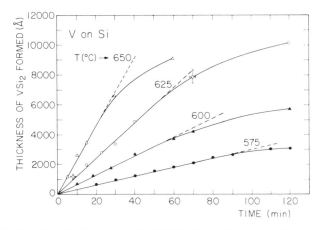

Fig. 2. Data for linear growth of VSi_2. (From Krautle *et al.*, 1974.)

For the second category of silicide formation, the growth occurs in a very narrow temperature range, usually within 10–30°C. These reactions are commonly referred to as nucleation-controlled. The formation of rare earth silicides provides typical examples of nucleation-controlled reactions. For example, in the formation of $TbSi_{\sim 1.7}$, no reaction was observed for Tb–Si samples below a temperature of ~320°C. As the temperature was raised to ~325°C, the silicide-forming reaction was completed within a few minutes. These growth characteristics, therefore, exhibit a threshold-temperature dependence. Since the reactions reach completion rapidly above the threshold temperature, the growth kinetics have not been measured systematically. There is, however, an indication that this type of silicide nucleations at local "weak" spots near the metal–Si interface (Baglin *et al.*, 1980). Once the silicide islands are nucleated, they proceed to grow at a very rapid rate. The threshold-temperature dependence is speculated to be due to a small driving force for nucleation (e.g., small ΔH values for the silicide-forming reactions).

From the discussion above, it should be apparent that solid state reactions for silicide formation under steady state condition are relatively well understood; and reliable contacts (Schottky or ohmic) for device application are made based upon this understanding. It should also be apparent that solid state reactions under steady state condition seldom lead to the formation of metastable phases. Some of the typical equilibrium silicides observed under steady state conditions are listed in Table I. The contrary is true for reactions occurring under pulsed laser irradiation. This subject will be discussed in the next section.

TABLE I

TYPICAL EXAMPLES OF SILICIDE FORMATION IN THE SOLID STATE

Silicide	Formation temperature (°C)	Activation energy (eV)	Growth rate
Ni_2Si	200–~350	1.5	$t^{1/2}$
Pd_2Si	100–~700	1.3–1.5	$t^{1/2}$
Pt_2Si	200–500	1.1–1.6	$t^{1/2}$
Co_2Si	350–500	1.5	$t^{1/2}$
PtSi	≥300	1.6	$t^{1/2}$
PdSi	≥700	—	—
NiSi	350–750	—	—
CoSi	425–500	1.9	$t^{1/2}$
HfSi	550–700	2.5	$t^{1/2}$
$CrSi_2$	450	1.7	t
$MoSi_2$	525	3.2	t

III.   Melting, Mixing, and Quenching

A.   Introduction

The research activity of metastable phase formation was first initiated by Pol Duwez in early 1960 (Duwez, 1967). Duwez and his co-workers obtained solid solubility extension and new metastable crystalline or amorphous phases in certain binary alloy systems by rapid cooling from the liquid to the solid state. These rapid solidification techniques are commonly known as *splat cooling* techniques. The quenching rate of splat cooling generally ranges between 10^4 and 10^6 K/sec. With these quenching rates, Duwez and others have found that metastable alloy phases can be formed near the eutectic composition in binary systems where deep eutectic troughs are present. In Duwez's original thinking, metastable phases are formed as a result of "fooling" the atoms by freezing them into unconventional positions by rapid cooling from the liquid state. Subsequently, Turnbull and others (Cohen and Turnbull, 1961; Davies and Lewis, 1975; Sinha *et al.*, 1976) have dealt with the theoretical and experimental aspects of splat cooling, and the field of rapid solidification has progressed in the interim to a very active, well-documented area of research.

With the advent of pulsed laser processing of materials, quenching rates on the order of 10^{10} K/sec are now achievable. These much faster rates open up new dimensions for the formation of metastable phases. We shall begin by briefly reviewing the concept of melting, mixing, and quenching induced by heat pulses. It should be emphasized here that heat flow calculations for pulsed laser irradiation essentially involve a one-dimensional problem and that the analysis can be performed with a relatively high degree of accuracy and leads to meaningful results that reflect experimental conditions.

The transformation of a sample through melting by a heat pulse into its final form is due to the interplay of heat flow, mass diffusion, and molecular rearrangements. Many aspects of these processes have been treated in previous chapters of this book. A special feature of the structures discussed in this chapter is that they are composed of two elements in *comparable proportions*. In the following, the focus is on heat pulse-induced melting and quenching in a *concentrated alloy regime*. The main aspects—heat flow, diffusion, and solidification—will, for clarity, be discussed separately, although they occur simultaneously and in close mutual interdependence. In a typical experiment, the sample is a planar, laterally uniform structure containing some distribution of two elements A and B (typically Si and a metal) as a function of depth. The sample surface is heated by a heat pulse with a duration in the range 10^{-9}–10^{-6} sec. Heat flow in this time regime is essentially one-dimensional, since the thickness of the heated region (on the order of micrometers or less) is small compared to the lateral dimensions of both the sample and the beam. The absorbed fluence is on the order of 1 J/cm^2, enough to melt several 100 nm of the sample while creating thermal gradients in the 10^6–10^8 K/cm range. The maximum surface temperature may exceed the melting point by several 100 K, but it should not approach the boiling point.

B. Heat Flow

The concepts of energy beam-induced creation and diffusion of heat have been introduced in Chapters 2 and 3. A special feature of the composite samples considered here is that their physical properties vary as a function of depth. In addition, chemical reactions between the elements may occur that influence the heat flow through their reaction enthalpies. The only way to include these phenomena is by numerical integration of the heat flow equation. It can be written (in one dimension):

$$\rho(\partial H/\partial t) = \partial/\partial z[\kappa(\partial T/\partial z)] + P(z,t,T) \tag{1}$$

where H is the enthalpy per unit mass, ρ the density, T the temperature, κ the thermal conductivity of the material, z the coordinate perpendicular to the sample surface (located at $z = 0$), and t time. P is the heat flux produced by absorption of the energy beam. The phase transition is reflected in the dependence of the enthalpy H on temperature:

$$H(T) = H(T_0) + \int_{T_0}^{T} C_p dT' + p(T, T_t) \, \Delta H_t \tag{2}$$

Here C_p is the heat capacity, and ΔH_t and T_t denote the latent heat and the transition temperature, respectively. The function $p(T, T_t)$ has the values 0 below and 1 above the transition temperature.

From a numerical solution of Eqs. (1) and (2), the transient temperature distributions in the heated film are obtained. However, as will be evident later, the most directly relevant information for the present discussion is not the temperature distributions as a function of space or time but the velocities of isothermal surfaces. If $T(z,t)$ is a temperature distribution, points of equal temperature $T(z,t) = T_0$ form a surface that moves through the material at a velocity v_0 given by

$$v_0 = \left(\frac{\partial T/\partial t}{\partial T/\partial z} \right)_{T_0} \tag{3}$$

i.e., by the ratio of the local heating or cooling rate to the local temperature gradient. If the isothermal surface coincides with the interface between the liquid and the solid phases, the condition for its velocity v_i is

$$v_i \rho H_i = [\kappa_s(\partial T_s/\partial z) - \kappa_l(\partial T_l/\partial z)]_{T_i} \qquad (4)$$

The subscripts l and s denote the liquid and the solid phase, respectively. The temperature gradients are evaluated at the interface, where $T_l = T_s = T_i$ and ΔH_i is the latent heat absorbed (during melting) or liberated (during solidification) at the l–s interface.

In general, the quantities in Eq. (4) are different for the melting and solidification processes. During melting, $T_i > T_m$ and ΔH_i is the heat of melting of the sample material at the position of the interface. During solidification, $T_i < T_m$. Moreover, ΔH_i for solidification may differ from ΔH_i for melting, because of the changes in local composition produced by melt diffusion and possibly because the solidifying phase has an enthalpy different from that of the original film material.

The described features of heat flow in a composite sample are best illustrated with an example, as shown in Fig. 3. The position of the liquid–solid interface is shown as a function of time, calculated for a film deposited on top of a substrate of much higher melting point (von Allmen

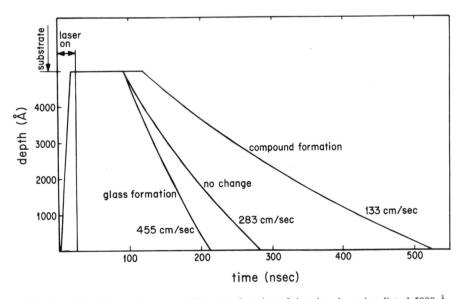

Fig. 3. Calculated melt front position as a function of time in a laser-irradiated 5000-Å alloy film on top of an inert substrate. The three curves result from different assumptions on the nature of the solidifying phase, as explained in the text. (From von Allmen *et al.*, 1981.)

et al., 1981). For this particular calculation, film thickness and laser pulse duration were chosen to be 500 nm and 25 nsec, respectively; the values for conductivity, specific heat, and melting heat were selected to be about halfway between those of Si and those of Pt. Since the function of the calculation was to obtain some insight into the relevance of various parameters (rather than to obtain accurate temperature values), the melting temperature was arbitrarily set at 1000°C. For the substrate, a constant conductivity κ of 0.24 W/(cm K) was assumed. The three different solidification curves in Fig. 3 result from three different assumptions about the solidification process:

(i)   For the curve labeled "no change," the same values for transition temperature and latent heat as for melting were used. This corresponds to the case of an elemental sample or, alternatively, to recrystallization of a mixture without a reaction between the components.

(ii)   For the curve labeled "compound formation," the latent heat for freezing was 1.5 times that for melting. This describes a case in which a compound is formed with a heat of formation of 50% of the average melting heat of the elements. (The corresponding value for formation of PtSi would be 46%; see Fig. 23 for the Pt–Si phase diagram.) Further, the freezing temperature of the compound was set at 800°C. The consequence of this reduction in liquid–solid interfacial velocity due to compound formation is discussed in more detail in Section IV.C.2 of this chapter.

(iii)   For the curve "glass formation," it was assumed that latent heat of two-thirds of the melting heat was liberated upon solidification. The freezing temperature was 1000°C in this case. (This procedure neglects the presence of the freezing interval.)

As is evident from Fig. 3, rather different values for the interface velocity during solidification can result under otherwise identical conditions, depending on the thermodynamics of the solidification process. As a rule, the interface velocity is roughly proportional to the inverse of the total latent heat liberated at the interface if all other parameters are left unchanged.

In order to further illustrate the quenching process represented in Fig. 3, Figs. 4 and 5 show temperature profiles at various times during cooling and equivalent cooling rates as a function of time, respectively, both for the "no change" case, as shown in Fig. 3.

With reference to Fig. 4, it can be seen that, while temperature gradients in excess of 10^6°C/cm are present in the melt before solidification, there is, because of the liberation of latent heat, virtually no temperature gradient in the melt once solidification has started. Although contrary to intuition, this observation has emerged as a quite general result under the

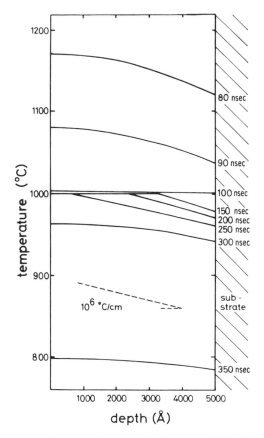

Fig. 4. Temperature profiles at various times for the curve labeled "no change" in Fig. 3. The freezing temperature is 1000°C.

assumption of a sharp transition temperature. (Allowance for a freezing interval results in finite, if comparatively small, temperature gradients in the melt.)

Cooling rates are often used in arguments to predict glass formation or to compare various techniques of melt quenching. Figure 5 is included here as a reminder that the concept of a cooling rate should be used with caution; shown is the cooling rate as a function of time at the film surface. (Similar curves represent the cooling rates at various depths within the film.) Just after the laser pulse ends, cooling rates on the order of 10^{11}°C/ sec are reached under the given circumstances. However, they occur in the superheated liquid and are of little relevance for the solidification process. As the freezing temperature is approached, the cooling rate

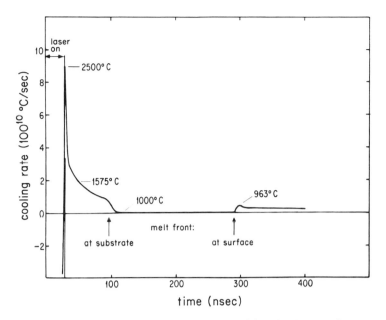

Fig. 5. Cooling rate at the film surface as a function of time for the case shown in Fig. 4.

sharply decreases and stays close to zero until solidification is completed. (This is, of course, connected with the vanishing of the temperature gradient in the melt.)

In the above example, the liquid–solid boundary was taken to be identical to the isothermal surface at which latent heat is absorbed or liberated. This is an idealization made for computing purposes. Even in the case of an elemental crystal growing from its melt, the *l*–s interface may be "rough" on an atomic scale (Jackson, 1975). If a compound crystal is formed, the *l*–s interface may bulge as a result of solidification instabilities, as discussed later in this section. Finally, if the solidifying phase is not crystalline but amorphous, freezing will occur over an extended temperature interval (Turnbull and Bagley, 1975) and there will be no well-defined *l*–s interface. The distinction between the amorphous phase and the undercooled melt is a matter of convention; it is usually made with the aid of the viscosity η and by defining an amorphous phase (a glass) as an undercooled liquid with a viscosity greater than 10^{15} P $= 10^{14}$ kg/msec. If the viscosity-versus-temperature relation of a material is known, the *l*–s interface can conveniently be taken as the isothermal surface at which $\eta(T) = 10^{15}$ P. This particular temperature is called the glass temperature T_g.

C. Mass Diffusion

Processing times in heat pulse annealing are usually small fractions of a second, thereby making the solid state diffusion negligible. However, the atoms in the melt have high mobilities, corresponding to diffusivities on the order of 10^{-4} cm^2/sec as compared to 10^{-9}–10^{-12} cm^2/sec in a solid. This allows mass transport over tens of nanometers even for heat pulses in the nanosecond range, and composite samples will necessarily undergo redistribution of the elements.

The diffusivities D in liquids depend on composition as well as on temperature (Takeuchi, 1973), but generally far less so than in solids. If D is taken to be a constant, concentration distributions due to diffusion in the melt can be calculated from

$$C(z,t) = \frac{1}{2\sqrt{\pi D t_l}} \int_0^\infty C_0(z) \left\{ \exp\left[-\frac{(z - z')^2}{4 D t_l} \right] \right.$$

$$\left. + \exp\left[-\frac{(z + z')^2}{4 D t_l} \right] \right\} dz' \tag{5}$$

where $C_0(z)$ is the initial distribution before melting and t_l is the time during which a melt exists.

Figure 6 gives concentration distributions for three different initial sample configurations often found in practice:

(i) The Gaussian (describing a profile obtained by ion implantation of a dopant B in a semiconductor A) (Fig. 6a),

(ii) The rectangular distribution (occurring in the case of a deposited layer B on top of a substrate A) (Fig. 6b), and

(iii) The square wave profile (describing a multilayer of alternating deposited films of A and B) (Fig. 6c).

Cases (ii) and (iii) will also be referred to as "unlimited supply" of and "limited supply" of the element A, respectively.

It can be seen that the initial concentration peaks are broadened by an amount of approximately $2\sqrt{D t_l}$. For more accurate calculations, t_l is allowed to vary within the molten layer as a function of depth, taking into account the finite velocity of the liquid–solid interface (Liau et al., 1979).

As the conditions governing heat flow during melting establish a boundary condition for atomic diffusion (via the duration of existence of the melt), the latter, in turn, results in an initial condition for solidification: It defines the local thermophysical properties (melting point, latent heat) of the molten layer at the instant of solidification and influences the structure of the solid, as well as the velocity of the l–s interface, as demonstrated in Fig. 3.

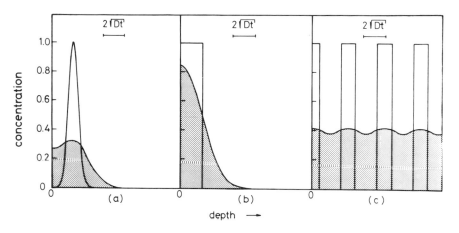

Fig. 6. Calculated diffusion profiles in the melt for three different initial concentration profiles. (a) Gaussian (as produced by ion implantation of an impurity into a substrate). (b) Rectangular profile (vapor-deposited thin film). (c) Square wave profile (alternating multilayer).

D. Solidification

Experiments on silicide and metastable phase formation are usually based on the sample configurations shown in Fig. 6b and c. Both result in a region of a concentrated mixture of A and B. In contrast, the configuration of Fig. 6a (implanted substrate) leads, in general, to very diluted mixtures. The structures resulting from solidification of diluted or concentrated melts are quite different: Generally, single-crystal regrowth with segregation or substitutional incorporation of the impurity is found in the diluted melt case, as explained in previous chapters. In the concentrated melt regime, solidification has been shown to result either in compound formation (usually polycrystalline and often microscopically inhomogeneous structures) or, alternatively, in amorphous phase formation. The regime of concentrated melts is briefly discussed in the following subsection.

1. NUCLEATION AND GROWTH IN CONCENTRATED MELTS

The basic laws of nucleation and growth of a crystal from its melt were introduced in Chapter 2. Figure 7 is a schematic of the general dependence of the velocity v of a crystal–melt boundary on temperature. For the case of an elemental crystal, it can be approximately described by the expression

$$v = u_0 \exp\left(-Q/kt\right)\left[1 - \exp(-\Delta H_m \, \Delta T_i / k T_m T_i)\right] \qquad (6)$$

452   MARTIN F. VON ALLMEN AND S. S. LAU

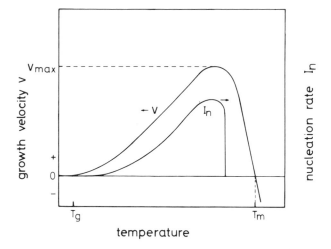

Fig. 7.   Schematic showing the velocity of a crystal boundary in contact with the melt as a function of temperature. ($v > 0$, solidification; $v < 0$, melting); T_m is the equilibrium melting temperature and T_g the glass temperature. Also shown is a schematic of the steady state nucleation rate I_n.

Here u_0 is a constant which depends on the detailed structure of the crystal–melt boundary, ΔH_m is the heat of melting of the crystal, and Q is the activation energy for the molecular rearrangement process (assumed here to be a simple thermally activated process). Q is sometimes identified with the activation energy for viscous flow (Jackson, 1975).

If two elements, A and B, are present in the melt, the composition of the solid C_s (expressed as an atomic fraction of A), is, in general, different from that of the liquid C_l in equilibrium with the solid. The compositions of the A–B phase diagram where the liquidus and the solidus temperatures coincide, e.g., eutectics or congruently melting phases, are exceptions.

The simplest model describing the growth of a compound crystal is based on the assumption that an expression of the form of Eq. (6) applies for each individual species but with a prefactor u_0 proportional to the concentration of this species in the melt (Chalmers, 1977). From this consideration alone, growth of a compound crystal is slowed down whenever its composition differs from that of the melt. If C_l and C_s are the concentrations of one species in the melt and the crystal, respectively, then the growth rate of a compound with $C_s > C_l$ is reduced by a factor

$$(1 - 1/C_l)/(1 - 1/C_s)_i$$

as compared to the case $C_s = C_l$. Here the compositions are, of course, those at the liquid–solid interface. Because of the rejection of excess atoms at the interface, the excess species is accumulated in the melt, gradually reducing the growth velocity further. Eventually, growth of a different compound with a composition closer to that of the melt may become thermodynamically more favorable.

In addition to the prefactors u_0 in Eq. (6), the enthalpy differences ΔH_m and the activation energies Q of the A and B atoms forming the compound crystal also are likely to differ because the corresponding binding forces need not be the same. The slower of the two growth processes of the A and B atoms then determines the overall growth rate of the compound.

The presence of a nucleus is required before a new crystal can grow. In samples of the unlimited supply type, shown in Fig. 6b, the single-crystal substrate serves as a nucleus when solidification starts. For the formation of compound phases, however, nucleation must occur in the melt. Nucleation is usually described in terms of a steady state nucleation rate I_n (see Chapter 4), which gives the number of supercritical nuclei produced per second per unit volume of a uniformly undercooled melt. A schematic of I_n as a function of temperature is also included in Fig. 7. A comparison of the nucleation and growth curves shows that, as a rule, a larger amount of undercooling is necessary to promote nucleation than is required for growth.

It is an inherent assumption in the concept of the steady state nucleation rate that the melt temperature is constant for a time sufficient to establish an equilibrium distribution of clusters in the melt (Chalmers, 1977). If a melt is undercooled suddenly, a finite time lag t_n is observed during which a steady state nucleation rate builds up (Kashiev, 1969).*

No exact theory describing nucleation from alloy melts seems to be currently available, although semiempirical formulas for I_n have been published (Uhlmann, 1972). However, it appears to be intuitively clear that nucleation tends to slow down when it is associated with a change in composition in a way similar to that discussed for growth.

2. GLASS FORMATION

It follows from the above discussion that, at least in principle, an alloy melt can be cooled too fast for either nucleation or growth to take place. If the undercooling is great enough, the melt will, in this case, retain its structure indefinitely and be called a glass (Turnbull, 1969). The question then is, What cooling rate (or what interface velocity) is required for glass formation from a melt of given properties?

---

* A time lag is also observed in the solid state growth of amorphous semiconductors (Köster, 1978).

Two cases can be distinguished depending on whether the liquid–solid transformation is nucleation-limited or growth-limited. For the *nucleation-limited* case, a condition for glass formation was formulated by Uhlmann (1972) in terms of a maximum crystallized volume fraction of the undercooled melt. The crystallized volume fraction X as a function of time can be expressed by

$$X = (\pi/3)\, I_n v^3\, t^4 \tag{7}$$

for small X. Equation (7) assumes that the melt is uniformly undercooled, i.e., that the nucleation rate and growth velocity are the same everywhere in the melt. The conditions for glass formation can now be formulated by specifying a maximum value for X, e.g., $X = 10^{-6}$, for which the solid cannot be distinguished from a glass. If the nucleation rate is taken to be time-independent, thus neglecting the time lag t_n, so-called time–temperature–transformation (TTT) curves can be constructed from Eq. (7), as illustrated in Fig. 8. The curve indicates the region where $X > 10^{-6}$. The critical cooling rate can be read off the curve as indicated by the dashed line. It is clear from the figure that, for glass formation, not only should the cooling rate exceed the critical one but also the final temperature should be low enough for the amorphous phase to be stable; i.e., it should not exceed the glass temperature.

If applied to heat pulse-induced melt quenching, the above treatment tends to overestimate the critical cooling rate, since it is based on the

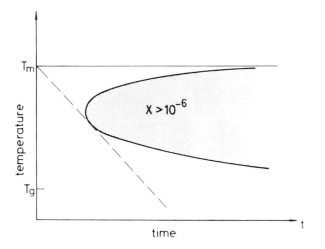

Fig. 8.   Schematic time–temperature–transformation diagram of an undercooled melt, showing the region where the crystalline volume fraction $X > 10^{-16}$. The dashed line indicates the minimum cooling rate for glass formation.

steady state nucleation rate. However, as demonstrated in Fig. 3, the lifetime of a liquid phase for nanosecond heat pulses is only on the order of 10^{-7} sec, and the time during which a given portion of the melt is undercooled,

$$\Delta t \cong \Delta T_i / (\partial T / \partial t)_i$$

is likely to be even smaller. In this time regime, the presence of a time lag for nucleation is most probably not negligible.

A *growth-limited* situation is present if a seed crystal is already in contact with the melt when solidification starts. In this case, the critical cooling rate for glass formation depends on the crystal growth velocity. As shown in Fig. 7, the crystal growth velocity as a function of undercooling has a maximum. From Eq. (6) we find that, for the temperature at which v is maximum,

$$T_{max} = T_m / [1 + (kT_m / \Delta H_m) \ln(1 + B)] \tag{8}$$

where $B = \Delta H_m / Q$. (Since Q in general is not independent of temperature, a value for B appropriate for T_{max} must be used.) The maximum growth velocity is then

$$v_{max} = v(T_{max}) = u_0 \exp(-Q/kT_m) \frac{B}{(1 + B)^{(1+1/B)}} \tag{9}$$

which must now be compared to the velocity $v(T_{max})$ of the corresponding isothermal surface. If the velocity of this isothermal surface is greater than v_{max}, no crystal can grow and the melt will freeze into an amorphous phase. Note that, in this case, the rate of liberation of latent heat will be reduced, since the heat of melting of a glass is less than that of the corresponding crystalline phase. This, as shown in Fig. 3, will even accelerate the velocity of the isothermal surface.

3. SOLIDIFICATION INSTABILITIES

The structures resulting from the solidification of concentrated alloy melts are often found to be inhomogeneous and to contain complicated microstructures. This is particularly true in samples of the unlimited supply type (Fig. 6b). These microstructures can result from a variety of causes, the most important of which seems to be the effect of *constitutional supercooling* (CSC).

The occurrence of CSC in solidification means that the undercooling $\Delta T = T_m - T$ is larger in the melt ahead of the *l*–s interface than at the interface itself, as illustrated in Fig. 9. This causes random protrusions in

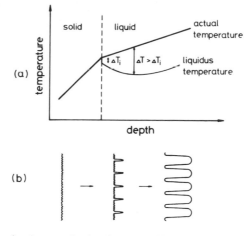

Fig. 9. Cell formation by constitutional supercooling. (a) Temperature profiles. (b) Sequence showing growth of random protrusions in the crystal boundary.

the crystal boundary to grow faster than their surroundings. As a result, small protrusions grow into columns, the shape of which depends on the details of the microscopic mass and heat flow (Chalmers, 1977). Part of the remaining melt is trapped between adjacent columns and eventually forms the "cell" walls. As an example, Fig. 10 shows cell formation in a laser-irradiated Pd film on Si. In this case, up to five different Pd–Si phases were detected in the same layer, and the Si in the columns was epitaxial (von Allmen *et al.*, 1980c).

CSC is a phenomenon well known in metallurgy. A necessary condition for its occurrence is that the liquidus temperature of the melt increase with the distance from the *l*–s interface. This is possible only in the presence of a compositional gradient in the melt which, in conventional metallurgy, is usually present only if segregation occurs. However, for the unlimited supply samples, large compositional gradients are necessarily present when solidification starts, and CSC can result even without segregation (von Allmen *et al.*, 1980c). This can be seen from Fig. 11 which presents a simple scheme for predicting if CSC is possible in a particular binary system: The liquidus temperature versus the composition as given by the phase diagram is combined with a plot of concentration versus depth, as shown in Fig. 6b, to obtain a plot of liquidus temperature versus depth $T_m(z)$. The actual temperature profile in the melt $T(z)$ at various times is also included in the plot. The condition of CSC is now fulfilled wherever the difference $T_m(z) - T(z)$ increases away from the *l*–s interface. In the example in Fig. 11 (and neglecting for a moment the presence

30 nsec laser pulse

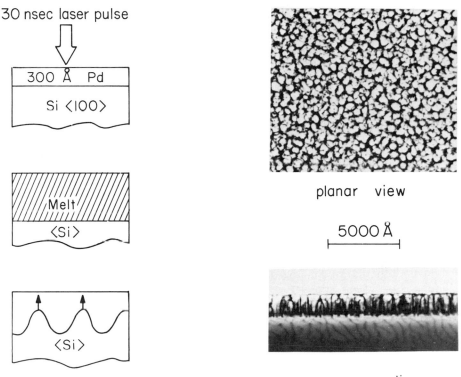

planar   view

5000 Å

cross   section

Fig. 10.   Schematic and TEM results showing the formation of a cellular structure by constitutional supercooling in a laser-irradiated Pd film on a Si substrate. (From von Allmen *et al.,* 1980c.)

of interfacial undercooling), no CSC is present for the temperature profile T_0 (initial profile) or T_1, but the dashed area shows CSC for the profile T_2. In other words, by the time the temperature is given by T_2 and the *l*–s interface has moved to z_2, unstable solidification is expected between z_2 and the surface. From the scheme in Fig. 11, it can be easily seen that CSC is more likely to occur if the melting point of the deposited film exceeds that of the substrate than if the reverse is true.

In addition to the structures caused by CSC, a different kind of cellular structure has been reported (van Gurp *et al.,* 1979), which was attributed to *hydrodynamic currents* in the melt driven by temperature or concentration gradients (Bénard cells). This kind of instability seems to occur mainly in the case of narrowly focused laser beams where appreciable lateral temperature gradients are created.

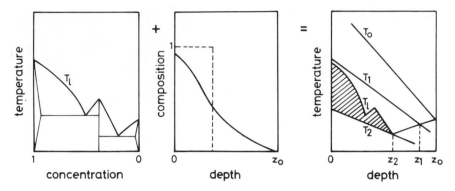

Fig. 11.   Schematic for predicting the occurrence of constitutional supercooling in unlimited supply-type samples from the phase diagram. (From von Allmen *et al.*, 1980c.)

Another phenomenon pertinent to this regime is the *formation of vapor bubbles* (von Allmen *et al.*, 1980c). This occurs when the melting point of a deposited metal layer is high enough for the substrate material to reach a substantial vapor pressure. An example of vapor bubbles in laser-irradiated W on Si is shown in Fig. 12. In this case, solidification seems to

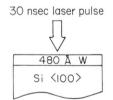

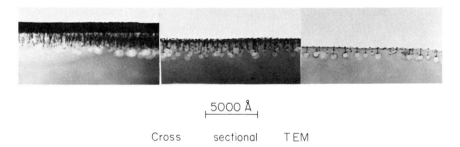

Fig. 12.   TEM micrograph showing bubble formation in laser-irradiated W on Si. The micrographs from left to right were obtained by progressively thinning the same sample area. (From von Allmen *et al.*, 1980c.)

start not only from the Si substrate but, independently, also from the surface which is rich in the refractory metal. At this instant, part of the underlying substrate is still above its equilibrium vapor point, a situation that could be called *constitutional superheating*. The mechanisms believed responsible for the structure in Fig. 12 are discussed in more detail in Section IV.B.3 of this chapter.

IV. Experimental Results

This section summarizes recent experimental results on phase transformation and compositional changes obtained from thin-film systems for three different initial sample configurations. The three typical initial concentration profiles include (a) Gaussian (or similar to a delta function) distribution of element B in substrate A, (b) rectangular distribution of element B on top of substrate A, and (c) square wave distribution of elements A and B on an inert substrate as shown in Fig. 6a–c.

A. Gaussian Profile

An initial Gaussian concentration profile is usually obtained by implanting element B in substrate A or by evaporating a thin layer (~5 Å) of element B buried in deposited substrate material A (Fig. 13). In the case of species implanted in Si, the situation has been fully discussed in previous chapters (see Chapters 5–7) with emphasis placed on implanted impurities that do not form compounds with the Si substrate. For implanted dopants of group III and V elements, it was found that, as a result of laser-induced melting, mixing, and quenching, Si amorphized by implantation regrew into a single-crystal film with the implanted species at substitutional sites. The concentration of the implanted impurity is usually low, on the order of 1 at.% or less. At sufficiently high regrowth velocities, there is little or no segregation for these elements (White *et al.*, 1980). The concentration profile after regrowth is often essentially the diffusion profile present in the molten layer before regrowth.

In high-dose implantation into Si of transition elements such as Pt, Cu, and Fe, (Cullis *et al.*, 1979) which form silicides in the solid state, pronounced segregation of the transition metals to the Si surface was observed. On the other hand, thermal annealing of the as-implanted samples showed little or no redistribution of elements in the case of Pt and Fe and deep in-diffusion of Cu into the Si bulk. Two interesting observations after

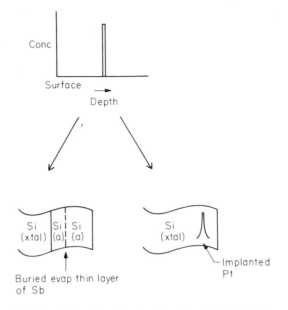

Fig. 13. Schematics showing the Gaussian distribution.

laser irradiation should be noted here: (i) The highly segregated Si surface shows a cellular structure with the implanted atoms locating at the cell walls and forming equilibrium silicides and/or metastable phases (Cullis *et al.*, 1979, 1980); and (ii) the solubility of the implanted species is significantly enhanced after laser irradiation (three orders of magnitude for Pt in Si). The amount of surface segregation, the cellular structure, and impurity substitutionality were found to be critically dependent on the solid–liquid interfacial velocity. Generally speaking, high regrowth velocity increases dopant substitutionality, decreases surface segregation, and discourages cellular structure formation (Cullis *et al.*, 1980; Baeri *et al.*, 1981). These phenomena for metals implanted in Si are discussed in detail in Chapter 6.

For buried thin deposited layers of Sb in Si (Liau *et al.*, 1979), it was found that the redistribution of Sb after laser irradiation followed closely a calculated liquid phase diffusion profile, taking the finite velocity of the liquid–solid interface and the melt depth into consideration. The microstructure of this type of sample has not been examined; however, the major advantage of this sample configuration is to allow a reasonably accurate estimate of the melt depth and surface melt time as a function of incident laser energy, since the buried thin layer can be placed at a depth not easily accessible by ion implantation (see Fig. 14).

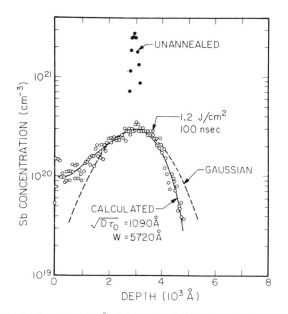

Fig. 14.   Redistribution of a ~10-Å Sb layer (buried in a deposited amorphous Si layer of 3600 Å thickness) after pulsed electron annealing. The asymmetry of the diffusion profile is due to a depth dependence of the diffusion time on the melt depth w. (From Liau *et al.*, 1979.)

B.   Rectangular Distribution

This type of sample configuration (deposited layer on semiconductor substrate) presents the most commonly used initial concentration profile for phase transformation experiments (Fig. 15). In this case, there is essentially an unlimited supply of substrate material (i.e., Si) for reaction with the deposited layer. The redistribution of the initial concentration, and therefore the phases formed, depends on the melt depth during laser

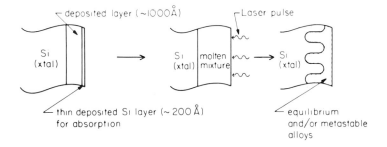

Fig. 15.   Schematics showing the rectangular distribution.

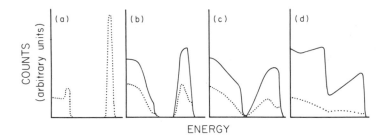

Fig. 16. Backscattering spectra from laser-processed Ge–Si heterojunctions. (a) A channeling orientation spectrum of the unprocessed sample. (b–d) Random and channeling orientation spectra for samples processed at 5.6, 8.1, and 12.1 J/cm^2, respectively (125 nsec Nd:YAG) and for which Ge χ_{min} values are 21, 19, and 9%, respectively. (From Leamy *et al.*, 1980.)

irradiation. For highly reflective surfaces (such as Au), it is sometimes necessary to deposit a thin amorphous semiconductor layer (e.g., a few hundred angstroms of Si) on top of the reflecting surface to facilitate laser energy absorption. Three types of material systems are described here: (i) systems in which complete solid solutions are formed (completely miscible systems), e.g., Ge–Si; (ii) simple eutectic systems, e.g., Au–Si; and (iii) systems in which compounds are formed, e.g., Pt–Si.

1. Completely Miscible Systems

The Ge–Si system presents one of the most interesting couples in this category because of the possibility of heterostructure and a tailored band gap in Ge–Si alloys. One of the first successful heterostructures of Ge on Si was obtained by depositing a layer of Ge on ⟨100⟩ Si substrate followed by pulsed laser and/or electron beam irradiation (Lau *et al.*, 1978, 1979). Subsequent investigation (Leamy *et al.*, 1980) showed that an increase in pulse energy density beyond the threshold resulted in increased crystal quality and decreased junction abruptness (Fig. 16). Cellular microstructures accompany the growth of concentrated Ge alloy as a result of constitutional supercooling, and the cells are surrounded by misfit dislocations (Fig. 17). It appears that Ge–Si heterojunctions of device quality cannot be fabricated by laser processing unless the intervention of cellular growth and consequent formation of misfit dislocation arrays during recrystallization can be eliminated. Recent experimental investigation of Ge–Si heterojunctions obtained in the solid state regime appear to be more promising for device application (Tsaur *et al.*, 1981a).*

* To the best of our knowledge, the first successful experiment of solid phase epitaxial growth of deposited Ge on Si was performed by Hung (1980).

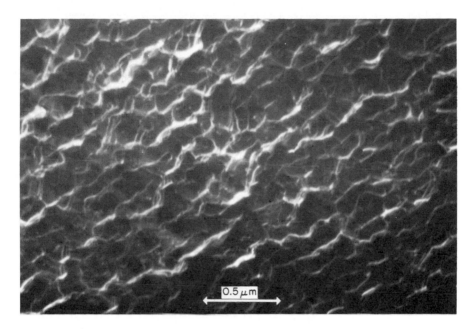

Fig. 17. [220], weak-beam, dark-field micrograph of defect structure of a Ge–Si sample after processing at 12.1 J/cm². (From Leamy *et al.*, 1980.)

2. EUTECTIC SYSTEMS

Since the first demonstration of an Au–Si metallic glass made by splat cooling in 1960, a multitude of metastable phases (amorphous or crystalline) have been obtained by rapid quenching of the melts (Duwez, 1976). With the extremely high cooling rate made available by short laser pulses, it was a natural consequence to investigate the glass-forming characteristics of classical deep eutectic systems under laser irradiation. The configuration of a metal layer deposited on a semiconductor substrate is suitable for the study of glass formation at the metal–semiconductor interface under unlimited supply conditions. Figure 18 shows the schematics and backscattering spectra for an Au–Si sample before and after laser irradiation (Lau *et al.*, 1981). The thin Si layer (~300 Å) on top of the Au layer (~2500 Å) was to facilitate the absorption of laser power (Nd–glass, 30-nsec pulse). After irradiation, the Au signal spread in width in both directions, indicating that the Au layer had reacted with crystalline as well as amorphous Si. From the decrease in the height of the Au signal near the Au–Si substrate interface, the mixed layer has a composition of $Au_{82}Si_{18}$ with an amorphous structure. It is interesting to note that the composition of the mixed layer, $Au_{82}Si_{18}$, is the eutectic composition of the Au–Si

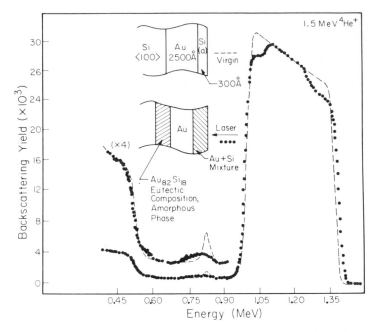

Fig. 18. Backscattering spectra of a Au–Si sample (unlimited supply) before and after pulsed laser irradiation. The top amorphous Si layer is used to facilitate absorption. The composition of the mixed layer near the crystal–Au interface is $Au_{82}Si_{18}$. (From Lau *et al.*, 1981.)

system. Investigations on three other metal–semiconductor systems (Au–Ge, Al–Ge, and Ag–Si) showed that mixing under unlimited supply conditions produced mixed layers with a composition very close to the eutectic composition of the system under investigation (Lau *et al.*, 1981). The structure of the mixed layer is always metastable (amorphous and/or crystalline). The results for these four metal–eutectic systems are summarized in Table II. We believe that laser irradiation induces melting, mixing, and fast quenching near the metal–semiconductor interface. Under the conditions of an unlimited supply of metal and semiconductor atoms, the system selects its eutectic composition to be the composition of the mixed layer probably due to the lowest melting temperature. After liquid phase interdiffusion, the mixed liquid solidifies very rapidly, resulting in a mixed layer with a metastable structure.

3. SILICIDE-FORMING SYSTEMS

Laser annealing as a means of forming metal silicide contacts on Si provides the advantage of transferring energy to only a very localized

TABLE II

PULSED LASER MIXING OF FOUR EUTECTIC SYSTEMS

Structure of specimen	Au–Si, eutectic composition $Au_{82}Si_{18}$, T_{eut}. 370°C		Au–Ge, eutectic composition $Au_{73}Ge_{27}$, T_{eut}. 356°C		Al–Ge, eutectic composition $Al_{70}Ge_{30}$, T_{eut}. 424°C		Ag–Si, eutectic composition $Ag_{87}Si_{13}$, T_{eut}. 855°C	
Before irradiation	Au–Si substrate Uniform $Au_{82}Si_{18}$ layer	Multilayers of $Au_{71}Si_{29}$ Uniformly mixed	Au–Ge substrate Uniform $Au_{60}Ge_{40}$ layer	Multilayers of $Au_{73}Ge_{27}$ Uniformly mixed	Ge–Al substrate Uniform $Al_{60}Ge_{40}$ layer	Multilayers of $Al_{70}Ge_{30}$ Uniformly mixed	Ag–Si substrate $Au_{86}Si_{14}$ layer	Multilayers of $Ag_{87}Si_{13}$ Mixed
After irradiation[a]	A	A	MX	MX	A plus trace of Al	A plus trace of Al	MX	MX

[a] A, Amorphous; MX, metastable crystalline.

region instead of heating up the whole wafer. Because of this practical aspect, a wealth of experimental investigations are reported in the literature. One of the first experiments of this kind involved pulsed laser irradiation of Si substrates covered with a Pd-deposited layer (von Allmen *et al.*, 1980c; Wittmer and von Allmen, 1979). Figure 10 shows the schematic and TEM results for Pd–Si (von Allmen *et al.*, 1980c). The mixed layer is nonhomogeneous laterally as well as in depth. The reacted structure is cellular and consists essentially of columns of pure Si, surrounded by thin walls of silicide (black in the micrograph). Backscattering analysis shows that the Si in the columns is partially epitaxial (channeling yield for the Si $\approx$30–70% of random yield). The overall Si content of the layer depends on the laser energy (melt depth). The silicide is polycrystalline and consists mainly of Pd_2Si, PdSi, and other silicides. The irradiated samples undergo phase transformation upon subsequent thermal annealing. There is a general tendency for the complicated multisilicide coexistence to return to a state where only one or two silicides exist, a condition usually found only in samples processed with thermal annealing. The postirradiation annealing situation is not entirely clear at present and is still an active area of research (Wittmer and von Allmen, 1981; Stacy *et al.*, 1980). For the Pt–Si system, the laser-induced interactions are very similar to those described for Pd–Si. The phases and resistivities of laser-induced and furnace-annealed Pd and Pt silicides are summarized in Table III.

It appears that the cellular structure of laser-induced silicides is a common feature of most silicide-forming systems investigated (i.e., Pd, Pt, W, Co, Fe, Mo, Stacy *et al.*, 1980; there was no evidence of cell structure for Mg, Wittmer *et al.*, 1979) and that cellular structure is a consequence of solidification instability due to constitutional supercooling (see Section

TABLE III

COMPOSITION AND RESISTIVITY OF LASER-INDUCED AND FURNACE-ANNEALED PALLADIUM AND PLATINUM SILICIDES

Metal	Annealed in furnace			Formed by laser irradiation Q-switched Nd : YAG		
	Annealing conditions	Com-pound	Resistivity ($\mu\Omega$ cm)	Laser fluence (J/cm^2)	Compounds	Resis-tivity ($\mu\Omega$ cm)
Evaporated Pd	450°C, 1 hr	Pd_2Si	25–35	2.1–2.4	Pd_5Si, Pd_4Si, Pd_3Si, Pd_2Si, PdSi	150–300
Evaporated Pt	600°C, 1 hr	PtSi	35–50	2.1–3.5	Pt_3Si, Pt_2Si, $Pt_{12}Si_5$, PtSi	250–600

III.D.3). Generally, the morphology of the cellular structure and the composition of the silicide layer depend upon experimental conditions such as the melt depth and the system. For the W–Si system, the morphological features of the reacted structure are different from those observed for Pd–Si and Pt–Si. Figure 12 is a cross-sectional TEM micrograph of a W–Si sample after laser irradiation. The reacted layer consists of thin needles of WSi_2 (the only compound observed) embedded in single-crystal Si where each ends in a spherical void. Apart from the voids, the appearance of WSi_2 needles is equivalent to the growth of Si columns in the case of Pd–Si, and we believe that it is due to constitutional supercooling. The striking difference from the Pd–Si and Pt–Si cases is the appearance of voids. The size of the voids (all perfectly spherical) shown in Fig. 12 ranges from 150 to 800 Å. The voids are believed to form as vapor bubbles in the overheated molten Si. At the melting point of W (3407°C), the vapor pressure of pure Si is about 2 atm and the maximum pressure reached is probably higher. However, at a vapor pressure on the order of 10 atm or less, spontaneous nucleation of vapor bubbles is unlikely to occur in the short lifetime of the superheated liquid. The fact that the bubbles are always connected to a "stem" of WSi_2 suggests that the latter acts as a nucleation center for them. The stems must therefore have grown from the already solidified W-rich surface layer into the molten Si. The decrease in the bubble size toward the surface implies that the layer bubbles nucleated earlier than small ones and therefore had more time to grow before they were frozen by the rapidly advancing Si solidification front.

The cellular structures obtained from an interaction between a Si single crystal and Fe, Co, and Ni are particularly interesting. These three elements form disilicide ($FeSi_2$, $CoSi_2$, and $NiSi_2$) skeletons which, although twinned, are coherent with the Si matrix. Figure 19 shows a dark-field micrograph of a laser-irradiated Co–Si sample with a $CoSi_2$ (200) reflection. Figure 20 shows the twinning relationship between the Si matrix and the $CoSi_2$ skeleton. It is believed that skeletons form because of the low surface energy associated with the coherent interface. The stability of these structures against postirradiated thermal annealing is probably also related to the interfacial energy (van Gurp et al., 1979; Stacy et al., 1980).

C. Square Wave Distribution

The square wave distribution approach is to limit the composition of the alloy by depositing a fixed amount of each component in sequence on an inert, nonabsorbing substrate (limited supply). This eliminates the influence of the melt depth and allows the investigation of phase transforma-

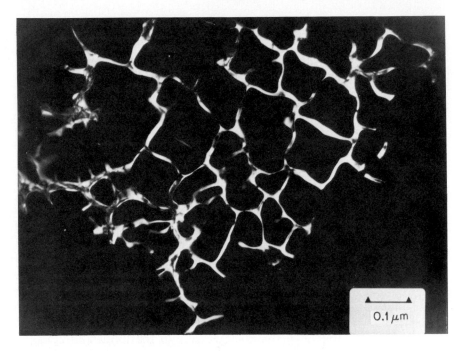

Fig. 19. Dark-field micrograph of the $CoSi_2$ skeleton in Si imaged with a (200) $CoSi_2$ reflection. (From Stacy *et al.*, 1980.) This figure was originally presented at the Spring 1980 Meeting of The Electrochemical Society, Inc., held in St. Louis, Missouri.

tion at a predetermined composition (Fig. 6c). Two types of systems are discussed here: (i) metal–semiconductor eutectic systems, and (ii) silicide-forming systems.

1. The Au–Si System

It is well known that Au and Si form a metallic glass near the eutectic composition under splat cooling conditions (Duwez, 1976) (quench rate $\sim 10^5$–10^6 °C/sec). With the limited supply approach, the extent of glass-forming ability of Au–Si can be investigated at extremely fast quench rates which are induced by pulsed laser irradiation and are not accessible with splat cooling techniques. Experiments (von Allmen *et al.*, 1980a) of this kind have been done using samples where multiple layers of Au and Si were vacuum-deposited on sapphire substrates. The thicknesses of the layers were adjusted such that the average film composition ranged between $AuSi_{10}$ and $Au_{10}Si$, i.e. from 9 to 91 at.% Au. The individual layer thickness was not more than a few hundred angstroms, with a total layer

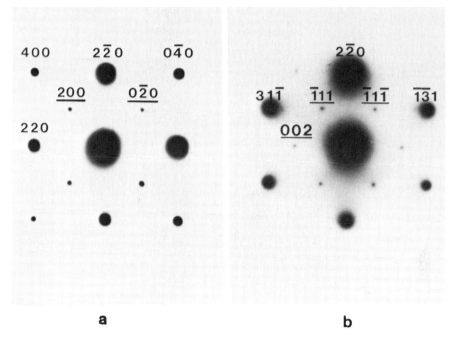

Fig. 20. Electron diffraction pattern; Si spots indexed *hkl*, CoSi$_2$ spots indexed *hkl*. (a) Region where matrix and skeleton have the same orientation; B = [001]. (b) Region where skeleton is twinned with respect to the matrix; B = [114]. (From Stacy *et al.*, 1980.) This figure was originally presented at the Spring 1980 Meeting of the Electrochemical Society, Inc., held in St. Louis, Missouri.

thickness of about 2000 Å. The surface layer was chosen to be Si and served the purpose of an antireflection coating. The samples were irradiated with pulses from a Nd–glass laser operated either in the free-running ($\tau \approx 300$ μsec) or in the Q-switched mode ($\tau \approx 30$ nsec). This produced melting and subsequent quenching of the films with estimated average cooling rates on the order of 10^6 or 10^{10}°C/sec, respectively.

The composition of the irradiated samples remained the same as the initial average film composition, as monitored by backscattering. As an example, Fig. 21 shows backscattering spectra for a sample with the composition AuSi$_5$ before and after irradiation with a 30-nsec pulse. In general, the layers mixed by 30-nsec pulses showed some residual waviness in composition versus depth, whereas those obtained with 300-μsec pulses were very uniform. This is a natural consequence of the liquid-mixing mechanism. The time available for melt interdiffusion is on the order of 10^{-7} sec for 30-nsec pulses and 10^{-3} sec for 300-μsec pulses,

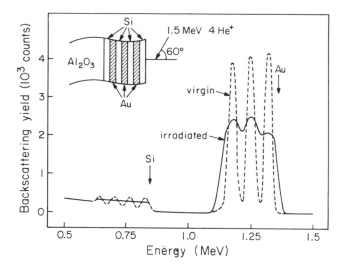

Fig. 21. Backscattering spectra of a $AuSi_5$ sample before and after irradiation with a Q-switched pulse of 30-nsec duration, Nd–glass laser. (From von Allmen *et al.*, 1980a.)

corresponding to diffusion lengths of a few hundred angstroms and a few micrometers, respectively, if a liquid diffusivity of 10^{-4} cm^2/sec is assumed.

X-ray diffraction analysis of films irradiated by 30-nsec pulses revealed an amorphous structure in all but the most Au-rich (91 at.%) samples, usually together with traces of a metastable Au–Si compound. The films irradiated with 300-μsec pulses were polycrystalline and consisted of the same metastable compound, generally together with precipitates of the predominant component.

The structure of the metastable polycrystalline silicide observed in all cases is different from that of previously reported metastable Au–Si phases (Suryanarayana and Anatharaman, 1979; Green and Bauer, 1970). The observed d spacings in our case can be satisfactorily fitted to a hexagonal structure with a crystalline/amorphous ratio of 1.24. The structure is similar to the one found by the thermal transformation of an ion beam-produced amorphous Au_5Si_2 alloy (Tsaur *et al.*, 1980).

All laser-irradiated films were unstable at room temperature. Their initial shiny silver color changed gradually to a gold color in a matter of hours (Au-rich films) to days (Si-rich films), indicating that phase separation into polycrystalline Si and Au had occurred. This explains the mixture of phases revealed by x-ray analysis which usually takes 4–8 hr.

Figure 22 shows a comparison of the stability of the laser-quenched Au–Si amorphous phases measured by the amorphous–crystalline

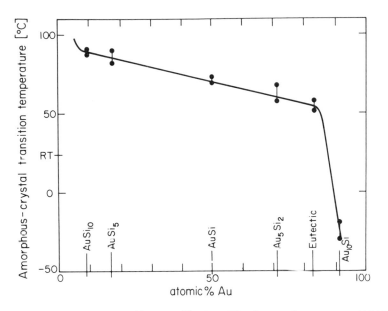

Fig. 22. Amorphous-metastable crystalline transition temperatures versus composition as obtained from the resistivity measure. (From von Allmen *et al.*, 1980a.)

(metastable) transition temperature (determined while heating the film at a rate of ~3°C/min) as a function of composition. It can be seen that, while the transition temperature decreases monotonically with increasing Au content, its slope is very different at different regions of the phase diagram. Pure amorphous Si is known to crystallize at ~650°C, and the presence of only 9 at.% Au decreases the temperature to about 80°C. In the medium compositional range, the transition temperature is nearly constant, but it decreases again dramatically near the Au-rich end.

The compositional range of Au–Si glasses reported here greatly exceeds previously established limits for glass formation by liquid quenching. Glass-forming ability is believed to be related to melting point depression (Donald and Davies, 1978), which is largest at the eutectic point. The only Au–Si glasses reported so far have eutectic compositions (Klement *et al.*, 1960). A depression of ≥20% of the actual melting temperature (relative to the weighted average of the melting temperatures of the components) has been found necessary to produce glasses in metallic systems for quenching rates in the 10^5–10^7°C/sec range. In contrast, the most Si-rich glasses reported here (17 and 9 at.% Au) have a melting point depression, as defined above, of only 6 and 2%, respectively. However, neither pure Si nor Au can be obtained in a glassy phase by irradiation with Q-switched pulses. To provide a more complete picture of the reactions of eutectic systems, four systems with a limited supply configuration are included in Table II.

A comparison of the results obtained from unlimited supply samples (Section IV.B) with those from limited supply samples of fixed composition shows that the eutectic composition is favored if the system is free to select the composition of the mixed layer, whereas the limited supply approach provides information on the compositional range of metastable phase formation.

2. SILICIDE-FORMING SYSTEMS

In this section, the application of the same limited supply techniques to the systems Pt–Si and Pd–Si (von Allmen *et al.*, 1980b) (both exhibit several equilibrium compounds) are discussed.

The samples consisted of multiple layers of metal and Si, alternately vapor-deposited onto sapphire substrates. Individual layers were no more than a few hundred angstroms thick, and the total film thickness was about 1000 Å. The samples were irradiated in air and at room temperature with pulses from a Q-switched Nd–glass laser ($\tau \approx 30$ nsec). The power density was adjusted to induce melting of the whole layer without damaging the surface.

Unlike the case of Au–Si, not all the compositions tested could be quenched into an amorphous state. Generally speaking, polycrystalline films resulted from samples with compositions at or close to those of the congruently melting compounds, Pt_2Si, PtSi, and Pd_2Si, PdSi, respectively. Figure 23 shows the phase diagrams of the two systems along with arrows marking the compositions of the films studied. The letters beneath the arrows give the structure of the as-irradiated films.

The two systems reported here show rather close similarities in their compositional range of glass-forming ability (GFA) (Donald and Davies, 1978). As Fig. 23 demonstrates, there are at least two distinct regions for GFA at the present cooling rate: one is at compositions near the metal-rich deep eutectic, and the other one extends from the Si-rich end of the phase diagram to compositions near the Si-rich eutectic. (Compositions between the congruently melting phases were not tested.)

Compositions considered to show GFA by conventional liquid quenching techniques are those of the metal-rich eutectics ($Pt_{77}Si_{23}$ and $Pd_{84}Si_{16}$) (Donald and Davies, 1978).

Intimate mixing of the components by laser irradiation was found to be essential in obtaining amorphous films; incompletely mixed films contain-

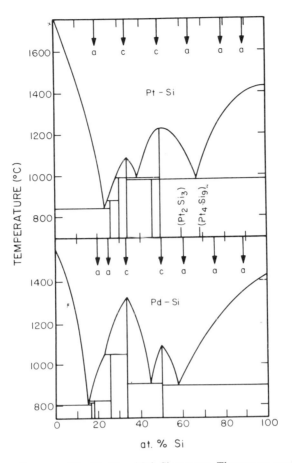

Fig. 23. Phase diagrams of the Pt–Si and Pd–Si systems. The arrows mark the compositions of films used in this study. Small letters indicate the structure of the As-quenched films
(a, amorphous; c, polycrystalline). (From von Allmen *et al.*, 1980b.) See note added in proof.

ing local compositions too close to that of a congruently melting compound were not amorphous after irradiation but rather showed the presence of the congruently melting compound. The reason for this could lie in
the liberation of the heat of formation of the compound, which reduces the
cooling rate as discussed in Section III.B. The fact of a composition-
dependent effective cooling rate notwithstanding, it seems clear that the
cooling rate necessary to quench a liquid binary mixture into a glassy state
is highest for compositions close to those of congruently melting compounds, as well as to those of the pure components.

The amorphous-to-crystalline transformation was measured by resistance change as a function of temperature (at a constant temperature rise of ~3°C/min). The results are summarized in Table IV.

With reference to Table IV the Pt-rich amorphous films (20 at.% Si), unlike all the others, did not show a detectable drop in resistivity; therefore an unambiguous amorphous–crystalline transition temperature cannot be defined in this case. X-ray analysis carried out for samples annealed to 170°C showed only the precipitation of pure Pt. The resistivity curve showed an abrupt *increase* at 200°C after which diffraction lines of Pt and Pt_3Si plus traces of an unidentified metastable phase were found. Samples annealed up to 550°C showed exclusively Pt and Pt_3Si in accordance with the equilibrium phase diagram. At ~400°C the Si-rich amorphous films all transformed into the metastable silicides Pt_2Si_3 or Pt_4Si_9 discovered recently (Tsaur *et al.*, 1979). No Si lines were detectable after transformation, indicating that any excess Si present was still amorphous at this stage.

While the films with 64 and 80 at.% Si showed only lines of one compound (Pt_2Si_3 and Pt_4Si_9, respectively) after transformation, the situation is not equally clear in the case of 90 at.% Si. In addition to the set of

TABLE IV

AMORPHOUS–CRYSTALLINE TRANSFORMATION TEMPERATURES AND CRYSTALLINE PHASES (AS IDENTIFIED FROM X-RAY ANALYSIS) OF Pt–Si AND Pd–Si FILMS

Composition (at.% Si)	T_{ac} (°C)	As-irradiated	Crystalline phases after a–c transition	At 550°C
Pt–Si film				
20	<170	None (amorphous)	Pt	—
	200	—	Pt, Pt_3Si[a]	Pt, Pt_3Si
33	—	Pt_2Si	—	Pt_2Si
50	—	PtSi	—	PtSi
64	390	None	Pt_2Si_3	Pt_2Si_3, PtSi, Si
80	400	None	Pt_4Si_9	PtSi, Si
90	400	None	Pt_2Si_3[a]	PtSi, Si[a]
Pd–Si film				
20	365	None (amorphous)	Pd, Pd_5Si, Pd_9Si_2, Pd_3Si	Pd, Pd_5Si, Pd_9Si_2, Pd_3Si
33	—	Pd_2Si	—	Pd_2Si
50	—	PdSi	—	PdSi
60	70	None	PdSi[a]	PdSi, Si
76	100	None	PdSi[a]	PdSi, Si
88	100	None	PdSi	PdSi, Si

diffraction lines identified with Pt_2Si_3, some extra lines were present, possibly indicating the presence of an additional, as yet unidentified, metastable silicide. Annealed to sufficiently high temperatures, all three metastable films decayed into PtSi and Si, although this equilibrium situation was not always reached at 550°C.

In spite of the similarities between the Pt–Si and Pd–Si systems in terms of GFA, the thermal decomposition study revealed significant differences in the stability of the respective amorphous films. In Table IV the Pd-rich film (20 at.% Si) is shown to transform at 365°C into a mixture of equilibrium phases. Identification of the rather complicated set of diffraction lines is based on a recent reevaluation of the Pd-rich part of the phase diagram (Duwez, 1980). The simultaneous presence of several phases in this case may be due to inhomogeneities in film composition because of incomplete mixing by the laser pulse. Surprisingly, the Si-rich amorphous Pd–Si films (60, 76, and 88 at.% Si) turned out to be very unstable and transformed at ∼100°C. After transformation, they all contained mainly PdSi, while precipitation of pure Si was detectable only at higher temperatures. (See note added in proof.)

In the Pd–Si samples, the monosilicide is found to precipitate at a temperature as low as 100°C, while in the Pt–Si case this happens only well above 400°C. This observation is in contrast to previous experience with solid state silicide growth. There it was found that PtSi could be grown at ∼400°C, while PdSi required ∼800°C (both grow by the reaction of Si with the metal-rich silicides formed at lower temperatures). The present situation reflects the growth of a compound from a homogeneous amorphous phase. In conventional solid state silicide growth, the kinetics are dominated by energy barriers associated with interfaces or mass transport. A difference is also found in the structure of the silicide films: All crystalline phases reported in Table IV appear to be randomly polycrystalline, while silicides formed by solid-state growth often have a textured structure.

In the dependence of thermal stability of the amorphous films (reflected by T_{ac}) on composition, some trends seem to emerge when the systems studied here are compared with Au–Si. In reference to the Si-rich amorphous films only, it is seen that T_{ac} tends to increase with increasing Si content (pure amorphous Si crystallizes at ∼650°C), and the same is true for Au–Si. In fact, for the parts of the Pt–Si and Pd–Si phase diagrams with ≥50 at.% Si, similar T_{ac}-versus-composition curves can be drawn as for the simple eutectic Au–Si system. On the other hand, there is no obvious correlation between the transformation temperatures of the Si-rich and the metal-rich amorphous films for a given system. This can be interpreted as indicating that the thermal stability of an amorphous binary film depends only on the nucleation and growth behavior of the available *neighboring* crystalline phase (metastable or stable) that first precipitates from it. Therefore, every part of a phase diagram bounded by two equilibrium phases is expected to behave in its own fashion as far as the thermal stability of amorphous mixtures is concerned.

D. Summary

Pulsed irradiation (laser or electron beam) typically induces melting, interdiffusion, and rapid solidification of near-surface regions. The composition and phases in the reacted layers depend on the pulse energy via the melt depth. The structure of the mixed layers is often laterally nonuniform and cellular in nature as a result of constitutional supercooling. Because of the very rapid solidification from the melt, not all alloys and compounds formed are necessarily thermodynamically stable. Alternatively, amorphous phases over extended compositional ranges may be obtained under appropriate conditions. Pulsed irradiation, therefore, permits a systematic search for, and the investigation of, metastable phases inaccessible by other means. These results are in marked contrast to those obtained with furnace or continuous-wave scanning laser annealing, where simple and uniform reactions take place in the solid phase.

Note added in proof: According to a recent reevaluation of the Pd–Si phase diagram, the monosilicide PdSi is not a congruently melting phase (Langer and Wachtel, 1981).

Acknowledgments

The authors are grateful to their colleagues in various laboratories for discussion and for providing figures. We are also grateful to the Böhmische Physical Society for inspiration and encouragement.

References

Baeri, P., Foti, G., Poate, J. M., Campisano, S. U., and Cullis, A. G. (1981). *Appl. Phys. Lett.* **38,** 800.
Baglin, J. E., d'Heurle, F. M., and Petersson, C. S. (1980). *Appl. Phys. Lett.* **36,** 594.
Chalmers, B. (1977). "Principles of Solidification." Krieger, Huntington, New York.
Cohen, M. H., and Turnbull, D. (1961). *Nature (London)* **189,** 132.
Cullis, A. G., Poate, J. M., and Celler, G. K. (1979). *In* "Laser–Solid Interactions and Laser Processing" (S. D. Ferris, H. J. Leamy, and J. M. Poate, eds.), p. 311. Am. Inst. Phys., New York.
Cullis, A. G., Webber, H. C., Poate, J. M., and Simons, A. L. (1980). *Appl. Phys. Lett.* **36,** 320.
Davies, H. A., and Lewis, B. G. (1975). *Scr. Metall.* **9,** 1107.

Donald, I. W., and Davies, H. A. (1978). *J. Non-Cryst. Solids* **30**, 77.

Duwez, P. (1967). *ASM Trans. Q.* **60**, 607.

Duwez, P. (1976). *Annu. Rev. Mater. Sci.* **6**, 83.

Duwez, P. (1980). Personal communication.

Green, A. K., and Bauer, E. (1970). *J. Appl. Phys.* **47**, 1284.

Hung, L. S. (1980). Personal communication.

Jackson, K. A. (1975). *In* "Treatise on Solid State Chemistry" (N. B. Hannay, ed.), Vol. 5, Chap. 5. Plenum, New York.

Kashiev, D. (1969). *Surf. Sci.* **14**, 209.

Klement, W., Willens, R. H., and Duwez, P. (1960). *Nature (London)* **187**, 869.

Köster, U. (1978). *Phys. Status Solidi A* **48**, 313.

Krautle, H., Nicolet, M.-A., and Mayer, J. W. (1974). *J. Appl. Phys.* **45**, 3304.

Langer, H., and Wachtel, E. (1981). *Z. Metallkd.* **72**, 769.

Lau, S. S., Tseng, W. F., Nicolet, M.-A., Mayer, J. W., Minnucci, J., and Kirkpatrick, A. R. (1978). *Appl. Phys. Lett.* **33**, 235.

Lau, S. S., Tseng, W. F., Golecki, I., Kennedy, E. F., and Mayer, J. W. (1979). *In* "Laser–Solid Interactions and Laser Processing" (S. D. Ferris, H. J. Leamy, and J. M. Poate, eds.), p. 503. Am. Inst. Phys., New York.

Lau, S. S., Tsaur, B. Y., von Allmen, M., Mayer, J. W., Stritzker, B., White, C. W., and Appleton, B. (1981). *Nucl. Instrum. Methods* **182/183**, 79.

Leamy, H. J., Doherty, C. J., Chiu, K. C. R., Poate, J. M., Sheng, T. T., and Celler, G. K. (1980). *In* "Laser and Electron Beam Processing of Materials" (C. W. White and P. S. Peercy, eds.), p. 581. Academic Press, New York.

Liau, Z. L., Tsaur, B. Y., Lau, S. S., Golecki, I., and Mayer, J. W. (1979). *In* "Laser–Solid Interactions and Laser Processing" (S. D. Ferris, H. J. Leamy, and J. M. Poate, eds.), p. 105. Am. Inst. Phys., New York.

Mohammadi, F. (1981). *Solid State Technol.* **24**, 65.

Muraka, S. P. (1980). *J. Vac. Sci. Technol.* **17**, 715.

Sinha, A. K., Giessen, B. C., and Polk, D. E. (1976). *In* "Treatise on Solid State Chemistry" (N. B. Hannay, ed.), Vol. 3, p. 1. Plenum, New York.

Stacy, W. T., van Gurp, G. J., Eggermont, G. E. J., Tamminga, Y., and Gijsbers, J. R. M. (1980). *In* "Laser and Electron Beam Processing of Electronic Materials" (C. L. Anderson, G. K. Celler, and G. A. Rozgonyi, eds.), p. 504. Electrochem. Soc., Pennington, New Jersey.

Suryanarayana, C., and Anatharaman, T. R. (1979). *Mater. Sci. Eng.* **13**, 73.

Takeuchi, S., ed. (1973). "The Properties of Liquid Metals," p. 521. Halsted, New York.

Tsaur, B. Y., Liau, Z. L., and Mayer, J. W. (1979). *Phys. Lett. A* **71A**, 270.

Tsaur, B. Y., Mayer, J. W., Nicolet, M.-A., and Tu, K. N. (1980). *In* "Ion Implantation Metallurgy" (C. M. Preece and J. K. Hirvonen, eds.), p. 142. Metall. Soc. AIME, New York.

Tsaur, B. Y., Fan, J. C. C., and Gale, R. P. (1981a). *Appl. Phys. Lett.* **38**, 176.

Tsaur, B. Y., Lau, S. S., Nicolet, M.-A., and Mayer, J. W. (1981b). *Appl. Phys. Lett.* **38**, 922.

Tu, K. N., and Mayer, J. W. (1978). *In* "Thin Films—Interdiffusion and Reactions" (J. M. Poate, K. N. Tu, and J. W. Mayer, eds.), Chap. 10. Wiley (Interscience), New York.

Tu, K. N., Chu, W. K., and Mayer, J. W. (1975). *Thin Solid Films* **25**, 403.

Turnbull, D. (1969). *Contemp. Phys.* **10**, 473.

Turnbull, D., and Bagley, B. G. (1975). *In* "Treatise on Solid State Chemistry" (N. B. Hannay, ed.), Vol. 5, Chap. 10. Plenum, New York.

Uhlmann, D. R. (1972). *J. Non-Cryst. Solids* **7**, 337.

478   MARTIN F. VON ALLMEN AND S. S. LAU

van Gurp, G. T., Eggermont, G. E., Tamminga, Y., Stacy, W. T., and Gijsbers, J. R. M. (1979). *Appl. Phys. Lett.* **35**, 273.
von Allmen, M., Lau, S. S., Mäenpää, M., and Tsaur, B. Y. (1980a). *Appl. Phys. Lett.* **36**, 205.
von Allmen, M., Lau, S. S., Mäenpää, M., and Tsaur, B. Y. (1980b). *Appl. Phys. Lett.* **37**, 84.
von Allmen, M., Lau, S. S., Sheng, T. T., and Wittmer, M. (1980c). *In* "Laser and Electron Beam Processing of Materials" (C. W. White and P. S. Peercy, eds.), p. 524. Academic Press, New York.
von Allmen, M., Affolter, K., and Wittmer, M. (1981). *In* "Laser and Electron Beam Solid Interactions and Materials Processing" (J. F. Gibbons, L. D. Hess, and T. W. Sigmon, eds.), p. 559. North-Holland Publ., New York.
Walser, R. M., and Bené, R. W. (1976). *Appl. Phys. Lett.* **28**, 624.
White, C. W., Wilson, S. R., and Appleton, B. R. (1980). *J. Appl. Phys.* **51**, 738.
Wittmer, M., and von Allmen, M. (1979). *J. Appl. Phys.* **50**, 4768.
Wittmer, M., and von Allmen, M. (1981). *In* "Laser and Electron Beam Solid Interactions and Materials Processing" (J. F. Gibbons, L. D. Hess, and T. W. Sigmon, eds.), p. 533. North-Holland Publ., New York.
Wittmer, M., Luthy, W., and von Allmen, M. (1979). *Phys. Lett. A* **75A**, 127.

Chapter 13

Factors Influencing Applications

C. HILL

Plessey Research (Caswell) Limited
Allen Clark Research Centre
Caswell, Towcester
Northamptonshire, England

I. Introduction

Conventional heat treatment techniques have served for semiconductor device fabrication very well up until now. Most of the physical processes

479

involved in fabrication are rate-determined by a solid state diffusion process (e.g., oxidation, dopant redistribution, anneal of implantation damage), and such processes have activation energies in the range 2–5 eV. The resulting sensitivity to temperature has required temperature control to ±1°C for modern semiconductor furnaces, and this is routinely met. What, then, is the place of radiant beam processing? This family of techniques offers a flexible alternative to the necessarily spatially uniform and temporally restricted heat treatment associated with furnacing. The spatial and temporal extent of the temperature profile inside the semiconductor can be controlled over a wide range of times (picoseconds to minutes) and distances (submicrometers to centimeters), including the isothermal option conventionally available. In preceding chapters, the exploitation of these new degrees of freedom in heat treatment to develop and understand new ways of changing semiconductor properties has been described.

The processing flexibility and new materials properties that have resulted from such work have exciting implications for semiconductor device fabrication. In considering the suitability of radiant beam processing for the reliable fabrication of useful devices, however, a number of additional features have to be considered. These are

(i) Is control of the total process adequate? This is determined by the required tolerances on the physical parameters of a particular structure, the sensitivity of the values of these parameters to variations in heat treatment, and the degree of control over the temperature profile–time cycle obtained with a particular radiant beam system and semiconductor material.

(ii) Does the beam processing induce secondary effects in the processed material and, if they are deleterious to the device, can they be sufficiently minimized by good process design?

(iii) Is the processing compatible with other parts of the device structure and processing sequence? For most applications it is essential to be able to heat-treat multiphase structures.

(iv) Can sufficiently large areas be heat-treated with adequate throughput?

(v) Is it cost-effective to introduce beam processing into the fabrication sequence for a particular device?

These factors have to be evaluated in each individual case, and no general approach is possible. However, enough experimental and theoretical data on beam processing of silicon are available to illustrate the influence of these factors in real situations.

In order to discuss these criteria for application systematically, the techniques for beam processing have been classified into three basic modes, determined by pulse length (Section II), and the materials changes that can be effected have been grouped into six basic categories (Section III). The factors involved in control of the heat treatment process are covered in Section IV. The occurrence of secondary effects and compatibility with other process steps and structures are dealt with in Section V (single-phase structures) and Section VI (multiphase structures). The problems of processing large areas of material rapidly and the economics of beam processing are considered in Section VII.

II. Modes of Beam Processing

There are three main modes of beam processing, characterized by the duration of the heating cycle responsible for changes in materials properties. These are referred to (Hill, 1981) as the adiabatic, thermal flux, and isothermal modes of beam processing. Typical experimental arrangements and resulting temperature profiles (both laterally and in depth over a silicon slice) are shown schematically in Fig. 1.

In the adiabatic mode the radiation pulse is sufficiently short (less than 200 nsec) that little heat is lost by diffusion from the absorption layer (typically 2 μm deep) during the pulse, and so nearly all the energy absorbed is utilized in raising the absorption layer temperature. The temperature profile in depth is consequently very shallow and steep. The short heat diffusion distance also causes the lateral temperature profile to follow the spatial distribution of energy in the beam: a uniform energy distribution yields a uniform temperature, as shown in Fig. 1. Because the pulse lengths are short compared with the times required for significant reordering in solid silicon (about 1 msec), most changes in materials properties are affected in this mode by forming a liquid silicon layer with much shorter reordering times ($\sim$1 nsec).

In the thermal flux mode, the pulse duration is sufficiently long for heat to flow completely through the thickness of the semiconductor. Heat is continuously supplied to the slice surface by a scanned focused beam and continuously removed from the back of the slice. In this way the effective heating pulse duration is controlled by the dwell time of the beam at any one spot, and dwell times are typically 1 msec. The temperature distribution is necessarily monotonically decreasing with depth and laterally decreases from the beam center at a rate determined by beam size (typically 100 μm) and heat diffusion distance (typically 100 μm). These profiles are

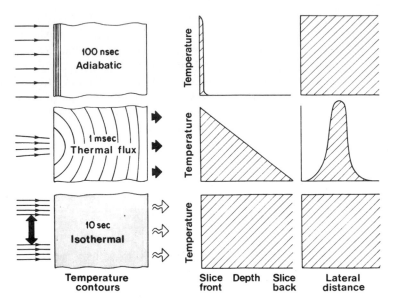

Fig. 1. Schematic diagram showing the characteristics of the three modes of beam processing. From left to right the diagram shows typical beam configurations used for each mode, typical heating pulse durations and temperature contours through the thickness of the semiconductor, heat loss at the opposite face of the semiconductor (adiabatic, none; thermal flux, loss equal to beam input by conduction to heat sink; isothermal, loss equal to beam input by radiation), temperature profiles perpendicular to the semiconductor surface, and temperature profiles in the plane of the sample surface. (From Hill, 1981.)

shown schematically in Fig. 1. Significant solid state reordering can occur in semiconductors in a millisecond, and materials changes can be effected through solid state mechanisms. The existence of the vertical temperature gradient allows the formation of surface liquid layers, and so liquid phase processing is also possible.

Isothermal processing involves pulse durations sufficiently long that heat is distributed uniformly through the semiconductor material. For a silicon slice 400 μm thick and 100 mm in diameter this implies pulse durations of greater than 10 sec for a nonuniform beam and greater than 0.1 sec for a uniform beam. Vertical and lateral temperature profiles are by definition uniform (Fig. 1). The slice is thermally isolated from its surroundings, so that heat loss is possible only by radiation, and the slice temperature is determined by the balance between input beam power and output radiated power. The uniform temperature distribution allows only solid phase processing to be utilized.

All these modes have been utilized for materials processing, and typical equipment used for each mode is shown schematically in Fig. 2. Note that

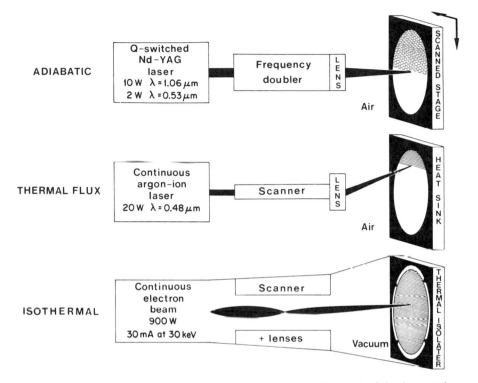

Fig. 2. Schematic diagram showing typical equipment used for each of the three modes of beam processing. From left to right the diagram shows the beam source and its characteristics, the beam handling components, the ambient, and the semiconductor slice and the characteristics of the slice mounting. The equipment shown utilizes a rapid scan of the focused beam to process the whole slice area (corresponding to the last column in Fig. 29). Typical scan rates for a 100-μm spot are adiabatic, 200 cm/sec; thermal flux, 10 cm/sec; and isothermal, 2 km/sec.

the heat must be generated near the front surface of the material for the adiabatic and thermal flux modes, so that, for silicon, visible or near-infrared lasers or very low-energy electron beams are necessary. In the isothermal mode energy can be deposited anywhere in the depth of the slice, and the deeply penetrating radiation from far-infrared lasers or higher-energy electron beams can be used.

III. Changes in Materials Parameters

The changes that can be brought about in a semiconductor structure by beam processing can be grouped into six categories: (i) reordering of

TABLE I

CATEGORIES OF MATERIALS CHANGES THAT CAN BE EFFECTED BY BEAM PROCESSING, THE MODES AND MECHANISMS AVAILABLE, AND EXAMPLES OF EACH COMBINATION[a]

Material change effected	Liquid phase process		Solid phase process	
	Adiabatic mode	Thermal flux mode	Thermal flux mode	Isothermal mode
Reordering of matrix material	Anneal of implant damage (5, 6, 7, 11) Anneal of diffusion damage (6) Increase in substitutionality of dopants (5)		Anneal of implant damage (10, 11)	Anneal of implant damage (10)
Disordering of matrix material	Introduction of point and line defects and associated electronic levels (6, 8) Amorphization (4, 6)		Introduction of point and line defects and associated electronic levels (10)	
Reordering of deposited material	Epitaxial regrowth of deposited Si on crystalline substrates (8)	Crystalline regrowth of deposited Si on amorphous substrates (8)	Epitaxial regrowth of deposited Si on crystalline substrates (8)	

Redistribution of dopants	Novel controlled redistribution of dopants in silicon (5, 6, 8, 13)	Uniform distribution of dopants in deposited layers (10, 13)	Minimal redistribution of doparts in silicon (10)	Small controlled redistribution of dopants in silicon (10)
Reaction of surface layers	Formation of silicides (12) Ohmic contacts to GaAs (11), Si (12) Synthesis of semiconductor films (11, 13) Surface cleaning (9) Surface oxidation (9)		Formation of silicides (10, 12) Surface oxidation (10)	
Redistribution of surface material	Shaping steps in silicon, silicon oxide (13) Smoothing poly Si (13) Formation of waves and ripples in Si (11, 13) Topographic changes in oxide-coated Si (13)	Long-range redistribution of Si deposited on amorphous substrates (13)		

^{a} Chapter numbers are shown in parentheses.

matrix material, (ii) disordering of matrix material, (iii) reordering of deposited material, (iv) redistribution of dopants, (v) reaction of surface layers, and (vi) redistribution of surface material. The full range of materials changes is available only if all three modes (adiabatic, thermal flux, and isothermal) and both mechanisms (solid and liquid phase regrowth) are utilized. The categories of materials changes and the modes and mechanisms that can be used to effect them are summarized in Table I, with important examples of each and the appropriate chapter reference. In any particular application of beam processing, one, two, or even three simultaneous materials changes may be necessary. For example, it may be required simply to anneal all the implantation damage in a particular region or, in addition, a particular redistribution of dopant may also be required; a third requirement might be to obtain greater than equilibrium substitutionality of dopant atoms. Table I shows that these simultaneous requirements can be met by utilizing adiabatic liquid phase processing. Conversely, when anneal of implant damage with minimal redistribution of dopant is required, thermal flux solid phase processing is appropriate. What may be a desired concomitant of annealing for one application may be an undesirable side effect for another. The materials changes listed in Table I under "Disordering of matrix material" and "Redistribution of surface material" are particularly likely to be unwanted side effects for most applications.

IV. Control of the Heat Treatment Process

A. *The Control Required*

1. THE ENERGY WINDOW

In any given application of beam processing there will be one or more physical parameters of the final device that will be required to fall within certain limits. These acceptable tolerances will determine the overall control required for the heat treatment process, through the chain of interactions shown schematically in Fig. 3. The device parameter tolerance will require a certain maximum variation in some materials property(ies), which in turn will permit only a particular variation in the heating cycle, which will demand a corresponding control over the beam parameters. The actual values of the tolerances will depend on the relationships between each of these pairs of parameters and will differ widely for different applications and methods of processing. The demand placed on the con-

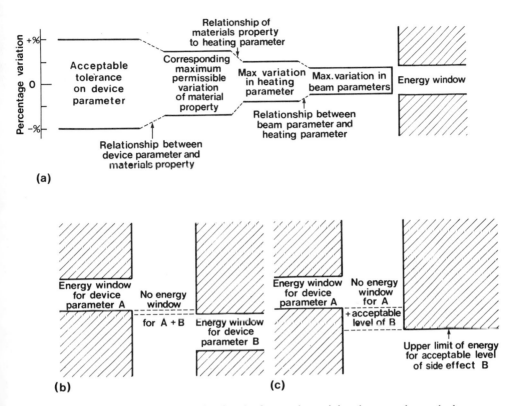

Fig. 3. Schematic diagrams showing the factors determining the control over the beam energy imposed by the requirements of a device fabrication process. (a) The relationship between the percentage variation allowable in a particular device parameter and the corresponding allowable energy window of the beam used to fabricate that device. The important factors determining the relationship are illustrated. (b) The situation where two different device parameters can be achieved to the desired tolerance individually but not simultaneously, because of a mismatch of the corresponding energy windows. (c) The situation where the desired tolerance on parameter A cannot be achieved without incurring an unacceptable level of a deleterious side effect.

trol of beam parameters can usually be expressed in terms of pulse energy or pulse energy per unit time (power), since in practice the other important beam parameters (wavelength and pulse time) can be sufficiently well controlled to render their effect on the heating cycle variation of second-order importance. Thus the parameter control desired defines an allowable energy window for the beam processing (Fig. 3a). As shown in Section III, it may be necessary to control more than one parameter in the final structure or to minimize other unwanted physical changes (side effects). The energy windows required by these simultaneous requirements may be quite different and, although individually they may allow the use of a wide energy window, together they may have no permissible window at all (see Fig. 3b and c). Much ingenuity and effort are required to create possible energy windows for incorporating beam processing into fabrication of a particular device, as will be seen in later sections.

In order to illustrate the main considerations that determine the size of the energy window in the three processing modes, the fabrication of a particular device structure with defined tolerances on its characteristics will be described. The device chosen is an integrated circuit bipolar transistor, and the beam processing is to be used to anneal and redistribute an arsenic implant to form the final emitter region of the transistor as shown in Fig. 4. The parameter to be controlled is the transistor gain, which is approximately proportional to the total number of boron atoms per square centimeter in the base region for this particular structure. As shown in Fig. 4, if the arsenic is redistributed without changing the base profile, a desired tolerance of $\pm 15\%$ on gain corresponds to $\pm 15\%$ control of base doping which requires a maximum permissible deviation from the designed junction depth of $\pm 4\%$ (i.e., ± 150 Å). In this case there is a rather strong relationship between the device parameter (gain) and the materials parameter (dopant redistribution) because of the shape of the base profile $[\Delta \text{ gain} \simeq (\Delta x_j)^2]$, demanding a high degree of control of the materials parameter. The requirement for anneal of the implant does not place any further restriction on processing control, because the threshold for anneal will always have been exceeded if significant redistribution occurs (Chapter 5).

The way in which the other two components of the chain in Fig. 3 (the relationship between dopant distribution and heating cycle and the relationship between heating cycle and beam parameters) translate the $\pm 4\%$ control of junction depth into a particular energy window will now be discussed for each of the three modes of processing. The overall relationship between the dopant distribution and the relevant beam parameters for the three modes is shown in Figs. 5–7.

2. ADIABATIC PROCESSING

In this mode, anneal and redistribution of dopant are controlled by the depth and duration of the liquid layer. For a given laser system (e.g., a 25-nsec pulse from a ruby laser of wavelength 0.69 μm) and material (e.g., silicon) both these parameters are uniquely determined by the heating cycle which is itself monotonically related to the total energy density of the incident pulse (Chapter 3). This gives rise to the single-valued rela-

tionship between junction depth and energy density given in Fig. 5. The
relationship between maximum melt depths and energy density for silicon
with an amorphous and a crystalline surface is also shown. These fairly
simple relationships arise because once the silicon surface is molten the
coupling between the beam and the silicon surface is fixed, and so the heat
generated in the surface layer is almost linearly related to the energy
density of the beam. This heat is nearly all converted into latent heat as
the melt front advances, so that the melt depth is also approximately
linearly related to pulse energy density. The dopant diffuses into the advancing liquid layer at a rate somewhat slower than that of the advancing

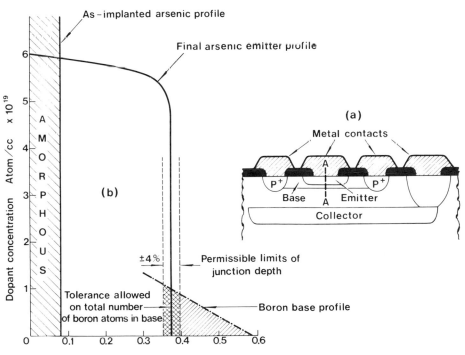

Depth below sample surface (micrometers)

Fig. 4. Typical bipolar transistor structure as used in a high-speed integrated circuit. (a)
Vertical section through the transistor, showing the heavily doped n-type emitter and collector regions and the more lightly doped p-type base region. (b) The dopant profiles along the
line A–A that could result from beam processing of a 40-keV 2×10^{15} arsenic ions/cm^2
implantation to form the emitter region. It is assumed that the previously fabricated boron
base profile does not redistribute during the arsenic redistribution. The maximum variation
in emitter junction depth shown is consistent with $+15\%$ control of transistor gain. The
desirable squarish emitter profile is achieved in adiabatic processing by multiple pulsing, and
in the other two modes by concentration-dependent solid state diffusion.

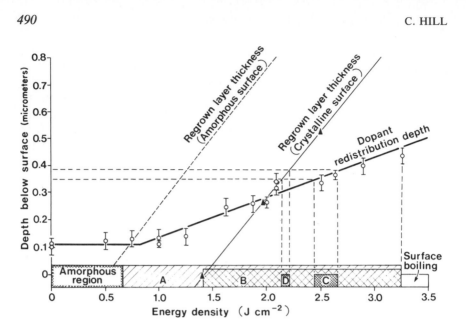

Fig. 5. Adiabatic processing: the relationship between processed depth and relevant beam parameter (total energy density) for anneal and redistribution of the arsenic implant shown in Fig. 4, using a 25-nsec ruby laser pulse. The three relationships shown are the maximum melt depth when processed as-implanted (a), the maximum melt depth when processed after prior anneal of the implant damage (b), the junction depth (in 5-Ω cm p-type material) when processed as-implanted (c). Energy windows for achieving anneal and redistribution to specified tolerances (see text) are shown. (From Hill, 1981.)

melt front, so that the junction depth is a less strong function of energy density than the melt depth. If a single pulse is used to anneal and redistribute the dopant, it can be seen from Fig. 5 that the $\pm 4\%$ control of x_j requires $\pm 4\%$ control of the energy density (energy window C). The use of multiple pulses allows the arsenic to diffuse right up to the maximum melt depth and, because of the eight orders of magnitude ratio between the diffusion rates of arsenic in the liquid and in solid silicon, allows desirable emitter profiles with extremely sharp boundaries to be formed (Hill, 1981; Hill *et al.*, 1982). If the correct energy is chosen, the specified junction depth can be obtained but, because of the stronger dependence of melt depth on energy density, an energy window of $\pm 1.5\%$ will be required to achieve the $\pm 4\%$ tolerance in x_j (window D). Note that, because the first pulse completely anneals the amorphous material, all subsequent pulses are incident on a single-crystal surface, and so the melt depth is determined by the crystalline surface line, not the amorphous one. Note also that, when only an anneal is required, the energy window is very wide, being determined at the lower limit by the threshold for melt-through of the amorphous material and at the upper limit by the onset of surface

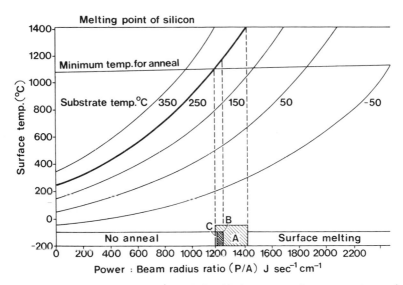

Fig. 6. Thermal flux processing: the relationship between surface temperature and relevant beam parameter (beam power/beam radius) for solid state anneal of silicon implanted as shown in Fig. 4 and annealed using a 1-msec pulse from a focused, scanned argon ion laser beam. The family of curves refer to different thermal sink (slice back) temperatures and are calculated after the method of Gold and Gibbons (1980). The minimum temperature for anneal of the implant damage is shown and also the energy windows for anneal and redistribution to specified tolerances (see text). (After Hill, 1981.)

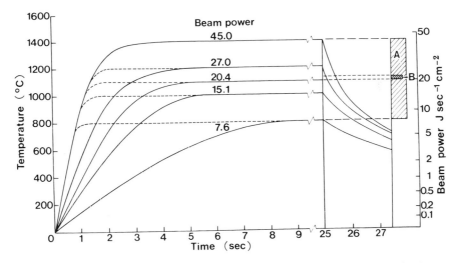

Fig. 7. Isothermal processing: the relationship between slice temperature and relevant beam parameters (beam power and time) for solid state anneal of a silicon slice 400 μm thick implanted as shown in Fig. 4 and annealed using a focused electron beam multiscanned for 25 sec. The rate of temperature rise and the final temperature are both determined by beam power. Energy windows for achieving a specific anneal and redistribution are shown (see text). (After Hill, 1981.)

boiling. The latter is a good example of a deleterious side effect limiting the allowable energy window. The energy windows for anneal are shown in Fig. 5 and are A, 0.7–3.3 J cm^{-2} (amorphous surface), and B, 1.5–3.3 J cm^{-2} (partially annealed surface). The factors determining the energy window for anneal and redistribution for this particular mode and application are summarized in Fig. 8.

3. THERMAL FLUX PROCESSING

Anneal of damage and redistribution of dopants in this mode occur by solid state diffusion, and the rate at which they occur is dependent on surface temperature; this temperature is determined by the power-to-radius ratio of the beam spot (P/A), the temperature at which the back of the slice is maintained, and the slice thickness and thermal conductivity (Chapter 10). For a given processing situation all the parameters except P/A can be fixed, and then surface temperature is uniquely defined by the

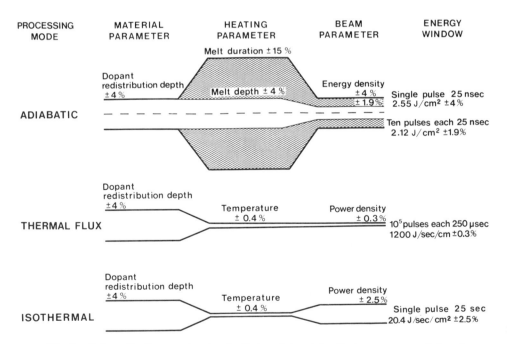

Fig. 8. Schematic diagram showing the factors determining the beam energy window for fabrication of the arsenic emitter structure to the tolerance specified in Fig. 4 by the three modes of beam processing. The alternative diagrams for adiabatic processing refer to fabrication of the desired junction depth by a single pulse and by 10 pulses (of a lower energy).

value of P/A, as shown in Fig. 6. These curves were calculated for a slice thickness of 400 μm and for total absorbed power after the method of Gold and Gibbons (1980). It can be seen that temperature is a weak exponential function of P/A, so that at 1100°C a 1% change in P/A causes a 1.4% change in the temperature to 1115°C. The relationship between the anneal rate R, the coefficient, and the temperature is also of exponential form, given by

$$R = R_0 t \, \exp(-E_a/kT) \tag{1}$$

for a heat treatment of duration t and temperature T, where the parameters R_0 and E_a depend on the dopant and its concentration (Cspregi *et al.*, 1978). Redistribution of dopants also occurs during this heat treatment with a characteristic diffusion length x determined by a diffusion coefficient D and a scaling parameter N,

$$x = N\sqrt{Dt} \tag{2}$$

and D is given by

$$D = D_0 \, \exp(-Q/kT) \tag{3}$$

N, D_0, and Q depend on the dopant and its concentration and may also be a function of the ambient gas and the presence of other dopants (Willoughby, 1981).

   To achieve only annealing it is enough to ensure that R is sufficiently high for the amorphous crystalline interface to traverse the amorphous layer completely during the dwell time of the beam spot. For the amorphous layer in Fig. 4 (750 Å thick, arsenic-doped) and the spot size and scan rate given in Fig. 6, a temperature of 1100°C causes R to be just sufficient for annealing. Further heat treatment causes no further anneal, so that the energy window for anneal at first sight appears to be quite wide, bounded at the lower edge by the P/A value required to achieve 1100°C (1200 J sec^{-1} cm^{-1}) and at the upper edge by the P/A value that causes melting (1420 J sec^{-1} cm^{-1}), as shown by window A in Fig. 6. A much more restricted window for anneal seems to be required, however, to avoid the introduction of electrically active damage into the silicon matrix by the combination of high temperatures and large thermal strains (Mizuta *et al.*, 1980), and the usable window is approximately ±3% of P/A, as shown by B in Fig. 6. In one scan, the redistribution of dopant is insignificant, but controlled redistribution can be effected by using a large number of sequential scans (Matsumoto *et al.*, 1980). About 25 sec total time at 1100°C is required to achieve the arsenic redistribution shown in Fig. 4, corresponding to 10^5 pulses of the chosen pulse time of 250 μsec. It is unlikely that this technique would be used where such large redistributions are needed. However, the thermal flux mode could well be used to advantage in obtaining small, well-controlled redistributions, and ±4% control of any redistribution implies the same percentage control of the diffusion coefficient. This parameter is a very sensitive function of temperature [Eq. (3)] and, in fact, for a typical activation energy of diffusion in silicon of 3 eV, ±4% control of the junction depth implies control of the temperature to 1100°C ± 4.5°C (or ±0.4%), which in turn requires control of P/A to ±0.3% (window C in Fig. 6). The actual distances of anneal and redistribution involve the duration of heating as well as the rate of the process. The thickness annealed is a linear function of the effective dwell time of the spot [Eq. (1)], whereas the redistribution distance depends on the square root of the same parameter [Eq. (2)].

The factors determining the energy window for anneal and redistribution in this particular process are summarized in Fig. 8.

4. ISOTHERMAL PROCESSING

Solid state mechanisms also control anneal and redistribution in this mode, and thus temperature and duration of anneal are the key parameters. The relationship between anneal rate and redistribution rate and temperature is the same as for thermal flux processing (Section IV.3). The relationship between beam power and temperature is, however, quite different, as can be seen in Fig. 7. Since the beam power heats the slice up uniformly, and the only heat loss is by radiation into a vacuum, the slice temperature rises to a limiting value set by the balance between input beam power and radiation loss given by Stefan's law:

$$\text{Radiated Power } P = K \, (\text{Absolute Temperature } T)^4 \qquad (4)$$

Final slice temperature is thus a very insensitive function of input power ($P^{1/4}$). The pulse duration is longer than in thermal flux processing, so that lower processing temperatures can often be used. This allows a very wide energy window for anneal (from the beam power to achieve 800°C, 7.6 W/cm^2, to that required to reach the melting point, 45 W/cm^2) (Window A Fig. 7). The temperature control (±0.4%) required for ±4% control of junction depth in Fig. 4 can be achieved with control of beam power to ±2.5% (Window B Fig. 7). Control of pulse duration is determined by the square root dependence of junction depth on time [Eq. (2)], and is thus in this case ±16%. The factors determining the energy window for anneal and redistribution for this particular mode and application are summarized in Fig. 8.

B. The Control Obtainable

1. UNIFORMITY AND REPRODUCIBILITY OF THE MATERIAL

It has been tacitly assumed throughout Section IV.A that the material being processed has no lateral variations in optical or thermal properties. Such variations would cause lateral variations in the induced temperature and its absolute value (Chapter 3) and destroy the unique relationships between beam parameters and heating parameters illustrated in Figs. 4–6. The uniformity of optical and thermal properties in silicon slices (with or without an overall implant) appears to be very good, since no reports of nonuniform processing from this cause have been reported. In fact, the evidence is that very exact control of, for example, anneal depth can be obtained (Chapter 6) in such material. However, the strong dependence on temperature of both the optical absorption coefficient for photons near band gap energy and the thermal conductivity in semiconductors can cause even small nonuniformities in these parameters to be responsible for large temperature differences between adjacent areas (Chapter 2). Such effects will be most marked in adiabatic processing where the rate of energy absorption is too great for thermal diffusion to smooth out temperature differences within the pulse time. In extreme cases this may lead to local annealing or even local melting, as suggested (Prussin and von der Ohe, 1980) for the anneal of low-dose boron and phosphorus implant damage in silicon, where the damage exists as discrete volumes of disordered material in an essentially crystalline matrix. The planar melting model used throughout this book in interpreting the results of adiabatic annealing is not applicable to such cases. However, such cases are only likely to arise when processing is close to the surface melting threshold because, once a liquid layer forms, the faster transport processes in the liquid and the rapid advance of the melt front (Chapter 3) tend to smooth out such variations. Where surface melting occurs, the effect of the nonuniformities is simply to increase the variation in materials parameters and so effectively worsen the control available in a given processing operation. In multiphase materials, very large differences in optical and thermal properties can, of course, occur, which give rise to some striking effects to be described in Section VI.

Processing with electron beams is much less sensitive to differential absorption effects, since the absorption coefficient is not sensitive to structure but only to atomic number (Chapter 2). Differential absorption will still be important where metal overlays are used (e.g., gold). Differences in thermal properties will affect the uniformity of processing in electron beam systems just as in laser systems.

2. UNIFORMITY AND REPRODUCIBILITY OF RADIANT BEAM SOURCE

a. The Raw Beam. Whatever equipment is used to generate the radiant beam there will be an inherent variation in beam parameters. The important parameters for control of beam processing are the total energy, total power, and wavelength of the beam, and the spatial and temporal variation of these (Chapter 2). The relative importance of these factors depends on the mode of processing and are discussed below.

b. Adiabatic Processing. The key parameters for control are the total pulse energy, the spatial distribution of this energy, pulse duration, and wavelength (Chapter 3). The Q-switched lasers used almost universally to generate the short pulses required by this mode of processing have sufficient control over the latter two parameters to make variations in them a second-order effect, and these parameters will not be considered further. However, future improvements in the other two parameters may then require better control of pulse length and wavelength. Currently available lasers with adequate output power are the ruby laser (1–2 W at 0.63-μm wavelength) and the Nd:YAG laser (2–20 W at 1.06-μm wavelength).

Total pulse energy in solid state Q-switched lasers is reproducible at present from ±3 to ±6% (Wright, 1980). The spatial distribution of energy is very nonuniform, being either Gaussian for lasers operated in the TEM_{00} mode or even less uniform (see Fig. 9) for the multimode operation required to extract the maximum energy per pulse from the laser. The additional nonuniformity arises from a fine-scale pattern of hot and cold spots generated by optical interference and heating effects between the slightly different wavelength modes of the beam and varies with time (Maracas *et al.*, 1978). Energy density uniformity over the useful part of the Gaussian beam is about ±30% and in a quasi-Gaussian beam can vary ±60%.

Adiabatic processing has been carried out using dual-wavelength lasers (Auston *et al.*, 1979; Daly, 1978), mixing the primary pulse (e.g., 1.06 μm) with its harmonically generated frequency-doubled component (0.53 μm) and thus increasing the efficiency of absorption of the longer-wavelength, higher-energy pulse by heating up the silicon with the lower-energy, shorter-wavelength pulse (which is more strongly absorbed at room temperature). In such systems both the partition of energy between the pulses and their relative timing must be accurately controlled. The pulse energy required to melt crystalline silicon in such a system (Newstein *et al.*, 1980) was found to increase from 0.7 to 3.2 J/cm² as the proportion of green

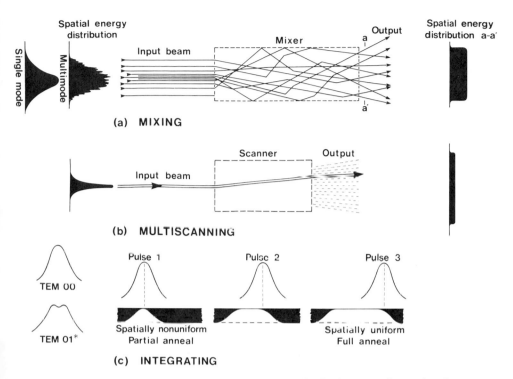

Fig. 9. Techniques for obtaining uniform energy distribution over the semiconductor surface using nonuniform radiant beams. On the left are shown the spatial energy distributions (across the beam cross section) characteristic of commonly available beam sources; on the right are shown the corresponding spatial energy distributions after beam homogenization. The three techniques available for homogenization are shown in the center of the diagram, and all rely on the principle of superimposing radiation from all parts of the beam to produce a uniform total irradiation energy. The superposition is achieved in (a) by optically splitting the beam and recombining the components, in (b) by rapidly scanning the whole beam over the anneal area, and in (c) by exposing the anneal area to a squence of identical pulses, each of which is displaced a fraction of the beam width laterally with respect to the previous pulse. The time–temperature cycles produced by these three techniques are very different (see text).

(0.53 μm) light was decreased from 100% to zero. It was also computed that a 10-nsec delay in the infrared pulse relative to the green pulse was equivalent to increasing the energy density of the infrared pulse by 18% in terms of the effect on the heating cycle. It is thought that with good systems design both these parameters are controllable to at least the same tolerance as present energy densities. Pulsed electron beams approach pulsed lasers in their reproducibility of total pulse energy (Kirkpatrick, 1980); spatial distribution of energy is also very nonuniform at present.

c. *Thermal Flux Processing.* The crucial parameters here are total beam power to beam spot radius ratio (P/A), its spatial distribution, and the heating pulse duration (Chapter 10). The high beam powers and millisecond pulses required by this mode of processing have so far been generated by focusing a continuous laser beam into a small spot (typically 100 μm) and the pulse duration controlled by scanning the beam relative to the semiconductor slice. The optical and mechanical scanning techniques give good control of the spot dwell time (t), hence of the duration of the heating pulse (t_{eff}), so this parameter will not be further discussed. [It should be noted, however, that the relationship between t_{eff} and t is not linear and involves other parameters (Gold and Gibbons, 1980).]

The parameter P/A is reported (Moore, 1980) to be controllable to between ± 1 and $\pm 3\%$ in a 20-W argon ion laser system, of the type illustrated in Fig. 2, by the use of feedback loops to control P, and careful design to minimize changes in A caused by alterations in effective focal length and refractive index during scanning. The spatial distribution of power in the beam is expected to be of only second-order importance, because heat flow distances in the semiconductor during the pulse time (about 1 msec) are comparable to the spot size (e.g., 100 μm). However, it is reported (Moore, 1980) that a significant improvement in the uniformity of laser-induced regrowth of amorphous silicon on oxide results from replacing the Gaussian spatial distribution of energy obtained in the TEM_{00} mode of laser operation by the more uniform distribution obtained in the TEM_{01*} mode (see Fig. 9). Control of the beam shape thus may be necessary, and this technique of vibrational mode selection allows some shaping of the raw beam to be effected (Fig. 9).

Focused electron beams can also be used for thermal flux processing (McMahon and Ahmed, 1979; Ratnakumer *et al.*, 1979). The control of P/A to the same tolerance as in laser systems is likely to be difficult if beam deflection over a large area is used, because of the problems of simultaneously controlling high power and small spot size with the intrinsically less flexible options available in electron lenses as compared to optical lenses. Scanning the slice under a fixed beam should allow comparable tolerances to be obtained. An additional problem is controlling the temperature of the back of the slice and removing the heat flowing through the slice. The use of atmospheric pressure to maintain good contact between slice and chuck (as used in laser systems) is not available in the necessarily vacuum environment of the electron beam system. No data are available as yet on the overall control of surface temperature that can be obtained with an electron beam system operated in the thermal flux mode.

d. Isothermal Processing.   The time–temperature cycle is entirely controlled by beam power and pulse duration. Spatial uniformity of the beam is unimportant, because the diffusion of heat laterally in the slice is nearly always faster than the rate of deposition of heat by the beam. There are two operating conditions where this criterion is not satisfied, which will be dealt with later. Historically, the requirement for thermal isolation for this mode of processing has favored its development in electron beam systems. The high scanning speeds and high beam powers available have allowed the power to be delivered to the slice uniformly in space and time by a continual rapid raster scan. Beam power is controllable to about 1% and pulse duration to at least 0.01% in systems specifically designed for isothermal processing (Plows, 1980). Currently, beam powers up to 1.5 kW are available.

There is no reason in principle why laser beams should not also be used for isothermal processing, and some preliminary experiments with a CO_2 continuous laser ($\lambda = 10.6 \ \mu m$) have been reported (Celler *et al.*, 1979). No data are available on isothermal processing by lasers where full thermal isolation (minimal contact in vacuum) was present.

Spatial distribution of energy in the beam can matter if the rate of absorption of energy from the beam is greater than the thermal flux inside the semiconductor, as mentioned above. In electron beam systems, where typically a small-area, high-power beam is being rastered over a large area, it is necessary to ensure that the scan is both rapid enough and uniform enough (e.g., avoiding Lissajous scan patterns) so that local hot spots of a transiently high-energy deposition rate do not occur. In CO_2 laser systems, where optical absorption is through nonlinear second-order effects (e.g., free-carrier absorption, both from dopant atoms and across the band gap by temperature rise), the efficiency of coupling between the beam and the material can be a very strong function of the integrated beam power and pulse duration (Chapter 2). Spatial variations in beam intensity (of the magnitude found in the normal TEM_{00} output beam) can thus give rise to adjacent regions of a very different absorption coefficient, and a runaway situation can develop where extreme temperature gradients transiently occur that can cause irreversible deformation of the material (Celler *et al.*, 1979; Webster, 1980).

3. METHODS OF IMPROVING BEAM UNIFORMITY

a. Beam Homogenization Techniques.   The nonuniform spatial distribution of energy and power in the radiant beams available from present sources (Section IV.B.2), the small area of the beams relative to many of the semiconductor slices requiring anneal (Section VII), and the narrow energy window required by many applications often necessitate the use of techniques for homogenizing the effective energy distribution over an area. There are three basic approaches to beam homogenization represented schematically in Fig. 9, and these methods are described in more detail in the next sections. All the approaches described involve splitting the beam either spatially or temporally into a number of components and recombining these components at the sample surface to produce an averaged effect, but the three techniques and their characteristics are quite different.

   b. Mixing.   The basis of this technique is to split up the initial beam spatially into many components, introduce angular and phase differences between the components, and recombine them at the surface to be processed (Fig. 9a). In this way the spatial and temporal variations in energy density can be smoothed out over the whole of the irradiated area and a uniform energy density can result. A very effective way of mixing the beam is to scatter it through a colloidal solution. Unfortunately the unavoidably large scattering angles cause a large fraction of the pulse energy to be lost. The pulse can be split up into a number of beams using angled mirrors that recombine it at the sample surface, and good results and low energy loss are claimed for this technique (cited in Grojean *et al.*, 1980). The rather small number of beams involved makes it unlikely, however, that energy density is uniform on a microscale, since the size of the speckle pattern generated by coherent beams is approximately inversely proportional to the number of combining wave fronts (Lahart and Marathay, 1975). A third technique (Cullis *et al.*, 1979), which combines good homogenization (~5% uniformity) with fairly low energy loss (~40%) is illustrated in Fig. 10. The beam is scattered into a small angle at the matte surface of a solid quartz waveguide, and the large number of scattered rays of many different angles follow different paths by total internal reflection through the guide to the polished output face. The large-radius, right-angle bend is reported (Cullis, 1979) to be required for efficient mixing, and the narrowing of the guide in the output section ensures that the energy density at the output face is at least as high as at the input face, despite losses from the guide. A modification of this design (Fig. 10b), involving a slight double bend in the output section, has been found (Hill, 1980) to make the uniformity of the energy density of the output less sensitive to the exact length of the output section than in the original design (Fig. 10a). In both designs, a uniformity of 5% in energy density over a circle almost the diameter of the output face can be obtained for samples placed closer than 1 mm to the output face. Uniformity rapidly worsens beyond this distance. An important characteristic of the

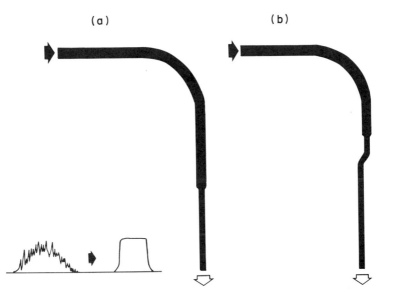

Fig. 10. Quartz rod laser beam homogenizers. (a) Original design (Cullis *et al.*, 1979) consisting of cylindrical 10-mm quartz rod with matte input face and polished output face, curved and tapered as shown. (b) Modified design (Hill, 1980) incorporating kink to ease manufacturing tolerances. Spatial energy density profiles for a typical multimode laser pulse and the same pulse after homogenization using either (a) or (b) are shown. (From Hill, 1980.) This figure was originally presented at the Spring 1980 Meeting of The Electrochemical Society, Inc., held in St. Louis, Missouri.

output beam is that, as well as being spatially uniform in energy density, it also consists of rays of many angles emerging from every point on the output face. Thus effects that are sensitive to beam angle (e.g., interference, diffraction), which can introduce large inhomogeneities into the effective energy density across the surface (Sections V and VI), are also averaged out by this technique. Beam mixing homogenization is applicable to all laser beam sources but so far has only been used in conjunction with Q-switched lasers in beam processing applications. A mixer suitable for CO_2 lasers ($\lambda = 10.6 \ \mu$m), based on the principle of the kaleidoscope, has been described (Grojean *et al.*, 1980) and is reported to have low energy loss (14%) and to result in good uniformity ($\pm 8\%$).

    c. Multiscanning. The energy in an inhomogeneous beam can be redistributed more uniformly over the sample surface by scanning the beam many times during the total pulse time over an area greater than the spot size (Fig. 9b). Thus every point on the sample surface is exposed to every part of the beam, and the total energy delivered to each point should be constant. The rate of delivery of this energy is, however, not constant and the technique is equivalent to dividing the continuous beam into a large number of smaller pulses separated by time intervals (Fig. 11a). It is evident that for effective homogenization the fraction of the total energy delivered in any one pulse should be small, otherwise the nonlinear relationships among optical absorption, thermal conductivity, diffusion coefficients, or phase changes and temperature will result in instabilities and nonuniformities in the heat treatment. The time interval between the pulses should also be short enough so that the induced temperature does not decrease significantly between pulses, otherwise the effective heat

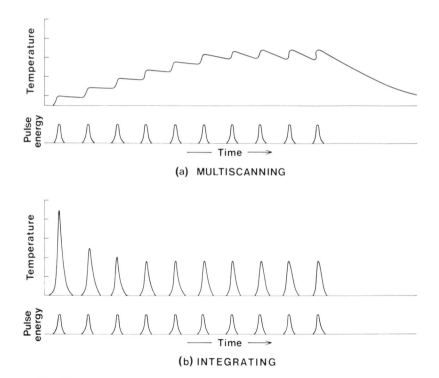

Fig. 11. The temperature–time profiles that can result from a regular sequence of radiation pulses of equal energy. If most of the absorbed energy is retained in the irradiated region in the interval between one pulse and the next, the temperature increases stepwise quasi-continuously during the sequence as in (a); this is characteristic of multiscanning. If most of the absorbed energy is lost from the irradiated region in the pulse interval, a series of discrete temperature–time cycles result as in (b); this is characteristic of integrating (see text). The higher temperatures shown in the first three pulses illustrate the likely situation that the material properties (hence coupling between beam and material) are likely to change very significantly in the first few temperature cycles.

treatment will not approximate to the envelope shown in Fig. 11a and the surface temperature over the scan area will not be uniform. The exact relationship among temperature uniformity, energy per scan, and scan interval will depend on the factors discussed in Sections IV.A.4 and IV.B, but the number of scans required to achieve 1% temperature uniformity is likely to lie between 10 and 1000 for most applications. This rules out scanning as a practical means of pulse homogenization for adiabatic processing, since it implies scan frequencies of greater than 1 GHz (25 pulses/25 nsec). In isothermal processing the much lower scan frequency required, about 100 kHz (100 pulses/1 msec), would be readily achievable with rotating prism deflector systems and in principle would allow homogenization of the Gaussian laser beams commonly used (Section IV.B.2.c). The comparatively slow raster scan used in thermal flux processing (~5–10 Hz) could be modulated with a small-amplitude, high-frequency modulation (100 kHz) of the spot about its mean position. This technique, although reported for beam homogenization in liquid phase thermal flux processing (Laff and Hutchins, 1974), has not been widely adopted, and this may be because control of spot size is a critical parameter in thermal flux processing. The dimensions of the larger spot generated by the rapid scan would need to be controlled to the same accuracy as the original beam spot (Section IV.A.3). In practice, the combination of inherent smoothing of temperature fluctuations by heat flow in the material (over about 100 μm/msec) and improvement of the uniformity of the raw beam (Section IV.B.2.c) may offer the best hope of obtaining a uniform beam in this mode. Beam homogenization by multiscanning is particularly well suited to isothermal processing because pulse times are long enough (typically greater than 1 sec) for low scan frequencies to be used, allowing homogenization by this method even over large areas. In a typical electron beam system (Plows, 1980) 0.1% of the total pulse energy can be delivered per scan (10^{-2} sec) over a 4-in. slice area in a total pulse time of 10 sec, so that 1000 scans per pulse are easily achieved. Uniformity of temperature is also assisted by the long heat flow distances over the time of a single scan (typically 3 mm) as compared to the size of the scanned spot (typically 0.2 mm).

d. Integration. Where beam homogenization is not possible by either spatial averaging of the beam energy (mixing) or temporal averaging (multiscanning), integration of the effects of multiple pulses offers a third approach to achieving effective beam uniformity (Fig. 9c). It was noted in the previous section that, when scanned multiple pulses are separated by small time intervals and each delivers only a small fraction of the beam energy, the overall heat treatment approximates to that from a large-area uniform beam. However, if the time intervals between the pulses are lengthened, although the total energy delivered to every point on the surface is still uniform, this energy is quantized in time and results in a series of separate sequential heat treatments. This difference between homogenization by multiscanning and by integrating is made clear in Fig. 11b. While the effect of multiscanning can be characterized by a single temperature–time cycle, the effect of integrating is the result of a number of sequential temperature–time cycles, not necessarily of equal intensity. This is because the change in materials properties effected by the first pulse may considerably alter the coupling between the second pulse and the material, and so on. Whether integrating is successful depends very much on the material and the materials change required. Integration can be used with both adiabatic and thermal flux processing and in fact always occurs where overlap techniques are used to process large areas with a smaller beam (Section VII). In such cases the final uniformity of the processed region depends on the success of the integration of the two or three sequential heat treatments that result from overlapping sequential pulses. Simply increasing the degree of overlap may not improve the final uniformity. An example of this (Aspnes *et al.*, 1980) is the adiabatic annealing (by melt and regrowth) of an amorphized silicon surface using the scanned focused beam from a Nd:YAG ($\lambda = 1.06~\mu$m) Q-switched laser emitting pulses of 140-nsec duration every 714 msec. The scan rate is such as to move the focused spot by about a quarter of its diameter each pulse, thus giving every point four sequential heart treatments of differing intensity.

The effect of the first and second pulses is shown schematically in Fig. 12. The first pulse creates a melt pool of smoothly varying depth according to the energy profile of the pulse, and a recrystallized zone is formed as shown. The second pulse is displaced with respect to the first and is incident on both amorphous and crystalline silicon. The absorption coefficients of these two materials are very different at 1.06-μm wavelength, thus there is a large difference in the energy absorbed, hence in the melt depth at the boundary of the two regions. This causes a meniscus line to form, and the final trace of the overlapped pulses contains a series of well-marked ridges corresponding to these menisci. The ridges can be prevented from forming not by increasing overlap but by using a wavelength where amorphous and crystalline silicon have comparable absorption coefficients. It was demonstrated (Aspnes *et al.*, 1980) that a completely flat surface was obtained when the same scan pattern was repeated with frequency-doubled Nd:YAG pulses ($\lambda = 0.53~\mu$m).

Adiabatic processing using pulsed electron beams apparently does not allow the use of overlapping pulses to achieve uniformity of heat treat-

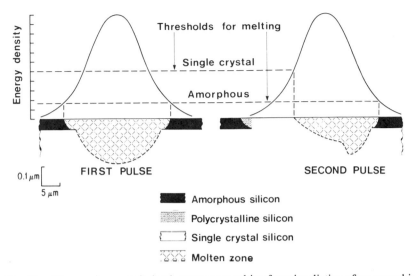

Fig. 12. The maximum melt depth contours resulting from irradiation of an amorphized silicon surface by a Q-switched laser pulse of Gaussian energy distribution and a subsequent identical pulse partially overlapping the region annealed by the first pulse. The different melting points of amorphous and single-crystal silicon are responsible for the small kinks in the otherwise smooth contour induced by the first pulse: the different energy thresholds for the melting of amorphous and crystalline silicon are responsible for the large step in the contour induced by the second pulse. [Based on the results of Aspnes *et al.* (1980) and Rozgonyi *et al.* (1978).]

ment by integration. While pulses incident on amorphized surface layers give satisfactory anneals, subsequent pulses on the annealed region give rise to microcracks (Greenwald *et al.*, 1979).

In the thermal flux mode applied to the anneal of implantation damage, integration by the use of overlapping scans has been successful in producing uniformly and well-annealed arsenic- and boron-implanted (100) silicon surfaces (Gat *et al.*, 1979). However, uniform annealing will not necessarily be achieved for all dopants and crystal orientations in silicon. The material scanned by the wings of the Gaussian will see a moderate temperature (600–800°C) anneal: Such anneals can give rise to stable line defects, or even polycrystalline silicon, which are more difficult to remove than the original damage (Williams, 1980). This difficulty is expected where regrowth rates are slow, as for (111) silicon, and with high doping levels, especially of oxygen or carbon.

Integration effects can be used to improve the control over a particular materials change if real-time assessment of the materials change is possible after each pulse. The effects of each pulse must be truly additive for this technique to be successful. Examples of such materials changes are the stepwise advance of junction depth during multipulse adiabatic processing (Section V.B.2) and the stepwise movement of the single-crystal–amorphous interface during successive scans of a laser beam in thermal flux processing (Chapter 10).

V.   Processing Single-Phase Structures

A.   Single-Phase Structures

These structures are defined, for the purpose of this chapter, as those in which the uniformity of the beam processing is not significantly affected by vertical or lateral inhomogeneities in the material. It includes, of course, material that is flat and completely uniform compositionally and structurally, but also material that is homogeneous laterally (e.g., containing a planar implant or a uniform diffusion) and free of very large changes in physical properties vertically (e.g., excluding structures containing metal or oxide overlayers or interlayers). The absence of discontinuities in the optical absorption reflection or heat flow greatly simplifies the processing of single-phase structures as compared to the multiphase structures described in Section VI. However, there can be problems in application, particularly in regard to secondary effects and compatibility with other process steps. If these processing problems can be overcome, there are a number of primary materials changes that can be effected in single-phase structures which either have been, or could be, incorporated into device structures. All these aspects are discussed in more detail below.

B.   Primary Materials Changes

1.   ANNEAL OF MATRIX DAMAGE

All three modes of beam processing (Section II) can be used to anneal a wide range of damage in the surface regions of a single-crystal silicon matrix, including amorphous layers and line defects introduced by implantation (Chapters 6, 7, 10, and 11) and by high dopant concentration (Chapter 6). Adiabatic melt phase regrowth yields the most perfect material, and the perfection is largely independent of dopants present or their concentration, except at very high levels, and independent of crystal orientation. Perfect material is also obtainable by thermal flux and isothermal solid phase anneal, but the perfection is reduced for some dopants, some crystal orientations, and high dopant levels. The dopants present are incorporated onto substitutional sites, and close to 100% electrical activation of dopants can be obtained for all three modes of anneal for dopant levels up to the solid solubility limit. Above this limit, behavior depends on the mode used and is described in Section V.B.3. At low dopant levels it appears that either mobility or dopant activation can be lower after adiabatic processing than for furnace processing (Prussin and von der Ohe, 1980).

In compound semiconductors, neither complete removal of line defects nor ideal carrier mobilities are obtained by any of the modes of processing unless loss of material by evaporation is prevented (e.g., by capping) (Chapter 11).

2. REDISTRIBUTION OF DOPANTS

The factors influencing dopant distribution during beam processing are described in Chapters 4, 5, 10, and 11. The wide range of heat treatments available allows distributions not previously obtainable by furnace processing. These include fully electrically activated dopant profiles not significantly redistributed from the as-implanted distribution (thermal flux and isothermal solid phase anneal) and Gaussian distributions of controllable depth from the as-implanted profile to large redistributions (adiabatic liquid phase anneal). Examples of the latter two types of profiles are shown in Fig. 13. The initial distribution of the arsenic dopant was a 40-keV 10^{16} ions cm^{-2} ion implant profile and coincided closely with the curve marked 1 J cm^{-2}, 10 zaps. The effect of a single 1-J cm^{-2} ruby laser pulse on the original implant was to anneal the amorphous implant damage completely and produce a carrier profile coinciding at low concentrations with the original implant. The melt front evidently penetrated very little beyond the implanted region. At a higher energy density (1 zap, 2.6 J cm^{-2}) full anneal is again achieved; in addition, considerable redistribution of arsenic occurs, caused by diffusion in the liquid phase, which penetrates much deeper in this case. These profile shapes are both reminiscent of furnace-produced redistributions, and this is because of Dt product (which determines the redistribution) is similar in both cases. For example, solid state redistribution in a furnace at ~1000°C gives typically $Dt = 10^{-4}$ cm^2 sec^{-1} × 1800 sec = 1.8 × 10^{-11} cm^2, and liquid phase redistribution during pulsed laser anneal has typically $Dt = 10^{-4}$ cm^2 sec^{-1} × 200 × 10^{-9} sec = 2 × 10^{-11} cm^2. An important difference, however, is that in liquid phase processing there is a boundary across which the diffusion coefficient changes abruptly by 8–10 orders of magnitude

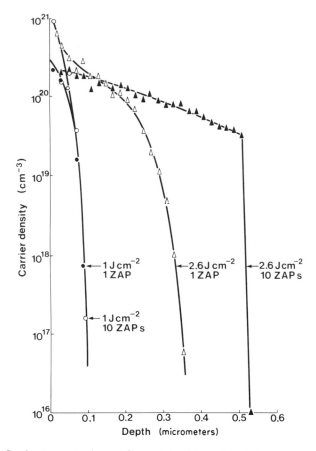

Fig. 13. Carrier concentration profiles obtained by adiabatic laser processing (25-nsec homogenized pulse from a ruby laser) of silicon implanted with 10^{16} arsenic ions/cm² at 40 keV. Profiles after irradiations with one pulse at either 1.0 J/cm² or 2.6 J/cm² are shown, together with profiles after irradiation with 10 sequential pulses (separated by 10-sec intervals) each of 1.0 J/cm² or 2.6 J/cm². (From Hill, 1981.)

(the liquid–solid interface). If the dopant is allowed to diffuse up to this boundary, a quite novel dopant distribution is obtained. This happens at long diffusion times, which can be effectively achieved by melting and regrowing the material a number of times. The effect of 10 pulses at 2.6 J cm⁻² is shown in Fig. 13. An almost uniform distribution of arsenic, bounded by an extremely sharp cutoff is obtained (Hill, 1980). The bounding slope is 1.5×10^{25} ions cm⁻⁴ and is a factor of 3 steeper than can be obtained by the best previous technique (concentration-dependent solid state diffusion) at this depth and is independent of the arsenic concentration (Hill *et al.*, 1982). Such sharply bounded regions, with independent control of depth and dopant concentration, are likely to find applications in small-geometry device structures. It should be noted that, because the thresholds for melting amorphous and crystalline silicon by ruby laser pulses are very different (Fig. 5), the melt depth obtained in subsequent pulses is always lower than for the first pulse. Where pulse energy is near threshold, subsequent pulses may not cause melting at all: this is the case for the 1 J cm^{-2} pulses in Fig. 13, where the redistribution is little affected by multiple pulses.

Other novel dopant redistributions can be obtained, exploiting the low segregation coefficients of some elements to accumulate them near the surface (Chapters 5 and 6). This has been suggested as a way to form a $p-n$ junction in one step, utilizing two dopants with different segregation coefficients (Stoneham, 1978). In any application of this sort, lateral segregation of impurities (Chapter 6) would need to be controlled, especially at high doping levels.

3. SUPERSATURATED SOLID SOLUTIONS

The mechanism of regrowth and the very rapid cooling rates associated with adiabatic melt phase processing of silicon allow solid solutions of substitutional dopants to be formed at much higher concentrations than in conventional furnacing (Chapter 5). The cooling rates in solid phase thermal flux annealing, while much lower, are still high enough for supersaturated solid solutions of some dopants to be formed (Chapter 10).

The solubilities obtained by adiabatic processing are in the range 10^{20}–10^{22} atom cm^{-3} and appear to be simply related to the equilibrium segregation coefficient K_0 by the expression (Fogarassy *et al.*, 1980)

$$C_{max} = 8.6 \times 10^{21} K_0^{0.51} \quad cm^{-3} \tag{5}$$

The electrical conductivity of these supersaturated solutions is also much higher than for conventionally produced doped layers, though the maximum carrier concentration is not necessarily the same as the maximum atomic solubility (Natsuaki *et al.*, 1981; Finetti *et al.*, 1981; Young *et al.*, 1978). The carrier concentration is equal to the atomic concentration for concentrations up to about 5×10^{21} cm^{-3} (P), 3×10^{21} (As), and 1×10^{21} (B), corresponding to material of resistivity 0.00014 Ω cm, 0.00016 Ω cm, and 0.00023 Ω cm, respectively. At higher dopant concentrations, the ratio of carrier concentration to atomic concentration falls below 1. The maximum atomic concentrations, carrier/dopant ratio, and resistivity for arsenic- and phosphorus-doped silicon are 9×10^{21} cm^{-3}, 0.4, 0.00013 and 9×10^{21}, 0.65, 0.00011, respectively; these may not, however, be usable in devices because of the large stored strains and tendency to precipitation. These effects are even more marked at high boron concentrations (Larson *et al.*, 1978) and can give rise to gross cracking of samples implanted with doses greater than 10^{16} cm^{-2} of boron and pulse laser-annealed (Natsuaki *et al.*, 1981). This restriction to lower than maximum solid solubility is not such a great disadvantage for device application as might at first appear, because even using the lower limits previously described, a high percentage (about 80%) of the maximum possible conductivity is achieved, giving minimum practical resistivities of about 0.00014 Ω cm (*n*-type silicon) and 0.00024 Ω cm (*p*-type silicon). The ineffectiveness of additional dopant in reducing resistivity is ascribable to electrical inactivation of the dopant by clustering (Chapter 6) and to a decrease in the electrical mobility with concentration, as shown in Fig. 14.

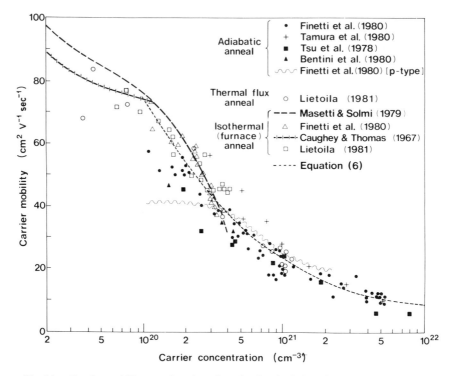

Fig. 14. Carrier mobility as a function of carrier density in heavily doped *n*-type silicon at room temperature. Results included are for published data on adiabatically annealed samples (pulsed laser), thermal flux-annealed samples (cw laser), and isothermally annealed samples (furnace). In all three modes, both phosphorus- and arsenic-doped silicon was measured, but the data have been plotted without distinguishing the dopant. Hole mobilities reported for adiabatically annealed boron-doped silicon are also plotted for comparison.

The compilation of available data (Tsu *et al.*, 1978; Tamura *et al.*, 1980; Bentini *et al.*, 1980; Finetti *et al.*, 1980) shows that, despite some scatter, there is good general agreement; mobility in *n*-type silicon falls steadily from the previous mobility limit of about 75 cm² V⁻¹ sec⁻¹ (Caughey and Thomas, 1967) at 10^{20} carriers cm⁻³ down to about 8 cm² V⁻¹ sec⁻¹ at 9 × 10^{21} carriers cm⁻³. The mobility in *p*-type silicon is apparently a much weaker function of doping level and becomes similar to that in *n*-type silicon for carrier concentrations greater than 4 × 10^{20} cm⁻³ (Finetti *et al.*, 1981).

Thermal flux processing can give supersaturated solid solutions of arsenic up to at least 10^{21} atom cm⁻³ with an equal carrier concentration (Lietoila *et al.*, 1979). The mobility values are plotted for comparison in Fig. 14 and are seen to fit the same curve as that for adiabatically processed *n*-type silicon and for isothermally annealed silicon (Masetti and Solmi, 1979; Finetti *et al.*, 1981; Lietoila, 1981). This suggests that the mobility is determined in both cases primarily by direct impurity effects and not dominated by the particular point defect structure obtained, which is different for the two modes (Chapter 6); see also Section V.C.1. The dependence of electron mobility on carrier concentration which empirically fits the data shown in Fig. 14 (for $C = 10^{20}$–10^{22} carriers cm⁻³) is

$$\mu = 7.5 \times 10^{11} C^{-1/2} \tag{6}$$

where the units of μ are square centimeters per volt-second, and of C are carriers per cubic centimeter. This dependence is quite different from the weak concentration-dependent mobility expected from simple theory (Ziman, 1972) in a degenerate semiconductor.

4. HOMOEPITAXY

The conversion of deposited films by beam processing to single-crystal layers structurally continuous with the substrate is described in Chapter 8. Applications suggested for this technique are growth of substrate materials (Kirkpatrick, 1980) and low-temperature epitaxy for bipolar transistor fabrication (Greenwald and Little, 1979). In both cases the expected advantage is reduced contamination of the grown material because of the much reduced time–temperature cycle and the smaller volume of heated material. For device applications, the films must be free of line and point defects; the factors controlling this and the big differences in solid and liquid phase epitaxy in this respect are described in Chapter 8. In particular, the lower density of deposited films and consequent void formation on recrystallization are likely to pose problems (Leamy *et al.*, 1980).

A particularly promising area of application is the selective regrowth of semiinsulating semiconductor films to produce conducting regions in an insulating matrix. For example, hydrogen-doped amorphous silicon is a good insulator and passivates $p–n$ junctions as well as (or even better than) thermally grown silicon dioxide (Pankove and Tarng, 1979). A layer of this material deposited on a single-crystal silicon substrate, and subsequent selective regrowth of certain areas to single crystal, would offer a very attractive approach to device isolation for integrated circuits and a truly planar technology. Unfortunately, the presence of high concentrations of gaseous impurities in deposited films hampers the regrowth process, and in particular heavily hydrogen-doped films have not, so far, been perfectly recrystallized (Pankove *et al.*, 1980; Peercy and Stein, 1980; Toulemonde *et al.*, 1980).

5. SURFACE CLEANING

The production of atomically clean surfaces by adiabatic laser processing in vacuum is fully described in Chapter 9. The most promising application for this technique in device processing appears to be in preparing surfaces for surface-sensitive processing, e.g., molecular beam epitaxy (MBE), and in creating a clean, well-characterized surface for metal deposition and subsequent Schottky diode fabrication. In the former application, the high-temperature transient induced by the laser offers a very attractive and efficient alternative to the present prolonged pre-heat treatment at the moderate temperature required to clean the semiconductor surface to the high standard essential for good-quality epitaxy (Ploog, 1979) and brings MBE a step nearer to being a production technique. This is important for very high-frequency devices, in both silicon and group III and V semiconductors, because MBE is one of the few techniques that offers the possibility of making the sequences of very shallow layers (2000 Å) with well-controlled doping levels and abrupt changes in doping required (~200 Å) by some of the oscillator and amplifier structures envisaged for operation at frequencies in excess of 200 GHz (Purcell, 1979). Schottky diodes are active devices very economical of space in integrated circuits and so are attractive for the very high packing densities hoped for in future integrated circuits (Lostroh, 1980). A major problem of such devices is that the metal–semiconductor interface forms the potential barrier, and so device operation is very susceptible to trace contamination of the semiconductor surface. Adiabatic processing of such surfaces prior to metal deposition may well minimize reproducibility problems in Schottky devices.

C. Secondary Materials Changes

1. ELECTRICALLY ACTIVE DEFECTS

The process of rapid heating and cooling introduces into the regrown silicon material, and into the adjacent substrate material, point defects that are electrically active (Chapter 10). The concentrations of these defects are high in adiabatic processing and thermal flux processing, where the cooling rates are high enough to trap the mobile point defects, and a wide variety of electronic levels are observed that are ascribed to these defects (see, e.g., Benton et al., 1980; Johnson et al., 1980). The recombination currents (in minority-carrier devices) and carrier generation rates (in majority-carrier devices) that these high (10^{13}–10^{15} cm^{-3}) densities of electronic levels cause would produce leakage currents in beam-processed material that would make it very unattractive for device fabrication. The density of electronic levels can, however, be reduced by at least two orders of magnitude by thermal annealing after beam processing. In inert ambients a temperature of about 700°C is required, in molecular hydrogen about 480°C, and in atomic hydrogen (plasma) about 200°C (Benton et al., 1980; Lindner, 1980). The point defects themselves are not removed by the low-temperature hydrogen anneals, as shown by the reappearance of the electronic levels after anneal in vacuum at 400°C (Benton et al., 1980). The molecular hydrogen passivation technique reduces the concentration of defect electronic levels sufficiently so that at least in bipolar devices the leakage currents are at least as low as those obtained in conventionally furnace-processed devices (Lindner, 1980). This treatment fits well into the fabrication sequence of many devices, because the last heat treatment stage is very often at 400–500°C in a reducing ambient to sinter the metallization to the silicon surface to produce good ohmic contacts.

The point defects introduced by adiabatic annealing into GaAs also cause electronic levels in the material. These are not so easily passivated as in silicon and are thought to be responsible for the low electrical activity obtained after anneal (Davies et al., 1980).

It is possible to introduce electrically active line defects also into the semiconductor by thermal flux processing (Mizuta et al., 1980). These are more stable than point defects, and it seems to be necessary to keep the processing temperatures as low as possible to avoid their formation. This requirement severely restricts the use of thermal flux processing and certainly reduces the acceptable energy window to about 5% in P/A (Section IV.A.3) for applications involving recrystallization of bulk material (Hill, 1981).

No electrically active defects are reported for isothermally processed silicon, and the successful fabrication of MOS and bipolar devices (McMahon *et al.*, 1980; Speight *et al.*, 1981) indicates that, if present, their concentration is very low. The high fluxes of electrons associated with these isothermal anneals might be expected to produce high point defect concentrations in semiconductors, but it is suggested (Ahmed, 1980) that the fraction of a second spent at high temperature immediately after the beam is switched off (see Fig. 7) is sufficient to allow the point defects to anneal out.

2.   TOPOGRAPHIC EFFECTS

In adiabatic processing a radiant beam sufficiently uniform to achieve certain tolerant energy window requirements for anneal (see Fig. 5) may yet cause deviations from planarity of the semiconductor surface. The occurrence and characteristics of these effects have been reviewed (Hill, 1980). In fact, for all adiabatic annealing not carried out with very uniform beams (5% on a microscale), surface ripple is observed, varying in height between 100 and 1000 Å. Beams in which interference, diffraction, or mode beating occur can give rise to much larger topographic effects, up to a micrometer in amplitude. An example of adiabatic processing modulated by diffraction effects is shown in Fig. 15a. In this case the modulation is sufficiently strong to be visible as periodic fluctuations in the perfection of the annealed material and in the surface level (shown in Fig. 15a as variations in TEM sample thickness). Topographic effects are probably insignificant for device fabrication if their amplitude can be kept below 1000 Å. They are often, however, associated with nonuniformities in materials properties (see Fig. 15a and Section VI.C.2) that are more serious.

3.   COMPOSITIONAL CHANGES

The high temperatures involved in beam processing, although transient, can allow in or out diffusion of atomic species between sample and ambient, which may be deleterious to device parameters. The loss of arsenic, which occurs by evaporation during adiabatic annealing of GaAs (Eisen, 1980), is one of the major factors that have prevented device-quality material from being produced by this technique in GaAs and other group III and V compounds. Most laser processing of Si has been carried out in an air ambient, and oxygen is reported to diffuse in during adiabatic processing (Hill, 1979; Hoh *et al.*, 1980) and to oxidize the surface during thermal flux processing (Gibbons, 1980).

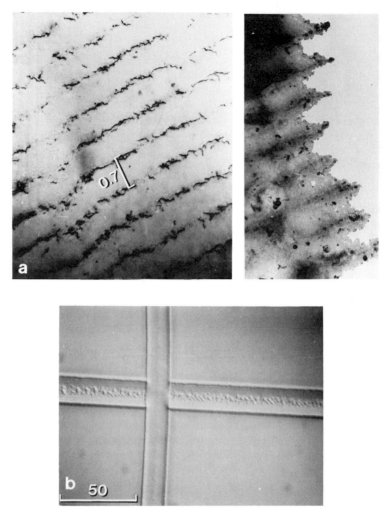

Fig. 15. Topographic nonuniformities in silicon surfaces amorphized by implantation with 10^{16} arsenic ions at 40 keV and melted and regrown by beam processing. (a) Transmission electron micrograph of this silicon surface annealed by a 25-nsec ruby laser pulse. The crystal structure is fully restored, except in narrow parallel zones regularly spaced where dislocations occur. These zones are also higher than the recrystallized surface, as shown by the greater thickness of the TEM foil (evident in the figure from the darker tone and the scalloped edge at the etched hole in the foil). The scale marker shows that the zone spacing is 0.7 μm, similar to the wavelength of the incident light (0.69 μm); the dislocated zones are thought to result from low-intensity regions of a diffraction pattern in the annealing beam (Hill, 1980). (b) Phase-contrast optical micrograph of the same amorphized surface annealed by the transits of a focused argon ion laser spot approximately 50 μm wide. First transit (horizontal) 8 W at 2.5 cm/sec, second transit (vertical) 12 W 10 cm/sec. The intensity at the center of the spot (Gaussian spatial energy distribution) was high enough to melt the silicon. The boundaries of the melted zone are delineated by a change in surface level ascribed to redistribution of silicon in the molten phase. Scale marker is in micrometers.

No contamination has been reported for isothermal processing, but the relatively prolonged heat treatment at high temperature makes the sample holder and the residual hydrocarbons in the vacuum ambient potential contamination sources.

D. Compatibility with Device Fabrication

1. PROCESSING DEPTH REQUIRED

a. Limitations. The thickness of the layer in which materials changes can be controllably induced varies widely with the mode of processing and sets a limit to the vertical size of the device that can be usefully considered for beam processing: this dimension varies between 0.1 μm and 1 cm dependent on the situation for each mode, as discussed below.

b. Adiabatic Processing. Because of the large difference in rates of diffusion and reactions between solid and liquid phases, the processed zone is very sharply defined by the maximum melt front boundary in adiabatic processing. The limiting layer thickness is determined by the maximum energy density that can be utilized without causing the surface to boil. This not only initiates surface damage but also removes heat as latent heat of vaporization and presents the melt front from penetrating further into the material (Mehrabian *et al.*, 1979). The occurrence of boiling anywhere on the surface is unacceptable for device structures, hence hot spots in the beam can severely limit the average energy density and the maximum layer thickness that can be processed. The use of a beam of very uniform energy density increases the usable energy density markedly. For example, homogenizing a ruby laser pulse allowed the threshold energy for surface damage in an amorphized silicon surface to be raised from 1.4 to 3.3 J cm^{-2}, corresponding to melt depths of about 0.3 and 0.9 μm (Hill, 1980, 1981). The higher energy density corresponds approximately to the whole surface reaching the boiling point; it is reported (Surko *et al.*, 1979) that a more exact correlation is the onset of damage in silicon and germanium with the temperature at which the vapor pressure reaches about 10 atm, and calculations show that this corresponds to an energy density about 25% higher than that required for boiling (1 atm). It is suggested that the formation and effects of a hot plasma in the ambient above the molten layer cause surface damage rather than simple loss of material by evaporation. With a uniform beam, the only way in which the melt depth can be further increased is by reducing the temperature gradient between surface and melt front (temperature fixed at the melting point). Some reduction can be effected by absorbing the radiation over a greater depth (e.g., by using a less strongly absorbed wavelength of light or by using electron beams that penetrate more deeply, Chapter 2). A good example of this effect is reported (though for a three-phase polysilicon–SiO_2–silicon matrix structure) where the processable depth of the top polysilicon layer was increased from 0.4 to 0.85 μm by increasing the wavelength from the strongly absorbed 0.53 μm to the weakly absorbed 1.06 μm (Kaplan *et al.*, 1980). This technique reaches its useful limit at about a micrometer for laser irradiation, because at these energy densities sufficient energy is available in the early part of the pulse to ensure that most of the energy of the pulse is incident on a melted surface (Baeri *et al.*, 1979) and the high absorption coefficient of liquid silicon confines the absorption of heat to the near-surface region. The temperature gradient can also be reduced by maintaining the substrate at an elevated temperature (Bloembergen, 1978), but this takes away one of the most attractive features of adiabatic processing, the opportunity for room temperature handling of slices during heat treatment of the surface.

It should be possible to select the usable melted layer depth by using electrons of energy appropriate to that absorption depth. However, although it is possible to vary the depth of the high-temperature region over the range 1–10 μm by this technique (Kirkpatrick, 1980; Neukermans and Saperstein, 1979), the stability of thick liquid layers under irradiation may not be good enough to allow planar undamaged structures to be fabricated. The energy densities involved would be similar to those used for inducing gettering damage, where craters are formed (Section VI.E.2). In addition, the required electron fluxes being absorbed in hot silicon below the melt front give rise to cracking (Wilson *et al.*, 1978; Kirkpatrick, 1980) and again limit the energies to those sufficient for melting to about 1 μm depth. Thus adiabatic melt phase processing of structures thicker than 1 μm is probably precluded, though of course much thicker structures can be built up by a sequence of deposit–melt–regrow operations (e.g., as in Kirkpatrick, 1980).

c. Thermal Flux Processing. In solid phase thermal flux processing the high-temperature region penetrates at least tens of micrometers into silicon (Gat *et al.*, 1979), and so inducement of material changes in layers deeper than 1 μm is possible, though not so far reported. The rate at which solid state processes occurs is, however, sufficiently slow that, at temperatures low enough to avoid damage (Mizuta *et al.*, 1980), the dwell time required might reduce throughput (Section VII.B.2) to an unacceptably low level. In addition, for annealing applications, the competitive hetero-

geneous nucleation of polycrystals during high-temperature anneals probably limits the amorphous layer thickness to 3000 Å (Gat *et al.*, 1979).

In liquid phase processing the low-temperature gradient that obtains in
the liquid layer is calculated to allow liquid layers of at least 10 μm thickness to be formed without surface boiling (Kokorowski *et al.*, 1981). However, the same hydrodynamic instabilities that will probably limit adiabatic melt phase processing to about 1 μm thickness may well limit thermal flux processing to a similar thickness. A nonflat surface always seems
to occur in deeply melted materials, as illustrated in Fig. 15b for a
scanned cw anneal of silicon, where melting has occurred.

d. Isothermal Processing. The processed depth in this mode is limited only by the uniformity of the temperature in depth required by the
process. Uniformity to 1°C is obtained in processing silicon slices 400 μm
thick (Section IV.B.2), and several millimeters is in principle feasible.

2. INTERACTION WITH EXISTING LAYERS

As well as effecting the primary materials change, beam processing
often alters the properties of previously fabricated layers that lie within
the processed zone. The most important interaction of this kind in existing
device schedules is the redistribution of dopants that occurs whenever a
heating cycle is carried out. Redistribution of dopants takes place during
beam processing too (Section V.B.2) and must be taken into account when
designing fabrication schedules. In adiabatic processing, the redistribution is confined to the melted zone (Chapter 5) and the boundary between
the redistributed region and the unchanged region beneath can be very
abrupt (Fig. 13). It is thus possible to preserve exactly dopant profiles
previously fabricated in the deeper part of a device structure although
effecting very large materials changes and dopant redistributions in the
immediately adjacent surface region (Hill *et al.*, 1982). In thermal flux
solid state processing the diffusion coefficients and processing times are
such that no significant redistribution of substitutional dopants occurs
(Chapter 10). This allows anneal of implantation damage (which requires
only very local reordering) without significant changes in the ion-
implanted profile and also without significant redistribution of preexisting
layers.

Thermal flux liquid state processing on single-phase material (e.g., silicon single crystal) is likely to give uniform distribution of substitutional
dopants within the melt zone and no redistribution outside it. No experimental data are available. Isothermal processing has a time scale over
which small (~200 Å) redistributions of dopants occur (McMahon *et al.*,

1980) and, since temperature is uniform, all preexisting layers throughout the device will see the same heating cycle.

3. EFFECT OF SUBSEQUENT HEAT TREATMENTS

If the beam processing operation is not the last in the fabrication schedule, then the processed material may be undergoing a subsequent high-temperature furnace stage. This can radically change the properties of beam-processed material which is often in a metastable state. The removal of the electrically active levels by subsequent hydrogen anneal has already been mentioned (Section V.C.1). Higher-temperature heat treatments cause the precipitation of dopant from supersaturated solutions. This precipitation begins at 400°C and reaches thermal equilibrium values after 2 min at 900°C in thermal flux-processed arsenic-implanted silicon (Lietoila *et al.*, 1979); in adiabatically processed material, isochronal (30 min) anneals caused a steady precipitation of supersaturated phosphorus as the temperature was increased from 400 to 700°C, at which temperature the equilibrium solid solubility appeared to have been reached (Miyao *et al.*, 1980). An important observation was that phosphorus-implanted layers that had been adiabatically processed at just above threshold for anneal gave supersaturated solutions with 100% electrically active dopant, but on subsequent thermal anneal at 600°C heavy precipitation occurred, leaving only about half the thermal equilibrium value of the dopant in solution. This effect was ascribed to precipitation on dislocation loops which were observed after near-threshold anneals.

This sensitivity of metastable solutions to subsequent moderate heat treatments considerably reduces their attractiveness in device fabrication. Metallization and bonding steps necessarily have to follow beam processing and typically involve 400–500°C heat treatments, so that some precipitation will be incurred. The use of even moderate temperature deposition techniques (e.g., low-pressure polysilicon, thermally densified deposited silicon oxide, both about 600–700°C) cannot be combined with metastable regions fabricated in previous steps.

Subsequent heat treatments can also change the carefully controlled dopant distributions previously fabricated by beam processing, for example, the novel dopant concentration profiles obtained by multiple-pulse adiabatic processing (Section V.B.2) or the as-implanted profiles resulting from thermal-flux annealing (Chapter 10). Such changes will only be significant if the $\sqrt{Dt}$ product of the heat treatment exceeds the tolerance allowable on the profile (typically 100–200 Å). For the common dopants in silicon, temperatures up to 800°C will be tolerable (15 min at 800°C gives $\sqrt{Dt} = 50$ Å for arsenic in silicon at moderate concentrations).

E. Devices

1. DEVICE STRUCTURES

The device structures that can be beam-processed as single-phase structures are necessarily very simple. They must have no lateral variations in materials properties, and no large vertical variations. The device must thus consist (at least at the beam processing stage) of a series of uniformly doped regions, one on top of the other, underlying and parallel with the flat semiconductor top surface. Very few devices are as simple as this. Integrated circuit slices based on bipolar technology look nearly as simple as this at the early epitaxial layer growth stage, and adiabatic beam processing for reduced contamination growth (Section V.B.4) of epitaxial silicon would approximate to single-phase processing. However, even then, the essential heavily doped and selectively patterned buried layer (that in final devices acts as the transistor collector) would introduce lateral variations in the absorption coefficient, hence temperature (Section VI.C.1). No devices so far have been reported in such regrown material. Two simple devices to which adiabatic beam processing has been applied are the solar cell and the impatt diode.

2. THE SOLAR CELL

Considerable effort has been put into improving the cost and efficiency of solar cells by incorporating beam processing into the technology (see, e.g., Young *et al.*, 1980; Kirkpatrick *et al.*, 1977; Muller *et al.*, 1980). A schematic diagram of a silicon solar cell at the stage of adiabatically processing the critical surface layer is shown in Fig. 16a. Nearly all the special characteristics of adiabatic processing have been utilized in attempting to improve the solar cell efficiency. The complete anneal of both implant and diffusion damage has resulted in greater collection efficiency of carriers in the heavily doped surface region; the sharply defined zone of materials changes has resulted in the lifetime, hence collection efficiency, of the bulk p-type silicon being preserved; the ability to form supersaturated solid solutions has been used to form highly conducting (but also thin) surface regions, resulting in lower resistive losses (Muller *et al.*, 1980). Cell efficiency was indeed increased, by about 1% as compared to conventionally processed cells. This may not be sufficient to justify incorporation of laser processing into manufacture.

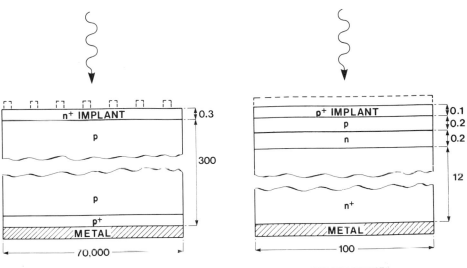

Fig. 16. Examples of devices that can be fabricated by beam processing single-phase structures. (a) Schematic diagram of a solar cell showing laser anneal of the implant forming the n^+ side of the n^+–p junction and ohmic contact. (b) Schematic diagram of a double-drift impatt diode showing laser anneal of the implant forming the p^+ ohmic contact region. Dimensions are in micrometers, and dotted regions indicate metal contacts fabricated after the laser anneal stage.

Adiabatic processing has also been applied in trying to lower the total cost of solar cell fabrication. These approaches (Young *et al.*, 1980; Muller *et al.*, 1980; Kirkpatrick *et al.*, 1977; Kirkpatrick, 1980) incorporate those already mentioned plus the formation of low-resistance metallic contacts with the front and rear of the cells and also the substitution of deposited metal films (~100 Å) for the more expensive ion implantation as a doping source. Complete processing lines where all heat treatments are adiabatically effected are envisaged (Kirkpatrick *et al.*, 1977). The economics of this approach have not been reported in detail, but the use of pulsed electron beams of high efficiency is intended to lower processing costs. The uniformity of such beams is not, so far, very good (Kirkpatrick, 1980) but may be adequate for solar cells if variations in junction depth can be minimized.

3. THE IMPATT DIODE

A schematic diagram of a double-drift impatt diode for the production of microwave oscillations in the 140-GHz range is shown in Fig. 15b.

These structures require narrow, uniformly doped regions separated by abrupt changes in doping level. Adiabatic processing has been applied to such structures (Hess *et al.*, 1980) to reduce the interdiffusion of the doped layers and to minimize contamination during anneal and activation of the heavily doped $p+$ layer. The devices are also required to have as low a total resistance as possible, and adiabatic processing allows the surface layer to be more highly doped, reducing both contact resistance and ohmic resistance. It appears that the main improvement in the final devices is not so much in the achievement of a new level of output power and efficiency (values obtained are typical of good conventionally processed impatt diodes) but in the reproducibility with which these values are obtained.

VI. Processing Multiphase Structures

A. Multiphase Structures

As their name suggests, these structures include those that contain two or more distinct phases with different chemical and physical constitutions: also included, because of their similar behavior during beam processing, are structures that are basically single phase but have sufficiently marked variations in properties laterally to cause significant lateral temperature variations during beam processing. All device structures are multiphase when completed (metal plus semiconductor at least), and most are multiphase at the stages in the fabrication sequence where beam processing is likely to be used. All the considerations discussed with regard to single-phase structures (Section V) apply to the individual single-phase regions of multiphase structures: where the phases are adjacent, or overlap, a number of additional possibilities (Section VI.B) and problems (Section VI.C) arise. Many of these additional aspects have been explored in, and are well illustrated by, the beam processing of structures containing silicon and silicon dioxide layers (Section VI.D). Some applications to device structures are described (Section VI.E).

B. Differential Processing

1. DIFFERENTIAL HEATING

In a multiphase structure the abrupt changes in physical and chemical properties that occur at phase boundaries can, and usually do, cause abrupt changes in the heating profile, both laterally and in depth, during beam processing. Thus different phases see very different heat treatments when irradiated with the same radiation pulse. These differences can arise from differential reflection, absorption, and heat flow.

2. DIFFERENTIAL REFLECTION

Surfaces may differ in reflectivity either because a different reflective material is exposed to the beam in different regions (e.g., metal, semiconductor) or because a different thickness of transparent film overlies different areas of an underlying reflective surface (e.g., oxide layers of different thicknesses on silicon). In the former case, a fraction of the beam energy is reflected away that is characteristic of the intrinsic optical properties of the material. In the latter case, the fraction reflected away depends on the thickness of the overlying film t and its refractive index n_1, and that of the substrate n_2, and is an optical interference phenomenon (Chapter 2). The fraction of incident power transmitted through the film is a periodic function of film thickness oscillating between a minimum value close to that of the bare substrate surface and a maximum value that can be close to 100% if the refractive indices n_1 and n_2 are related to the refractive index of the ambient n_0 by (Born and Wolf, 1964):

$$n_1^2 = n_0 n_2 \tag{7}$$

Thus in a planar structure where metal and dielectric films overlie selected regions of a semiconductor substrate, the fraction of the incident power transmitted into the surface at room temperature can vary typically between 5 and 15% (metal), 50 and 70% (bare semiconductor), and 50 and 100% (dielectric-coated semiconductor) (von Allmen, 1980; Hill, 1979; Tamura *et al.*, 1979). In addition, these parameters can change as the temperature rises; the reflectivity of metals tends to fall, whereas the reflectivity of semiconductors rises as the materials change from cold solid to hot solid to liquid (von Allmen, 1980). The accompanying change in the refractive index of semiconductors can also change the fraction of interference-controlled transmitted light in dielectric-coated semiconductor regions as the temperature rises. The periodicity of the power transmission in such regions depends on meeting an exact interference criterion and therefore is much less marked when a wide range of beam angles (e.g., from a Cullis homogenizer, Section IV.B.3) or wavelengths (e.g., white light) is used. In such cases an average enhanced power transmission, independent of dielectric thickness, is obtained (Hill and Godfrey, 1980).

The differential reflectivity effects described above apply equally to all optical beam systems irrespective of pulse length. Electron beam systems exhibit weak differential reflectivity effects, the transmission of power of a 30-keV beam varying from 65 to 100% as the atomic number of the material changes from 100 to 1 (von Allmen, 1980).

3. DIFFERENTIAL ABSORPTION

A very wide range of optical absorption coefficients can be found in multiphase structures: typical values are 10^{-4} cm^{-1} (silicon oxide), 10^1–10^4 cm^{-1} (crystalline semiconductor), 10^4–10^5 cm^{-1} (amorphous semiconductor), and 10^6 cm^{-1} (metallic) at room temperature. These values can change markedly as the temperature rises and can also be a strong function of the beam wavelength. For instance, for a 1.06-μm wavelength incident on silicon, as the temperature rises from 0 to 1400°C, the absorption coefficient increases from 1.5×10^1 to 6.8×10^3 cm^{-1} (crystalline silicon) and from 2.2×10^4 to 5.0×10^4 cm^{-1} (amorphized silicon) (Godfrey *et al.*, 1981). The influence of free-carrier absorption on the absorption coefficient becomes very important as the photon energy decreases below the band gap energy (Chapter 2). For example, in crystalline silicon (band gap energy 1.0 eV), the absorption coefficients of near-intrinsic (10^{16} carriers cm^{-3}) and heavily doped (10^{20} carriers cm^{-3}) material as photon energy decreases are, respectively: 1.9×10^4 and 1.9×10^4 cm^{-1} (2.54 eV, argon ion laser; Godfrey, *et al.*, 1981), 3.0×10^1 and 6.0×10^2 cm^{-1} (1.17 eV, Nd:YAG laser; cited in Fowler and Hodgson, 1980), and 5×10^0 and 1.5×10^4 cm^{-1} (0.12 eV, CO_2 laser; Schuman *et al.*, 1971). Metals and liquid semiconductors have high absorption coefficients (about 10^6 cm^{-1}) which are insensitive to temperature.

These differential absorption effects apply equally in all optical beam processing systems irrespective of pulse length. Differential absorption occurs in electron beam irradiation but is a much weaker function of the material and insensitive to the temperature (Chapter 2).

4. DIFFERENTIAL HEAT FLOW

Although the rate of deposition of heat at every point in a multiphase structure is entirely determined by the energy transmission and absorption parameters just described, the resulting temperature profile also depends on the rate of loss of heat from the absorption zone (Chapter 3). This is mainly determined by the thermal conductivity of the adjacent material. The range of values likely to be met with increases from 0.015 (silicon dioxide) through 0.29 (amorphous silicon) and 1.3 (single-crystal silicon) to 5.0 W cm^{-1} °C^{-1} (aluminum) (Godfrey *et al.*, 1981). The temperature dependence of this parameter depends very much on material structure and can be positive or negative. For the first three materials quoted above, the percentage change in thermal conductivity between 0 and 1000°C is +250, −20, and −80%, respectively (Godfrey *et al.*, 1981). Thermal conductivity in liquid metals and semiconductors is high, 6–10 W cm^{-1} °C^{-1}.

Differential heat flow effects occur independently of the method of energy deposition (photon or electron beam), but their importance strongly depends on the pulse duration. In adiabatic processing, the short heat diffusion time and consequent similarity of the heat flow distances and processed layer thicknesses can cause differential heat flow effects, giving rise to large temperature differences in the different regions of a multiphase structure. An extreme example is given by the adiabatic processing of a 0.5-μm polycrystalline silicon film deposited uniformly over an oxide-coated silicon substrate in which windows have previously been photoengraved in the oxide. This gives a multiphase structure with two types of layer structure exposed to the beam: a polysilicon–oxide–silicon substrate and a polysilicon–silicon substrate. The large difference in thermal conductivity between silicon and silicon dioxide gives such a large difference in temperature during irradiation with a pulsed ruby laser that the whole film can be evaporated from the oxide, although the silicon islands directly in contact with the substrate remain undamaged.

In thermal flux processing, differential heat flow effects are rather weak because of the long heat flow distances as compared with the optical absorption depth. However, in liquid phase regrowth of crystals on amorphous films (Section VI.C.1) control of heat flow is an important parameter, and amorphous layers with different thermal conductivities may influence crystal quality (Biegelson *et al.*, 1981).

5. DIFFERENTIAL MATERIAL BEHAVIOR

Regions of a multiphase structure that experience the same time–temperature cycle may nevertheless respond very differently. A difference in melting points can mean that solid and liquid regions exist side by side or one on top of the other. This can occur when there are merely structural differences between phases—amorphous and crystalline silicon have been shown (Baeri *et al.*, 1980) to melt at 897 and 1412°C, respectively. Other material properties likely to be involved in differential material behavior are boiling point, vapor pressure, volume change on melting, mechanical strength, thermal expansion coefficient, and Young's modulus. Obviously, if the differential material behavior affects the differential heating mechanisms previously described (Section IV.B.1), e.g., through a phase change occurring during irradiation, the temperature profiles in the two regions will rapidly diverge. Differential material behavior is responsible for many of the effects described in Section VI.D.

6. CONTROL OF DIFFERENTIAL EFFECTS

It is essential to have adequate control over differential processing effects if multiphase device structures are to be satisfactorily fabricated. In structures containing a sequence of uniform layers of very different reflection and absorption properties, the energy window for control of the process can often be made much wider by attention to the order and thickness of the layers. The principle is to ensure that the temperature–time cycle of the process is controlled by the layers that have a weak temperature dependence of the relevant parameter and is described in detail by von Allmen (1980). This approach is particularly well suited to reacting deposited layers, since the layer order can often be altered (Section VI.C.2).

More often, the need to control differential heating arises because of lateral variations in the structure. A particular process is being carried out in region A, and unwanted processes occur in the differing regions B, C, and D. If the energy window for process A gives rise to unacceptable levels of materials changes in other regions of the structure, various approaches are possible. The process can be redesigned to give a lower energy threshold for process A by increasing the transmission, absorption, or containment of energy in this region as compared with the other regions. An example of this approach is the successful melt and regrowth of polysilicon islands on oxide-coated silicon substrates without melting the silicon substrate material where it was exposed to the same adiabatic processing induced by a ruby laser pulse (Yaron et al., 1980a). The differential heating was greatest in the polysilicon layers because of the higher absorption coefficient (fine grain size), lower heat loss (lower thermal conductivity of the underlying SiO_2), and lower reflectivity (selected polysilicon thickness) as compared with the coated silicon substrate material. Regrowth of an amorphous silicon layer on the silicon substrate (Fowler and Hodgson, 1980) in selectively controlled areas was achieved by doping the silicon in these areas (before amorphous silicon deposition) with diffused arsenic. The enhanced absorption (due to free-carrier absorption) at a wavelength of 1.06 μm allowed selective melt and regrowth of the amorphous material to be effected only in the doped regions, with a resolution of 2.5 μm.

If the energy threshold for the desired process cannot be brought below those for the unwanted processes, selective heating must be used. This technique is applicable only to adiabatic annealing because of the confinement of the heating effect to the irradiated area. Selectivity can be achieved either by imaging the pulse (e.g., as in Fig. 30) or by protecting the sensitive areas with a reflecting overlayer. For high-resolution requirements the latter approach is preferred, since penumbra and diffraction effects limit resolution in the imaged pulse to about 5 μm, whereas standard photoengraving techniques and self-alignment procedures enable the protective overlayer edge and processed region edge to coincide within 500 Å (Hill, 1980). Details of the use of protective overlayers in silicon dioxide–silicon structures are given in Sections VI.D.2 and VI.E.1.

C. Primary Materials Changes

1. REORDERING DEPOSITED LAYERS

With the use of adiabatic and thermal flux processing techniques for melting and regrowing, semiconductor layers deposited on foreign substrates can be converted to polycrystalline or even single-crystal structure. These techniques and their characteristics are fully described in Chapter 8. The driving force for this work is the attractiveness in many applications of a starting semiconductor material that consists of an active layer, no thicker than the device thickness required, grown onto a substrate whose properties are optimized for the particular application. The high-quality silicon slices currently available are not optimized in this way. The devices made in the top few micrometers of the slice are very satisfactory, but for solar cells, for example, the substrate is far too expensive and wasteful of semiconductor material; for integrated circuits, the substrate is too conductive and thus creates a parasitic capacitance which slows down the operation of MOS devices. Much of the work in this field so far has concentrated on exploring and characterizing the properties of beam-processed amorphous films, usually using the convenient thermally oxidizing silicon slice as a convenient flat and chemically pure amorphous substrate. Solar cell work has been mainly concerned with attempts to change the structure of hydrogenated amorphous silicon films to obtain higher electrical conductivity and longer carrier lifetime, also preserving the high optical absorption coefficient of amorphous silicon so that efficient thin layer solar cells become possible (Pankove *et al.*, 1980; Sussmann *et al.*, 1980). Some success has been achieved, but the basic difficulty appears to be that the hydrogen is easily lost during the heat treatment and is essential to the structural integrity of the α-S$_i$.

For integrated circuit applications the prime requirement of the regrown material is that it have uniform properties on both a macro- and micro-scale; the matching of the thousands of components in such a circuit demands this. The least demanding requirement for regrown amorphous material is for use as a passive element, e.g., a resistor or conductor in a circuit. The control problem is then simply to obtain a uniform resistivity at a required value. High-value polysilicon resistors have an immediate application as load devices in integrated circuits (O'Connell *et al.*, 1977), whereas low-resistance polyresistors could extend the use of polysilicon as a first-level metallization to smaller structures without incurring unacceptable voltage drops across the circuit. Polysilicon films of controllable high resistance (10^3–10^8 Ω per square) have been fabricated by thermal flux processing (with a scanned laser) of phosphorus-implanted, low-pressure vapor-deposited polysilicon films on silicon nitride-coated silicon (Yaron *et al.*, 1980b). As compared with furnace-annealed films, the beam-processed films have a more uniform resistivity much less sensitive to phosphorus doping variations. This difference is ascribed to the lower density of trap states at grain boundaries in the beam-processed material. Polysilicon films of uniform low resistivity (10 Ω per square) that are stable during subsequent 400°C anneals have been obtained by adiabatic processing of arsenic-implanted 0.4-μm-thick polysilicon layers (Wada *et al.*, 1980). To attain the necessary stability during 400°C passivating anneals, the grain size of the polysilicon had to be increased by phosphorus doping and furnace anneal before the arsenic implantation. The improvement over small-grain films was ascribed to a reduction in the available sites for precipitation of the metastable solid solution of arsenic.

A more demanding application for regrown semiconductor films is in acting as the matrix in which active circuit components, e.g., diodes and transistors, are fabricated. Not only resistivity but also carrier mobility and lifetime will, in general, need to be high and uniform. The uniformity requirement is paramount for integrated circuits, and two approaches have been adopted. One is to achieve uniformity by producing a polysilicon layer with uniform grain size and improved mobility due to reduced traps at grain boundaries (Lee *et al.*, 1979) by a technique similar to that described above (Yaron *et al.*, 1980b). The uniformity depends on averaging the effects of a number of grains over a device. Average mobility and resistivity of this material were about a factor of 2 lower than for bulk single crystal. A second approach, which has been pursued using a wide variety of techniques, is to attempt to produce single-crystal material on the amorphous substrate. Ordered crystallization is induced by "seeding" the regrowth from selective regions where the polycrystalline film makes direct contact with the substrate (lateral epitaxy), by favoring the growth of fast-growing crystal planes by shaping the silicon into islands (Biegelson *et al.*, 1981) or by shaping the amorphous substrate (Geis *et al.*, 1980). All these techniques have had some success, but none so far has succeeded in producing defect-free material in which the crystal orientation is exactly controlled and reproducible. For integrated circuits this is essential. The islands need not be very large, however; probably the most satisfactory size would coincide with the individual chip area, typically 1–10 mm square. The next most convenient size would be that of the individual component, typically 10–40 μm square, but this would be acceptable only if the nonactive areas between components could be kept to a small fraction of the whole (less than 25%).

2. REACTING DEPOSITED LAYERS

As well as reordering one layer on another, beam processing can effect reactions between layers, either reacting them completely to form a new phase or partially reacting them to produce an intermediate layer of a new phase. This technique has been used to produce silicide layers from elements (Chapters 12 and 10), to synthesize semiconductors from elements, e.g., AlSb (Andrew *et al.*, 1979), and to form thin reacted layers between semiconductor substrates and overlying metal films to produce ohmic contacts on silicon (Allen *et al.*, 1980; Wittmer, 1980) and Group III–V compounds (Eckhardt, 1980). A very wide variety of beam sources and pulse lengths have been used for these studies, since the metal component absorbs all wavelengths equally well. The difference between the absorption coefficients of silicon (about 50) and copper (about 10^6) at a wavelength of 10.6 μm has been exploited to bond silicon directly to a copper heat sink by an adiabatic pulse through the thickness of the semiconductor (Allen *et al.*, 1980). The control of this process is very good, since the heat is generated only where it is required, at the metal–semiconductor interface. The advantages to be obtained in devices from reacting deposited layers are not clear as yet. Silicide layers for low-resistance metallization can be formed without subjecting the whole slice to a furnace anneal: yet a great advantage of polycides (a silicide underlaid by silicon is called a polycide) requires that they be thermally oxidized in a furnace to form automatically a dielectric spacer layer before the second level of metallization is deposited. Ohmic contacts and underlying p–n junctions have been formed in one operation by adiabatic pulse processing of Al on Si structures (Harper and Cohen, 1970).

3. CONTOURING DEPOSITED LAYERS

The melting that occurs during adiabatic processing of deposited layers changes the topography of the surface on both a macro- and a microscale. The surface tension of the liquid film tends to smooth out sharp changes in the direction of the layer surface. This has been exploited to round off the sharp etched edges in polycrystalline silicon films (Wu and Schnable, 1979; Hess *et al.*, 1981) and in silicon oxide (Runge, 1980). The much smaller asperities characteristic of low-temperature-deposited films are also removed, and thermal oxide grown on the beam-processed silicon is found to be of much higher quality for capacitor fabrication (Yaron *et al.*, 1980a). These techniques are very attractive for integrated circuit fabrication because, as circuit components become smaller, the design rules require the vertical–horizontal aspect ratio to increase, necessitating vertical etching techniques such as plasma etching, which produce very sharp edges. Subsequently deposited layers cannot cover these steps without cracking, and various techniques are employed to smooth over the steps. Beam processing has the advantages that the threshold for smoothing is low (so other parts of the circuit are unaffected; Yaron *et al.*, 1980b), and no additional layer is required.

A contouring technique that does not necessarily involve melting depends on changing the material properties in selected regions and then applying a selective etch to remove the processed (or unprocessed) material. Mesas have been fabricated by etching unprocessed amorphous silicon (Hess *et al.*, 1979) and processed AlGaAs (Salathe *et al.*, 1979). The latter structure was used as a 2.5-μm-wide waveguide, and such mesa structures may find application in the growing field of optical integrated circuits.

D. *Silicon Dioxide Films on Silicon*

1. SILICON DIOXIDE

The oxide that forms on silicon during high-temperature oxidation is an almost perfect material for every aspect of integrated circuit fabrication and operation and is largely responsible for the preeminence of silicon in this field. Thermally grown silicon oxide is strongly bonded to the silicon, is an effective mask against high-temperature diffusion of dopants and against ion implantation of dopants, and is completely resistant to all chemicals except those containing hydrofluoric acid. Electrically it is a very good insulator, has high dielectric strength, and can be grown with less than 10^{11} electronic charges per cm^2. The silicon–silicon dioxide interface can be routinely produced with interface states below 10^{10} per cm^2. One consequence of this is that silicon oxide tends to be present throughout the processing of integrated circuits (and indeed all low-current silicon devices), often being used in a multiple role. If beam processing is to be applied to integrated circuit manufacture, the problems of controllably irradiating silicon coated with selective areas and different thickness of silicon dioxide have to be solved.

2. ADIABATIC PROCESSING

The structure shown in Fig. 17 is a basic building block of an integrated circuit. The window in the thermal oxide covering acts as both an implantation mask and a junction edge passivation for the selectively arsenic-doped region. In Section V.B it was shown that the implant damage could be annealed and the dopant controllably redistributed by adiabatic processing. The result of irradiating this structure with an unhomogenized ruby laser pulse is shown in Fig. 18a. The bevel section reveals that the dopant has been redistributed and electrically activated, and also that the

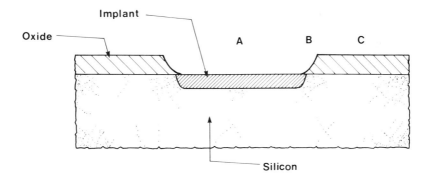

Fig. 17. Schematic diagram of a vertical section through a bipolar diode structure, a basic building block of silicon integrated circuits. The oxide coating C (typically 0.5 μm thick) acts both as a mask against ion implantation and so defines the doped region A, and also as a passivating dielectric layer which greatly reduces electrical leakage currents in region B where the p–n junction formed by annealing the implanted ions (n-type silicon is hatched; p-type substrate is dotted) meets the silicon surface. (From Hill, 1980.) This figure was originally presented at the Spring 1980 Meeting of The Electrochemical Society, Inc., held in St. Louis, Missouri.

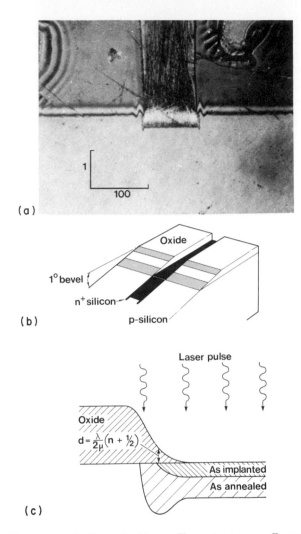

(a)

(b)

(c)

Fig. 18. Photoengraved silicon dioxide on silicon structures—effects of differential reflection. (a) Optical micrograph of a bevel section through a structure of the type illustrated in Fig. 17; the window region was implanted with 10^{16} arsenic ions/cm² at 40 keV and adiabatic anneal carried out using an unhomogenized multimode 25-nsec pulse from a ruby laser. The bevel section has been stained to delineate the heavily doped n-type region which results from the redistribution of arsenic during melt and regrowth. (b) Schematic diagram of the bevel section clarifying the geometry of the section. Note that the effective vertical magnification is approximately 60 times the horizontal magnification when a continuous linear structure is beveled at a 1° angle, so that the planar junction depth of 0.25 μm and the more deeply penetrating "spikes" on this junction (0.30 μm) are clearly seen, even though the dimensions are smaller than the wavelength (0.59 μm) of the illumination used. (c) Schematic diagram of the window edge region illustrating how the deeper penetration of the arsenic results from enhanced local heating where the thickness of the overlying oxide meets the criterion for minimum reflection of the laser light. (From Hill, 1980.) This figure was originally presented at the Spring 1980 Meeting of The Electrochemical Society, Inc., held in St. Louis, Missouri.

$p-n$ junction is not planar but has "spikes" at the window edges. The schematic diagram (Fig. 18b) makes the geometry of the sectioning clearer. The cause of the spikes was shown to be (Hill, 1979) enhanced power transmission at the sloping edges of the chemically etched oxide window (Fig. 18c), a localized example of the interference effect discussed in Section VI.B.1.

A second effect that is noticeable when a structure such as that in Fig. 17 is adiabatically processed is that the window width becomes narrower (Hill, 1980). Also, the oxide-covered areas are no longer flat but are covered with slight ripples as shown in Figs. 19 and 20. Ripples in oxide-coated structures after adiabatic anneal have also been reported for laser irradiation by Stephen et al. (1980) and Naryan (1980) and for electron beam irradiation by Leas et al. (1980). A systematic study of these effects (Hill and Godfrey, 1980) showed that the dimensional change in the windows and the ripple were different manifestations of the same basic effect. The differential heating, between the amorphized silicon in the window and the oxide-coated silicon elsewhere, is not sufficient to allow process-

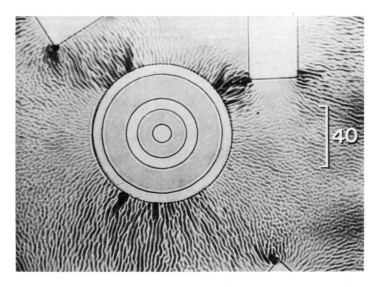

Fig. 19. Photoengraved silicon dioxide-on-silicon structures: effects of differential material behavior on oxide islands free to move in two dimensions. The optical micrograph shows the topographic effects that occur when a silicon surface coated with 0.18 μm of oxide, photoengraved to form annular islands, is irradiated with a homogenized ruby laser pulse ($\tau = 25$ nsec, $E = 1.5$ J/cm^2). The initially flat oxide surface becomes rippled in large areas but remains flat in the small islands (darker tone) separated by gaps through which the underlying silicon can be seen (lighter tone). For explanation see text. Scale marker in micrometers.

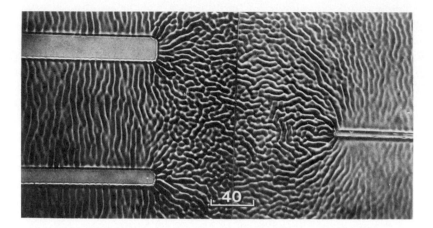

Fig. 20. Photoengraved silicon dioxide-on-silicon structures: effects of differential material behavior on oxide islands free to move in one dimension. Oxide thickness and anneal conditions as for Fig. 19. The optical micrograph shows the randomly oriented ripple, characteristic of large areas of oxide after adiabatic anneal, in the center region of the photograph. At the left-hand edge, the ripple is aligned approximtely at right angles to the long edges of the parallel windows through which the underlying silicon (uniform tone) can be seen. For explanation see text. Scale marker in micrometers.

ing of the window region without melting the silicon under the oxide. This oxide is now floating free on liquid silicon and is apparently under stress, because it buckles, forming the ripple patterns observed. Sectioning of rippled surfaces shows that the surface topography is faithfully followed by the underlying silicon and that there are no voids (Fig. 21). The total energy of the system can be further lowered if the localized buckling is converted to a lateral strain over the whole oxide island. This can, of course, occur only while the underlying silicon is still molten, and the speed of propagation of the deformation is certainly limited to a maximum of 5×10^5 cm sec^{-1}, the speed of sound in silicon dioxide. In a typical 150-nsec melt time, this gives a propagation distance of 750 μm. Thus oxide islands significantly larger than 1500 μm might be expected to retain their ripple structure, whereas much smaller islands should become flat and increase in width. Inspection of Figs. 19 and 20 shows that this is exactly what happens. The circular islands in Fig. 19 are less than 40 μm from edge to edge and are completely flat after exposure to a Q-switched ruby laser pulse, in contrast to the oxide-coated regions outside the circle, which form part of a 3000-μm^2 raft and are heavily rippled. Where the oxide is equally constrained by surrounding oxide in all directions (as in much of Fig. 20), random ripple results. Where one dimension of the oxide raft is much smaller than the critical value, the buckling strain is relieved

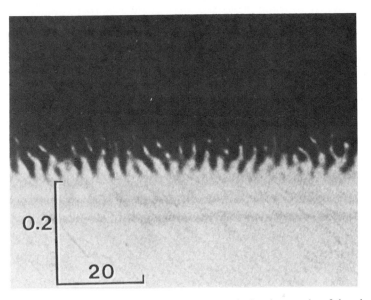

Fig 21. Silicon dioxide-on-silicon structures: optical micrograph of bevel section through rippled oxide on silicon (after adiabatic anneal) showing that the underlying silicon (light tone) is also rippled in step with the overlying oxide (dark tone) and that the oxide–silicon interface remains continuous with no voids. Scale marker in micrometers.

in that direction, but not in the orthogonal direction, resulting in the ordered ripple structure in Fig. 20. Measurements of the buckling strain and lateral strain showed that they were indeed equivalent in sign and magnitude and were proportional to the oxide land width. Thermal expansion stress induced by the pulse was originally thought to be responsible (Hill and Godfrey, 1980), but more recent work (Hill and Godfrey, 1982) makes it almost certain that the compressive stress built into the oxide during cool-down from the original thermal oxidation causes the oxide movements. The degree of movement is found to be independent of oxide thickness (0.2–1.0 μm) but dependent on oxide growth temperature (700–1100°C).

In structures such as Fig. 17, where the window region is coated with an oxide of different thickness, very ordered ripple effects can occur when the lateral movement of the large oxide area buckles the thinner oxide inside the window (Fig. 22). The surface topography becomes reproduced in the melt front, hence in the position of the $p–n$ junction (formed by dopant diffusion in the melt), as shown in Fig. 23 (Hill, 1980).

All the deleterious effects described can be avoided by covering the oxide areas during the laser processing with a reflection mask. A summary of the effects obtained with protected and unprotected oxide is given in

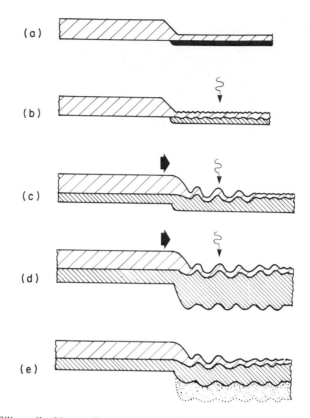

Fig. 22. Silicon dioxide-on-silicon structures. Schematic diagram of the development of topographic changes that occur during pulsed adiabatic anneal of an implanted region coated with thin oxide and bounded by a thick oxide region. (a) Starting structure. (b) Early part of the laser pulse; the heavily doped silicon in the window melts first, and the thin oxide buckles slightly. (c) Later in the pulse; the melt region has penetrated further under the thin oxide, and melting has now occurred in the undoped silicon under the thick oxide, which can now relieve its built-in compressive stress (see text) by expanding into the window; the thin oxide begins to buckle. (d) The buckling propagates further into the window region, and the altered heat flow through the melted region results in a variation in melt depth corresponding to the surface topography. (e) The laser pulse is finished, and the melt front moves back toward the surface as cooling occurs; the variation in melt depth and melt time is recorded as a variation in the dopant redistribution front. Implanted region: solid; oxide: lightly hatched; melt zone: heavily hatched; doped region: dotted. (From Hill, 1980.) This figure was originally presented at the Spring 1980 Meeting of The Electrochemical Society, Inc., held in St. Louis, Missouri.

Fig. 24. One technique adopted (Hill, 1980) is to use an aluminum mask, with the processing sequence given in Fig. 25. It is very important that self-alignment techniques be used, as here, because the implant, oxide edge, and reflection mask must coincide to within 1000 Å to prevent oxide

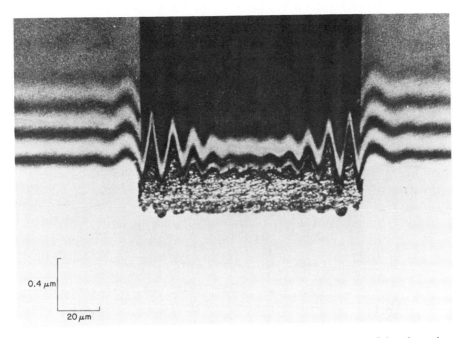

Fig. 23. Silicon dioxide-on-silicon structures. Optical micrograph of bevel section through an oxide-coated silicon window of the type shown in Fig. 22. Thick oxide, 0.7 μm; thin oxide, 0.3 μm; implant, 10^{16} arsenic ions/cm^2, 40 keV; anneal, 25-nsec ruby laser pulse; energy, 2.0 J/cm^2. The reproduction of the surface topography in the penetration of the arsenic-doped melted and regrown region (stained dark gray) into the substrate silicon (white) can be clearly seen. Note the regular buckling of the thin oxide, the amplitude of which decreases steadily toward the center of the window. Scale markers are in micrometers.

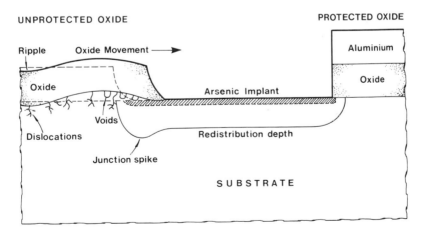

Fig. 24. Silicon dioxide-on-silicon structures. Schematic diagram summarizing the effects observed during adiabatic processing of unprotected and protected oxide–silicon structures. (From Hill, 1981.)

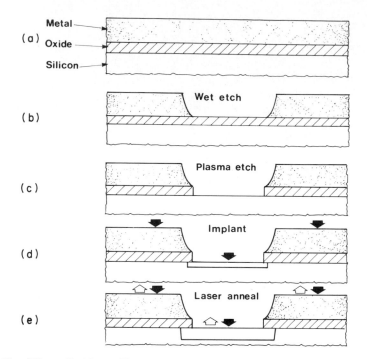

Fig. 25. Silicon dioxide-on-silicon structures. Diagrammatic process schedule for producing a self-aligned metal mask for oxide protection during adiabatic laser processing. (a) Starting structure; (b) window photoengraved and etched in metal; (c) metal window defines oxide window formed by plasma etching; (d) and (e) metal mask protects oxide during implant and laser anneal. (From Hill, 1980.) This figure was originally presented at the Spring 1980 Meeting of The Electrochemical Society, Inc., held in St. Louis, Missouri.

damage and yet allow the junction to be terminated under the oxide by lateral diffusion. A junction fabricated by this technique is shown in Fig. 26, and the planarity of the junction and perfection of the oxide are evident (cf. Fig. 18).

3. THERMAL FLUX PROCESSING

The solid state annealing process avoids the problems described above. However, the enhanced transmission of radiant power caused by interference effects (Section VI.B.1) can cause melting to occur under oxide-covered regions if the oxide thickness is close to that for maximum transmission. If the oxide thickness is not variable, reflection techniques have to be used.

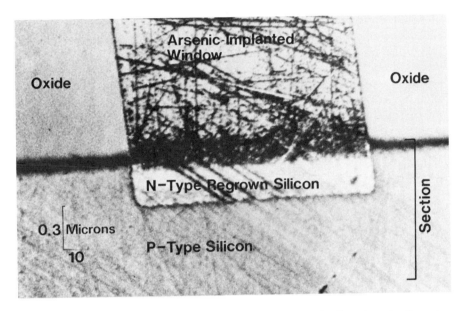

Fig. 26. Silicon–silicon dioxide structures. Optical micrograph of bevel section through a planar n^+–p junction fabricated using the schedule shown in Fig. 25. Oxide thickness, 0.15 μm; implant, 10^{16} arsenic ions at 40 keV; anneal, 25 nsec; ruby laser pulse, homogenized, 2.3 J/cm². The metal masking film has been removed, and the integrity of the oxide and the planarity of the junction are evident.

4. ISOTHERMAL PROCESSING

Since this mode is usually employed with electron beams, differential reflectivity is not important. However, if laser beams are used, the effect is simply to change the overall balance between input and output power, hence to alter the beam power needed to attain a certain temperature.

E. *Devices*

1. INTEGRATED CIRCUIT COMPONENTS

Relatively few devices have been made by beam processing so far. Those that have show a steady improvement in properties over the first two years. Adiabatic processing has been used to make oxide-passivated diodes (Lindner, 1980; Kamins and Rose, 1979), MOS transistors (Koyanagi *et al.*, 1979), and bipolar transistors (Natsuaki *et al.*, 1981). The overall picture is that the bulk of the semiconductor and the $p-n$ junctions fabricated in it can be satifactorily annealed but that where the junction meets the Si–SiO$_2$ interface electronic levels are introduced that are reduced but not removed by a 30-min treatment at 415°C in H$_2$. These levels can be prevented from forming by protecting the junction region with a metal masking film (Lindner, 1980). No improvement over furnace-processed devices has been reported except for short-channel MOSFETs, where the active channel length was 0.6 μm longer than in furnace-annealed samples, thus improving control of the threshold voltage (Koyanagi et al., 1979). The improvement is ascribed to the minimal penetration of the source and drain regions under the gate (as shown in Fig. 27), itself a consequence of the shallow structure required to obtain the same sheet resistance and the selective heat treatment. In this case, regions other than source and drain were protected from the pulsed radiation by the polysilicon gate layer.

MOSFET devices have also been fabricated (by conventional techniques) in silicon layers on amorphous substrates, regrown by adiabatic processing (Shah et al., 1981) and by liquid phase thermal flux processing (Lee et al., 1979). In the latter case, satisfactory devices resulted but channel mobilities were about half those obtained in bulk single crystal. A small integrated circuit was also fabricated in thermal flux regrown material and worked, though again with about half the bulk mobility and a higher supply voltage (Kamins et al., 1980). MOS devices fabricated in adiabatically regrown silicon-on-sapphire were found to have up to 30% improvement in channel mobility, as against those in unprocessed SOS (Yaron and Hess, 1980). In this case, severe deterioration of transistor performance resulted when the melt front was allowed to reach the silicon–sapphire interface; ideally regrowth took place only in the upper half of the silicon layer. Structural studies on adiabatically processed SOS (Cullis et al., 1981) show that the defect structure changes when the melt front reaches the silicon–sapphire interface. Solid state thermal flux processing has been used to anneal MOS threshold implants with good uniformity and characteristics as good as those of furnace-annealed transistors (Zimmer, 1979). It has also been used to reduce surface state densities in capacitor Si–SiO$_2$ structures from 5×10^{11} cm^{-2} to 6×10^{10} cm^{-2}, approaching values typical of good thermal oxide, 2×10^{10} cm^{-2} (Gibbons, 1980).

Isothermal processing has been used to fabricate $n+p$ and $p+n$ diodes, bipolar transistors (McMahon et al., 1980), and MOS transistors (Speight et al., 1981). The devices appear in all important respects to be as satisfactory as furnace-annealed devices and to have, in some cases, even lower leakage currents.

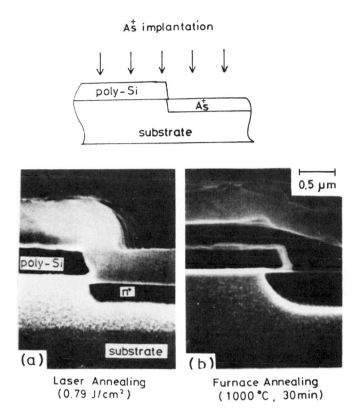

Fig. 27. Scanning electron micrographs of cleaved sections through an integrated circuit MOS transistor structure fabricated by laser annealing of the source and drain regions (a) and furnace annealing of these regions (b). The device requires a very high concentration of electrically active dopant in the source and drain regions (one of which is labeled As$^+$ in the schematic diagram) combined with minimal lateral penetration of this doped region under the polycrystalline silicon gate (labeled "poly-Si"). The prolonged furnace treatment required to achieve the electrical properties results in a marked lateral penetration of the n$^+$ region under the poly-Si (b), whereas the adiabatic laser anneal using a ruby laser of pulse length 25 nsec, energy 0.79 J/cm^2, gives the same electrical properties with very small lateral penetration (a). (From Koyanagi *et al.*, 1979.)

2. GETTERING

The technique of gettering impurities into noncritical regions of a silicon integrated circuit is essential in obtaining the high yields required for satisfactory operation of the large number of components in a circuit. Laser-induced damage is found to provide an efficient gettering site for impurities in silicon (Hsieh *et al.*, 1973; Rozgonyi *et al.*, 1975; Pomerantz,

Fig. 28. Effect of introducing laser damage into the back of a silicon slice on the generation lifetime of the silicon in the front surface. Lifetime was measured by capacitance–time measurements in MOS capacitors fabricated on the front surface; the horizontal axis shows the lateral position of the capacitors relative to the regions of the back surface containing laser damage (gettered) and those untreated (ungettered). During heat treatments subsequent to the laser damage (focused pulses of 1630 J/cm² from a Nd : YAG laser) dislocation networks form at which mobile electrically active ions preferentially precipitate, rendering the overlying silicon purer and resulting in the three orders of magnitude improvement in generation lifetime shown. (From Hayafuji *et al.*, 1980.) Reprinted by permission of the publisher, The Electrochemical Society.

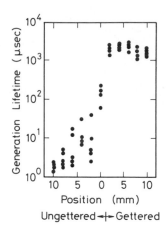

1976; Petroff *et al.*, 1977; Schwuttke *et al.*, 1977; Hayafuji *et al.*, 1980) and is at least as effective as other mechanical damage techniques. Many experiments have been made to elicit the improvement in electrical parameters that results from the introduction of adjacent gettering sites. An example is shown in Fig. 28, where the parameter (minority-carrier lifetime) improves by two orders of magnitude when adiabatically introduced gettering damage is present (Hayafuji *et al.*, 1980). The advantages over other gettering damage techniques are that no contamination is introduced, the technique is completely compatible with every processing stage, and it is reproducible and quick. Gettering damage can be introduced into both the front and back of the wafer as required, but there is some raising of the front surface around the damage area, which may interfere with some contact printing techniques.

VII. Processing Large Areas

A. *Introduction*

Although all the process control factors and changes in materials parameters described in previous sections can be demonstrated and achieved using quite small areas of a semiconductor (e.g., 2 mm square), the application of beam processing to device fabrication requires the capability to process large areas of material at an acceptable rate. What is acceptable will depend on the semiconductor material and its application. Probably the most demanding situation will be the processing of silicon slices for integrated circuits, since processing rates of 100 3-in. slices per hour are already common and 4- and 5-in. slices are being introduced. The different possible approaches to achieving uniform processing over such large areas, the factors influencing the rate of processing obtainable, and the economic considerations are discussed below in terms of beam processing 3-in. silicon slices.

B. Whole Slice Processing

1. Techniques

a. The Problem. The basic problem in whole-slice processing is to achieve the uniformity of irradiation required by a particular application over an area much greater than the uniform beam area. Eight possible techniques for achieving this uniformity are illustrated schematically in Fig. 29a–h.

There are three distinguishable approaches depending on the relative size of the beam and slice (the three columns in Fig. 29), and each of these approaches can be applied in the three modes of processing with one exception. The uniformity of heat treatment is obtained by one of (or a combination of) the three methods, mixing, multiscanning, or integrating, described in Section IV.B.3. The currently available equipment available for beam processing 4-in. silicon slices is based on techniques (c), (f) and (h) and achieves area uniformity by integrating [(c) and (f)] and multiscanning (h), respectively. The other techniques in Fig. 29 are all in principle usable, however, and some applications will require them. The characteristics of each technique and its present viability will now be discussed in more detail.

b. Adiabatic Processing. Where a beam area of adequate intensity and greater than the slice area is available, uniform area processing can be achieved if the pulse is spatially homogenized by mixing (Fig. 29a). To date pulses of the required uniformity ($\pm 5\%$) and energy density (1–3 J cm^{-2}) have been produced in beam diameters of 1 in. (Cullis *et al.*, 1981) or less (Hill, 1980; Cullis, 1979). Although it is feasible with current standard laser technology to generate larger beams, the simultaneous requirement of adequate pulse rate, high uniformity, and reasonable cost are not likely to be met. Successful implementation of giant pulse adiabatic processing of 3- and 4-in. slices will probably require a purpose-built laser homogenizer system, optimize for high power, uniformity, and reproducibility at the expense of the normally desirable laser parameters of high coherence, narrow band width and low beam divergence. A currently more feasible approach uses the smaller homogenized (by mixing) beams

REGIME

ADIABATIC

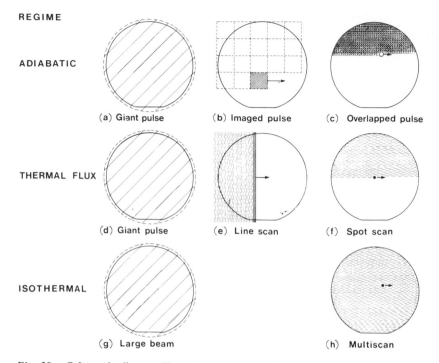

(a) Giant pulse   (b) Imaged pulse   (c) Overlapped pulse

THERMAL FLUX

(d) Giant pulse   (e) Line scan   (f) Spot scan

ISOTHERMAL

(g) Large beam   (h) Multiscan

Fig. 29. Schematic diagram illustrating eight possible approaches to processing large-area slices. The radiant beam shape and size, and its motion, are shown superimposed on a typical workpiece, a 3-in. silicon slice. The columns are characterized by very large, intermediate, and very small beam areas (from left to right) and the rows refer to the applications of the different beam sizes to the three modes of processing.

currently available and is illustrated in Fig. 29b. The spatial extent of the pulse is controlled by imaging to a shape that tesselates, so that the slice area can be processed by a stepped sequence of imaged homogenized pulses which just abut. This approach is possible in adiabatic processing because of the confinement of the heated region to the irradiated region (Section IV.A.2). Because of heating variations at the edge of the image due to diffraction, stepping tolerances, and heat flow, the abutting regions need to be confined to noncritical regions of the slice. Such regions already conveniently exist in integrated circuits as the rectangular grid of scribing channels that separates individual chips. Use of these channels as abutment zones for processing chips or groups of chips requires confinement of edge effects to less than the channel width (typically 100 μm). That this is quite feasible is shown in Fig. 30 which shows the effect of imaging a ruby laser pulse through a photoengraving mask onto a silicon

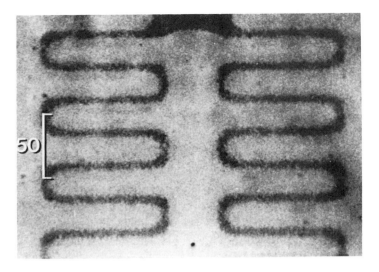

Fig. 30.   Selective annealing of an amorphized silicon surface by imaging a homogenized 25-nsec ruby laser pulse through a photoengraving mask. The optical micrograph shows the areas amorphized by ion implantation of 10^{16} arsenic ions/cm² at 40 keV as a light tone (due to the higher reflectivity of amorphous silicon) and the selectively annealed regions as a dark wavy track of width about 5 μm. (From Hill, 1981.)

surface amorphized by an arsenic implant. The silicon has been selectively annealed with a resolution of about 5 μm (Hill, 1981). The stepping accuracy is easily achievable by current photoengraving equipment. The approach illustrated in Fig. 29c is the one most easily adopted utilizing existing Q-switched laser systems and is available commercially. Large-area processing is achieved by overlapping the melt zones of a large number of focused nonuniform pulses emitted at a high rate (e.g., 10^4 Hz) from a continuously pumped Q-switched laser. Pulse diameter is typically 100 μm (Kaplan *et al.*, 1980). The degree of uniformity of processing depends entirely on the averaging effect within the material of integration processes (Section IV.B.3), since the stability of the Nd:YAG laser used in this way is very good ($\pm 3\%$). The wavelength emitted couples rather poorly into silicon, but efficient coupling can be ensured by mixing in a proportion of strongly absorbed 0.53-μm light generated by frequency doubling (Newstein *et al.*, 1980). However, this overlapped pulse technique will not be suitable for all applications, particularly those with a narrow energy window or very nonlinear integration effects, because of the spread of energy density across the beam and the necessity for overlap.

c. Thermal Flux Processing.   A giant uniform pulse can, in principle, be used for thermal flux processing (Fig. 29d). The total pulse energy requirement is very large, however; calculation is based on the data shown in Fig. 6, scaled up from a spot size of 100 μm to a 3-in. slice and adjusted for the planar heat flow in the giant pulse case (as against the spherical heat flow in the scanned spot) shows that about 1000 J has to be absorbed in a millisecond to raise the slice temperature to 1100°C. Existing pulsed lasers have output energies typically two orders of magnitude smaller than this. A more suitable radiation source for giant pulse thermal flux processing appears to be the xenon flash tube (Bomke et al., 1978; Cohen et al., 1978; Lue, 1980). Annealing of implantation damage by this technique has been achieved over areas of 1 cm², using two 800-μsec pulses separated by 0.5 sec. The energy absorbed by the silicon from each pulse was about 25 J (Correra and Pedulli, 1980), in good agreement with the calculated value for a 3-in. slice quoted above. The double pulse is apparently necessary to preheat the slice (to about 250°C) to ensure full anneal on the second pulse and suggests that the energy density of the lamp used is only marginally adequate. The use of multiple arrays of such tubes and improvement in flash output would make this technique very feasible. It may, however, be difficult to obtain sufficient uniformity or irradiation for applications with a narrow energy window because of the merely local smoothing of nonuniformities by heat flow (about 100 μm) and the present requirement for having the flash tubes very close to the sample (0.5 mm). The technique is attractive for processes that are self-limiting (e.g., anneal of implant damage) where moderate uniformity of irradiation is adequate, because the temperature gradients are normal to the slice surface, in which direction the slice is mechanically strong. Beam-induced damage should thus be minimal. An alternative and commercially available technique for achieving the high power densities required is to focus continuous argon ion laser beam into a small (100 μm) spot and control the pulse time by scanning (Section IV.B.2). Large-area coverage can then be effected by covering the slice in overlapping raster scans as shown in Fig. 29f. As in (c), the uniformity of the processing depends on how successful integration in the material is in smoothing out nonuniformities due to the beam nonuniformity and the presence of overlap regions. These disadvantages could be much reduced by a single line scan across the wafer as shown in (e), the line being of uniform power density achieved by either mixing and focusing a beam into a line source or multiscanning a spot along a line. Thermal flux processing with a laser using this technique has not yet been reported. It would require a high beam power in the region of 3 kW to process a 3-in. silicon slice with a typical front-to-back temperature difference of 800°C and a dwell time of 1 msec. The line-scan technique has, however, been implemented with electron beams (Soda *et al.*, 1981) and blackbody radiant strips (Fan *et al.*, 1980; Maby *et al.*, 1981). In both cases the slice temperature was high so that lower input power was required. A serious disadvantage of both (e) and (f) for large-area processing is the presence of lateral temperature gradients which can introduce severe damage into single-crystal substrates, particularly if substrate temperatures are high.

d. Isothermal Processing. By definition there are insignificant temperature gradients in materials being processed isothermally. Thus there are only two options available in covering large areas: using a giant homogenized pulse (Fig. 29g) or multiscanning a beam rapidly over the entire wafer (h). Sufficient power is available in continuous CO_2 lasers to use technique (g): As can be seen in Fig. 7, a beam power of about 900 W (20.4×44.2) is required to achieve 1100°C in a 3-in. slice. However, preliminary experiments (Celler *et al.*, 1979; Webster, 1980) show that slice preheating (to perhaps 400°C) is essential if laser damage due to large lateral temperature gradients (Section IV.B.1) is to be avoided. Processing by technique (g) using diffuse radiation has been successfully effected on small samples by placing them on a strip heater and heating them by blackbody radiation emitted when a current pulse is passed through the heater (Scovell, 1981). An improved version of this approach, with better control of the thermal and chemical environment, might well offer a cheap, satisfactory method of isothermal processing. An alternative source of diffuse radiation is gas discharge lamps. Adequate power is available from both these sources. A disadvantage of the diffuse nature of the radiation is that it will be very difficult to avoid heating of the slice supports and containing vessel by the primary radiation and so increase contamination problems. Use of transparent media may reduce this problem to acceptable levels.

The multiscanning technique (h) offers precise control over the heat treatment and is the basis of present commercial electron beam equipment for isothermal annealing (Section IV.B.2). The uniformity and reproducibility of the heating cycle are adequate for even exacting applications (Section IV.B.2), and the beam power of 1 kW available (Plows, 1980) is adequate for processing up to at least 1100°C in 3-in. silicon slices.

The temperature cycle characteristic of multiscanned samples (Fig. 11) approximates to a very smooth curve in electron beam systems because of the high scan rates available. Typical control of beam current to better than 1% and the use of 1000 scans limits deviations from the smooth curve by less than ±2°C when an equilibrium temperature of 1100°C is used. For shorter heating times, where the slice temperature is rising during the heat treatment, the slice temperature is still uniform to within 10°C throughout a 400-μm-thick silicon slice for pulse times longer than 1 msec (Shah *et al.*, 1981).

The large amount of reradiated energy (allowing temperature rise of the surroundings) and the relatively long heat treatment times (normally seconds) may require more attention to contamination problems in isothermal processing of large slices than in the adiabatic and thermal flux techniques.

2. THROUGHPUT

Beam processing in general will form part of a sequential fabrication process, hence must be capable of matching the throughput of the other stages in the process if it is not to form a bottleneck. Currently, in many integrated circuit production lines, a throughput of 100 3-in. slices/hr would be required. The maximum possible throughput in any beam processing technique is set by the ratio of the time-averaged power of the beam divided by the beam energy required to process one slice. This latter energy is very dependent on the mode of processing, as can be seen in Fig. 31. For typical pulse times presently available for each of the three modes (adiabatic 25 nsec, thermal flux 1 msec, isothermal 10 sec) and the requirement to effect materials changes in a $\frac{1}{2}$-μm-thick surface layer of a 3-in. silicon slice (Fig. 1) shows that absorbed energy densities of 0.005, 4, and 4 kcal are required, respectively. The average beam powers for typical equipment in these categories is 1 W (Q-switched ruby laser), 20 W (continuous argon ion laser), and 1000 W (electron beam), giving maximum possible throughputs of 144, 4, and 214 3-in. slices/hr (allowing for some reflection losses in the laser systems). These figures can, however, be strongly modified by the process control requirements (Section IV.A) of the structure being fabricated. In adiabatic processing, if a narrow energy window is required, throughput will depend on the spatial uniformity of the beam. This is illustrated in Fig. 32, which shows the fraction of the beam that can be used for annealing when process control requires ±4% tolerance on total beam energy and the shot-to-shot variation of the laser is 3%. If the beam has a Gaussian spatial distribution, only 16% of the energy is usable, wheras 60% of the homogenized beam energy is available (assuming a typical 40% loss in the homogenizer). Throughput is thus reduced from 144 slices/hr to 86 (homogenized beam) and 23 (Gaussian beam). Some processes require higher beam energies (e.g., for deep distribution of dopant), and this can reduce the throughput to about 30 and 10 slices/hr, respectively. Processes requiring multiple zaps (Section V.B

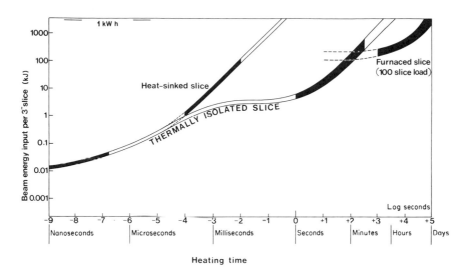

Fig. 31. Total beam energy input required to raise the surface temperature of a 400-μm-thick, 3-in.-diameter silicon slice to 1100°C (lower boundary) and 1400°C (upper boundary) as a function of heating time. The experimentally accessible regions are shown in black and correspond from left to right to the adiabatic, thermal flux, and isothermal modes of processing. For comparison, the total input energy per 3-in. slice for isothermal processing of a 100-slice load in a ramped furnace is also shown. (From Hill, 1981.)

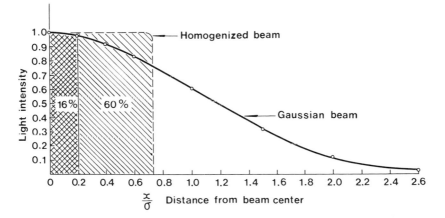

Fig. 32. Effect of the spatial distribution of energy in an adiabatic pulse on the efficiency of utilization of the total beam energy for processing requiring the use of narrow energy windows (see text for explanation).

would have even lower throughputs, perhaps 3 and 1 slice/hr, respectively. Where final uniformity depends on integrating techniques (Fig. 29c, e, and f), small process tolerances often require a large overlap of scans, and this can reduce throughput greatly. A requirement for 75% overlap would reduce throughput in the examples taken from 144–36 slice/hr (adiabatic, scanned pulse) and from 4–1 slice/hr (thermal flux, spot scan). The low throughput of the thermal flux mode is due to the high thermal conductivity of silicon, which requires a high heat flux to maintain a large temperature difference (typically 800°C) between the front and back of the slice. Throughput in this mode can be considerably increased by increasing the substrate temperature, but this is not feasible for all processes. The throughput of the isothermal mode is likely to be reduced only by delays incurred through the necessary vacuum environment (e.g., slow cooling and air-lock transfer of slices) and should be capable of throughputs of 100 slices/hr in one machine. Long (e.g., 1 min) heat treatments will obviously reduce throughput. Higher throughputs are, of course, to be expected where only a fraction of the slice area requires processing (e.g., introducing gettering damage, Section VI.E.2), or where the energy density required is much lower than for processing the matrix material, either because of more efficient absorption of the light and better thermal isolation from the substrate [e.g., polysilicon films separated from the silicon matrix by silicon dioxide (Section VI.C.1)] or because of a lower processing temperature requirement (formation of metal eutectic contacts to silicon (Allen *et al.*, 1980).

3. ECONOMICS

The cost of introducing beam processing into a production line is determined by the total cost of capital equipment, replacement parts, and power consumed over the life of the equipment as compared with the total number of slices processed. It is assumed that no increased labor costs are incurred in running and maintaining the machine as compared with a furnace. It is evident that the capital cost per wafer will be very throughput-dependent and that the running cost will depend very much on the efficiency of conversion of electrical power to beam power. For the equipments considered below these efficiencies are approximately 30% (electron beam), 0.1% (ruby laser), and 0.01% (argon laser). In Table II, the approximate costs of beam processing by the currently most efficient available techniques in each of the three modes are calculated for the range of throughputs discussed in the previous section. For comparison,

TABLE II

ESTIMATED COST OF BEAM PROCESSING IN THE THREE THERMAL MODES[a]

Mode	Equipment	Units of equipment	Cost per hour			
			Capital ($)	Parts ($)	Power ($)	Wafer (¢)
Adiabatic	Q-switched ruby	1	2.86	0.04	0.18	3
	laser, homogenized	24	68.6	0.96	4.2	74
	beam-imaged pulses					
Thermal flux	Continuous argon ion	8	22.9	17.0	22.8	63
	laser, 60% overlap	25	71.4	55.0	70.0	196
	scan					
Isothermal	Focused electron beam,	1	2.86	0.3	0.5	4
	rapid multiscan	5	14.3	1.5	2.5	20
	Furnace	1	—	—	—	4.9[b]

[a] Throughput, 100 slices/hr (3 in.); equipment utilization, 80%; amortization time, 5 years; unit equipment cost, $100,000; electric power, 7¢/kWh.
[b] From Kaplan *et al.* (1980).

the cost per slice of processing a 100-slice batch in a furnace operation (Kaplan *et al.*, 1980) is also included. It can be seen that the cost per wafer of beam processing is very process- and mode-dependent, but that for heat treatment of the silicon material itself it is unlikely to displace furnacing on cost grounds alone.

The increased cost of beam processing may be willingly accepted if the overall cost of the total fabrication process is thereby reduced. Processing stages near the end of a fabrication sequence may deal with slices worth $20 each: an improvement in yield of 15% effected by beam processing would amply cover the $0.6–$1.9 increase in processing cost for this particular stage. If a viable large-area technique for converting polysilicon on insulators to single crystal is developed using beam processing in the thermal flux mode, as a better alternative to silicon on sapphire, the difference in the cost of the starting wafers would cover the largest processing cost ($1.96) in Table II. For some applications involving selective processing of only part of the slice area (e.g., gettering damage production) the high throughputs possible of 200–300 slices/hr in one machine (*Quantronix*, 1980) may reduce the cost of the gettering damage well below that of alternative techniques, namely, ion implantation, furnacing in a POCl$_3$ atmosphere, and mechanical abrasion.

A composite diagram illustrating the various applications of beam processing to integrated circuits so far reported is shown in Fig. 33.

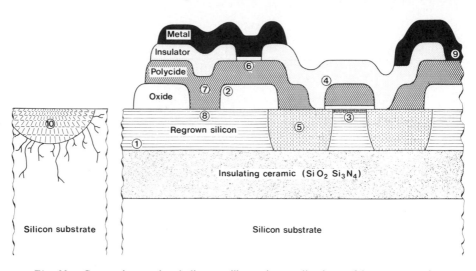

Fig. 33. Composite sectional diagram illustrating applications of beam processing to integrated circuit structures currently being explored. (1) Regrown single crystal on amorphous dielectric; (2) contoured oxide edge; (3) implantation damage annealed without redistribution of dopant; (4) contoured step in polycrystalline silicon and silicide layer (polycide); (5) redistributed supersaturated solid solution of dopant; (6) smoothed polycrystalline silicon to give high-quality oxide for capacitor use; (7) reacted polycrystalline silicon plus metal to give silicide; (8) cleaned surface for Schottky contact, (9) ohmic contact; (10) beam-induced damage to form gettering sites for impurities.

Acknowledgments

I am greatly indebted to my colleagues, Derek Godfrey and John Daly, for their support and enthusiasm in our joint experimental work on which much of this chapter is based, and for permission to use some of our unpublished results. I also thank Dr. Miyao and Dr. Hayfuji for allowing me to reproduce their diagrams and for providing excellent original copies. The chapter owes much to the typing skill and patience of Marion Welbourn, the draftsmanship of Roy Finch, and the encouragement and help of my wife in preparing the manuscript. Much of the work on which this chapter is based has been carried out with the support of the Procurement Executive, Ministry of Defence, sponsored by DCVD, and thanks are due to them and the Plessey Company Limited for permission to publish.

References

Ahmed, H. (1980). Personal communication.
Allen, S. D., von Allmen, M., and Wittmer, M. (1980). *In* "Laser and Electron Beam Processing of Electronic Materials" (C. L. Anderson, G. K. Celler, and G. A. Rozgonyi, eds.), p. 514. Electrochem. Soc., Pennington, New Jersey.
Andrew, R., Ledezma, M., Lovato, M., Wautelet, M., and Laude, L. D. (1979). *Appl. Phys. Lett.* **35**, 418.

Aspnes, D. E., Celler, G. K., Poate, J. M., Rozgonyi, G. A., and Sheng, T. T. (1980). *In* "Laser and Electron Beam Processing of Electronic Materials" (C. L. Anderson, G. K. Celler, and G. A. Rozgonyi, eds.), p. 414. Electrochem. Soc., Pennington, New Jersey.
Auston, D. H., Golovchenko, J. A., and Venkatesan, T. N. C. (1979). *Appl. Phys. Lett.* **34**, 558.
Baeri, P., Campisano, S. U., Foti, G., and Rimini, E. (1979). *J. Appl. Phys.* **50**, 788.
Baeri, P., Foti, G., Poate, J. M., and Cullis, A. G. (1980). *Phys. Rev. Lett.* A **46A**, 2036.
Bentini, G. G., Galloni, R., Merli, P. G., Pedulli, L., Vecchi, I., and Zigneni, F. (1980). *In* "Laser and Electron Beam Processing of Materials" (C. W. White and P. S. Peercy, eds.), p. 272. Academic Press, New York.
Benton, J. L., Doherty, C. J., Ferris, S. D., Kimerling, L. C., Leamy, H. J., and Celler, G. K. (1980). *In* "Laser and Electron Beam Processing of Materials" (C. W. White and P. S. Peercy, eds.), p. 430. Academic Press, New York.
Biegelson, D. K., Johnson, N. M., Bartelink, D. J., and Moyer, M. (1981). *In* "Laser and Electron Beam Solid Interactions and Materials Processing" (J. F. Gibbons, L. D. Hess, and T. W. Sigmon, eds.), p. 487. North-Holland Publ., New York.
Bloembergen, N. (1979). *In* "Laser–Solid Interactions and Laser Processing" (S. D. Ferris, H. J. Leamy, and J. M. Poate, eds.), p. 1. Am. Inst. Phys., New York.
Bomke, H. A., Berkowitz, H. L., Harmatz, M., Kronenberg, S., and Lux, R. (1978). *Appl. Phys. Lett.* **33**, 955.
Born, M., and Wolf, E. (1964). "Principles of Optics," p. 64. Pergamon, Oxford.
Caughey, D. M., and Thomas, R. E. (1967). *Proc IEEE* **55**, 2192.
Celler, G. K., Bonitta, R., Brown, W. L., Poate, J. M., Rozgonyi, G. A., and Sheng, T. T. (1979). *In* "Laser–Solid Interactions and Laser Processing" (S. D. Ferris, H. J. Leamy, and J. M. Poate, eds.), p. 381. Am. Inst. Phys., New York.
Chiu, T. L., and Ghosh, H. N. (1971). *IBM J. Res. Dev.* **15**, 472.
Cohen, R. L., Williams, J. S., Feldman, L. C., and Wost, K. W. (1978). *Appl. Phys. Lett.* **33**, 751.
Correra, L., and Pedulli, L. (1980). *Appl. Phys. Lett.* **37**, 55.
Cspregi, L., Kennedy, E. F., Mayer, J. W., and Sigmon, T. W. (1978). *J. Appl. Phys.* **49**, 3906.
Cullis, A. G. (1979). Personal communication.
Cullis, A. G., Webber, H. C., and Bailey, P. (1979). *J. Phys.* E **12**, 688.
Cullis, A. G., Webber, H. C., Chew, N. G., Hill, C., and Godfrey, D. J. (1981). *In* "Microscopy of Semiconducting Materials" (A. G. Cullis and D. Joy, eds.), p. 95. Inst. Phys., London.
Daly, R. (1978). *Proc. Laser Eff. Ion Implanted Semiconduct., Univ. Catania* p. 97.
Davies, D. E., Lorenzo, J. P., and Ryan, T. G. (1980). *Appl. Phys. Lett.* **37**, 612.
Eckhardt, G. (1980). *In* "Laser and Electron Beam Processing of Materials" (C. W. White and P. S. Peercy, eds.), p. 467. Academic Press, New York.
Eisen, F. (1980). *In* "Laser and Electron Beam Processing of Materials" (C. W. White and P. S. Peercy, eds.), p. 309. Academic Press, New York.
Fan, C. C. J., Geis, M. W., and Tsaur, B.-Y. (1980). IEDM *Tech. Dig.—Int. Electron Devices Meet.* Dec., p. 845.
Finetti, M., Lotti, R., Nobili, D., and Solmi, S. (1981). *J. Electrochem. Soc.* **128**, 1313.
Fogarassy, E., Stuck, R., Grob, A., and Siffert, P. (1980). *In* "Laser and Electron Beam Processing of Materials" (C. W. White and P. S. Peercy, eds.), p. 117. Academic Press, New York.
Fowler, A. B., and Hodgson, R. T. (1980). *Appl. Phys. Lett.* **36**, 914.
Gat, A., Leitoila, A., and Gibbons, J. F. (1979). *J. Appl. Phys.* **50**, 2926.

Geis, M. W., Antoniadis, D. A., Silversmith, D. J., Mountain, R. W., and Smith, H. I. (1980). *Appl. Phys. Lett.* **37**, 454.

Gibbons, J. F. (1980). *In* "Laser and Electron Beam Processing of Electronic Materials" (C. L. Anderson, G. K. Celler, and G. A. Rozgonyi, eds.), p. 1. Electrochem. Soc., Pennington, New Jersey.

Godfrey, D. J., Hill, A. C., and Hill, C. (1981). *J. Electrochem. Soc.* **128**, 1798.

Gold, R. B., and Gibbons, J. F. (1980). *In* "Laser and Electron Beam Processing of Materials" (C. W. White and P. S. Peercy, eds.), p. 77. Academic Press, New York.

Greenwald, A. C., and Little, R. G. (1979). *Solid State Technol.* **22**, 143.

Greenwald, A. C., Kirkpatrick, A. R., Little, R. G., and Minucci, J. A. (1979). *J. Appl. Phys.* **50**, 783.

Grojean, R. E., Feldman, D., and Roach, J. F. (1980). *Rev. Sci. Instrum.* **51**, 375.

Harper, F. E., and Cohen, M. I. (1970). *Solid-State Electron.* **13**, 1103.

Hayafuji, Y., Yanada, T., and Aoki, Y. (1980). *J. Electrochem. Soc.* **128**, 1975.

Hess, L. D., Forber, R. A., Kokorowski, S. A., and Olson, G. L. (1979). *Proc. Soc. Photo-Opt. Instrum. Eng., San Diego, Calif.* Paper H6.

Hess, L. D., Olson, G. L., Ito, C. R., and Nakaji, E. M. (1980). *In* "Laser and Electron Beam Processing of Materials" (C. W. White and P. S. Peercy, eds.), p. 621. Academic Press, New York.

Hess, L. D., Kokorowski, S. A., Olson, G. L., and Yoron, G. (1981). *In* "Laser and Electron Beam Solid Interactions and Materials Processing" (J. F. Gibbons, L. O. Hess, and T. W. Sigmon, eds.), p. 307. North-Holland Publ., New York.

Hill, C. (1979). *In* "Laser–Solid Interactions and Laser Processing" (S. D. Ferris, H. J. Leamy, and J. M. Poate, eds.) p. 419. Am. Inst. Phys., New York.

Hill, C. (1980). *In* "Laser and Electron Beam Processing of Electronic Materials" (C. L. Anderson, G. K. Celler, and G. A. Rosgonyi, eds.) p. 26. Electrochem. Soc., Pennington, New Jersey.

Hill, C. (1981). *In* "Laser and Electron Beam Solid Interactions and Materials Processing" (J. F. Gibbons, L. D. Hess, and T. W. Sigmon, eds.) p. 361. North-Holland Publ., New York.

Hill, C., and Godfrey, D. J. (1980). *J. Phys. (Orsay, Fr.)* **41**, C479.

Hill, C., and Godfrey, D. J. (1982). to be published.

Hill, C., Butler, A. L., and Daly, J. A. (1982). *In* "Laser and Electron Beam Interactions with Solids" (G. K. Celler and B. R. Appleton, eds.), p. 579. North-Holland Publ., New York.

Hoh, K., Koyama, H., Uda, K., and Miura, Y. (1980). *Jpn. J. Appl. Phys.* **19**, L375.

Hsieh, C. M., Matthews, J. R., Seidel, H. D., Pickar, K. A., and Drum, C. M. (1973). *Appl. Phys. Lett.* **22**, 238.

Johnson, N. M., Batelink, D. J., Moyer, M. D., Gibbons, J. F., Leitoila, A., Ratnakumer, K. N., and Regolini, J. L. (1980). *In* "Laser and Electron Beam Processing of Materials" (C. W. White and P. S. Peercy, eds.), p. 423. Academic Press, New York.

Kamins, T. I., and Rose, P. H. (1979). *J. Appl. Phys.* **50**, 1308.

Kamins, T. I., Lee, K. F., Gibbons, J. F., and Saraswat, K. C. (1980). *IEEE Trans. Electron Devices* **ED-27**, 290.

Kaplan, R. A., Cohen, M. G., and Liu, K. C. (1980). *In* "Laser and Electron Beam Processing of Electronic Materials" (C. L. Anderson, G. K. Celler, and G. A. Rozgonyi, eds.), p. 58. Electrochem. Soc., Pennington, New Jersey.

Kirkpatrick, A. R. (1980). *In* "Laser and Electron Beam Processing of Electronic Materials" (C. L. Anderson, G. K. Celler, and S. A. Rozgonyi, eds.), p. 108. Electrochem. Soc., Pennington, New Jersey.

Kirkpatrick, A. R., Minnucci, J. A., and Greenwald, A. C. (1977). *IEEE Trans. Electron Devices* **ED-24**, 429.

Kokorowski, S. A., Olson, G. L., and Hess, L. D. (1981). *In* "Laser and Electron Beam Solid Interactions and Materials Processing" (J. F. Gibbons, L. D. Hess, and T. W. Sigmon, eds.), p. 139. North-Holland Publ., New York.

Koyanagi, M., Tamura, H., Miyao, M., Hashimoto, N., and Tokuyama, T. (1979). *Appl. Phys. Lett.* **35**, 621.

Laff, R. A., and Hutchins, G. L. (1974). *IEEE Trans. Electron Devices* **ED-21**, 743.

Lahart, M. J., and Marathay, A. S. (1975). *J. Opt. Soc. Am.* **65**, 769.

Larson, B. C., White, C. W., and Appleton, B. R. (1978). *Appl. Phys. Lett.* **32**, 801.

Leamy, H. J., Rozgonyi, G. A., Sheng, T. T., and Celler, G. K. (1980). *In* "Laser and Electron Beam Processing of Electronic Materials" (C. L. Anderson, G. K. Celller, and G. A. Rozgonyi, eds.), p. 333. Electrochem. Soc., Pennington, New Jersey.

Leas, J. M., Smith, P. J., Nagarajian, A., and Leighton, A. (1980). *In* "Laser and Electron Beam Processing of Materials" (C. W. White and P. S. Peercy, eds.), p. 645. Academic Press, New York.

Lee, K. F., Gibbons, J. F., Saraswat, K. C., and Kamius, T. I. (1979). *Appl. Phys. Lett.* **35**, 173.

Lietoila, A. (1981). Ph.D. Thesis, Stanford Univ., Palo Alto, California.

Lietoila, A., Gibbons, J. F., Magee, T. J., Peng, J., and Hong, J. D. (1979). *Appl. Phys. Lett.* **35**, 532.

Lindner, M. (1980). *Phys. Status Solidi A* **57**, 263.

Lostroh, J. (1980). *In* "Solid State Devices" (J. E. Carroll, ed.), p. 51. Inst. Phys., London.

Luc, J. (1980). *Appl. Phys. Lett.* **36**, 73.

Maby, E. W., Geis, M. W., Le Coz, L. Y., Silversmith, D. J., Mountain, R. W., and Antoniadis, D. A. (1981). *IEEE Electron Devices Lett.* **EDL-2**, 241.

McMahon, R. A., and Ahmed, H. (1979). *J. Vac. Sci. Technol.* **16**, 1840.

McMahon, R., Ahmed, H., Speight, J. D., and Dobson, R. M. (1980). *In* "Laser and Electron Beam Processing of Electronic Materials" (C. L. Anderson, G. K. Celler, and G. A. Rozgonyi, eds.), p. 130. Electrochem. Soc., Pennington, New Jersey.

Maracas, G. N., Harris, G. L., Lee, C. A., and McFarlane, R. A. (1978). *Appl. Phys. Lett.* **3**, 453.

Masetti, G., and Solmi, S. (1979). *IEE J. Solid-State Electron Devices* **3**,(3), 65.

Matsumoto, S., Gibbons, J. F., Deline, V., and Evans, C. A., Jr. (1980). *Appl. Phys. Lett.* **37**, 821.

Mehrabian, R., Kou, S., Hsu, S. C., and Munitz, A. (1979). *In* "Laser–Solid Interactions and Laser Processing" (S. D. Ferris, H. J. Leamy, and J. M. Poate, eds.), p. 129. Am. Inst. Phys., New York.

Miller, G. L. (1980). *In* "Laser and Electron Beam Processing of Electronic Materials" (C. L. Anderson, G. K. Celler, and G. A. Rozgonyi, eds.), p. 83. Electrochem. Soc., Pennington, New Jersey.

Miyao, M., Itoh, K., Tamura, M., Tamura, H., and Tokuyama, T. (1980). *J. Appl. Phys.* **51**, 4139.

Mizuta, M., Sheng, N. H., Merz, J. L., Leitoila, A., Gold, R. B., and Gibbons, J. F. (1980). *Appl. Phys. Lett.* **37**, 154.

Moore, J. (1980). Personal communication.

Muller, C. M., Fogarassy, E., Salles, D., Stück, R., and Siffert, P. M. (1980). *IEEE Trans. Electron Devices* **ED-27**, 815.

Narayan, J. (1980). *In* "Laser and Electron Beam Processing of Materials" (C. W. White and P. S. Peercy, eds.), p. 397. Academic Press, New York.

Natsuaki, N., Miyazaki, T., Ohkura, M., Nakamura, T., Tamura, M., and Tokiyama, T. (1981). *In* "Laser and Electron Beam Solid Interactions and Materials Processing" (J. F. Gibbons, L. D. Hess, and T. W. Sigmon, eds.), p. 375. North-Holland Publ., New York.

Neukermans, A., and Saperstein, W. (1979). *J. Vac. Sci. Technol.* **16**, 1847.

Newstein, M. C., Liu, K. C., and Kaplan, R. (1980). *In* "Laser and Electron Beam Processing of Materials" (C. W. White and P. S. Peercy, eds.), p. 303. Academic Press, New York.

O'Connell, T. R., Hartman, J. M., Errett, E. B., and Leach, G. S. (1977). *IEEE J. Solid-State Circuits* **SC-12**, 497.

Pankove, J. I., and Tarng, M. L. (1979). *Appl. Phys. Lett.* **34**, 156.

Pankove, J. I., Wu, C. P., Magee, C. W., and McGinn, J. T. (1980). *J. Electron. Mater.* **9**, 905.

Peercy, P. S., and Stein, H. J. (1980). *In* "Laser and Electron Beam Processing of Materials" (C. W. White and P. S. Peercy, eds.), p. 411. Academic Press, New York.

Petroff, P. M., Rozgonyi, G. A., and Sheng, T. T. (1976). *J. Electrochem. Soc.* **125**, 565.

Ploog, K. (1980). *Cryst.: Growth Prop. Appl.* **3**, 73.

Plows, G. (1980). Personal communication.

Pomerantz, D. (1976). *J. Appl. Phys.* **38**, 5020.

Prussin, S., and von der Ohe, W. (1980). *J. Appl. Phys.* **51**, 3853.

Purcell, J. J. (1979). *Radio Electron. Eng.* **49**, 347.

Quantronix Laser News (1980). **1**(3), 2.

Ratnakumer, K. N., Pease, R. F. W., Bartelink, D. J., Johnson, N. M., and Meindl, J. D. (1979). *Appl. Phys. Lett.* **35**, 463.

Rozgonyi, G. A., Petroff, P. M., and Read, M. H. (1975). *J. Electrochem. Soc.* **122**, 1725.

Rozgonyi, G. A., Leamy, H. J., Sheng, T. T., and Celler, G. K. (1978). *In* "Semiconductor Characterization Techniques" (P. A. Barnes and G. A. Rozgonyi, eds.), p. 492. Electrochem. Soc., Pennington, New Jersey.

Runge, H. (1980). Personal communication.

Salathe, R. P., Gilgen, H. H., Rytz Froidevaux, Y., Lüthy, W., and Weber, H. P. (1979). *Appl. Phys. Lett.* **35**, 543.

Schuman, P. A., Keenan, W. A., Tong, A. H., Gegenworth, H. H., and Schneider, C. P. (1971). *J. Electrochem. Soc.* **118**, 145.

Schwuttke, G. H., Yang, K., and Krappert, H. (1977). *Phys. Status Solidi A* **42**, 553.

Scovell, P. (1981). *Electron. Lett.* **17**, 403.

Shah, R. R., and Crosthwait, L. (1981). *In* "Laser and Electron Beam Solid Interactions and Materials Processing" (J. F. Gibbons, L. D. Hess, and T. W. Sigmon, eds.), p. 471. North-Holland Publ., New York.

Soda, K. J., De Jule, R. M., and Streetman, B. G. (1981). *In* "Laser and Electron Beam Solid Interactions and Materials Processing" (J. F. Gibbons, L. D. Hess, and T. W. Sigmon, eds.), p. 353. North-Holland Publ., New York.

Speight, J. D., Glaccum, A. E., Machin, D., McMahon, R. A., and Ahmed, H. (1981). *In* "Laser and Electron Beam Solid Interactions and Materials Processing" (J. F. Gibbons, L. D. Hess, and T. W. Sigmon, eds.), p. 383. North-Holland Publ., New York.

Stephen, J., Smith, B. J., and Blamires, N. G. (1980). *In* "Laser and Electron Beam Processing of Materials" (C. W. White and P. S. Peercy, eds.), p. 639. Academic Press, New York.

Stoneham, A. M. (1978). Report AERE M.3001 UKAEA, Harwell, England.

Surko, C. M., Simons, A. L., Auston, D. H., Golovchenko, J. A., Slusher, R. E., and Venkatesan, T. N. C. (1979). *Appl. Phys. Lett.* **34**, 635.

Sussman, R. S., Harris, A. J., and Ogden, R. (1980). *J. Non-Cryst. Solids* **35/36**, 249.

Tamura, H., Miyao, M., and Tokuyama, T. (1979). *J. Appl. Phys.* **50**, 3783.

Tamura, M., Natsuaki, N., and Tokuyama, T. (1980). *J. Appl. Phys.* **51**, 3373.

Toulemonde, M., Muller, J. C., Siffert, P., and Bowdon, B. (1980). *In* "Laser and Electron Beam Processing of Materials" (C. W. White and P. S. Peercy, eds.), p. 417. Academic Press, New York.

Tsu, R., Baglin, J., Tan, T., Tsai, M., Park, K., and Hodgson, R. (1979). *In* "Laser–Solid Interactions and Laser Processing" (S. D. Ferris, H. J. Leamy, and J. M. Poate, eds.), p. 344. Am. Inst. Phys., New York.

von Allmen, M. (1980). *In* "Laser and Electron Beam Processing of Materials" (C. W. White and P. S. Peercy, eds.), p. 6. Academic Press, New York.

Wada, Y., Tamura, M., and Tokuyama, T. (1980). *In* "Laser and Electron Beam Processing of Materials" (C. W. White and P. S. Peercy, eds.), p. 266. Academic Press, New York.

Webster, J. (1980). Personal communication.

Williams, J. S. (1980). *In* "Laser and Electron Beam Processing of Electronic Materials" (C. L. Anderson, G. K. Celler, and G. A. Rozgonyi, eds.), p. 249. Electrochem. Soc., Pennington, New Jersey.

Wilson, S. R., Appleton, B. R., White, C. W., and Narajan, J. (1979). *In* "Laser–Solid Interactions and Laser Processing" (S. D. Ferris, H. J. Leamy, and J. M. Poate, eds.), p. 481. Am. Inst. Phys., New York.

Wittmer, M. (1980). *In* "Laser and Electron Beam Processing of Electronic Materials" (C. L. Anderson, G. K. Celler, and G. A. Rozgonyi, eds.), p. 485. Electrochem. Soc., Pennington, New Jersey.

Wright, J. K. (1980). Personal communication.

Wu, C. P., and Schnable, G. L. (1979). *RCA Rev.* **40**, 339.

Yaron, G., and Hess, L. D. (1980). *IEEE Trans. Electron Devices* **ED-27**, 573.

Yaron, G., Hess, L. D., and Kokorowski, S. A. (1980a). *IEEE Trans. Electron Devices* **ED-27**, 964.

Yaron, G., Hess, L. D., and Olson, G. L. (1980b). *In* "Laser and Electron Beam Processing of Materials" (C. W. White and P. S. Peercy, eds.), p. 626. Academic Press, New York.

Young, R. T., White, C. W., Clark, G. J., Narajan, J., Christie, W. H., Murukami, M., King, P. W., and Kramer, S. D. (1978). *Appl. Phys. Lett.* **32**, 139.

Young, R. T., Wood, R. F., Narayan, J., White, C. W., and Christie, W. H. (1980). *IEEE Trans. Electron Devices* **ED-27**, 807.

Ziman, J. M. (1972). "Principles of the Theory of Solids." Cambridge Univ. Press, London and New York.

Zimmer, G. (1979). *Electron. Lett.* **15**(6), 184.

Index